Recent Advances in Materials for Energy Harvesting and Storage

Online at: https://doi.org/10.1088/978-0-7503-5749-4

Recent Advances in Materials for Energy Harvesting and Storage

Edited by
Suresh C Pillai
Department of Environmental Sciences, Atlantic Technological University, Sligo, Ireland

Daniel M Mulvihill
James Watt School of Engineering, University of Glasgow, Glasgow, UK

Aswathy Babu
Nanotechnology and Bio-Engineering Research Group, Department of Environmental Sciences, Atlantic Technological University, Sligo, Ireland

IOP Publishing, Bristol, UK

ISBN 978-0-7503-5749-4 (ebook)
ISBN 978-0-7503-5747-0 (print)
ISBN 978-0-7503-5750-0 (myPrint)
ISBN 978-0-7503-5748-7 (mobi)

DOI 10.1088/978-0-7503-5749-4

Version: 20240801

IOP ebooks

British Library Cataloguing-in-Publication Data: A catalogue record for this book is available from the British Library.

Published by IOP Publishing, wholly owned by The Institute of Physics, London

IOP Publishing, No.2 The Distillery, Glassfields, Avon Street, Bristol, BS2 0GR, UK

US Office: IOP Publishing, Inc., 190 North Independence Mall West, Suite 601, Philadelphia, PA 19106, USA

To those who tirelessly explore the frontiers of science and engineering.

Contents

Preface

This book encapsulates the collective efforts of researchers and scientists who have explored the realms of batteries, supercapacitors, piezoelectricity, new electrolytes, photovoltaics, and triboelectric nanogenerators (TENGs) to revolutionize the landscape of energy technologies. With each turn of the page, the reader will embark on a journey through the latest breakthroughs, methodologies, and applications that promise to redefine how we harness and store energy. From the intricacies of materials synthesis to the practical implications for real-world implementation, this book serves as a testament to more sustainable technologies.

Chapter 1's primary focus is the examination of current research trends that employ sustainable polymers as effective materials across a range of high-performance energy storage and harvesting systems. These systems encompass TENGs, photovoltaics, batteries, supercapacitors, and hydrogen energy, highlighting the versatility and potential of sustainable polymer-based solutions in advancing energy technologies.

Chapter 2 introduces the materials used in TENGs and delivers an elaborate examination of various TENG materials. These encompass conventional polymers, 2D materials, natural materials, porous crystalline materials, ferroelectric materials, and textile materials. The chapter concludes by emphasizing the potential of two-dimensional materials, showcasing their capability to function dually as electrodes and tribolayers.

Chapter 3 offers valuable insights into TENGs propelled by 2D materials, exploring their fundamental material characteristics, their integration into TENG manufacturing processes, and the broad spectrum of applications they enable. By consolidating the current advancements in this field, the chapter serves as a guidepost, outlining potential research directions and laying a foundation for future investigations in this dynamic domain.

Chapter 4 explores the fundamental concept of piezoelectricity and its underlying mechanisms. It then examines various classes of piezoelectric materials in detail. The discussion culminates in a systematic review of several case studies, providing a comprehensive overview of the research advancements in this domain and shedding light on the potential application scope of piezoelectric materials.

Chapter 5 offers an extensive overview of the latest progress in hybrid materials involving layered double hydroxides (LDHs), MXenes, and LDH/MXenes combined with conductive polymers within the supercapacitor domain. It encompasses composite fabrication, electrode materials, symmetric supercapacitors (SSCs), and asymmetric supercapacitors (ASCs) and delves into various facets of the subject matter. Despite the considerable research efforts directed towards MXenes and conductive polymers, certain obstacles remain, hindering their practical utilization.

Chapter 6 outlines a novel solar-driven energy-storage-based hybrid desalination system. The quest for sustainable water desalination technologies has led to the development of innovative solutions such as this novel solar-driven energy storage-based hybrid desalination system. This system represents a pioneering approach that integrates solar energy harvesting with energy storage technologies to power the desalination process. As a component of a hybrid multi-effect distillation and adsorption (MED-AD) cycle water treatment system, a MgO-based solar-heat-storing scheme is also proposed in this chapter.

Chapter 7 discusses the developments of sulfur host cathode materials, electrolytes, separators, and binders employed in current lithium-ion battery technology. In addition, it explores the evolution of separator technologies, elucidating their crucial role in optimizing ion transport and preventing detrimental side reactions within batteries. Through a comprehensive examination of recent developments, this chapter provides valuable insights into the ongoing efforts aimed at pushing the boundaries of lithium–sulfur battery (LISB) technology, ultimately paving the way for next-generation energy storage solutions.

Chapter 8 provides an overview of the latest advancements in the electrode materials, electrolytes, and separators utilized in sodium-ion batteries (SIBs). The essential classes of anode materials in SIBs include insertion-type, conversion-type, and alloying-type materials. Meanwhile, cathode materials predominantly consist of layered metal oxides, polyanionic compounds, and Prussian blue analogs. Through an examination of these key components, this chapter sheds light on the evolving landscape of SIB technology and its potential applications in energy storage systems.

Chapter 9 presents the fundamental aspects of the ionic conductivity of solid-state electrolytes (SSEs), their types, and their expected contributions to battery technologies. Furthermore, it explores the chemistry and synergistic interactions within their composites. It also addresses the challenges inherent in current SSEs and proposes trajectory strategies aimed at accelerating their commercial adoption and recognition.

Chapter 10 offers a comprehensive examination of recent progress in electrocatalyst development tailored for anodic reactions directed towards hydrogen (H_2) production. The evolving landscape of electrocatalyst development imposes stringent criteria encompassing high selectivity, activity, and durability. These electrocatalysts comprise a diverse array of materials, ranging from noble metals to non-noble metals, nonmetals, transition metals, and their alloys. Extensive efforts are dedicated to enhancing the electrochemical activity of these catalysts, with a particular emphasis on understanding the relationships between catalyst structure and activity.

Chapter 11 presents an overview of both established and emerging technologies in hydrogen production and storage. A primary emphasis lies in addressing challenges pertaining to physical infrastructure within technology advancement efforts. This chapter also reviews the use of nanomaterials for hydrogen storage, such as carbon nanotubes, other carbon-based nanomaterials, and metal–organic frameworks.

Each chapter unfolds a narrative of scientific discovery, unveiling novel materials, preparation techniques, and theoretical frameworks that promise to reshape the landscape of energy harvesting and storage. From the nanoscale of materials design to macroscopic applications poised to revolutionize renewable energy systems, this book serves as a vital resource for researchers, engineers, undergraduates, postgraduates, and policymakers alike, illuminating pathways towards a more sustainable and efficient energy future.

Suresh C Pillai
Daniel M Mulvihill
Aswathy Babu

Acknowledgements

We would like to acknowledge Atlantic Technological University (ATU) and University of Glasgow (UoG) for their unwavering support and invaluable resources in the completion of this book.

Editor biographies

Suresh C Pillai

Professor Suresh C Pillai obtained his PhD from Trinity College, Dublin and completed his postdoctoral research at the California Institute of Technology (Caltech, USA). He currently heads the Nanotechnology and Bio-Engineering Research Group at the Atlantic Technological University, Ireland. His research interests include the synthesis of nanomaterials for energy and environmental applications. He is the recipient of a number of awards, including the Boyle-Higgins Award 2019, the Linus Pauling Lecture Award 2020, etc. Suresh was also a recipient of the Industrial Technologies Award 2011 for licensing functional coatings to Irish companies. He received the Hothouse Commercialisation Award 2009 and also the Enterprise Ireland Research Commercialization Award 2009. He was also nominated for the One to Watch award 2009 for commercializing R&D work (Enterprise Ireland).

Daniel M Mulvihill

Prof. Daniel M. Mulvihill is Professor of Materials Engineering and Tribology at the James Watt School of Engineering, University of Glasgow, UK. He is also a faculty member of the Materials and Manufacturing Research Group (MMRG). He obtained his doctorate at the University of Oxford (UK) in 2012 and then completed postdoctoral work at the University of Limerick, EPFL Switzerland, and the University of Cambridge prior to joining the University of Glasgow in 2016. His research interests are focused on tribology, experimental mechanics, and materials for energy applications. His work has included contributions on: friction, wear, surface topography, triboelectrification, adhesion, composite materials, and aerospace metal alloys. He is a 'novel materials editor' for the Chemical Engineering Journal (Elsevier) and also an editor for Results in Engineering (Elsevier). He has been a chartered engineer in the UK (C.Eng.) since 2017 and was elected fellow of the Institution of Mechanical Engineers (FIMechE) in 2023. Dr Mulvihill is also a former Institution of Mechanical Engineers Tribology Trust Bronze Medallist (2013). Much of his present work is focused on electromechanical problems, especially those of triboeletrification and triboelectric nanogenerators.

Aswathy Babu

Dr Aswathy Babu obtained her PhD in chemistry from the University of Kerala, India in 2015 and subsequently undertook postdoctoral work at the CSIR-NIIST, India, and Atlantic Technological University, Sligo, Ireland. She worked on the development of functional plasmonic, fluorescent, and bio-nanomaterials for energy and environmental applications. Currently, she is an assistant lecturer at the Department of Life Sciences, Atlantic Technological University, Sligo. Her recent research interest is the area of energy materials, especially the development of textile triboelectric nanogenerators (T-TENGs) for wearable electronic applications and the fabrication of functional nanomaterials for advanced oxidation processes.

List of contributors

Hasna M Abdul Hakkeem
Materials Science and Technology Division, CSIR-National Institute for Interdisciplinary Science and Technology (CSIR-NIIST), Thiruvananthapuram 695019, Kerala, India

S Wazed Ali
Department of Textile and Fibre Engineering, Indian Institute of Technology Delhi, Hauz Khas, New Delhi 110016, India

Satyaranjan Bairagi
Materials and Manufacturing Research Group, James Watt School of Engineering, University of Glasgow, Glasgow, G12 8QQ, UK

Swagata Banerjee
Department of Textile and Fibre Engineering, Indian Institute of Technology Delhi, Hauz Khas, New Delhi 110016, India

Qian Chen
Institute for Ocean Engineering, Shenzhen International Graduate School, Tsinghua University, Shenzhen 518055, China

Sheng Dai
School of Chemical and Process Engineering, University of Leeds, Leeds, LS2 9JT, UK

Priyanka Ganguly
Chemical and Pharmaceutical Sciences, Applied Chemistry and Pharmaceutical Technology (ADAPT), School of Human Sciences, London Metropolitan University, London, N7 8DB, UK

Zhanhu Guo
Construction Engineering, Faculty of Engineering and Environment, Northumbria University, Newcastle Upon Tyne, NE1 8ST, UK

M Muhammad Imran
Department of Mechanical, Biomedical and Design Engineering, Aston University, Birmingham B4 7ET, UK

Nida Imtiaz
Mechanical Engineering Department, Universiti Teknologi Malaysia, Johor Bahru, Malaysia

Shaista Jabeen
Nanotechnology and Bio-Engineering Research Group, Department of Environmental Science, Atlantic Technological University, ATU Sligo, Ash Lane, Sligo, F91 YW50, Ireland

Muhammad Ahmad Jamil
Mechanical and Construction Engineering Department, Northumbria University, Newcastle Upon Tyne, NE1 8ST, UK

E J Jelmy
Inter University Centre for Nanomaterials and Devices, Cochin University of Science and Technology, Cochin-22, Kerala, India

Yinzhu Jiang
School of Materials Science and Engineering, Zhejiang University, Hangzhou 310027, China

Honey John
Department of Polymer Science and Rubber Technology, Cochin University of Science and Technology, Cochin-22, Kerala, India
Inter University Centre for Nanomaterials and Devices, Cochin University of Science and Technology, Cochin-22, Kerala, India

Jithu Joseph
Department of Chemistry, Indian Institute of Space Science and Technology, Thiruvananthapuram 695547, India

Sherin Joseph
Department of Polymer Science and Rubber Technology, Cochin University of Science and Technology, Cochin-22, Kerala, India

Gaurav Khandelwal
Materials and Manufacturing Research Group, James Watt School of Engineering, University of Glasgow, Glasgow, G12 8QQ, UK

K S Krishnendu
Department of Chemistry, Indian Institute of Space Science and Technology, Thiruvananthapuram 695547, India

Yong Ma
School of Material Science and Engineering, Shandong University of Science and Technology, Qingdao 266590, P R China

J Mary Gladis
Department of Chemistry, Indian Institute of Space Science and Technology, Thiruvananthapuram 695547, India

A Peer Mohamed
Materials Science and Technology Division, CSIR-National Institute for Interdisciplinary Science and Technology (CSIR-NIIST), Thiruvananthapuram 695019, Kerala, India

Daniel M Mulvihill
Materials and Manufacturing Research Group, James Watt School of Engineering, University of Glasgow, Glasgow, G12 8QQ, UK

Kim Choon Ng
Water Desalination and Reuse Center, King Abdullah University of Science and Technology, Saudi Arabia

Nikoleta D Nikolova
Chemical and Pharmaceutical Sciences, Applied Chemistry and Pharmaceutical Technology (ADAPT), School of Human Sciences, London Metropolitan University, London, N7 8DB, UK

Sreedhanya Pallilavalappil
Nanotechnology and Bio-Engineering Research Group, Department of Environmental Science, Atlantic Technological University, ATU Sligo, Ash Lane, Sligo, F91 YW50, Ireland

Rajagopalan Pandey
Department of Smart Manufacturing, NAMTECH, Gujrat 382355, India

Saju Pillai
Materials Science and Technology Division, CSIR-National Institute for Interdisciplinary Science and Technology (CSIR-NIIST), Thiruvananthapuram 695019, Kerala, India

Gopika Preethikumar
Materials Science and Technology Division, CSIR-National Institute for Interdisciplinary Science and Technology (CSIR-NIIST), Thiruvananthapuram 695019, Kerala, India

Keith Sirengo
Nanotechnology and Bio-Engineering Research Group, Department of Environmental Science, Atlantic Technological University, ATU Sligo, Ash Lane, Sligo, F91 YW50, Ireland

K Sreekala
Department of Chemistry, Indian Institute of Space Science and Technology, Thiruvananthapuram 695547, India

Muhammad Wakil Shahzad
Mechanical and Construction Engineering Department, Northumbria University, Newcastle Upon Tyne NE1 8ST, UK

Chirantan Shee
Department of Textile and Fibre Engineering, Indian Institute of Technology Delhi, Hauz Khas, New Delhi 110016, India

Mathew Sunil
Department of Polymer Science and Rubber Technology, Cochin University of Science and Technology, Cochin-22, Kerala, India

J Tanaya Dutta
Department of Chemistry, Indian Institute of Space Science and Technology, Thiruvananthapuram 695547, India

Adithya A Venugopal
Materials Science and Technology Division, CSIR-National Institute for Interdisciplinary Science and Technology (CSIR-NIIST), Thiruvananthapuram 695019, Kerala, India

Ben B Xu
Mechanical and Construction Engineering, Faculty of Engineering and Environment, Northumbria University, Newcastle upon Tyne NE1 8ST, UK

List of abbreviations

PHAs	Bacterial polyhydroxyalkanoates
CNCs	Cellulose nanocrystals
CNFs	Cellulose nanofibers
CVD	Chemical vapor deposition
HER	Hydrogen evolution reaction
LDHs	Layered double hydroxides
LIB	Lithium-ion battery
LISBs	Lithium–sulfur batteries
MOFs	Metal–organic frameworks
OER	Oxygen evolution reaction
PAN	Poly(acrylonitrile)
PDMS	Poly(dimethylsiloxane)
PeSCs	Perovskite solar cells
PENGs	piezoelectric NGs
PEG	Poly(ethylene glycol)
PEO	Poly(ethylene oxide)
PMMA	Poly(methyl methacrylate)
PVDF	Poly(vinylidene fluoride)
PVDF-HFP	Poly(vinylidene fluoride-co-hexafluoropropylene)
PDMS	Polydimethylsiloxane
PLA	Polylactic acid
KNN	Potassium sodium niobate
PWE	Pure water electrolysis
rGO	Reduced graphene oxide
RO	Reverse osmosis
SEM	Scanning electron microscopy
SIBs	Sodium-ion batteries
SEI	Solid electrolyte interphase
SSE	Solid-state electrolyte
SCs	Supercapacitors
SPEs	Sustainable polymer electrolytes
SPs	Sustainable polymers
THF	Tetrahydrofuran
TGA	Thermogravimetric analysis
TENG	Triboelectric nanogenerator
XPS	X-ray photoelectron spectroscopy

IOP Publishing

Recent Advances in Materials for Energy Harvesting and Storage

Suresh C Pillai, Daniel M Mulvihill and Aswathy Babu

Chapter 1

Sustainable polymers for energy harvesting and storage

Adithya A Venugopal, Hasna M Abdul Hakkeem, Gopika Preethikumar, A Peer Mohamed and Saju Pillai

The emergence of sustainable materials and technologies has led to innovations in high-performance energy harvesting and storage systems that produce minimal carbon emissions. Energy harvesting and storage system components are often made of synthetic petroleum-based hydrocarbons, metals, and inorganic compounds. However, these conventional resources might not be able to keep up with the growing demands for environmental sustainability. This chapter mainly focuses on the recent trends in research into the use of sustainable polymers as efficient materials for various high-performance efficient energy storage and harvesting systems, including nanogenerators, photovoltaics, batteries, supercapacitors, and hydrogen energy systems. Sustainable polymers (SPs) are very versatile, adaptable, robust, structurally strong, and abundant, making it possible for them to fulfill the requirements of emerging environmental challenges. Improved energy density, consistent efficiency over time, and more sustainable end-of-life behavior are now required for energy storage devices. Anode breakdown caused by phenomena such as dendrite growth frequently affects battery performance, and due to difficulties in charging and discharging, capacitors frequently struggle to reach functional energy densities. Nanogenerators are regarded as a promising alternative to conventional rigid and heavy energy storage devices, as they are flexible, lightweight, self-powered gadgets. The main focus of this chapter is to emphasize the potential energy harvesting and storage uses of sustainable polymer materials, including cellul.ose, lignin, starch, chitosan, and alginates. To further expand environmentally benign triboelectric nanogenerator (TENG) technology, more study is required to improve sustainable polymers' mechanical properties, triboelectric properties, and surface roughness characteristics.

doi:10.1088/978-0-7503-5749-4ch1

1.1 Introduction

The performance of energy harvesting and storage devices can be greatly enhanced by choosing the right materials for the device. While the performance characteristics of conventional materials are adequate, their synthesis, use, and disposal cause significant environmental issues. Many conventional materials and products raise considerable concerns about the management of their detrimental environmental effects. As a result, sustainable polymers (SPs) are gaining popularity in an effort to overcome the issues produced by conventional materials. Sustainable polymers are materials that can be recycled, biodegraded, or composted at the end of their service life and are made from renewable, recycled, and waste carbon resources [1]. Polysaccharides, cellulose, starch, lignin, proteins, polylactic acid (PLA), bacterial polyhydroxyalkanoates (PHAs), etc. are a few examples of naturally occurring or bio-based renewable SP resources. Since they are environmentally friendly, SPs have been recommended as feasible substitutes for plastic-based energy harvesting and storage devices, which currently use synthetic, non-degradable polymers. This chapter is mainly concerned with SP-based materials utilized for energy storage and harvesting applications. They are excellent choices for the creation of high-performance composites because of their structural features as well as their inherent biological capabilities. Recent reports of SP-based energy devices with remarkable performance suggest that the most efficient way to develop energy harvesting and storage systems would be to rely on SPs as compared to attempting to improve synthetic polymers for such applications. [2–4].

The use of environmentally friendly SPs that minimize detrimental impact has been the subject of a variety of articles. Compared with liquid and synthetic-polymer-based electrolytes, SP-based electrolytes have drawn much attention for use in electrochemical devices, since they have numerous advantages, such as their ability to develop thin films, ease of handling, the absence of solvent loss, and satisfactory conductive properties [5, 6]. The first section of this chapter discusses the various SP-based energy harvesting devices, while the different SPs reported for the development of lithium-ion battery (LIB) materials (including separator membranes, electrolytes, electrode materials, binders etc.) and supercapacitor materials are discussed in the second section. The sustainable polymer materials mainly discussed in this field include cellulose [7], keratin [8], starch [9, 10], lignin [11], alginate [12], gelatin [13, 14], agar [15], and chitosan [16]. The majority of natural polymers are chain-like and have repeated units, such as gelatin, silks, alginate, cellulose, chitosan, chitin, and others.

In the ever-changing technological landscape, the convergence of textiles and electronics has given rise to an innovative approach known as electronic textiles (e-textiles). As we approach the dawn of an entirely new century, the integration of sustainability concepts into this domain has become imperative. This leads to the revolutionary path of sustainable polymer-based electronic textiles. We highlight their potential for scalable commercialization with a special emphasis on the incorporation of energy storage and harvesting technologies. An immersive revolution, the incorporation of energy harvesting and storage materials into wearable electronic textiles adds an additional level of functionality to these state-of-the-art fabrics. By powering electronic components with energy derived from naturally-derived polymers and storing that energy inside the textile matrix, not only do these textiles become more functional and self-sufficient but they also allow the development of a wide range of sustainable, self-sufficient systems. Materials advances can pave the path for effective user adoption and the construction of a sustainable circular economy that realizes the transformative effects of wearable e-textiles. Recently, Shi *et al* proposed a systematic design framework that incorporates material selection and fabrication ideas to combine sustainable practices, market viability, supply-chain resilience, and user experience quality. They predicted that eco-friendly sustainable polymer-based materials might have a prominent position in the e-textile wearable materials library. Figure 1.1 shows the schematic representation of 4R design principle and the progress of wearable e-textile development [17].

Nevertheless, there are some key challenges in the use of sustainable polymers for energy storage and harvesting applications. Sustainable polymers are more prone to degradation, particularly in the severe environments of energy storage and harvesting systems. Over time, maintaining their structural integrity and performance becomes a major concern. Compared to more conventional materials such as metals or synthetic polymers, many sustainable polymers exhibit lower electrical conductivity. This reduces their effectiveness when used in components such as supercapacitors or the electrodes of energy storage devices, which need high electrical conductivity. For wider adoption, it is essential to reduce production costs while preserving performance. It may take a lot of research and development to improve the characteristics and architectures of sustainable polymers to achieve the

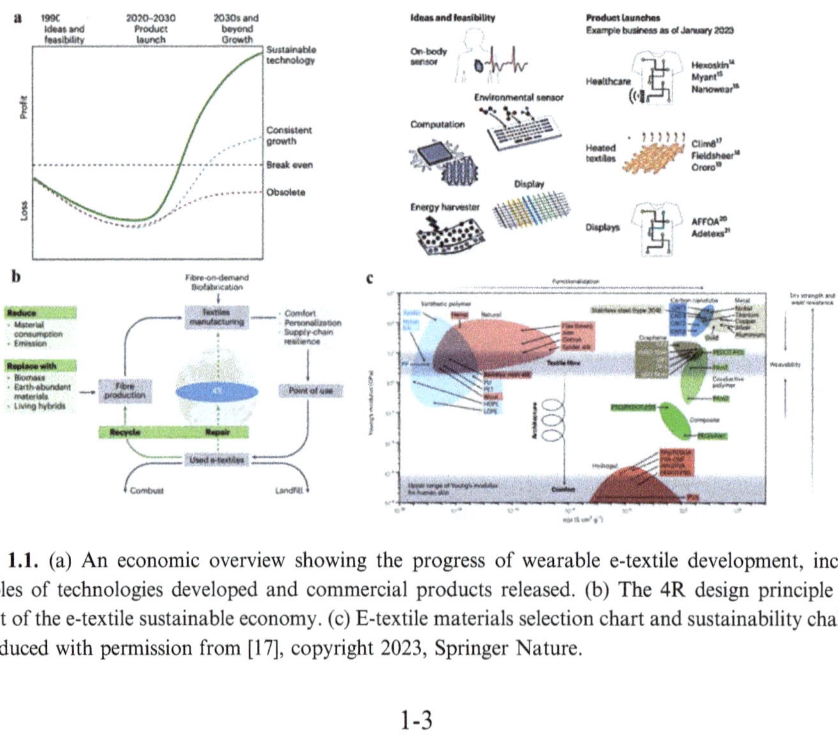

Figure 1.1. (a) An economic overview showing the progress of wearable e-textile development, including examples of technologies developed and commercial products released. (b) The 4R design principle in the context of the e-textile sustainable economy. (c) E-textile materials selection chart and sustainability challenge. Reproduced with permission from [17], copyright 2023, Springer Nature.

appropriate energy storage and harvesting capability. Current research developments that aim to overcome these challenges are discussed in this chapter.

1.2 Sustainable polymers for energy harvesting

1.2.1 Bio-triboelectric nanogenerators

Triboelectric nanogenerators (bio-TENGs) are miniature electronic energy harvesting systems that work by combining the triboelectrification process with electrostatic induction. They convert low-frequency mechanical motions in their host environment into electrical energy. There are different kinds of nanogenerators, such as piezoelectric NGs (PENGs) [18], pyroelectric NGs (PyENGs) [19], thermoelectric NGs, electromagnetic NGs, and triboelectric NGs. Triboelectric nanogenerators were first demonstrated by professor Zhong Lin Wang in 2012 [20]. TENGs are known to be promising candidates for flexible, lightweight, self-powered devices as opposed to conventional rigid and heavy energy storage devices. They dominate the whole nanogenerator category because of their relatively high output voltage and wide range of electrode source materials. TENGs produced from sustainable polymers, known as bio-TENGs or eco-friendly TENGs, have attracted a lot of interest due to their biocompatibility and stability. Bio-TENGs have previously demonstrated lower energy output and stability than synthetic polymer-based TENGs. More studies are currently being conducted to increase the capacity of TENGs for energy harvesting through functional modifications of sustainable polymers and the use of additives.

1.2.1.1 The triboelectric effect

The triboelectric effect is a contact-induced electrification phenomenon in which materials become electrically charged. It takes place when two different materials rub each other. An electron transfer takes place between the two materials, in which one donates electrons and other receives them. These materials are referred to as the positive side and negative sides of the TENG, respectively. When the materials cease contact, these electrons produce triboelectric charges on the surfaces of the materials. Generally, materials with strong triboelectrification effects are less conductive or are even insulators.

1.2.1.2 Electrostatic induction

Electrostatic induction is a method of inducing static electricity in a neutral material by bringing an electrically charged object close to it. This phenomenon was first demonstrated by professor John Canton in 1753 and later studied in detail by professor John Calville. TENGs utilize the synergic effect of both the triboelectric effect and electrostatic induction for energy generation [21]. Initially, electric charges accumulate on the surface of materials by the triboelectric effect; then, the generation of electrical energy takes place via electrostatic induction between the materials [22]. The basic device structure of TENGs includes two materials that are connected to electrodes at the back. The four well-known operational modes of TENGs, as indicated in figure 1.2, are lateral sliding, vertical contact–separation,

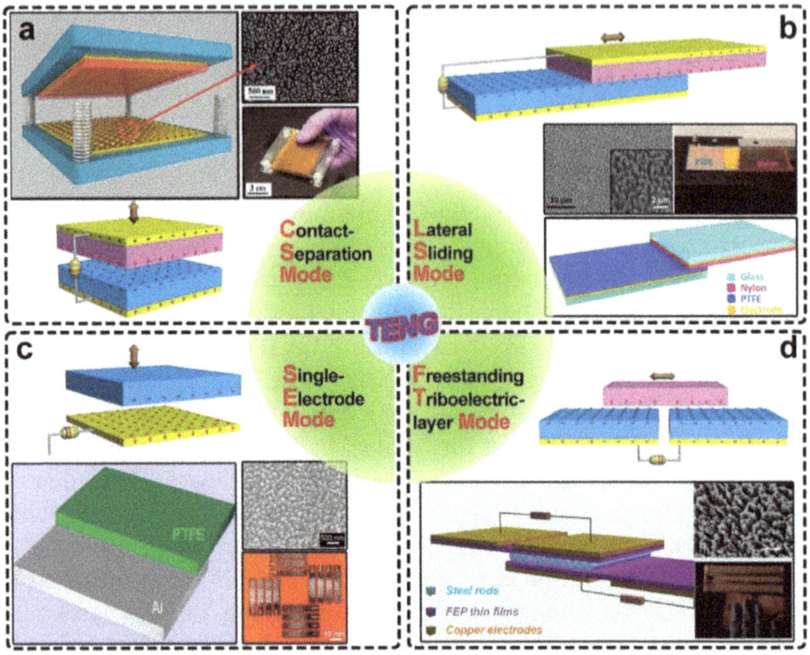

Figure 1.2. TENGs' four primary functioning modes: (a) the vertical contact–separation mode, (b) the lateral sliding mode, (c) the single-electrode mode, and (d) the freestanding triboelectric layer mode. Reproduced with permission from [24], copyright 2023, John Wiley & Sons.

single-electrode, and a freestanding triboelectric layer [23, 24]. The selection of materials for both the positive and negative sides of TENGs is purely based on the triboelectric series shown in figure 1.3 [25]. Almost all materials have triboelectric effects, including metals, polymers, etc. TENGs mainly work through three consecutive steps comprising charge generation, charge separation, and charge flow. Therefore, the selection of materials that have large differences in triboelectric charge density improves charge generation [26]. This can be further augmented by modifying material surfaces, enlarging the area of contact, and changing the surface morphology (figure 1.2) [27].

Bio-TENGs that are made of sustainable polymers such as cellulose, chitosan, lignin, gelatin, starch, etc. are good candidates in terms of their biocompatibility and biodegradability. To overcome the inadequate energy conversion capability and mechanical stability of sustainable polymers, these areas still require more focus.

1.2.2 Polysaccharide-based TENGs

Polysaccharides are a major class of sustainable polymers that has a specific position in the triboelectric series (figure 1.4). Among them, cellulose is predominant due to its high positive triboelectric charge, which can be modified or altered by functionalization. These functional groups play an important role in changing the electron-donating or -withdrawing nature of sustainable polymers, especially cellulose [28, 29]. This can

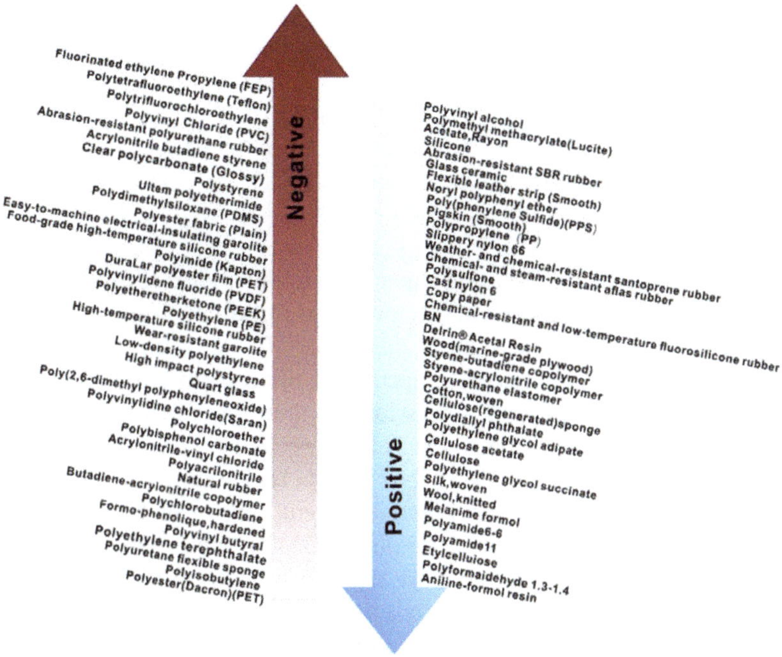

Figure 1.3. The triboelectric series from negative to positive tribopolarity [27] John Wiley & Sons. © 2020 Wiley-VCH GmbH.

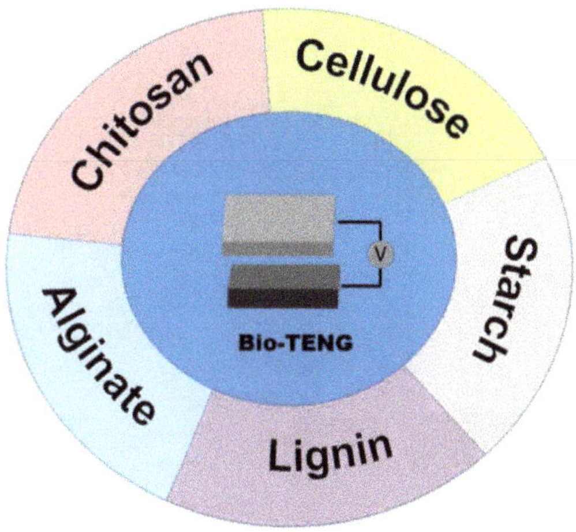

Figure 1.4. Sustainable polymer-based triboelectric nanogenerators (bio-TENGs).

be considered as one of the major advancements in polysaccharide-based TENGs. In regard to the other performance enhancement factors of TENGs, surface area and morphology play an important role, since TENGs function due to the contact between two different materials. The total contact surface area can be increased by processes such as electrospinning, plasma etching [30], electron-beam lithography, etc. For example, Varghese *et al* achieved a 400 V output voltage, a 3 mA m^{-2} short-circuit current, and a 0.9 W m^{-2} peak power density using electrospun cellulose acetate (CA) as the positive side and polydimethylsiloxane (PDMS) with micropyramidal arrays fabricated by photolithography as the negative side of a TENG [31, 32]. In addition, this sustainable polymer also can be modified to have micro- or nanostructures with different morphologies and increased surface area.

1.2.2.1 Cellulose-based TENGs

Cellulose has a high electron-donating capacity, making it ideal for the positive side of TENGs. In the case of cellulose, well-known nanosized cellulose categories include cellulose nanocrystals (CNCs), cellulose nanofibers (CNFs), and bacterial cellulose (BC). CNCs are spindle- or needle-shaped crystalline forms of nanocellulose that have more exciting properties and processibilities than those of bulk cellulose. CNCs are well suited for use in bio-TENGs due to their mechanical properties, transparency, etc [33]. Peng *et al* fabricated a PDMS/CNC TENG and attained a higher open-circuit voltage (OCV) of 320 V and a short-circuit current density of 5 μA cm^{-2}, which is almost ten times higher than that of its pure PDMS counterpart (figure 1.5a) [34]. CNFs are a fibrous form of nanocellulose that has a width in the range of 3–25 nm. CNF is also utilized in TENG constructions that provide more advantages of nanosized cellulose [35–37]. Zhang *et al* fabricated a gear-like TENG using CNF and fluorinated ethylene propylene film (figure 1.5b) [38]. The triboelectric properties of the CNF were modified by introducing amino groups by mixing the fibers with a polyethyleneimine (PEI) and silver nanoparticle coating. They achieved a high OCV of 286 V and a power density of 0.43 W m^{-2} using the gear-like structure, which directly powered 60 light-emitting diodes and performed as a self-powered sensor (figure 1.5).

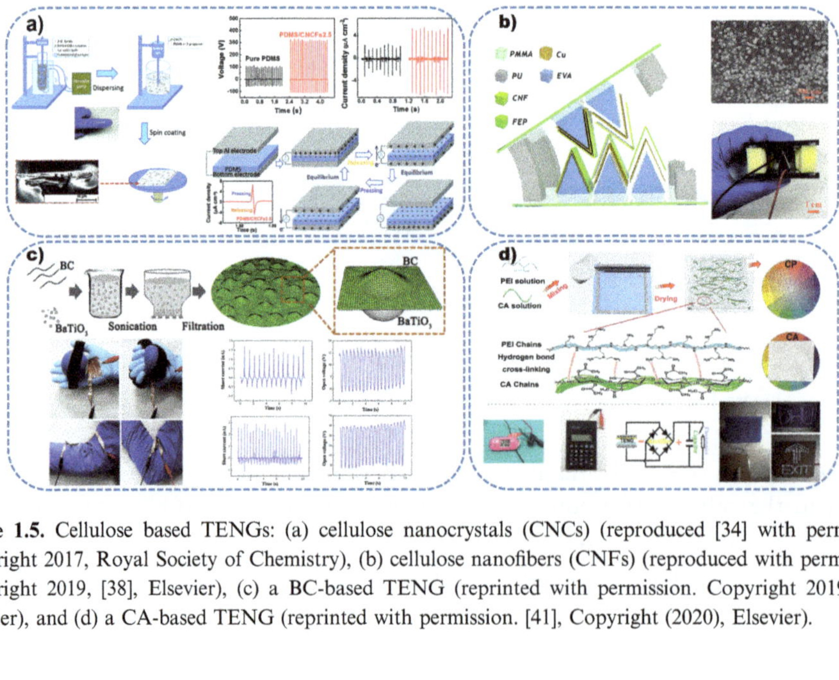

Figure 1.5. Cellulose based TENGs: (a) cellulose nanocrystals (CNCs) (reproduced [34] with permission Copyright 2017, Royal Society of Chemistry), (b) cellulose nanofibers (CNFs) (reproduced with permission. Copyright 2019, [38], Elsevier), (c) a BC-based TENG (reprinted with permission. Copyright 2019, [40], Elsevier), and (d) a CA-based TENG (reprinted with permission. [41], Copyright (2020), Elsevier).

BC is a pure form of natural CNF that has a width in the range of 20–100 nm and is generated by bacteria. Pure BC-based bio-TENGs can only provide small open-circuit output voltages, but the output can be improved by adding additives that enhance the output [39]. Shao *et al* incorporated dielectric particles of $BaTiO_3$ into BC by vacuum infiltration and placed it against PDMS in a TENG device. They obtained an OCV of 181 V and a peak power density of 4.8 W m^{-2} (figure 1.5c) [40]. CA can provide greater output than CNF, CNC, and BC in triboelectric nanogenerators due to the highly electron-donating nature of the acetate group [31, 32]. For example, Bai *et al* reported a high OCV of 478 V, a short-circuit current density of 6.3 μA cm^{-2}, and a power density of 2.21 mW cm^{-2} for a TENG constructed from a CA/PEI composite against vulcanized silicone rubber (figure 1.5d) [41].

1.2.2.2 Starch-based TENGs

The direct application of starch in the production of TENGs has been observed to be difficult to implement due to the high solubility of starch in water and its weak mechanical properties. The reinforcement of starch by additives can improve its mechanical properties and roughness. In 2017, Bao *et al* fabricated a starch/lignin composite film for use in TENGs [42]. This was considered to be the first attempt to produce a starch composite film for a starch-based TENG electrode material. In comparison to a starch film, the mechanical properties and roughness of the film were both improved. Bao *et al* obtained an average short-circuit current output of 3.96 nA·cm^{-2}, an OCV of 1.04 V, and a power density of 173.5 nW·cm^{-2}. Subsequently, Zhu *et al* demonstrated a bio-TENG that utilized starch paper to sense human perspiration [47]. Starch's affinity for water was advantageous in this application. The biodegradability of this starch paper outperformed that of the entire electronic device and could be degraded in water in just 4 min. A combination of micro-nanostructured starch and cellulose ester were also reported to have been used in a bio-TENG [48]. The same research group also demonstrated the fabrication and operation of a polymer electrolyte starch-based bio-TENG which was produced by incorporating $CaCl_2$ salt into the starch. They achieved about a threefold enhancement in the voltage output [49]. Sarkar *et al* successfully produced a bio-TENG using starch and PDMS as active materials (figure 1.6a) [43]. An average output voltage of ~560 V, an output current density of ~120 mA m^{-2}, and an output power density of 17 W m^{-2} were achieved, and the device was able to power at least 100 commercial blue LEDs. In addition, Sarkar *et al* demonstrated its usefulness as a self-powered speedometer and pedometer. Khandelwal *et al* enhanced the mechanical stability of starch by introducing an edible laver filler to introduce some degree of hydrophobicity into the system for use in a TENG [50]. A 1 wt.% starch/laver composite exhibited a contact angle of 107° and was utilized for biomechanical energy harvesting.

1.2.2.3 Lignin-based TENGs

The first lignin-based bio-TENG was demonstrated by Bao *et al* in 2017. They prepared a composite of lignin and starch. This combination exhibited better mechanical and triboelectric properties than those of starch alone. Subsequently,

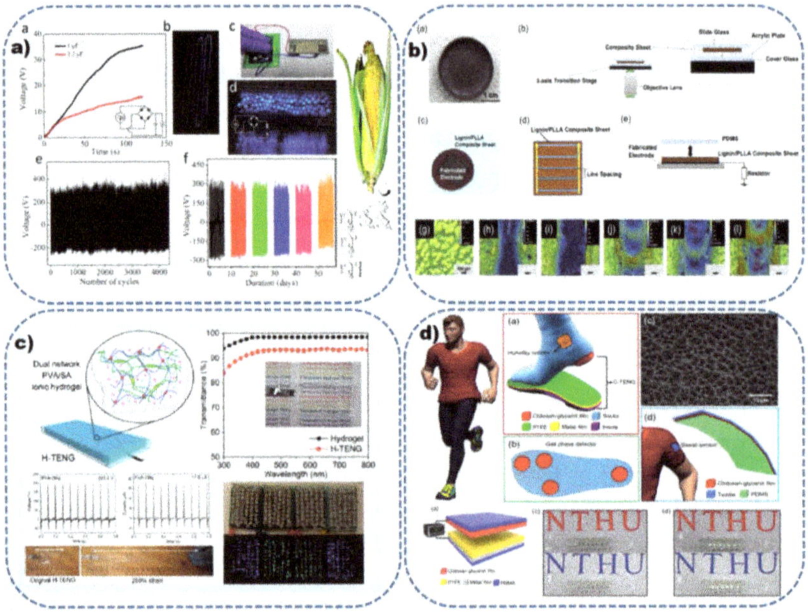

Figure 1.6. TENGs based on sustainable polymers including: (a) starch (reprinted with permission from [43], Copyright (2019), American Chemical Society), (b) lignin (reprinted with permission from [44], Copyright (2023), American Chemical Society), (c) alginate (reprinted with permission from [45], Copyright (2020), American Chemical Society), and (d) chitosan (reprinted from [46], Copyright (2018), Elsevier).

Funayama *et al* fabricated a lignin/polylactic acid composite for use in a biodegradable TENG. They successfully achieved a power density of about 1.98 mW m^{-2} at a load resistance of 200 MΩ (figure 1.6b) [44]. In 2018, An *et al* fabricated an eco-TENG from solution-blown nanofiber (NF) mats of soy protein and lignin [51]. According to their report, this was the first nanotextured plant-derived bio-TENG to use fibers on the few-hundred nanometer scale that can be extensively and safely used for self-powered biomedical devices. Yu *et al* developed a novel strategy for fabricating an insoluble and infusible biomass-based TENG made up of lignin, chitosan, and cellulose [52]. They successfully achieved an OCV of 308 V and a power density of up to 5.93 W m^{-2}. Feng *et al* successfully developed a self-healing hydrogel-based TENG consisting of sulfonated lignin and Fe^{3+} ions [53]. This eutectic gel TENG was capable of withstanding temperature as low as −80 °C and was also transparent, stretchable, and self-healing. The lignin-based eutectic gel TENG delivered an OCV of 105 V, a short-circuit current of 0.5 μA, and a power density of up to 53 mW m^{-2}. Tanguy *et al* used natural lignocellulosic nanofibrils as tribonegative materials in a bio-TENG [54]. They also explored the enhancement to the TENG provided by lignin by comparing the performance of cellulose nanofiber against that of an aluminum electrode. The high contact potential difference of lignocellulosic nanofibers compared to that of CNFs confirmed the polarizability of lignin that led to pronounced electric output. They also fabricated a cascade lignocellulosic nanofibril nanopaper-based TENG which could even power energy-intensive wireless communication nodes and other sustainable self-powered IoT devices. By taking advantage of polarizability, the strong electron-donating ability of lignin, and the enhanced TENG performance due to nanofibers, Wang *et al* constructed a lignin electrospun nanofiber-based TENG [55]. This waste-derived bio-TENG can be utilized for energy harvesting and self-powered pressure sensing. Sun *et al* reported the fabrication of a Cu^{2+}–alkali lignin organogel-based TENG that could withstand different extreme temperature conditions [56]. This organogel-based TENG can be used in commercial electronics and for the movement monitoring of human joints. Edburg *et al* also tried to demonstrate the triboelectric properties of sustainable lignin and nitrocellulose by fabricating a bio-TENG. They obtained a maximum OCV of 232 V, a short-circuit current of 17 mA m^{-2}, and a power density of up to 1.6 W m^{-2} [57]. In 2023, Jo *et al* also utilized an electrospinning method to fabricate a lignin/polycaprolactone NF-based TENG [58]. As a result of the high specific area of the fibers and the electron transfer of the composite, the TENG delivered an OCV of 95 V at a low tapping force of 9N and a frequency of 9Hz. Nanthagal *et al* utilized lignin, enhancing the performance of a natural-rubber-based TENG by about 500% [59]. The power output density was even improved to about 1300% by modifying the particle size and polar functionality of lignin by sonicating it in a basic medium. Subsequently, they grafted natural rubber with polyacrylamide to eliminate the reduction in electrical output caused by lignin aggregation in the natural rubber matrix. Most recently, Chen *et al* constructed a bio-TENG by using Fe-ion coordinated lignin in a carboxymethyl cellulose (CMC) matrix as the positive side [60]. They attained an OCV of 110 V, a short-circuit current of 8.97 μA, and a maximum output power density of 147.19 mW m^{-2}. An expanded electron cloud on the lignin skeleton and its interaction with transition-metal ions were the reasons for this enhanced output.

1.2.2.4 Alginate-based TENGs

Alginate is widely used in the food and biomedical industry due to its biocompatibility and biodegradability. In 2015, Valentini *et al* fabricated a bio-TENG from a sodium alginate graphene oxide nanocomposite [61]. A solvent casting method was implemented to produce sodium alginate (Al)–graphene oxide (GO) films, which were used as one of the advantages of TENGs compared to polyethylene terephthalate (PET). They obtained a short-circuit current density of 10^{-4} mA cm^{-2}, an OCV of up to 1.3 V, and a maximum power density of 1.33 mW m^{-2}. The whole device was also deployed as a pressure sensor. Liang *et al* constructed a bio-TENG device that consisted of sodium alginate and polyvinyl alcohol (PVA) as triboelectric materials that completely degraded when they came into contact with water [62]. Instead of a sodium alginate film, calcium alginate films were prepared by ion exchanging the sodium alginate with $CaCl_2$ for use in a TENG [63]. They achieved an OCV of 38 V, a short-circuit current of 245 nA, and a power density of 3.8 mW m^2. The device was easily destroyed by immersing it in salt water. Jing *et al* fabricated a PVA and sodium-alginate-based hydrogel TENG with elasticity that could be tuned by varying the amount of cross-linking with sodium alginate (figure 1.6c) [45]. The fabricated transparent hydrogel-based TENGs (H-TENG) produced a peak output voltage of 203.4 V, a current of 17.6 mA, and a power density of 0.98 W m^{-2}. They demonstrated the simultaneous lighting of 240 green and blue LEDs and powered a digital timer and pedometer.

1.2.2.5 Chitosan-based TENGs

Chitosan exhibits a strong electron-donating nature due to the presence of amine groups which can be successfully utilized in bio-TENGs [64]. Jao *et al* used the electron-donating capability of chitosan in a TENG by fabricating a nanostructured chitosan–glycerol film (figure 1.6d) [46]. In addition to transparency, it also had a humidity-resistant nature. Compared to starch-based bio-TENGs, the chitosan-based TENG performed outstandingly without being affected by humidities ranging from 20% to 80%. Porous surfaces always produce greater triboelectric outputs than dense polymer films. The first completely porous TENG electrode was reported by Zheng *et al.* They used a porous aerogel film of chitosan and polyimide, obtaining an extremely high voltage of 60.6 V and a current of 7.7 μA, which is more than sufficient to light 22 blue LEDs [65]. Wang *et al* studied the morphology and triboelectric output of different nanocomposite films of chitosan [66]. Among chitosan–acetic acid, chitosan, chitosan–lignin, chitosan–starch, and chitosan–glycerol composite films, a chitosan–acetic acid (10%) film provided the maximum output as a result of the highly electron-donating nature of acetic acid. In 2021, Pongampai *et al* fabricated a flexible chitosan-based bio-TENG with an OCV of 111.4 V, a short-circuit current of 21.6 A cm^{-2}, and a power density of 756 μW cm^{-2} by incorporating piezoelectric $BaTiO_3$ nanorods (BT-NRs) (table 1.1) [67]. A summary of the performances of various sustainable polymers used in TENG devices, are shown in table 1.1.

1.2.3 Photovoltaic cells

1.2.3.1 Dye-sensitized solar cells

The affordable photoelectrochemical devices known as dye-sensitized solar cells (DSSCs) enable the direct conversion of sunlight into energy. Natural pigments or dyes absorb visible light and inject electrons into the TiO_2 conduction band. A DSSC is a cost-effective photovoltaic cell composed of a dye-sensitized photoanode such as (I-/I3-), a counter electrode, and an electrolyte. Lignin, a sustainable polymer, can be employed in DSSCs. Its photophysical properties permit the use of various phenols as photosensitizers in DSSCs. Using a phase-separation technique, lignophenols were synthesized directly from natural lignin [68]; they exhibited considerable visible-light absorbance and fluorescent emission at 400–600 nm.

1.2.3.1.1 Sustainable polymer electrolytes

Solid ion conductors called sustainable polymer electrolytes (SPEs) are formed when salts are dissolved in high-molecular-weight polymers. SPEs have high energy densities, ionic conductivities of up to 104 S cm^{-1}, wide electrochemical windows, and solvent-free, lightweight properties, making them useful for device applications including DSSCs. They may be made as solids or semisolids using simple trustworthy

Table 1.1. A summary of the performances of various sustainable polymers used in TENG devices.

Materials (SP in bold)	Open-circuit voltage output (V)	Short-circuit current output (μA)	Power density	References
SA/GO/ITO/PET vs ITO/PET	1.3	—	1.33 mW m^{-2}	[41]
Au vs **CNF** film	32.8	35	—	[69]
317L stainless steel vs **EC**	245	50	—	[70]
BC vs Cu	13	3	4.8 mW m^{-2}	[39]
Cellulose/PDMS vs Al	28	2.8	64 μW cm^{-2}	[33]
CNC/PDMS vs Al	320	—	0.76 mW cm^{-2}	[34]
AZO-coated CNF vs **AZO-coated CNF** + $TiCl_4$	7	0.7	—	[71]
CMC/Al/PET/polyimide (PI)	153	3.9	—	[72]
CNF/phosphorene vs Au/PET	5.2	—	9.36 μW cm^{-2}	[73]
SA/Al vs **PVA**/Li	1.47	3.9 $\times 10^{-3}$	3.8 mW m^{-2}	[62]
CMF/**CNF**/Ag vs fluorinated ethylene propylene	21.9	0.17	7.68 μW cm^{-2}	[74]
Porous EC vs Sn-doped In_2O_3/PDMS/poly(methyl methacrylate) (PMMA)	45	—	1.2 mW m^{-2}	[75]
CNF/PEI aerogel vs PVDF nanofibers	106.2	9.2	13.3 W m^{-2}	[35]
CA vs polytetrafluoroethylene (PTFE)	7.3	9.1	—	[76]
SH paper (single-electrode mode)	~14	—	—	[47]
Calcium alginate vs Al	33	150 $\times 10^{-3}$	—	[63]
CS aerogel vs PI aerogel	60.6	7.7	2.33 W m^{-2}	[65]
CS vs Kapton	~1.6	~40 $\times 10^{-3}$	17.5 μW m^{-2}	[66]
CS/glycerol vs PTFE/Al	130	~15	—	[46]
Wood vs PTFE/Cu	81	1.8	57 mW m^{-2}	[77]
Cellulose/$BaTiO_3$ aerogel vs PDMS/Al	48	—	—	[78]
Polypyrrole (PPy)-coated **cellulose paper**, **Cellulose paper**, **Nitrocellulose membrane**	60	8.8	0.83 W m^{-2}	[79]

Crepe cellulose paper vs **Nitrocellulose membrane**/Cu	196.8	31.5	16.1 W m^{-2}	[80]
Polycaprolactone (PCL)/GO vs **Cellulose paper**/Au	120	4	72.5 mW m^{-2}	[81]
CNF-PEI-Ag vs FEP/Cu	286	4	0.43 W m^{-2}	[38]
CNF aerogel vs PDMS/Ag	55	0.94	29 mW m^{-2}	[37]
BC/$BaTiO_3$ vs PDMS/Cu	181	21	4.8 mW m^{-2}	[40]
BC/$BaTiO_3$/Ag nanowires (NWs) vs PTFE/Al	170	9.8	180 μW cm^{-2}	[82]
SH vs cellulose ester/Cu	0.3	—	—	[48]
SH/$CaCl_2$ vs PTFE/Al	1.2	—	170 mW m^{-2}	[49]
Thermoplastic **SH** vs PDMS	~560	180	17 W m^{-2}	[43]
CS hydrogel/Ag NW/Cu^{2+}/PDMS (single-electrode mode)	218	—	2 W m^{-2}	[83]
Wood vs PTFE/Cu	220	5.8	158.2 mW m^{-2}	[84]
Cellulose/CS aerogel vs PTFE/Al	242	—	—	[64]
Cellulose/alginic acid aerogel vs PTFE/Al	~80	—	—	[64]
Cellulose/$BaTiO_3$ nanoparticles (NPs) vs PDMS/Ag NPs	88	8.3	—	[85]
Cellulose aerogel vs PTFE/Al	65	1.86	127 mW m^{-2}	[64]
Cellulose/PVDF/PA/$BaTiO_3$ vs PTFE	20.15	6	—	[86]
CMC aerogel/PDMS vs PDMS/Kapton/Al	~14	~0.22	—	[87]
CNF/Diatom frustules vs PTFE	388	18.6	85.5 mW m^{-2}	[88]
Allicin-grafted **CNF** vs PVDF	7.9	5.13	10.13 μW cm^{-2}	[89]
BC/ZnO NPs/Teflon/indium tin oxide (ITO) (single-electrode mode)	49.6	4.9	—	[90]

(Continued)

Table 1.1. (*Continued*)

Materials (SP in bold)	Open-circuit voltage output (V)	Short-circuit current output (μA)	Power density	References
CA/PEI vs LTV Conductive fabric	478	—	2.21 mW cm^{-2}	[41]
SA/PVA hydrogel/$CaCl_2$/borax vs PDMS/Al	203.4	17.6	0.98 W m^{-2}	[45]
CS vs PDMS/Al	306	—	—	[52]
CS vs FEP/Al	150	1.02	15.7 mW m^{-2}	[91]
BC/Ag NWs/$BaTiO_3$ vs PTFE	87	7.1	75 μW cm^{-2}	[92]
BC/carbon nanotubes (CNTs)/PPy vs BC	29	0.6	—	[93]
BC/CS vs PDMS/Cu NPs	23	0.5	—	[94]
CS/$BaTiO_3$ nanorods (NRs) vs PTFE	111.4	21.6	756 μW cm^{-2}	[67]
CS/egg-shell membrane vs PTFE	55	10	—	[95]
CS/silk fibers vs PTFE	77	13	—	[95]
Cellulose fibers/Ag NPs vs PTFE	142.6	31	—	[96]
Epoxidized **natural rubber/CS**/amino-triethoxysilane-treated **CNCs** vs Teflon	107.7	10.6	156 mW cm^{-2}	[97]
CA nanofibers vs PDMS micropyramidal arrays	400	3×10^3	0.9 W m^{-2}	[31]
Cellulose fibers/Ag NPs vs PTFE/Ag	142.6	31.2	—	[96]
EC/poly-ε-caprolactone vs polyvinylidene fluoride (PVDF)/graphene	44	2	157.17 mW m^{-2}	[98]
Porous bamboo cellulose/polyaniline		2.9	1.1 W m^{-2}	[99]
CMC/MXene $Ti_3C_2T_x$ (single-electrode mode)	54.73	1.22	402.94 mW m^{-2}	[100]
CS/SH/FEP/PET/ITO (Single-electrode mode)	1080	16.9×10^3	5.07 W m^{-2}	[101]
Regenerated cellulose vs **cellophane**	300	2.6×10^3	300 W m^{-2}	[102]
Calcium alginate fibers/vermiculite vs PDMS	270	1	—	[103]
SA reinforced polyacrylamide/xanthan gum—Cu^{2+}/LiCl/PDMS	150	1.2	—	[104]

Ethyl cellulose—**EC**, cellulose acetate—**CA**, cellulose microcrystals—**CMCs**, cellulose microfibers—**CMFs**, cellulose nanofibers—**CNFs**, cellulose nanocrystals—**CNCs**, starch—**SH**, sodium alginate—**SA**, chitosan—**CS**.

techniques. Since their introduction by O'Regan and Gratzel, DSSCs have been one of the most explored solar-powered systems. Due to stability concerns caused mainly by electrolyte leakage, quasi gel electrolytes were developed, even though liquid electrolytes have the highest conversion efficiency. However, due to poor ionic mobility and gel entrapment in metal oxide pores, these electrolytes could not overcome the stability issue and caused efficiency losses. These electrolytes have uses in DSSCs due to their thermal stability, biodegradability, and high conductivity.

A DSSC is composed of three parts: the anode, the cathode, and the electrolyte. Mesoporous metal oxides have been used to manufacture anodes. The cathode is made by depositing a transparent oxide on platinum. A TiO_2 electrode was stained overnight and loaded with electrolyte. The current was generated by an electrode connected to an external load and the cell. Polysaccharides are the most commonly recommended recyclable polymers for the fabrication of the DSSC's electrolyte. Figure 1.7 demonstrates the structures of various polysaccharides [105], KI, and ionic liquids (ILs) in dimethyl sulfoxide (DMSO). Considering that IL dispersion produces more ions and also assists in lowering crystallinity, the IL-doped sustainable polymer-KI matrix displays an additional improvement in conductivity.

It has been shown that it is possible to build cationic polymers derived from aqueous electrolyte-based DSSCs which have more than 7% efficiency [107]. DSSCs

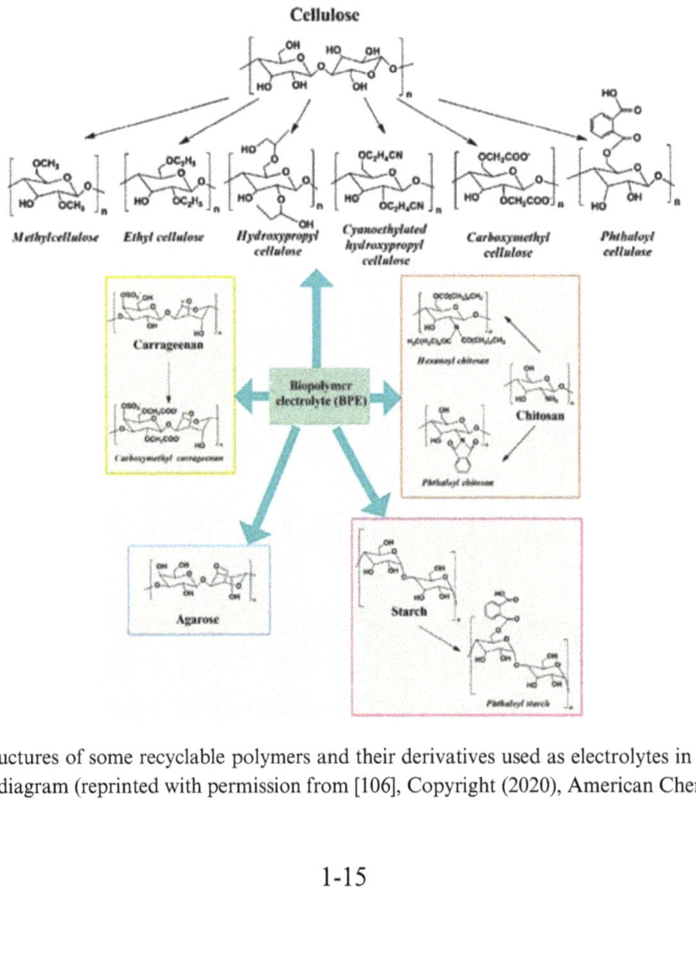

Figure 1.7. Structures of some recyclable polymers and their derivatives used as electrolytes in DSSCs shown in a schematic diagram (reprinted with permission from [106], Copyright (2020), American Chemical Society).

based on polymers, on the other hand, have lower power conversion efficiencies than DSSCs based on liquid electrolytes. Although the current greatest efficiencies are only around 10%, this meets the long-term viability requirement for DSSCs, and there are numerous research investigations on this subject. Other possible sensitizers utilized in DSSCs are natural pigments taken from fruit, flower petals, leaves, and vegetables (chlorophyll, betalain, carotenoids, flavonoids) [108]. In the last few decades, various photosensitizers for DSSC applications have been researched. They are classified into the groups described below. Examples include dyes made from ruthenium quantum dot sensitizers, which are derived from natural resources.

1.2.3.1.2 Cellulose and starch-based electrolytes

Microcrystalline, nanocrystalline, and nanofibrillated cellulose have all been proposed as gel electrolyte solutions for use in DSSCs. Bella *et al* designed a cell with a nanoscale microfibrillated cellulose (NMFC) sustainable polymer electrolyte and a fiber-based photoanode derived from cellulose. DSSCs developed with cellulose-based electrodes and electrolytes ensure sunlight conversion efficiencies as high as 3.55 and 5.2% at simulated light intensities of 1 and 0.2 sun, respectively, and an outstanding efficiency retention of 96% after 1000 h of accelerated aging test [109]. Poskela *et al* proposed a more efficient assembly method for large-scale DSSC manufacture using bio-based cryogels made of BC and CNF which were soaked in electrolyte. Their comprehensive investigation found that the residual lignin components in DSSCs made from BC resulted in higher efficiency and long-term viability than those of DSSCs made using CNF. In contrast to cellulose-based DSSCs, the overall stability of both BC- and CNF-based DSSCs declined significantly over time, demanding further optimization of this assembly approach [110]. Because of cellulose's high crystallinity, DSSC electrolytes have low conductivity and efficiency. Chemical alterations, plasticization, grafting, and combination with another polymer can all help to reduce this crystallinity [111]. There are several polymeric materials which are soluble in water; organosoluble cellulose derivatives have been used in the synthesis of DSSC electrolytes. Vasei *et al* used LiI, I_2, 4-tert-butylpyridine (TBP), and tetra-n-butyl ammonium iodide (TBAI) as liquid electrolytes that were then gelled by the addition of the cellulose derivative ethyl cellulose (EC) [112]. When EC was added, the gel's viscosity increased by 300 times compared to that of the pure electrolyte, which caused the DSSC's photovoltaic performance to drop from 6.5% to 5.8%. Even though the DSSC's stability increased with a greater EC concentration compared to other cells with a higher viscosity, its efficiency decreased. Furthermore, a cellulose polymer substrate grafted with acrylic acid (AA) with 1-butyl-3-methylimidazolium iodide ionic liquid as a medium might produce comparable results [113]. Kaschuk *et al* [105] created a DSSC using 1-propyl-3-methylimidazolium iodide (PMII), guanidinium thiocyanate, and nanofibers that were electrospun from both CA and deacetylated cellulose acetate (DCA). Because the CA-based DSSC had many more acetyl groups than the DCA-based ones, it had a greater J_{SC} and very high efficiency. The photovoltaic performance of CA and DCA was superior to that of the liquid-electrolyte-based DSSC used as a reference. Due to their ability to minimize electrolyte loss during operation, these cellulose derivatives disseminated the electrolyte evenly and stabilized performance for over 500 h.

Starch-based electrolytes exhibited mechanical stability and provided mobile ions with minimal steric hindrance. Starch, a sustainable polymer derived from carbohydrates, has gained considerable interest for use as an electrolyte in DSSC. An increase in amylose concentration modulates the stability and ease of ion transport. As a result, amylose-rich starches, such as starches derived from arrowroot, potato, and sago, have been frequently used in DSSCs. Numerous investigations have determined that the DSSC efficiency obtained using rice starch (RS) is only 2.09%, which is inadequate. Since potato starch (PS) has a greater conductivity than that of pure RS, its feasibility as a DSSC electrolyte was examined further. Yogananda *et al* synthesized electrolytes that contained both PS and PS nanocrystals. Their DSSC in DMSO with sodium iodide demonstrated much greater performance than that of any pure RS-based DSSC. A DSSC with improved performance and 3.3% efficiency was obtained using an electrolyte based on PS nanocrystals. It exhibited greater conductivity, J_{SC}, and V_{OC} than those obtained using PS. Even though PS nanoparticles had more intense crystallinity than PS, their higher conductivity could be due to the improved ability of the heterogeneous network of PS nanocrystals to facilitate ion movement. The frequency of electron injection and consequently higher J_{SC} and accuracy were attributed to the nanoparticles' deeper penetration into the porous TiO_2 photoanode [114].

1.2.3.2 Perovskite solar cells and quantum dot-sensitized solar cells

Perovskite solar cells (PeSCs) are regarded as the most economical developments in photovoltaics due to their ease of manufacture, cost-effectiveness, and superior efficacy. The quality of the perovskite film and the device interfaces, which are the primary contributors to non-radiative recombination losses in PeSCs, have a substantial effect on the power conversion efficiency and stability of this technology. 'Trihalides derived from organometallics,' of which $CH_3NH_3PbI_3$ is the most utilized, is another name for perovskites. The fundamental advantage of perovskite sensitizers is their narrow bandgap from visible light to IR, which allows them to absorb a broad spectrum of solar radiation and to be used in the production of DSSCs. Currently, inorganic–organic perovskites derived from lead compounds display a maximum cell efficiency of 25.2%. In this chapter, we examine the most recent developments and techniques for achieving excellent performance using PeSCs. In this part, we also cover the use of bio-based sustainable polymers to enhance the performance of perovskite films, as well as the development of stretchy, biocompatible, and biodegradable electrodes made from natural biomaterials.

Long-chain sustainable polymers with several functionalities can provide more interactions and a larger constriction force that can change their morphology. Yang *et al* added wood-based polymer ethyl cellulose (EC) [115] to an antisolvent to produce high-quality perovskite film. It was demonstrated that the EC sustainable polymer prevented the perovskite layer from crystallizing. It was obvious that the Lewis acid–base interaction generated by the EC durable polymer slowed down the crystallization of the perovskite layer. The slower crystallization provided more time for grain development, resulting in perovskite films that had denser, smoother, and bigger grains. The long-chain EC also served as a scaffold, alleviating the lattice strain caused by the annealing stage and stabilizing the perovskite structure of the crystal. Furthermore, the modified device demonstrated exceptional durability in the environment, sustaining 80% of its initial power conversion efficiency (PCE) even after 30 days. To create flexible PeSCs and eventually use them in wearable electronic devices, stretchable electrodes are required. Because of its low cost, relatively light weight, flexibility, biocompatibility, and total biodegradability, cellulose paper is an appealing substrate for flexible electronics as an efficient way to use a sustainable polymer. It has been used in devices such as adaptable sensors and organic photovoltaic cells. Gao *et al* [116] developed an anode material derived from cellulose paper which was modified with carbon and were the first to prepare hole-transport-layer-free (HTL-free) flexible PeSCs which exhibited better conductivity, energy level alignment with the perovskite layers, and resulting efficient charge extraction. Using this type of substrate, namely bio-based substrates, they achieved a PCE of 9.05% for the HTL-free flexible PeSCs in addition to high flexibility and excellent mechanical durability. Zou and colleagues [117] developed PeSCs that had a PCE of up to 11.68% by producing a transparent, elastic electrode made of bamboo. Large quantities of highly polymerized cellulose bamboo fibers were used to produce bamboo carbon nanofiber (B-CNF) substrates (cellulose nanofibril substrates). The presence of carboxylate groups in the bamboo carbon nanotubes (B-CNTs) and the B-CNF substrate enhanced transmission over the visible spectrum. A B-CNT/indium zinc oxide electrode demonstrated excellent transmittance, flexibility, and conductivity.

Quantum dot-sensitized solar cells (QDSSCs) are third-generation solar cells that use quantum dots (QDs) as a sensitizer or light-harvesting technology. As their nanoparticle sizes can be adjusted to fit the desired absorption spectra, QDs have exceptional properties that lead to a configurable bandgap. The construction of QDSSCs involves sandwiching a redox electrolyte containing sulfide/polysulfide (S^{2-}/S^{n-}) between a counter electrode (CE) and a photoanode (which is covered in a high bandgap semiconductor sensitized with QDs). The principle of this solar cell is identical to the DSSC principle. Photons of light are absorbed by electrons in the QDs, energizing electrons in the valence band and exciting them into the conduction band (CB). The excited electrons are subsequently introduced into the semiconductor's CB from the CB of the QD. The QDs oxidize as a result of this occurrence. The electrons then proceed via an external circuit to the CE. Subsequently, S^{2-} receives the electrons at the CE and changes to S^{2-} in the polysulfide electrolyte. S^{2-} then restores the oxidized QD to its initial state, continuing the process [118]. It should be noted that the iodide/tri-iodide (I^-/I^{3-}) redox couple, which is frequently employed in DSSCs, cannot be used in QDSSCs since it rapidly diminishes photocurrent by destroying the QD sensitizer. As a result, the PCE is also reduced. The highest PCE to date of more than 12% was achieved by a QDSSC based on a liquid polysulfide (Sn^{2-}/S^{2-}) electrolyte. However, because of leakage and volatilization, it suffered from long-term instability. Hence, in order to enhance the performance of DSSCs, cost-effective biopolymer-based electrolytes have been developed. A sustainable polymeric material, sodium-carboxymethyl-cellulose (Na-CMC) film, has beneficial characteristics such as nontoxic properties, an outstanding ability to produce thin films, transparency, and high mechanical strength. Baharun *et al* developed a Na-CMC film using a simple solution-casting approach. Because of the carboxyl and hydroxyl groups that allow hydrogen bonds to form in the polymer chains, Na-CMC is a naturally occurring anionic polysaccharide, and the three-dimensional network of this material has great water absorption and holding capabilities. With an activation energy value of 0.38 eV, the most conductive SPE has the lowest activation energy. The SPE's free ion concentration (n), ionic mobility (μ), and ionic coefficient of diffusion (D) were all calculated. Recently synthesized SPEs are being employed as electrolytes in QDSSCs with the configuration FTO/TiO_2/CdS/ZnS/SPE/Pt/FTO. Under 1000 W m^{-2} illumination, a QDSSC constructed using CMC-68 SPE had an optimum PCE of 0.90% [119]. According to electrochemical impedance spectroscopy (EIS) results, QDSSCs with high J_{SC} and PCE values have longer electron lifetimes and lower electron recombination rates.

1.2.3.3 Hydrogen production

As a carbon-neutral method of producing hydrogen from biomass or renewable resources, biohydrogen production from cellulose has garnered considerable interest. In this part, we examine the utilization of a two-step process to extract hydrogen from lignocellulose. The method includes a dark-fermentation step to increase the production of hydrogen, carbon dioxide, acetic, formic, succinic, and lactic acids as well as ethanol from a pretreated lignocellulosic biomass and an electrohydrogenesis step to generate hydrogen gas from the remaining volatile fatty acids (VFAs) and alcohols. Biochar-based materials can also be used to produce hydrogen, primarily through the natural generation of H_2, electrocatalysis, and the development of photocatalytic H_2. Fermentation (anaerobic digestion) and microbial photosynthesis are two biological processes that can produce hydrogen gas (H_2) from readily available and affordable organic waste and biomass. When organic matter is decomposed during anaerobic digestion, microbes can form biofilms that are grafted to the surfaces of particles, enhancing the efficiency of H_2 generation. Biochar can be used as a supplement to increase anaerobic H_2 production because it reduces acid and ammonia inhibition and promotes biofilm formation. According to Sharma *et al*, the addition of biochar to an anaerobic digestion system improved the rate of H_2 production to 96.63 ml per 2.8 g of carbon, compared to 22.55 ml per 2.2 g of carbon without biochar [120]. In addition to biological H_2 generation, the water-splitting electrocatalytic method offers an alternative to current systems based on fossil fuels for the creation of highly pure H_2. During the hydrogen production process, electrocatalysts play an essential role in lowering the overpotential and generating high catalytic current densities in the hydrogen evolution reaction (HER). Efficient catalytic hydrogen evolution from water employing biochar-based materials as catalysts is another significant way to manufacture clean and sustainable H_2 fuel. It should be possible to create high-performance HER electrocatalysts through the addition of catalytically reactive species to biochar [121].

Although highly significant, the efficiency of H_2 production through aqueous phase reforming (APR) needs to be further increased. The mixed-metal-oxide

(MMO)-supported catalytic production of hydrogen through the APR of cellulose over defect-rich nickel was reported by Zhang *et al* to produce a 70% hydrogen yield [122]. The supercritical water gasification (SCWG) process, which transforms biomass into biorenewable hydrogen, is a promising technology. In this process, a wet biomass is converted into hydrogen, carbon dioxide, methane, and carbon monoxide. The catalytic activity and hydrogen selectivity of the catalysts Ni/-Al_2O_3, Ni/hydrotalcite, Raney nickel, Ru/C, and Ru/-Al_2O_3 were assessed in the production of hydrothermal hydrogen from lignocellulosic biomass. Glucose, cellulose, fructose, xylan, pulp, lignin, and bark were the primary feedstocks used. The experiments were done in a batch reactor at 380 °C with a 2 wt.% concentration of feed [123].

Zou *et al* produced hydrogen from cellulose via catalytic gasification using a CeO_2/Fe_2O_3 catalyst. One issue for catalytic biomass gasification is the production of more tar compounds by the thermal–chemical conversion of biomass, which results in the formation of coke on the surface of the catalyst. To avoid these issues, a steam reforming two-stage pyrolysis–catalytic system was developed in which pyrolysis occurred at the first stage. The second stage of biomass gasification utilized a catalyst to catalytically convert the generated vapors into hydrogen. Catalysts play a significant role in hydrogen generation. The creation of hydrogen via the steam gasification of cellulose is best accomplished using a CeO_2/Fe_2O_3 catalyst with a Ce:Fe ratio of 3:7. The catalyst encouraged the gasification of volatile steam as the temperature increased from 500 °C to 900 °C, which enhanced the amount of gas produced: 28.58 mmol g^{-1} hydrogen was produced from the cellulose at 800 °C. However, because the iron promoted the reverse water shift reaction at 900 °C, the production of hydrogen and carbon monoxide was reduced at this temperature [124].

1.3 Sustainable polymers for energy storage

1.3.1 Battery materials

Due to their remarkable adaptability, eco-friendliness, and inherent efficiency, sustainable polymers are at the forefront of scientific research and technological innovation as we approach a green energy revolution. The use of natural sustainable polymers in future batteries that are environmentally friendly and sustainable is gaining popularity. Even though they are being explored for nearly every element of battery cell development, various studies have been conducted to investigate their application as an electrolyte component. These studies have addressed an extensive variety of topics, including their potential use for electrolyte wetting and as thermally versatile separators, as a substitute for synthetic polymers in gel and solid polymer electrolytes, and as functional membranes to prevent species deterioration (in order of elctroactivity) in different battery chemistries [125]. Here, we delve into the fascinating field of sustainable polymers to learn about their revolutionary impact on the development of various types of battery technology, including sodium-ion batteries (SIBs), potassium-ion batteries, and LIBs. Figure 1.8 summarizes the sustainable biopolymers which are commonly used in battery applications.

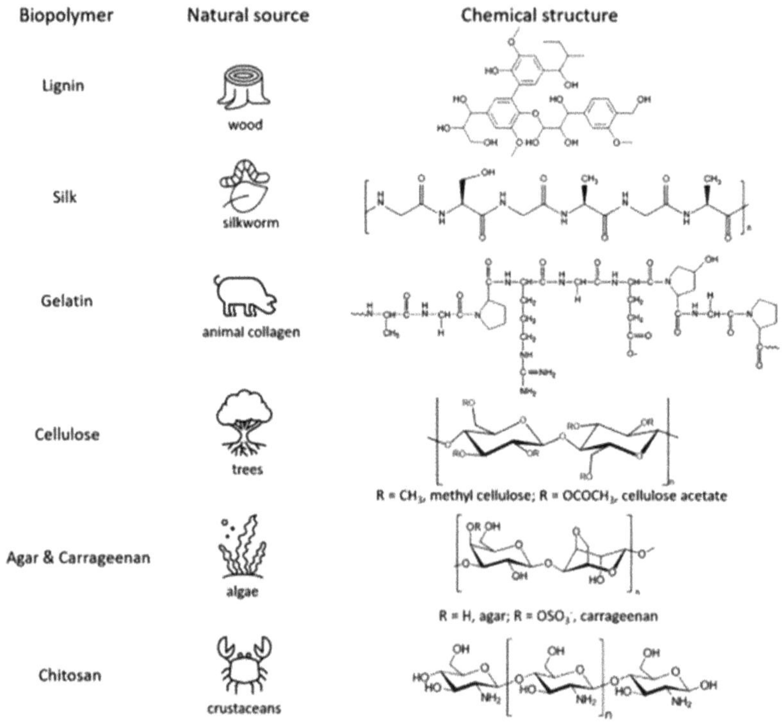

Figure 1.8. Schemes including a summary of the most commonly used sustainable polymers ([125] John Wiley & Sons, © 2020 Wiley-VCH GmbH).

1.3.1.1 Lithium-ion batteries

LIBs are widely employed to power portable devices, electric vehicles (EVs), and grid-scale energy storage systems (ESSs). As a result, the creation of economical, eco-efficient, and environmentally compatible LIBs is essential to enable the widespread usage of energy from renewable sources by future generations. A typical LIB consists of an anode, a cathode, a separator that has optimized wettability by an electrolyte (the major function of the separator is to avoid thermal runaway), and an electrolyte that facilitates Li^{+} transference between the cathode and anode during charging/discharging. The performance of LIBs is dependent on their design. Both electrodes are composed primarily of polymeric binders which promote particle cohesion. The three main categories of binders used in LIBs are organic-solvent-soluble binders, water-soluble binders, and binderless systems. This section also includes detailed studies of bio-based binders.

Separators, as a major component of LIBs, have a significant impact on battery security, lifespan, and electrochemical stability, which is also critical for the long-term sustainability of energy storage devices. Separators provides routes for lithium-ion migration between two electrodes, involving high Li-ion conductivity in liquid electrolytes. Because of their exceptional insulating qualities, high mechanical strength, and great electrochemical stability, polyolefin separators such as polyethylene (PE)- and polypropylene (PP)-based separators are today constantly employed in commercial LIBs. On the other hand, polyolefin separators generated from finite fossil oil are not eco-efficient, environmentally harmless, or sustainable. Furthermore, LIBs' low-temperature stability has an enormous effect on their safety and lifespan. Viable polymers have recently garnered a lot of interest as a replacement material for battery separators due to their good qualities, such as their outstanding thermal stability, excellent mechanical performance, non-toxicity, portability, and outstanding moisture retention capacity for the electrolyte.

1.3.1.1.1 Battery electrolytes and separator membranes

These beneficial characteristics, as well as their resistance to organic solvents and acids, have prompted a great number of studies of lignin-based electrolytes and separators for LIBs. Wang *et al* investigated a casting method that combined poly (N-vinyl imidazole)-co-poly(ethylene glycol) methyl ether methacrylate with lignin biopolymer to produce a freestanding gas-permeable electrolyte with a lithium transference number of 0.6 mS cm^{-1}, a tensile strength of 4 MPa, and an electrochemical stability of 7.5 V. This gel polymer electrolyte (GPE) was extremely effective at limiting the growth of lithium-ion dendrites. Although it worked well in a Li/$LiCoO_2$ cell, the lignin GPE was also added to a Li/$LiFePO_4$ which had a specific capacity of 150 mAh g^{-1} and was stable for up to 450 cycles. In a different method, a GPE with an electrolyte uptake of 230% and an ionic conductivity of 3.73 mS cm^{-1} was reported as an efficient GPE that had better compatibility with a lithium anode. It was not possible to trigger electrolyte loss when it was heated to 100 °C, implying significant contact between the ester and carbonyl groups in the electrolyte and the phenolic hydroxyl groups of the lignin, which prevented liquid from evaporating. These interactions resulted in a remarkable lithium transference number (tLi^+) of 0.85 as well as the ability to dissociate the anion from the lithium salt and facilitate the migration of Li ions. When added to a Li/$LiFePO_4$ cell, the GPE had a capacity of 129 mAh g^{-1} and 95% capacity retention. A GPE produced by combining polyethylene glycol with lignocellulose and 1 molar lithium hexafluorophosphate ($LiPF_6$) in ethylene carbonate, dimethyl carbonate, and diethyl carbonate (EC-DMCDEC) attained a maximum electrolyte absorption of 267%, an ionic conductivity of 3.22 mS cm^{-1}, and a tLi^+ of 0.81, in contrast to Celgard 2400's subpar tLi^+ of 0.26. The novel GPE had good compatibility with a lithium anode, as shown in figure 1.9 [126].

Zhao *et al* developed a nonwoven membrane separator by electrospinning lignin and polyacrylonitrile, which showed a porosity of 74%, an ionic conductivity of 1.24 mS cm^{-1}, and no shrinkage when subjected to a temperature of 150 °C for 15 min. The separator delivered a capacitance of 149 mAh g^{-1} after 50 cycles when placed in a Li/$LiFePO_4$ battery. In addition, lignin was applied to PVA nanofiber membranes to enhance the electrolyte uptake, resulting in electrolyte uptake values of 488%, 508%, and 533% for 1 m $LiPF_6$, 1 m $LiPF_6$, and 1 m $LiBF_4$ in EC–ethyl methyl carbonate (EC-EMC), EC–diethyl carbonate–fluoroethylene carbonate (EC-DECFEC), and EC–dimethyl carbonate (EC-DMC), respectively, as opposed to 121%, 146%, and 108% for Celgard 2400 [127]. The discharge capacity of the cells obtained very high efficiency compared with that of the Celgard separator.

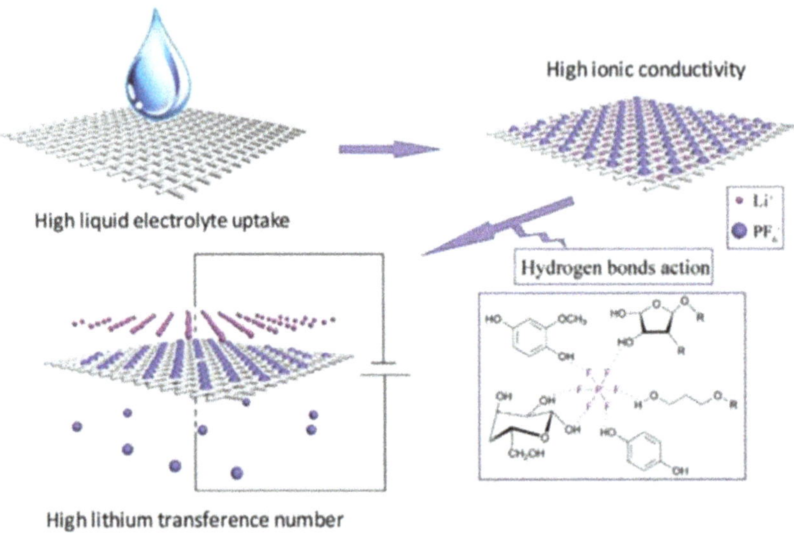

Figure 1.9. A schematic showing the preparation of GPE using a lignocellulosic matrix (reprinted from [126], Copyright (2017), with permission from Elsevier).

Cellulose, the most prevalent natural polymer, is distinguished by its composition and properties. It is insoluble in most solvents due to its strong intramolecular hydrogen bonds, and its hydroxyl groups become reactive. To produce a porous cellulosic separator membrane, cellulose was extracted from wood pulp and acetylated. The raw material was then dissolved in a suitable solvent, *N,N*-dimethylacetamide/acetone (DMAc-acetone), after which it was hydrolyzed using an LiOH–ethanol solution and electrospun to generate nonwoven cellulosic with an average diameter of 300 nm, 75% porosity, excellent dimensional stability, and electrochemical stability up to 5.0 V [128].

Cellulosic fibers that have diameters in the nanometer range make up nano-cellulose. Nanocellulose has a surface area of up to 500 m^2 g^{-1} and a three-dimensional network-like morphology. Hänsel *et al* constructed a single Li-ion conductor out of cellulose for the first time, successfully avoiding the production of an ion concentration variation throughout charging and discharging [129]. To eliminate the silica phase after the self-assembly of transition-metal oxides (TMOs) and CNCs, lithium hydroxide has been used as an alternative to sodium hydroxide solution, replacing the proton present in $-OSO_3H$ groups on the cellulose crystal surface with Li^+. A remarkable tLi^+ of 0.93 and an ionic conductivity of 0.02 mS cm^{-1} were produced, enabling uniform Li metal deposition and avoiding the growth of dendrites. Figure 1.10 illustrates the fabrication method for single-ion conductive mesoporous membrane using cellulose and water-dispersed CNCs.

Several studies have demonstrated that CNFs have the potential to serve as battery separators. There have been various attempts to improve the porosity of CNF separators in order to improve performance. Nanocellulose fibers have been infused with SiO_2 nanoparticles, sodium alginate, and nitrogen- and phosphorous-containing

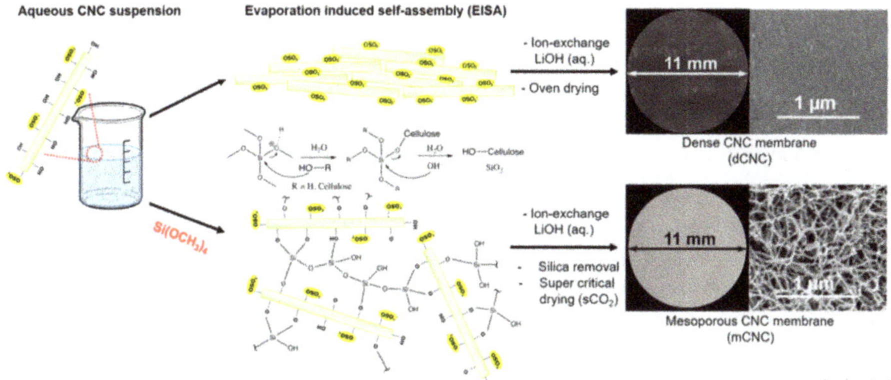

Figure 1.10. (a) A schematic showing the fabrication method used to produce a single-ion conductive mesoporous membrane using cellulose and water-dispersed CNCs (reprinted with permission from [129], Copyright (2019), American Chemical Society).

compounds to improve the efficacy and functionality of LIBs. Compared to commercial polypropylene separators, these separators have a higher oxygen index of 40% and a greater thermal stability (150 °C). Furthermore, a $LiCoO_2$/graphite separator had a specific capacity of 107 mAh g^{-1}, which was higher than that of a polypropylene (PP) separator. After calendering, a CNF/polysulfonamide membrane with a nanoscale pore diameter was generated [130].

LIB separators and GPEs have also been developed using a variety of sustainable polymers. To develop chitosan-based SPEs, chitosan was disseminated in 1% ethanoic acid, then lithium triflate ($LiCF_3SO_3$) salts, ethylene carbonate (EC) and propylene carbonate (PC) solvents, and lithium acetate ($LiCH_3COO$) were added. $LiCF_3SO_3$ has a greater ionic conductivity than that of $LiCH_3COO$ (5.0×10^{-6} Scm^{-1} versus 6.1×10^{-7} Scm^{-1}) because it dissociates more readily. Furthermore, monochloroacetic acid and chitosan were combined to create carboxymethyl chitosan [123]. Even though several LIB separators for high-performance LIBs have been built using sustainable polymer fibers, a few issues remain to be addressed in the future, such as the requirements for an easily tuned porous structure, hygroscopicity, flammability, thermal shutdown, and large-scale fabrication.

1.3.1.1.2 Electrode materials and binders

LIBs store charge using Li ions; the cathode of an LIB is commonly $LiCoO_2$ and the anode is graphite, which has a theoretical specific capacity of 372 mAh g^{-1}. To fulfill the expectations of future innovations such as electric and hybrid vehicles, it is necessary to research the fabrication of electrode materials that are cost-effective and have high capacity. Graphite is the most common electrode material, but biomass-derived porous carbon with a large surface area is an alternative [131]. Murali *et al* developed porous carbon with a graphene framework which had a high specific surface area (526 m^2 g^{-1}) by activating peanut shells with KOH, yielding higher specific capacity after 100 cycles. The framework morphology can improve and intercalation effect [132].

Binders are used during electrode fabrication, which is when the slurry for anode components is prepared and cast onto the current collector. The underlying cause and mechanism that allows a binder to link surfaces can be interfacial binding forces, mechanical interlocking, or a combination of the two. Mechanical interlocking is based on the penetration of a binder into imperfections in the surface and the subsequent interlocking of the surfaces as a result of the binder system's response mechanism, such as the polymerization/drying illustrated in (figure 1.11). In this section, we highlight current advances in the research, development, and utilization of biodegradable binders for the fabrication of LIB electrodes, emphasizing their importance for long-term sustainability. Polymeric binders have been employed in both electrodes of a battery to boost particle population cohesiveness. Binders also ensure that the slurry has the proper viscosity for homogeneous deposition and adherence to the metal current collector [133]. The establishment of enormous manufacturing facilities around the world has occurred in response to a substantial spike in LIB production due to rising automotive industry demand. This has increased the consumption of unsustainable fossil-based materials used to make electrode formulations during slurry preparation. Additional battery research is required into the use of metals such as nickel, manganese, and lithium as cathode materials and graphite as an anode material. Bio-based binder solutions for LIB electrodes are critical to permit a sustainable global changeover in LIB manufacture and to eradicate dependency on nonrenewable resources. Among the bioderived binders that can improve battery performance are cellulose, lignin, alginate, gums, and starch. This section focuses primarily on the current development and progress of bio-based polymeric binders and electrode materials for LIB applications.

The utilization of both functional and polar surface groups, such as COOH and OH, has the potential to improve electrode processing. Furthermore, these surface groups have a variety of functional features. The class of bio-based cathode and anode binders is a subset of a broader class of sustainable polymers. Figure 1.12 is a schematic representation of bio-based binders [136].

Cellulose is a biocompatible, biodegradable, and cost-effective polymer that is one of the most readily available sustainable polymers. Cellulosic binders include

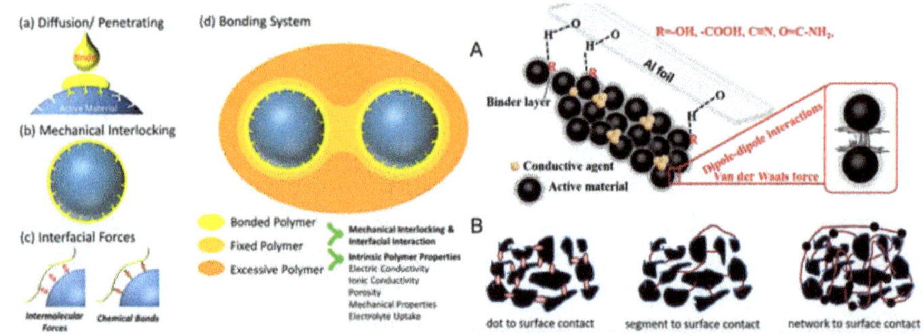

Figure 1.11. A schematic illustration depicting the mechanical interlocking and interfacial force binding systems. Panel (a) depicts a schematic diagram of a chemical connection (reprinted with permission from [134], Copyright (2018), American Chemical Society). Panel (b) depicts a schematic view of interactions (reprinted from [135], Copyright (2019), with permission from Elsevier).

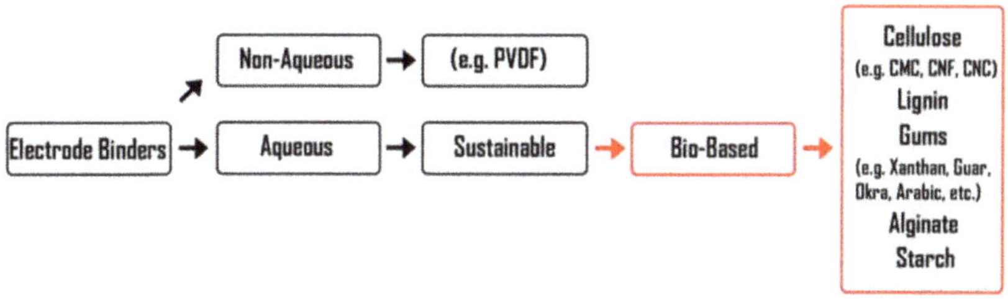

Figure 1.12. A schematic demonstrating the bio-based binder class that is the focus of this review, which includes cellulose, lignin, alginate, and starch-based binders used for the manufacture of LIB electrode material slurries (reproduced from [136], CC BY 4.0).

CNFs with a high aspect ratio, CNCs, CMC microparticles, and microscale pulp fibers. CMC, which is mainly used in electrode formulations, shows promising results due to its low cost, sustainability, and water solubility in comparison to that of PVDF. CMC is typically added with a rubber that improves the mechanical properties of the slurry deposited on the current collector (often styrene–butadiene rubber, SBR). The application of Na-CMC in combination with a fossil-based SBR copolymer results in a very cost-effective anode binder solution for LIBs. In a comparison of binder properties, a CMC-based binder demonstrated a higher specific capacity of 1153 mAh g^{-1} over 35 cycles [137]. In addition, for a graphite anode in $LiFePO_4$, a better capacity of 141 mAh was demonstrated for CMC modified with sulfobetaine methacrylate as opposed to 108 mAh g^{-1} for pure CMC. The discharged capacity loss was 6.6% for CMC modified with sulfobetaine methacrylate and 12.2% for a CMC binder during the first three cycles of the solid electrolyte interphase (SEI) layer formation process [138]. The main disadvantage of CMC binders is the possibility of electron trapping in the carbon coating, which reduces electrical conductivity and may impair coulombic efficiency as $-CH_2COOLi$ groups form. Another issue that must be rectified before fully bio-based anode slurries can be made available is the need to use fossil-based SBR in conjunction with CMC.

Nanocellulose, a cellulose derivative, can be used as the primary component of electrodes, electrolytes, and membrane separators. Cellulose is one of the most important bio-based polymers. As an alternative to synthetic-polymer-based electrode binders, various electroactive compounds have been combined with nanocellulose. Various nanocellulose-based composite electrodes have been created utilizing mixed CNF/carbon suspensions and solvent removal techniques such as vacuum filtration, supercritical drying, solvent exchange, etc. Reduced graphene oxide (rGO) and CNFs were combined to create an aerogel by Shao and colleagues. The CNF served as a nanospacer for the reduced GO. The CNF in this composite hydrogel significantly decreased rGO–rGO stacking interactions and prevented rGO aggregation [139]. Cellulose nanofiber was an efficient alternative electrode binder to the most advanced synthetic polymer binders due to its good mechanical compliance and high aspect ratio. PVDF is the most common polymer binder used for LIB electrodes. However, the production of PVDF-based electrodes necessitates the use of hazardous chemical solvents such as *N*-Methyl-2-pyrrolidone (NMP) as well as expensive and intricate drying machines. PVDF, on the other hand, is inadequate to withstand the applied external deformation. Beneventi and colleagues proved that CNF can be used as a binder in the production of flexible graphite anodes via a water evaporation approach. The freshly produced graphite anodes exhibited a well-developed porosity structure and retained their mechanical flexibility [140]. In contrast to PVDF, these systems provide better lithium-ion intercalation into graphite and better improvement in the binding strength due to the presence of hydroxyl groups (SEI). To develop multifunctional and hybrid binder systems, further investigations have been carried out in materials design for next-generation LIBs.

Lignin, another sustainable polymer found in woody biomass, is the subject of intensive research in many relevant fields and has numerous industrial applications. Lignin is a macromolecule with excellent electrochemical performance. When lignin is grafted with polyacrylic acid and copolymeric binders, a next-generation anode containing Si nanoparticles has proven to be beneficial. The stable specific capacity of the silicon microparticle anode was limited to 800 mAh g^{-1} throughout 940 cycles; its maximum specific capacity was 1914 mAh g^{-1}, and its columbic efficiency increased to 91%. After 100 cycles, a grafted lignin binder on silicon/graphite-based electrodes achieved an unaltered specific capacity of 492 mAh g^{-1} [133]. Furthermore, polyacrylic acid (PAA) copolymeric binders (L-co-PAA) have demonstrated promising outcomes when employed to modify lignin (L) binders. It has been demonstrated that silicon-based electrodes can maintain a constant capacity of 939 mAh g^{-1} for over thousand cycles [141].

Starch, a biodegradable and cost-effective water-based binder for LIBs, is an additional attractive binder option. The surface of starch can be modified in various ways to improve its properties. After 100 cycles, silicon-based anodes with starch additions had discharge capacities of less than 500 mAh g^{-1}, but fluorinated starch had a more stable and boosted discharge capacity of 1864–2874 mAh g^{-1}. Using fluorinated starch as a binder in Si–graphite anodes, a discharge capacity of around 600 mAh g^{-1} was achieved after 200 cycles [142]. The modification of the starch surface with polyethylene glycol is advantageous for Si-based anode materials, and the specific capacity of such modified material after 300 cycles at 99.9% PCE reaches approximately 1100 mAh g^{-1}, in contrast to pure starch, which exhibits only 59% PCE [143]. After 200 cycles, cross-linked starch with maleic anhydride displayed an enhanced charge capacity of around 2106 mAh g^{-1}, but non-cross-linked starch displayed a gradual reduction in capacity to less than 500 mAh g^{-1} for Si-based anodes [144].

When a lignin-based binder was used as a binder material in a LiFePO4 cathode, it showed a reversible capacity of 148 mAh g^{-1} at a rate of 0.1 C and 117 mAh g^{-1} at the higher rate of 1 C (a C-rate is a measure of the rate at which a battery is discharged relative to its maximum capacity. A 1C rate means that the discharge current will discharge the entire battery in 1 hour). A lignin-based binder was applied to an NMC111 cathode, which performed similarly to a PVDF binder and had an approximate capacity of 140 mAh g^{-1} at 0.1 C. As the C rate increased, the specific capacity of the lignin-based binder system decreased significantly. In the NMC111 cathode system, increasing C rates resulted in more stable efficacy with the CMC/

lignin binder [145]. Because of their exceptionally high power density while retaining exceptional efficiency and cyclability at low cost, LIBs have established themselves as the dominant energy technology during the last few decades. The quest is growing for enhanced LIBs and 'beyond lithium-ion' alternatives that are less expensive and more adaptable while yet preserving sufficient battery life and power density to suit application requirements. As a result, numerous promising options, such as solid-state, lithium-ion, sodium, multivalent ion, and lithium–sulfur batteries, have been developed.

LIBs have grown in popularity as a result of these qualities, and the entire energy market size has increased by approximately six times. LIBs accounted for practically the whole market for extremely energetic batteries in 2019, and this trend appears set to continue as governments throughout the world favor the usage of electric vehicles and sources of renewable energy. LIBs have increased in cost since 1991, as compared to the \$80/kWh target set by the US State Department of Energy's Energy Storage Grand Challenge. Furthermore, battery applications accounted for over 70% of all global lithium use, and the steady rise in demand has sparked issues about potential resource-depletion-related price increases in the future [146, 147]. Beyond this, sodium ions, being more affordable and abundant in the Earth's crust, are particularly beneficial as insertion ions because, like lithium, sodium is an alkali metal and has very similar chemical properties.

1.3.1.2 Lithium–sulfur (Li–S) batteries

Li–S batteries have significant potential due to their extremely high theoretical energy density of 2600 Wh kg^{-1}. Because of their significant theoretical specific capacity (1672 mAh g^{-1}) and reversible interconversion of S and Li_2S via polysulfide intermediates, lithium–sulfur batteries (Li–S) are excellent contenders for many applications in energy storage. Unfortunately, lithium polysulfide intermediates, which are soluble in ordinary liquid organic electrolytes, generate a polysulfide shuttle between the electrodes during cycling, polluting the entire system and reducing cell reliability and overall cell performance. Because polysulfide breakdown in organic electrolytes causes sulfur volume expansion, the battery's cyclic stability diminishes.

1.3.1.2.1 Electrodes

In Li–S battery development, the mitigation of shuttle mechanisms is one of the primary areas of research. In order to overcome volume expansion of sulfur, it is incorporated into carbon matrices derived from biomass, which can accommodate the volume expansion (figure 1.13). Carbon derived from biomass is mainly selected for Li–S batteries due to its special microstructure, porous nature, and cost-effectiveness [148].

The above figure illustrates how the shuttle effect is inhibited by the biomass-derived porous carbon hybrid material, thus improving specific capacity. The cathode generates lithium polysulfides (Li_2S_x), which dissolve easily in most solvents and shuttle between the cathode and anode during charging and discharging. In this instance, the porous carbon's tremendous surface area provides a physical desorption site, trapping the polysulfides on the cathode. These adsorption sites are, however, insufficient for full entrapment. To further limit the shuttle, porous carbon–metal oxide/sulfide composites or heteroatom doping can be utilized, which produce reliable chemical interactions while conserving more Li ions. This can

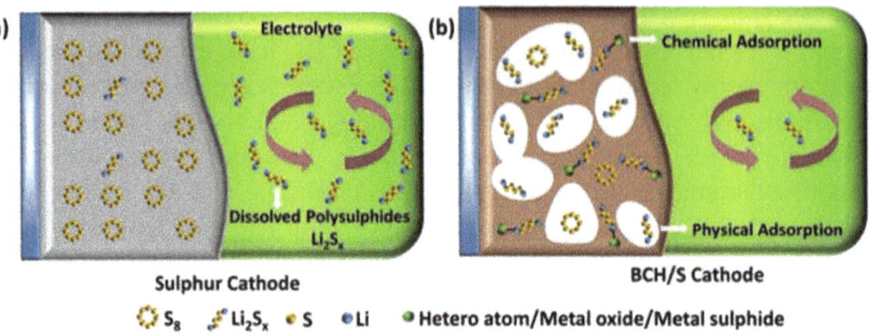

Figure 1.13. (a) Due to the limited adsorption of lithium polysulfides on sulfur, the electrolytes used with sulfur cathodes contain substantial amounts of dissolved lithium polysulfides. (b) A biomass carbon hybrid (BCH)/ sulfur cathode with less dissolved lithium polysulfide due to both physical and chemical adsorption (reproduced from [149], CC BY 4.0).

improve the efficiency as well as the capacity of the cell. Wang *et al* developed porous carbon with a large surface area (1712.6 m^2 g^{-1}) from reed blossoms for a sulfur cathode which prevented sulfur loss for polysulfide formation, effectively accommodating sulfur within its pores. The material had very high efficiency and showed a retention capacity of 908 mAh g^{-1} after 100 cycles [150]. Other efficient methods are the doping of bioextracted carbon with heteroatoms such as nitrogen or sulfur and dual doping with two heteroatoms, which provides additional electrochemical active sites and improved conductivity. Nitrogen doping of porous carbon may result in substantial chemical adsorption of polysulfides, inhibiting polysulfide dissolution. Lignin-derived nitrogen-doped carbon has been utilized, which improved the structural integrity of the electrode and enhanced the electrochemical characteristics by increasing conductivity, reducing the shuttle effect, and hastening ion diffusion. This improved the electrolyte wettability and the kinetics of the reaction. This electrode material showed a high capacity retention of 647.2 mAh g^{-1} [151]. Sulfur-doped aerogels had a very large surface area of 4037 m^2 g^{-1} and a high reversible capacity of 798 mAh g^{-1} after 200 cycles. Notably, the 3D hierarchical structure of the pores was controlled by Fe^{3+}, which favors smooth ion transport and provides a large space for sulfur accommodation. Despite the sulfur host's huge specific surface area, its discharge capacity is poor, which may be attributable to sulfur sites that are impenetrable to Li ions, indicating that just one region cannot help improve efficiency in energy use and storage [152].

1.3.1.2.2 Separator membranes and electrolytes

Lithium–sulfur batteries have a high theoretical capacity (1672 mAh g^{-1}), which is due to the reversible interconversion of sulfur and Li_2S through polysulfide intermediates, thus rendering them suitable for massive amounts of energy storage. As functional separators or polysulfides are immobilized and prohibited at the sulfur cathode side, interlayers can successfully control their shuttle effect. Because the hydroxyl group binds sulfur-containing compounds with significant strength, sustainable polymers may entrap polysulfide ions.

Chitosan is one of the sustainable polymers that can be used to modify membrane separators. Due to the presence of hydroxyl and amine groups in chitosan, which trap polysulfide through strong polysulfide chemisorption, sulfur loss in the electrolyte is diminished. By coating a carbon membrane with chitosan, Chen *et al* constructed Li–S batteries with a specific capacity of 830 mAh g^{-1} after 100 cycles, compared to the electrochemical performance of conventional separators. The improvement was attributed to the chitosan layer because it effectively caught and chemically immobilized polysulfides in the cathode compartment, preventing the shuttle effect. Protein-directed self-assembly was used to create a gelatin-based polysulfide nanofilter, which was inspired by air-filtration principles in which filters must trap pollutants with minimal flow resistance. Polysulfides, unlike the pollutants in air-filtration systems, may be captured by the nanofilter's distinctive porous structure, which also facilitates the movement of Li^+ ions between electrodes. After 200 cycles, the discharge capacity of a commercial Celgard separator coated with the nanofilter increased from 550 to 730 mAh g^{-1} [153].

To avoid polysulfide transfer, the traditional liquid electrolyte and separator can be replaced with a similar gel or solid electrolyte. The electrochemical stability of poly(ethylene oxide) (PEO) and PVDF-HFP polymer membranes used to create electrolytes is low due to their poor ionic conductivity. To enhance their performance, SPEs are the subject of ongoing research and development. SPEs derived from sucrose and starch have been created. An SPE was derived from sucrose after 257 cycles of combining sucrose, borane-tetrahydrofuran complex, PEO, and lithium bis (trifluoromethanesulfonyl) imide (LiTFSI) salt. This SPE exhibited a specific capacity of 620 mAh g^{-1} [154]. Li *et al* developed an SPE from food-quality starch by cross-linking it via a coupling agent (silane agent, KH-560), which resulted in an SPE that could accommodate the minimum amount of LiTFSI (due to the –C–O–C– groups). This starch, which had ten times the dielectric constant and cyclic polar subunits of PEO, stimulated LiTFSI dissociation, increasing Li^+ conductivity to 0.339 mS cm^{-1}. Throughout 100 cycles at a rate of 0.1 C, the resulting lithium–sulfur batteries achieved a minimum discharge capacity of 864 mAh g^{-1}, paving the path for high-performance solid-state batteries.

Fluorine has piqued the curiosity of scientists due to its substantial redox potential as well as its low atomic weight. However, the potential energy density of fluoride ions is still lower than those of magnesium, lithium, and aluminum, and the system confronts significant hurdles in the selection of appropriate electrolytes [146]. In addition, sodium ions, being more affordable and abundant in the Earth's crust, are particularly beneficial as insertion ions because, like lithium, sodium is an alkali metal and has very similar chemical properties to it.

1.3.1.3 Sodium-ion batteries

Na-based energy storage systems have become the focus of current research due to increasing concerns regarding renewable energy technologies and devices. Sodium is abundant in the Earth's crust (2.36 wt.%), is low cost, and is one of the best alternatives to the less abundant lithium. Na-ion-based energy storage systems has garnered significant attention for use in the next generation of sustainable batteries. Na ions have comparable chemical properties and performance to those of Li ions, making them a viable alternative for use in large-scale energy storage systems.

Compared to Li ions, the structural integrity of Na ions is significantly compromised due to their larger radii (1.02 Å), which reduces their interlayer interactions and causes structural collapse during cycling. This problem can be overcome by various modifications to the system, including electrodes, separator membranes, binders, and the electrolyte. Sustainable biomass-derived polymers can be used to improve the performance of SIBs. SIBs have less potential than LIBs to cause global warming, fossil resource depletion, eutrophication, and adverse effects to humans, according to life cycle assessments. In the following sections, we discuss current research into the separators, electrolyte materials, and electrodes of SIBs.

1.3.1.3.1 Separator membranes and electrolytes

Despite being considered a passive component of a battery, the separator–electrolyte pair is critical in determining the battery's electrochemical stability and safety. In SIBs, the function of the separator membranes is to prevent direct contact between the electrode materials while permitting the passage of sodium ions. The separator must comprise a porous, structurally stable material. Various sustainable polymers are utilized to produce more effective separator membranes. Several research studies have focused on developing eco-friendly, highly efficient polymer separator membranes. Using the electrospinning technique, Chen *et al* developed a CA membrane with extremely high surface porosity. The acetyl groups on the surface of CA limit its solubility in electrolytes. Therefore, 0.06 m NaOH was used to convert its acetyl functionalities into hydroxyl moieties, producing a membrane with greater affinity for different electrolytes, including EC/propylene carbonate (PC), EC/dimethylene carbonate (DMC), diglyme, and triglyme electrolytes. At 220 °C, CA separators exhibited good tensile strength (11 MPa), 9% elongation at break, and a capacity retention of 93.78% after 10 000 cycles. Figure 1.14 shows a schematic illustration representing the preparation and surface functionalization of the CA membrane for improved performance in $Na/Na_3V_2(PO_4)_3$ half cell [153].

Zhang *et al* developed SIB separators with adjustable pores that were made from prawn shell chitin nanofiber membranes using a salt-templating method [154]. This method generated dense porous chitin membranes with nanometer dimensions during self-assembly, whereas sodium dihydrogen citrate (SDCA) hindered the rigid

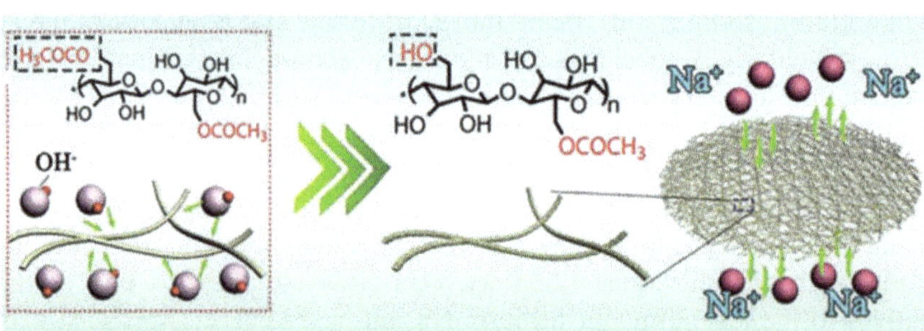

Figure 1.14. Schematic illustration representing the preparation and surface functionalization of the cellulose acetate (CA) membrane for improved performance in $Na/Na_3V_2(PO_4)_3$ half cell (reprinted with permission from [153], Copyright (2018), American Chemical Society).

packing of chitin fibers, resulting in mesoporous structures. This was accomplished via a vacuum drying process that utilized sodium dihydrogen citrate (SDCA). The chitin separators outperformed commercial PP separator membranes in terms of performance, thermal stability at 150 °C, mechanical stability, and electrochemical stability window.

In conventional SIBs, a fiber separator or polyolefin derived from petroleum is soaked in an organic electrolyte containing sodium salts dissolved in ethylene carbonate or propylene carbonate. The ion transference and electrolyte wettability of the separator are reduced due to its low porosity, resulting in high internal resistance and leakage that causes short circuits. To address this issue, sustainable polymers with polar functions are required. Polysaccharides, which include natural polymers (cellulose, chitin, gelatin, and silk) with $CONH_2$, OH, SO_3H, and NH_2 functionalities play an important role in increasing salt dissolution and cation transference within the electrolyte [155].

A sustainable polymer GPE for SIBs was first developed by Mittal *et al.* CNCs with 3D structures keep electrolytes from leaking or collapsing. At room temperature, combining CNFs with Na^+ transfer enhancers results in an effective transfer number of 0.63. A battery containing $Na_2Fe(SO_4)_3$/Na half cells achieved a stable specific capacity after 25 cycles [156]. By bringing together mechanically stiff nanocrystals of cellulose, flexible CNFs, and glutaraldehyde, Mittal *et al* successfully created a gel electrolyte with a mesoporous structure. Figure 1.15 represents the bottom-up and top-down approach for the formation of mesoporous gel electrolytes from biomass.

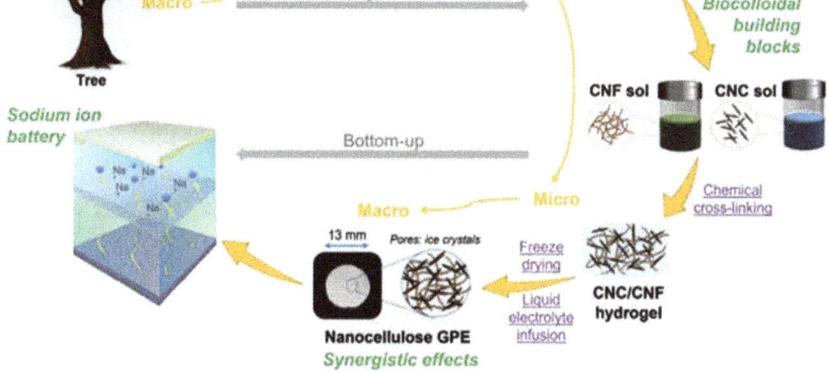

Figure 1.15. A schematic showing the fabrication of a gel electrolyte from biomass: CNCs and CNFs are extracted from biomass using a top-down approach and a bottom-up approach that combines self-assembly and sol–gel chemistry to produce a CNC/CNF cryogel (reproduced from [157], CC BY 4.0).

1.3.1.3.2 Battery electrodes

Two approaches are recommended and may be used to address the lower cyclic stability and columbic efficency in future battery chemistries: either Li-based systems are replaced by more abundant alternatives such as Na or inorganic nonrenewable components are replaced by organic analogs that can be developed from biomass resources. The search is ongoing for renewable, low-cost electrode materials with which to develop sustainable electrochemical energy storage systems. In the extensive prior research into SIBs, metals, metal oxides/sulfides, alloys, and organic compounds have all been considered as potential anode materials. It is crucial to design an electrode material to overcome issues such as low cycle life, poor efficiency, low coulombic efficiency, and high hysteresis. Carbonaceous materials generated from biomass are preferred for Na^+ storage and transport due to their hierarchical pore architectures and higher interlayer lattice distance, and as a result, they are used as sustainable electrode materials. Different morphologies of biomass-derived carbon, including spheres, tubes, fibers, and sheets, can be doped with heteroatoms such as B, N, and S [158] and these morphologies can be changed to highly porous structures which allow the fast diffusion of Na ions. Such doping can increase the number of active sites, which can improve the Na-ion absorption capacity during the charging process. Diverse sustainable polymers can be converted to carbonaceous materials and have been used to develop efficient anode materials. Chitin, which consists of nitrogen-containing N-acetyl groups and is the second most abundant polymer, is one of the most important sustainable polymers due to its interconnected network structure, sustainability, and availability.

By pyrolyzing chitin, Hao *et al* created nitrogen-doped amorphous carbon nanofibers (figure 1.16). Their electrode provided a large reversible capacity of 320.6 mAh g^{-1} in addition to excellent rate capability and extended cyclability. Unique one-dimensional mesoporous nanofibers which facilitated electron/electrolyte transmission and an N-doped amorphous nanostructure, which boosted electrical conductivity and the number of active sites, were primarily responsible for its superior electrochemical performance [155].

Lignin-based anode materials are one of the most promising candidates for SIBs, and research is currently being conducted using lignin derived from various sources (rice rusk, pitch, oak leaves, peat moss, etc.) as an anode material. As previously

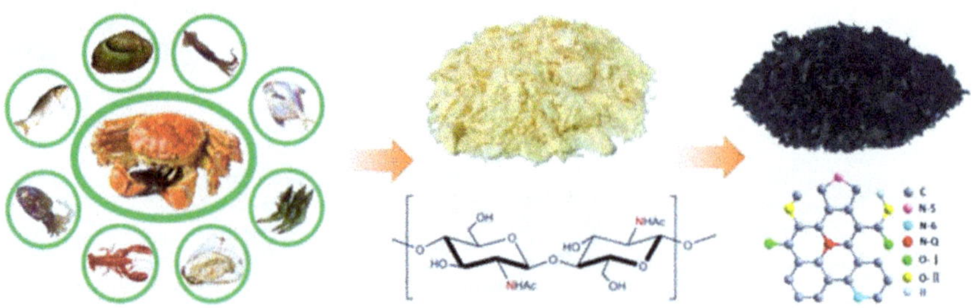

Figure 1.16. A schematic illustration of the synthesis of nitrogen-doped amorphous carbon nanofibers (NACFs) derived from biowaste chitin (reprinted from [155], Copyright (2018), with permission from Elsevier).

discussed, doping can enhance the cycling performance of lignin-based carbon anodes. Numerous scientific studies have indicated that nitrogen doping of porous carbon structures improves battery performance. By introducing SiO_2 nanoparticles with a particle size of 100 nm to the precursor and etching the material after pyrolyzing it in N_2, Du *et al* created an N-doped lignin porous carbon structure. Thus, a uniform nanoporous honeycomb-structured N-doped material with low ion diffusion resistance and a short ionic transport distance was developed. Consequently, after 1100 cycles at a high current density of 1 A g^{-1}, a high reversible capacity of 100 mAh g^{-1} was attained, and the corresponding capacity decay was only 8% [159]. Xu *et al* developed nitrogen-doped porous carbon microspheres for use as an anode material in SIBs. The formation of N-doped hierarchical porous carbon microspheres (NHPCSs) from a nickel metal–organic framework (MOF) integrated heteroatom doping and porous structures into a single material (figure 1.17). It was demonstrated that this hierarchical porous structure delivered more active Na^+ storage sites, quick Na^+ diffusion pathways, and volume reduction during the sodiation process. This material exhibited a high reversible capacity of 291 mAh g^{-1}, an exceptional rate performance of 256 mAh g^{-1} at a high current of 1000 mA g^{-1}, and a cycling stability of 204 mAh g^{-1} for more than 200 cycles [160].

Yang *et al* described a two-stage process for producing activated hard carbon from walnut shells that included carbonization and hydrothermal degradation. The resulting carbon acted as negatively charged electrode material and had a capacity of 244.6 mAh g^{-1} after 100 cycles. The carbon had 3.8 layer spacing and short-range pseudo graphitic domains. The inclusion of mesopores (2–20 nm) improved the charge storage behavior within the expandable carbon, which was associated with improved electrochemical performance [161]. Thus, these techniques can enhance the functionality and cyclic stability of sodium-ion electrode materials.

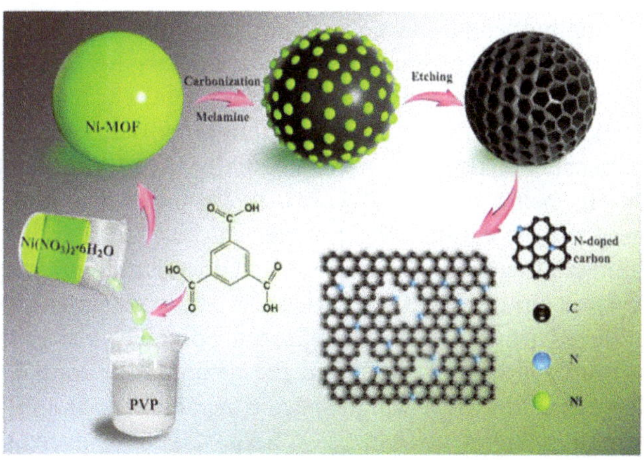

Figure 1.17. A schematic representation of the fabrication of N-doped hierarchical porous carbon microspheres (NHPCSs) (reproduced from [160], CC BY 4.0).

1.3.1.3.3 Supercapacitor materials

Because of their extended life cycle, high energy and power density, and outstanding reversibility, supercapacitors, also known as electric double-layer capacitors (EDLCs), offer an innovative option for the storage of clean and renewable energy. EDLCs store energy by establishing an electric double layer between the electrode and electrolyte at the inner interface. One of the most difficult challenges is storing and converting energy into various usable forms in order to meet the rising global demand for energy. The current excessive use of fossil fuels creates significant environmental challenges. It is essential to design a sustainable and renewable electrical energy storage system to meet the demands of energy-driven applications. The electrolyte, electrode material, and separator membrane influence supercapacitor performance. Sustainable polymer-based supercapacitor materials protect the environment and improve the electrochemical performance of supercapacitors. Compared to traditional carbon materials, raw biochar is frequently rich in oxygenated functional groups, such as OH, C=O, and COOH, which can improve the performance of supercapacitors. By changing the pyrolysis conditions, such as the temperature and heating rate, the oxygen content of biochar can be altered. The final biochar material can be functionalized to improve the performance of the pseudo capacitance. Traditional biomass consists of cellulose, lignin, hemicellulose, oil or fat, protein, and chitosan, among other components. Due to its availability, low cost, high carbon output, and product porosity, lignin-rich biomass is among the most often employed precursors. Heteroatom doping with nonmetallic elements such as N, O, B, S, and P is another approach that has recently been discovered to increase the performance of porous carbon for supercapacitors. In addition to the use of biochar and doping procedures, the surface functional groups of carbon influence the charge storage capacity of supercapacitor electrodes.

1.3.1.3.4 Supercapacitor electrolytes

The electrolyte is a critical component of a supercapacitor. Devices that store power, such as supercapacitors, are usually configured with a liquid electrolyte between two electrodes. There are many issues faced by liquid electrolyte supercapacitors, such as electrolyte flammability, gas production upon overcharging and discharging, and thermal runaway. These issues can be solved using solid electrolytes. Nevertheless, the use of hazardous, nonbiodegradable materials causes serious environmental problems. Thus, in order to develop environmentally friendly products, various bio-based materials are being developed [162]. A flexible GPE film with excellent conductivity based on the eco-friendly polymer carboxylated chitosan (CCH) was created by solution coating and hydrochloric acid cross-linking. Using HCl to phase separate carboxylated chitosan hydrogel, an eco-friendly gel polymer electrolyte film was initially produced from sustainable chitosan polymer. The principal chemical for cross-linking film synthesis is HCl, and the technique is simple, fast, and non-polluting, resulting in a film with a high specific capacitance of 45.9 F g^{-1} at 0.5 A g^{-1} and a maximum energy density of 5.2 Wh kg^{-1}. The produced film exhibited a maximum electrolyte absorption rate of 742.0 wt.%, exceptional flexibility, and a maximum ionic conductivity of 8.69×10^{-2} Scm^{-1} (figure 1.18) [163].

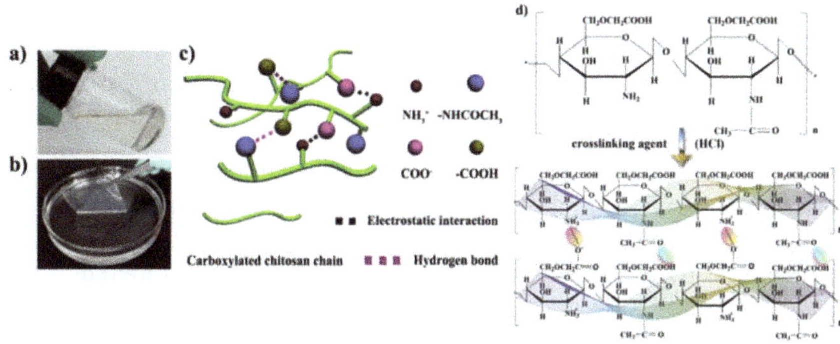

Figure 1.18. (a) Photographs of carboxylated chitosan solution, (b) cross-linked CCH film, and (c) and (d) schematic illustration of internal cross-linking for CCH film (reprinted from [163], Copyright (2019), with permission from Elsevier).

Energy storage devices for portable and wearable electronics have been made possible by combining the electrolyte and separator into a single polymer membrane. A polyethylene glycol (PEG)-grafted poly(arylene ether ketone) membrane matrix containing a $LiClO_4$ gel based on chitosan was created to produce a microporous polymer electrolyte. When used in a solid-state EDLC, the polymer electrolyte demonstrated good mechanical properties, minimal leakage, and great cycling stability. However, traditional polymer electrolytes are often difficult to prepare or are neither renewable nor biodegradable, limiting their broad application. It would be desirable, if challenging, to create a straightforward method for manufacturing multifunctional electrolytes based on polymers with greater ionic conductivity which are recyclable and compostable and have satisfactory mechanical properties. Cellulose is the most plentiful, renewable, and biodegradable sustainable polymer that can fulfill the demand for green/sustainable electrolytes. Cellulose-based membranes, notably CA, methylcellulose, and cellulose-based composites, have been utilized as separation membranes in energy storage devices [164]. Membranes with superior electrolyte retention and enhanced ionic conductivity, which may be obtained with a separator that has greater porosity, create less heat and lose less energy. According to previous studies, CNFs and electrospun cellulose mats have been utilized to develop nanoporous cellulose electrolyte membranes with high porosity. A flexible and transparent renewable mesoporous cellulose (mCel) membrane was created using a phase-inversion technique and an ionic liquid (IL) as the solvent. A solution of 1-butyl-3-methylimidazolium chloride ([Bmim]Cl) can be used to dissolve cellulose. This solution can subsequently be used to regenerate cellulose into a film using pure water as a nonsolvent. To create mesopores ($\approx$24.7 nm) and obtain a high porosity of 71.78% during the drying process, two PTFE millipore membranes were employed as templates. Furthermore, a highly integrated planar-type microsupercapacitor (MSC) can easily be built without the use of complex devices by simply placing the electrode materials onto a mCel-membrane-based polymer electrolyte. The resulting MSC had a high capacitance of 153.34 mF cm^{-2} and a volumetric capacitance of 191.66 F cm^{-3}, which is one of the highest values among all carbon-based MSC devices. This flexible, mesoporous, renewable cellulose membrane holds considerable promise for use in flexible, solid-state, portable energy storage devices other than supercapacitors [165].

1.3.1.3.5 Electrode materials

Typical biomass consists of hemicellulose, lignin, cellulose, oil, protein, chitosan, and other substances. Due to its accessibility, cost, high carbon output, and product porosity, lignin-rich biomass is the most preferred precursor. The composition, microstructure, and metallic and nonmetallic constituents of the different precursors influence the attributes of carbon-based products. The biomass structure strongly influences the microstructure of a product, while nonmetallic components (nitrogen and sulfur) are used as heteroatom dopants and metallic elements in the biomass (sodium and potassium) can serve as nascent activation agents during synthesis. Figure 1.19 shows TEM images of carbon materials derived from different biomass precursors.

Since the procedure used for synthesis has been found to have a significant impact on the end product's physicochemical features, it is necessary to process the carbon compounds made from plant matter or biowaste to obtain those features. Significant techniques and processes have been identified and documented to date. These approaches emphasize their distinctiveness by being simple, efficient, and inexpensive. The pyrolitic method involves breaking down organic materials at extreme temperatures and in the absence of oxygen to create charcoal. Hemicelluloses, cellulose, and lignin constitute the majority of plant biomass. Activation is the most widely used technique for creating holes and expanding the total surface area of carbon-based materials. It can be broadly classified into three types: chemical

Figure 1.19. Transmission electron microscopy (TEM) images of carbon materials derived from biomass precursors (reprinted from [166], Copyright (2020), with permission from Elsevier).

activation, physical activation, and self-activation processes. To create peanut shell nanosheet carbon (PSNC), Ding *et al* utilized a carbonization process with a KOH activation method and an air activation technique. Compared to air activation, the activation technique yielded graphene sheets with a remarkable specific surface area of 2396 $m^2\ g^{-1}$ [167]. Doping with heteroatoms is one approach for improving the overall electrochemical performance of batteries. The insertion of foreign atoms into the carbon-sourced matrix of the biochar resource is another successful way to boost capacitance and alter the pore structure and area. Nitrogen is the heteroatom that most frequently generates pseudocapacitive faradaic reactions in carbon materials; hence, nitrogen-doped biochar materials are widely used as electrode materials for supercapacitors.

Combining precarbonization with KOH activation, Ma *et al* produced multilayer porous carbonaceous materials using sakura petals as the starting material. The material's electrochemical efficacy as a supercapacitor electrode was enhanced by its substantial specific surface area, beneficial pore size distribution, low percentage of oxygen-dense groups and N functionalities, and partially graphitized phase. A supercapacitor was produced during the activation process when the mass-to-volume ratio of KOH to sakura carbon was 4. In addition to having a lower density and greater flexibility compared to those of conventional metal-based support materials, cellulose-based foams and aerogels also have a number of benefits that improve electrochemical performance, such as hydrophilic surfaces, rich absorptive sites, and an abundance of hierarchical pores for charge storage [168].

In addition to metal oxides, aerogel and cellulose–graphene hybrid foam materials may be employed with conductive polymers such as PPy and polyaniline (PANI). Citric acid A–Fe^{3+} complexes have been employed as oxidant precursors to control the deposition of PPy and generate flexible CNF/rGO/PPy aerogel electrodes with clearly defined three-dimensional porous topologies. Wang *et al* developed a small, porous, flexible supercapacitor electrode comprising graphene oxide (GO) and CNFs using a hydrothermal process and freeze-drying. Symmetrical supercapacitors perform admirably in terms of double-layer capacitance, with a specific capacity of 125.5 F g^{-1} at scan rates comparable to those of CNF/graphene foam electrodes [169]. After 5000 cycles, the CNF/graphene foam had a specific capacitance retention of up to 98.3%.

Adjusting the preparation methods may alter the characteristics of carbon electrodes based on lignin. Lignin-sourced carbons can be classified into three classes depending on their topologies and processing techniques: activated, templated, and composite carbons. Lignin undergoes high-temperature pyrolysis to become lignin-derived char during the process of carbonization. The next step is to activate lignin char to create porous compounds based on activated carbon. Physical and chemical activation are the two primary categories of lignin activation. The physical method involves activating the lignin char using CO_2 or steam at elevated temperatures. Chemical activation, on the other hand, involves combining char with chemical activators (KOH, K_2CO_3, etc) and then heating the combination in an inert environment [170].

Chitosan has garnered the greatest interest among these sustainable polymers because of its superior mechanical strength and high ionic conductivity. Due to the presence of lone pair electrons in the $-NH_2$, –OH, and C–O–C (glycosidic linkage) groups in its structure, it is able to conduct ions. Chitosan is a renewable and sustainable resource that can be used to create porous carbons more effectively, affordably, and sustainably. A newly developed double-doped porous carbon nanosheet made from N-doped chitosan and a nickel nitrate precursor demonstrated extremely high specific capacitance (614.6 F g^{-1}) and a cycle stability of 85.61% capacitance retention after 10 000 cycles. This electrode has the greatest specific capacitance of all porous carbon electrodes currently under development. These findings provide compelling evidence for the use of chitosan in the manufacture of supercapacitor electrodes, which has a promising future. Table 1.2 summarizes an overview of current sustainable polymer-based battery and supercapacitor materials and their electrochemical performances [171].

1.4 Perspectives and future directions

Looking to the future, the recent technological advancement in the field of sustainable polymers for energy harvesting and storage is gaining significant momentum in mitigating climate change and enabling the transition to a more sustainable energy system through the reduction of reliance on fossil fuels. These polymers play a key role in various technologies, such as solar cells, batteries, supercapacitors, and energy storage materials. Here are some perspectives and future directions for sustainable polymers in the field of energy harvesting and storage:

- Advances in the design and synthesis of sustainable polymers are vital. Researchers are focused on producing revolutionary polymers with improved performance, stability, and environmental sustainability. This includes the development of biodegradable polymers, organic polymers, and polymers derived from renewable resources.
- More focus is being placed on the development of biodegradable polymers for energy storage devices. These polymers, primarily in single-use and recyclable electronics applications, can reduce the environmental impact of batteries and supercapacitors. It is feasible to reduce the amount of hazardous waste produced and the reliance on nonrenewable resources by integrating biopolymers into energy system designs. It is indeed becoming more important to apply circular economy concepts. Waste and resource consumption can be reduced by recycling and upcycling the polymer materials used in energy harvesting and storage systems.
- Green chemistry principles, such as the use of nontoxic solvents and reduced energy consumption, are essential for more sustainable manufacturing processes. Sustainable polymers are essential for improving the performance and safety of energy storage systems, such as LIBs. Polymers are used as electrolytes, separators, and even in the electrode materials. Future developments will focus on increasing energy density, cycle life, and thermal stability and implementing efficient, environmentally friendly production techniques such as roll-to-roll manufacture and 3D printing to enable the large-scale

Table 1.2. An overview of current sustainable polymer-based battery and supercapacitor materials along with their performances.

Sustainable polymer-based system	Function	Performance	References
Carbonized silk fibroin nanofiber film	Anode and anode interlayers in Li/S batteries	Reversible capacity of 1913 mAh g^{-1} with 97.3% efficiency and 69% capacity retention after 200 cycles	[172]
Keratin-derived carbon added to $TiNb_2O_7$	Anode material for LIBs	Reversible capacity of 356 mAh g^{-1} with 55.0% efficiency and 85% retention after 50 cycles	[173]
Cellulose nanocrystal membrane	Separator membrane for LIBs	Reversible capacity of 122 mAh g^{-1} with 100% efficiency after initial decay and more than 90% retention	[174]
Cross-linked BC gel	Electrolyte for LIBs	Reversible capacity of 141.2 mAh g^{-1} with 89.46% efficiency at 0.5 C and 104.2% retention after 150 cycles	[175]
Lignocellulose-based gel	Electrolyte for LIBs	Reversible capacity of 1186.3 mAh g^{-1} and 55.1% capacity retention after 100 cycles at 20 mA g^{-1}	[176]
Nitrogen-doped carbon nanowire array coating	Electrodes for supercapacitors	Specific capacitance: 275.6 F g^{-1} at 0.5 A g^{-1}	[177]
Pyrolyzed BC aerogel	Electrodes for supercapacitors	Specific capacitance: 560.8 F g^{-1} at 1 A g^{-1}	[178]
Self-codoped N and O porous biochar	Electrodes for supercapacitors	Specific capacitance: 228 F g^{-1} at 1 A g^{-1}. Cyclic stability of 88% capacitance retention and stable at 5000 cycles	[179]
CNFs derived from PI and lignin precursors	Electrodes for supercapacitors	High specific capacitance of 83.6 F g^{-1} at 1 Ag^{-1}, capacitance retention of ~100% after 1000 cycles	[180]
Shrimp chitosan-derived carbon (CCS) and CCS/NiO@$Ni(OH)_2$ (CSSN) aerogel	Electrodes for supercapacitors	High capacitance of 316 mAh g^{-1} at 1.0 A g^{-1}, 84% capacitance retention after the 10 000 cycles	[181]
Lignin-grafted with sodium polyacrylate, PAL–NaPAA	Binder for LIBs	Specific capacity of 1914 mAh g^{-1} after 100 cycles	[182]
Lignin PAA copolymeric binders (L-*co*-PAA)	Binders for silicon anode	Specific capacity of 939 mAh g^{-1} with 84% initial columbic efficiency	[141]

production of polymer-based batteries and supercapacitors for commercial applications.

- Sustainable polymers offer opportunities for flexible and lightweight wearable energy harvesting devices. Researchers might explore the integration of biopolymer-based nanogenerators with flexible and wearable electronics, enabling self-powered wearable devices that can harvest energy from body movements. This integration could lead to the development of self-powered health monitoring systems and electronic textiles. Further exploration may be conducted into developments for powering implantable medical devices. These devices could harvest energy from natural body movements, eliminating the need for external power sources or frequent battery replacements, thus enhancing the longevity and functionality of implantable devices.
- Sustainable polymers could cause a paradigm shift in emerging technologies such as wearable electronics, additive manufacturing/3D printing, Internet of Things (IoT) devices, and smart grids. These applications require lightweight, flexible, biocompatible, recyclable, sustainable, and low-cost energy solutions. Sustainable polymer-based energy harvesters, such as piezoelectric and triboelectric generators, can be integrated into IoT systems to harness energy from ambient vibrations, mechanical movements, and environmental changes, contributing to the development of self-powered and sustainable IoT networks.
- Government policies and regulations are likely to play a significant role in shaping the future of sustainable polymers for energy harvesting and storage. Research and innovation will be propelled by incentives and regulations aimed at mitigating the environmental effect of energy storage technology. Interdisciplinary research between materials scientists, chemists, engineers, biologists, and environmental experts is crucial. Interdisciplinary research will help to identify new sustainable materials and technologies that can revolutionize the energy sector.

1.5 Summary

In this chapter, the use of sustainable polymer materials for energy harvesting and storage systems was briefly reviewed, with an emphasis on current research developments. The structural and mechanical properties, high surface area, and the suitability of the desired chemical functionalities of sustainable polymer materials have enabled remarkable improvements in the performance of these materials as electrodes, electrolytes, separator membranes, binders, etc. for both energy storage and harvesting systems. This chapter reviewed advancements in battery research, including Li-ion batteries, Li–S batteries, SIBs, supercapacitors, hydrogen generation, photovoltaic solar cells, and nanogenerators. Wearable triboelectric nanogenerators made from sustainable polymers decrease environmental impact and may be disposed of after use. However, these bio-TENGs are still in their infancy in terms of output, and their applicability in the real world has not yet been demonstrated. More studies must be conducted to boost the triboelectric, mechanical, and surface roughness properties of sustainable polymers for use in TENGs. These innovations are adaptable and advantageous to wearable electronics, biomedical sensors, body-implanted bio-TENGs, and IoT-based healthcare systems. In conclusion, sustainable polymers have the potential to revolutionize the field of energy harvesting and storage, making it more environmentally friendly and efficient. Ongoing research and development efforts, along with a focus on material innovation, recycling, and sustainable manufacturing, will pave the way for a cleaner and more sustainable energy future.

References and further reading

[1] Mohanty A K, Wu F, Mincheva R, Hakkarainen M, Raquez J-M, Mielewski D F, Narayan R, Netravali A N and Misra M 2022 *Nat. Rev. Methods Primers* **2** 46

[2] Aziz S B, Brza M A, Dannoun E M, Hamsan M H, Hadi J M, Kadir M F and Abdulwahid R T 2020 *Molecules* **25** 4503

[3] Chen Y, Qiu L, Ma X, Dong L, Jin Z, Xia G, Du P and Xiong J 2020 *Carbohydr. Polym.* **234** 115907

[4] Mansouri S A, Nematbakhsh E, Ahmarinejad A, Jordehi A R, Javadi M S and Matin S A A 2022 *J. Energy Storage* **50** 104206

[5] Alexandre S A, Silva G G, Santamaría R, Trigueiro J P C and Lavall R L 2019 *Electrochim. Acta* **299** 789–99

[6] Aziz S B 2013 *Iran. Polym. J.* **22** 877–83

[7] Zhu W-B, Yi F-L, Huang P, Zhang H, Tang Z-H, Fu Y-Q, Wang Y-Y, Huang J, Dong G-H and Li Y-Q 2021 *J. Mater. Chem.* A **9** 26758–66

[8] Bragulla H H and Homberger D G 2009 *J. Anat.* **214** 516–59

[9] Angel N, Li S, Yan F and Kong L 2022 *Trends Food Sci. Technol.* **120** 308–24

[10] Kebabsa L, Kim J, Lee D and Lee B 2020 *Appl. Surf. Sci.* **511** 145313

[11] Chen H, Yu F, Wang G, Chen L, Dai B and Peng S 2018 *ACS Omega* **3** 4724–32

[12] Peng K, Wang W, Zhang J, Ma Y, Lin L, Gan Q, Chen Y and Feng C 2022 *Carbohydr. Polym.* **278** 118927

[13] Wang G, Lin Z, Jin S, Li M and Jing L 2022 *J. Energy Storage* **45** 103525

[14] Guo Y, Wang T, Chen X and Wu D 2021 *J. Power Sources* **507** 230252

[15] Ababneh H and Hameed B 2021 *Int. J. Biol. Macromol.* **186** 314–27

[16] Shu Y, Bai Q, Fu G, Xiong Q, Li C, Ding H, Shen Y and Uyama H 2020 *Carbohydr. Polym.* **227** 115346

[17] Shi H H, Pan Y, Xu L, Feng X, Wang W, Potluri P, Hu L, Hasan T and Huang Y Y S 2023 *Nat. Mater.* **22** 1294–303

[18] Wang Z 2014 *Farad. Disc.* **176** 447–58

[19] Cao X, Jie Y, Wang N and Wang Z L 2016 *Adv. Energy Mater.* **6** 1600665

[20] Tian G, Deng W, Gao Y, Xiong D, Yan C, He X, Yang T, Jin L, Chu X and Zhang H 2019 *Nano Energy* **59** 574–81

[21] Lingam D, Parikh A R, Huang J, Jain A and Minary-Jolandan M 2013 *Int. J. Smart Nano Mater.* **4** 229–45

[22] Fan F-R, Tian Z-Q and Wang Z L 2012 *Nano Energy* **1** 328–34

[23] Wu C, Wang A C, Ding W, Guo H and Wang Z L 2019 *Adv. Energy Mater.* **9** 1802906

[24] Luo J, Xu L, Tang W, Jiang T, Fan F R, Pang Y, Chen L, Zhang Y and Wang Z L 2018 *Adv. Energy Mater.* **8** 1800889

[25] Chen A, Zhang C, Zhu G and Wang Z L 2020 *Adv. Sci.* **7** 2000186
[26] Liu Y, Mo J, Fu Q, Lu Y, Zhang N, Wang S and Nie S 2020 *Adv. Funct. Mater.* **30** 2004714
[27] Lee B-Y, Kim D H, Park J, Park K-I, Lee K J and Jeong C K 2019 *Sci. Technol. Adv. Mater.* **20** 758–73
[28] Ma C, Gao S, Gao X, Wu M, Wang R, Wang Y, Tang Z, Fan F, Wu W and Wan H 2019 *InfoMat* **1** 116–25
[29] Yao C, Yin X, Yu Y, Cai Z and Wang X 2017 *Adv. Funct. Mater.* **27** 1700794
[30] Kim M-K, Kim M-S, Kwon H-B, Jo S-E and Kim Y-J 2017 *RSC Adv.* **7** 48368–73
[31] Varghese H, Hakkeem H M A, Chauhan K, Thouti E, Pillai S and Chandran A 2022 *Nano Energy* **98** 107339
[32] Varghese H, Hakkeem H M A, Farman M, Thouti E, Pillai S and Chandran A 2022 *Results Eng.* **16** 100550
[33] Chandrasekhar A, Alluri N R, Saravanakumar B, Selvarajan S and Kim S-J 2017 *J. Mater. Chem.* C **5** 1810–5
[34] Peng J, Zhang H, Zheng Q, Clemons C M, Sabo R C, Gong S, Ma Z and Turng L-S 2017 *Nanoscale* **9** 1428–33
[35] Mi H-Y, Jing X, Zheng Q, Fang L, Huang H-X, Turng L-S and Gong S 2018 *Nano Energy* **48** 327–36
[36] Sriphan S, Charoonsuk T, Maluangnont T, Pakawanit P, Rojviriya C and Vittayakorn N 2020 *Adv. Mater. Technol.* **5** 2000001
[37] Qian C, Li L, Gao M, Yang H, Cai Z, Chen B, Xiang Z, Zhang Z and Song Y 2019 *Nano Energy* **63** 103885
[38] Zhang C, Lin X, Zhang N, Lu Y, Wu Z, Liu G and Nie S 2019 *Nano Energy* **66** 104126
[39] Bai Z, Xu Y, Zhang Z, Zhu J, Gao C, Zhang Y, Jia H and Guo J 2020 *Nano Energy* **75** 104884
[40] Kim H-J, Yim E-C, Kim J-H, Kim S-J, Park J-Y and Oh I-K 2017 *Nano Energy* **33** 130–7
[41] Shao Y, Feng C-P, Deng B-W, Yin B and Yang M-B 2019 *Nano Energy* **62** 620–7
[42] Funayama R, Hayashi S and Terakawa M 2023 *ACS Sustain. Chem. Eng.* **11** 3114–22
[43] Jing X, Li H, Mi H-Y, Feng P-Y, Tao X, Liu Y, Liu C and Shen C 2020 *ACS Appl. Mater. Interfaces* **12** 23474–83
[44] Jao Y-T, Yang P-K, Chiu C-M, Lin Y-J, Chen S-W, Choi D and Lin Z-H 2018 *Nano Energy* **50** 513–20
[45] Bao Y, Wang R, Lu Y and Wu W 2017 *APL Mater.* **5** 074109
[46] Sarkar P K, Kamilya T and Acharya S 2019 *ACS Appl. Energy Mater.* **2** 5507–14
[47] Zhu Z, Xia K, Xu Z, Lou H and Zhang H 2018 *Nanoscale Res. Lett.* **13** 6
[48] Ccorahua R, Cordero A, Luyo C, Quintana M and Vela E 2019 *MRS Adv.* **4** 1315–20
[49] Ccorahua R, Huaroto J, Luyo C, Quintana M and Vela E A 2019 *Nano Energy* **59** 610–8
[50] Khandelwal G, Joseph Raj N P M, Alluri N R and Kim S-J 2021 *ACS Sustain. Chem. Eng.* **9** 9011–7
[51] An S, Sankaran A and Yarin A L 2018 *ACS Appl. Mater. Interfaces* **10** 37749–59
[52] Yu Z, Wang Y, Zheng J, Xiang Y, Zhao P, Cui J, Zhou H and Li D 2020 *Nano Energy* **68** 104382
[53] Feng Y, Yu J, Sun D, Dang C, Ren W, Shao C and Sun R 2022 *Nano Energy* **98** 107284
[54] Tanguy N R, Rana M, Khan A A, Zhang X, Tratnik N, Chen H, Ban D and Yan N 2022 *Nano Energy* **98** 107337

[55] Wang J, Chen Y, Xu Y, Mu J, Li J, Nie S, Chen S and Xu F 2022 *Sustain. Energy Fuels* **6** 1974–82
[56] Sun D, Feng Y, Sun S, Yu J, Jia S, Dang C, Hao X, Yang J, Ren W and Sun R 2022 *Adv. Funct. Mater.* **32** 2201335
[57] Edberg J, Mulla M Y, Hosseinaei O, Alvi N u H and Beni V 2022 A Forest-Based Triboelectric Energy Harvester *Global Challenges* **2022** 2200058
[58] Jo H, Park D, Joo M, Choi D, Kang J, Ha J, Kim K H, Kim K H and An S 2023 *EcoMat* **5** e12413
[59] Nanthagal K, Khoonsap S, Harnchana V, Suphasorn P, Chanlek N, Sinthiptharakoon K, Lapawae K and Amnuaypanich S 2022 *ACS Sustain. Chem. Eng.* **11** 1311–23
[60] Chen W, Li C, Tao Y, Lu J, Du J and Wang H 2023 *Nano Energy* **116** 108802
[61] Valentini L, Rescignano N, Puglia D, Cardinali M and Kenny J 2015 *Eur. J. Inorg. Chem.* **2015** 1192–7
[62] Liang Q, Zhang Q, Yan X, Liao X, Han L, Yi F, Ma M and Zhang Y 2017 *Adv. Mater.* **29** 1604961
[63] Pang Y, Xi F, Luo J, Liu G, Guo T and Zhang C 2018 *RSC Adv.* **8** 6719–26
[64] Zhang L, Liao Y, Wang Y C, Zhang S, Yang W, Pan X and Wang Z L 2020 *Adv. Funct. Mater.* **30** 2001763
[65] Zheng Q, Fang L, Guo H, Yang K, Cai Z, Meador M A B and Gong S 2018 *Adv. Funct. Mater.* **28** 1706365
[66] Wang R, Gao S, Yang Z, Li Y, Chen W, Wu B and Wu W 2018 *Adv. Mater.* **30** 1706267
[67] Pongampai S, Charoonsuk T, Pinpru N, Pulphol P, Vittayakorn W, Pakawanit P and Vittayakorn N 2021 *Compos. B: Eng.* **208** 108602
[68] Hasan M M, Islam M D and Rashid T U 2020 Biopolymer-based electrolytes for dye-sensitized solar cells: a critical review *Energy Fuels* **34** 15634–71
[69] Yao C, Hernandez A, Yu Y, Cai Z and Wang X 2016 *Nano Energy* **30** 103–8
[70] Wang M, Li W, You C, Wang Q, Zeng X and Chen M 2017 *RSC Adv.* **7** 6772–9
[71] Yang B, Yao C, Yu Y, Li Z and Wang X 2017 *Sci. Rep.* **7** 4936
[72] Uddin A I and Chung G-S 2017 *Sens. Actuators* B **245** 1–10
[73] Cui P, Parida K, Lin M F, Xiong J, Cai G and Lee P S 2017 *Adv. Mater. Interfaces* **4** 1700651
[74] He X, Zou H, Geng Z, Wang X, Ding W, Hu F, Zi Y, Xu C, Zhang S L and Yu H 2018 *Adv. Funct. Mater.* **28** 1805540
[75] Šutka A, Ruža J, Järvekülg M, Linarts A, Mālnieks K, Jurķāns V, Gorņevs I, Blūms J, Rubenis K and Knite M 2018 *J. Electrostat.* **92** 1–5
[76] Srither S R, Shankar Rao D S and Krishna Prasad S 2018 *ChemistrySelect* **3** 5055–61
[77] Luo J, Wang Z, Xu L, Wang A C, Han K, Jiang T, Lai Q, Bai Y, Tang W and Fan F R 2019 *Nat. Commun.* **10** 5147
[78] Shi K, Huang X, Sun B, Wu Z, He J and Jiang P 2019 *Nano Energy* **57** 450–8
[79] Shi X, Chen S, Zhang H, Jiang J, Ma Z and Gong S 2019 *ACS Sustain. Chem. Eng.* **7** 18657–66
[80] Chen S, Jiang J, Xu F and Gong S 2019 *Nano Energy* **61** 69–77
[81] Parandeh S, Kharaziha M and Karimzadeh F 2019 *Nano Energy* **59** 412–21
[82] Oh H, Kwak S S, Kim B, Han E, Lim G H, Kim S W and Lim B 2019 *Adv. Funct. Mater.* **29** 1904066
[83] Wang L and Daoud W A 2019 *Nano Energy* **66** 104080

[84] Hao S, Jiao J, Chen Y, Wang Z L and Cao X 2020 *Nano Energy* **75** 104957
[85] Shi K, Zou H, Sun B, Jiang P, He J and Huang X 2020 *Adv. Funct. Mater.* **30** 1904536
[86] Sun Z, Yang L, Liu S, Zhao J, Hu Z and Song W 2020 *Sensors* **20** 506
[87] Beyranvand M and Gholizadeh A 2020 *Curr. Appl. Phys.* **20** 226–31
[88] Rajabi-Abhari A, Kim J-N, Lee J, Tabassian R, Mahato M, Youn H J, Lee H and Oh I-K 2020 *ACS Appl. Mater. Interfaces* **13** 219–32
[89] Roy S, Ko H-U, Maji P K and Kim J 2020 *Chem. Eng. J.* **385** 123723
[90] Jakmuangpak S, Prada T, Mongkolthanaruk W, Harnchana V and Pinitsoontorn S 2020 *ACS Appl. Electron. Mater.* **2** 2498–506
[91] Kim J-N, Lee J, Go T W, Rajabi-Abhari A, Mahato M, Park J Y, Lee H and Oh I-K 2020 *Nano Energy* **75** 104904
[92] Yu H, Shao Y, Luo C, Li Y, Ma H-Z, Zhang Y-H, Yin B, Shen I-B and Yang M-B 2021 *Compos. A: Appl. Sci. Manuf.* **151** 106646
[93] Zhang J, Hu S, Shi Z, Wang Y, Lei Y, Han J, Xiong Y, Sun J, Zheng L and Sun Q 2021 *Nano Energy* **89** 106354
[94] Huang J, Hao Y, Zhao M, Qiao H, Huang F, Li D and Wei Q 2021 *Compos. A: Appl. Sci. Manuf.* **146** 106412
[95] Charoonsuk T, Pongampai S, Pakawanit P and Vittayakorn N 2021 *Nano Energy* **89** 106430
[96] Lin C, Lan J, Yu J, Hua Z, Huang H, Ma X and Cao S 2022 *Mater. Today Commun.* **30** 103208
[97] Somseemee O, Sae-Oui P and Siriwong C 2022 *Cellulose* **29** 8675–93
[98] Fan C, Huang J, Mensah A, Long Z, Sun J and Wei Q 2022 *Cell Rep. Phys. Sci.* **3** 101012
[99] Zhao J, Zhang W, Liu T, Liu Y, Qin Y, Mo J, Cai C, Zhang S and Nie S 2022 *Small Methods* **6** 2200664
[100] Cheng Y, Zhu W, Lu X and Wang C 2022 *Nano Energy* **98** 107229
[101] Zheng Z, Yu D, Wang B and Guo Y 2022 *Chem. Eng. J.* **446** 137393
[102] Zhang R, Dahlström C, Zou H, Jonzon J, Hummelgård M, Örtegren J, Blomquist N, Yang Y, Andersson H and Olsen M 2020 *Adv. Mater.* **32** 2002824
[103] Liu Y, Shen Y, Ding W, Zhang X, Tian W, Yang S, Hui B and Zhang K 2023 *Npj Flex. Electron.* **7** 21
[104] Li T, Wei H, Zhang Y, Wan T, Cui D, Zhao S, Zhang T, Ji Y, Algadi H and Guo Z 2023 *Carbohydr. Polym.* **309** 120678
[105] Singh R, Jadhav N A, Majumder S, Bhattacharya B and Singh P K 2013 *Carbohydr. Polym.* **91** 682–5
[106] Hasan M M, Islam M D and Rashid T U 2020 *Energy Fuels* **34** 15634–71
[107] Bella F, Porcarelli L, Mantione D, Gerbaldi C, Barolo C, Grätzel M and Mecerreyes D 2020 *Chem. Sci.* **11** 1485–93
[108] Hosseinnezhad M, Gharanjig K, Yazdi M K, Zarrintaj P, Moradian S, Saeb M R and Stadler F J 2020 *J. Alloys Compd.* **828** 154329
[109] Bella F, Pugliese D, Zolin L and Gerbaldi C 2017 *Electrochim. Acta* **237** 87–93
[110] Poskela A, Miettunen K, Borghei M, Vapaavuori J, Greca L, Lehtonen J, Solin K, Ago M, Lund P and Rojas O 2019 *ACS Sustainable Chem. Eng.* **7** 10257–65
[111] Zainuddin N, Saadiah M, Abdul Majeed A and Samsudin A 2018 *Int. J. Polym. Anal. Charact.* **23** 321–30
[112] Vasei M, Tajabadi F, Jabbari A and Taghavinia N 2015 *Appl. Phys.* A **120** 869–74

[113] Li P, Zhang Y, Fa W, Zhang Y and Huang B 2011 *Carbohydr. Polym.* **86** 1216–20

[114] Yogananda K, Ramasamy E and Rangappa D 2019 *Ionics* **25** 6035–42

[115] Yang J, Xiong S, Qu T, Zhang Y, He X, Guo X, Zhao Q, Braun S, Chen J and Xu J 2019 *ACS Appl. Mater. Interfaces* **11** 13491–8

[116] Gao C, Yuan S, Cui K, Qiu Z, Ge S, Cao B and Yu J 2018 *Sol. RRL* **2** 1800175

[117] Zhu K, Lu Z, Cong S, Cheng G, Ma P, Lou Y, Ding J, Yuan N, Rümmeli M H and Zou G 2019 *Small* **15** 1902878

[118] Chebrolu V T and Kim H-J 2019 *J. Mater. Chem.* C **7** 4911–33

[119] Baharun N, Mingsukang M, Buraidah M, Woo H, Teo L and Arof A 2020 *Ionics* **26** 1365–78

[120] Sharma P and Melkania U 2017 *Int. J. Hydrog. Energy* **42** 18865–74

[121] Lin Y, Pan Y and Zhang J 2017 *Electrochim. Acta* **232** 561–9

[122] Zhang J, Zhu Y, An Z, Shu X, Ma X, Song H, Wang W, Zhao J, Cao X and He J 2020 *J. Mater. Chem.* A **8** 14697–705

[123] Navaratnam S *et al* 2015 Transport mechanism studies of chitosan electrolyte systems *Electrochim. Acta* **175** 68–73

[124] Zou J, Oladipo J, Fu S, Al-Rahbi A, Yang H, Wu C, Cai N, Williams P and Chen H 2018 *Energy Convers. Manage.* **171** 241–8

[125] Lizundia E and Kundu D 2021 *Adv. Funct. Mater.* **31** 2005646

[126] Song A, Huang Y, Liu B, Cao H, Zhong X, Lin Y, Wang M, Li X and Zhong W 2017 *Electrochim. Acta* **247** 505–15

[127] Uddin M-J, Alaboina P K, Zhang L and Cho S-J 2017 *Mater. Sci. Eng.:* B **223** 84–90

[128] Zhang J, Liu Z, Kong Q, Zhang C, Pang S, Yue L, Wang X, Yao J and Cui G 2013 *ACS Appl. Mater. Interfaces* **5** 128–34

[129] Hänsel C, Lizundia E and Kundu D 2019 *ACS Appl. Energy Mater.* **2** 5686–91

[130] Xu Q, Kong Q, Liu Z, Wang X, Liu R, Zhang J, Yue L, Duan Y and Cui G 2014 *ACS Sustain. Chem. Eng.* **2** 194–9

[131] Hernández-Rentero C, Marangon V, Olivares-Marín M, Gómez-Serrano V, Caballero Á, Morales J and Hassoun J 2020 *J. Colloid Interface Sci.* **573** 396–408

[132] Murali G, Harish S, Ponnusamy S, Ragupathi J, Therese H A, Navaneethan M and Muthamizhchelvan C 2019 *Appl. Surf. Sci.* **492** 464–72

[133] Lingappan N, Kong L and Pecht M 2021 *Renew. Sustain. Energy Rev.* **147** 111227

[134] Chen H, Ling M, Hencz L, Ling H Y, Li G, Lin Z, Liu G and Zhang S 2018 *Chem. Rev.* **118** 8936–82

[135] Ma Y, Ma J and Cui G 2019 *Energy Storage Mater.* **20** 146–75

[136] Dobryden I, Montanari C, Bhattacharjya D, Aydin J and Ahniyaz A 2023 *Materials* **16** 5553

[137] Munao D, Van Erven J, Valvo M, Garcia-Tamayo E and Kelder E 2011 *J. Power Sources* **196** 6695–702

[138] Li W-C, Lin C-H, Ho C-C, Cheng T-T, Wang P-H and Wen T-C 2022 *J. Taiwan Inst. Chem. Eng.* **133** 104263

[139] Gao K, Shao Z, Li J, Wang X, Peng X, Wang W and Wang F 2013 *J. Mater. Chem.* A **1** 63–7

[140] Jabbour L *et al* 2013 Cellulose-based Li-ion batteries: a review *Cellulose* **20** 1523–45

[141] Yuan J-M, Ren W-F, Wang K, Su T-T, Jiao G-J, Shao C-Y, Xiao L-P and Sun R-C 2021 *ACS Sustain. Chem. Eng.* **10** 166–76

[142] Jin B, Wang D, Song L, Cai Y, Ali A, Hou Y, Chen J, Zhang Q and Zhan X 2021 *Electrochim. Acta* **365** 137359

[143] Hapuarachchi S N, Wasalathilake K C, Nerkar J Y, Jaatinen E, O'Mullane A P and Yan C 2020 *ACS Sustain. Chem. Eng.* **8** 9857–65

[144] Huang J, Liu B, Zhang P, Li R, Zhou M, Wen B, Xia Y and Okada S 2021 *Solid State Ionics* **373** 115807

[145] Bryntesen S N, Finne P H, Svensson A M, Shearing P R, Tolstik N, Sorokina I T, Vinje J, Lamb J J and Burheim O S 2023 *J. Mater. Chem.* A **11** 6483–502

[146] Haidl P, Buchroithner A, Schweighofer B, Bader M and Wegleiter H 2019 *Sustainability* **11** 6731

[147] Arshad F, Lin J, Manurkar N, Fan E, Ahmad A, Wu F, Chen R and Li L 2022 *Resour. Conserv. Recycl.* **180** 106164

[148] Peng H J, Huang J Q, Zhao M Q, Zhang Q, Cheng X B, Liu X Y, Qian W Z and Wei F 2014 *Adv. Funct. Mater.* **24** 2772–81

[149] Baskar A V, Singh G, Ruban A M, Davidraj J M, Bahadur R, Sooriyakumar P, Kumar P, Karakoti A, Yi J and Vinu A 2023 *Adv. Funct. Mater.* **33** 2208349

[150] Wang Z, Zhang X, Liu X, Zhang Y, Zhao W, Li Y, Qin C and Bakenov Z 2020 *J. Colloid Interface Sci.* **569** 22–33

[151] Yeon J S, Park S H, Suk J, Lee H and Park H S 2020 *Chem. Eng. J.* **382** 122946

[152] Li D, Chang G, Zong L, Xue P, Wang Y, Xia Y, Lai C and Yang D 2019 *Energy Storage Mater.* **17** 22–30

[153] Chen W, Zhang L, Liu C, Feng X, Zhang J, Guan L, Mi L and Cui S 2018 *ACS Appl. Mater. Interfaces* **10** 23883–90

[154] Zhang T-W, Shen B, Yao H-B, Ma T, Lu L-L, Zhou F and Yu S-H 2017 *Nano Lett.* **17** 4894–901

[155] Hao R, Yang Y, Wang H, Jia B, Ma G, Yu D, Guo L and Yang S 2018 *Nano Energy* **45** 220–8

[156] Nie A, Gan L y, Cheng Y, Tao X, Yuan Y, Sharifi-Asl S, He K, Asayesh-Ardakani H, Vasiraju V and Lu J 2016 *Adv. Funct. Mater.* **26** 543–52

[157] Mittal N, Tien S, Lizundia E and Niederberger M 2022 *Small* **18** 2107183

[158] Yang J, Zhou X, Wu D, Zhao X and Zhou Z 2017 *Adv. Mater.* **29** 1604108

[159] Alvin S, Yoon D, Chandra C, Cahyadi H S, Park J-H, Chang W, Chung K Y and Kim J 2019 *Carbon* **145** 67–81

[160] Xu K, Pan Q, Zheng F, Zhong G, Wang C, Wu S and Yang C 2019 *Front. Chem.* **7** 733

[161] Yang L, Hu M, Zhang H, Yang W and Lv R 2020 *J. Colloid Interface Sci.* **566** 257–64

[162] Sudhakar Y N, Selvakumar M and Bhat D K 2018 *Biopolymer Electrolytes: Fundamentals and Applications in Energy Storage* (Amsterdam: Elsevier)

[163] Yang H, Liu Y, Kong L, Kang L and Ran F 2019 *J. Power Sources* **426** 47–54

[164] Zhu H, Luo W, Ciesielski P N, Fang Z, Zhu J, Henriksson G, Himmel M E and Hu L 2016 *Chem. Rev.* **116** 9305–74

[165] Zhao D, Chen C, Zhang Q, Chen W, Liu S, Wang Q, Liu Y, Li J and Yu H 2017 *Adv. Energy Mater.* **7** 1700739

[166] Priya M S, Divya P and Rajalakshmi R 2020 *Sustain. Chem. Pharm.* **16** 100243

[167] Ding J, Wang H, Li Z, Cui K, Karpuzov D, Tan X, Kohandehghan A and Mitlin D 2015 *Energy Environ. Sci.* **8** 941–55

[168] Zhang Y, Shang Z, Shen M, Chowdhury S P, Ignaszak A, Sun S and Ni Y 2019 *ACS Sustain. Chem. Eng.* **7** 11175–85
[169] Wang Y-Y, Fu Q-J, Ning X, Chen G-G and Yao C-L 2020 *BioResources* **15** 677–90
[170] Wu Y, Cao J-P, Hao Z-Q, Zhao X-Y, Zhuang Q-Q, Zhu J-S, Wang X-Y and Wei X-Y 2017 *Int. J. Electrochem. Sci.* **12** 7227–39
[171] Chen K, Liu J, Bian H, Wei J, Wang W and Shao Z 2020 *Nanotechnology* **31** 335713
[172] Wu K, Hu Y, Cheng Z, Pan P, Jiang L, Mao J, Ni C, Gu X and Wang Z 2019 *J. Membr. Sci.* **592** 117349
[173] Thiyagarajan G B, Shanmugam V, Wilhelm M, Mathur S, Moodakare S B and Kumar R 2021 *Open Ceramics* **6** 100131
[174] Goncalves R, Lizundia E, Silva M M, Costa C M and Lanceros-Mendez S 2019 *ACS Appl. Energy Mater.* **2** 3749–61
[175] Xu D, Wang B, Wang Q, Gu S, Li W, Jin J, Chen C and Wen Z 2018 *ACS Appl. Mater. Interfaces* **10** 17809–19
[176] Song A, Huang Y, Zhong X, Cao H, Liu B, Lin Y, Wang M and Li X 2018 *J. Membr. Sci.* **556** 203–13
[177] Yan B, Feng L, Zheng J, Zhang Q, Dong Y, Ding Y, Yang W, Han J, Jiang S and He S 2023 *Appl. Surf. Sci.* **608** 155144
[178] Xia Y, Guan J and Du X 2023 *J. Energy Storage* **72** 108776
[179] Husain Z, Raheman A S, Ansari K B, Pandit A B, Khan M S, Qyyum M A and Lam S S 2022 *Mater. Sci. Energy Technol.* **5** 99–109
[180] Park G-T, Jeon H-B, Kim S-Y, Gang H-E and Jeong Y G 2022 *Mater. Sci. Eng.:* B **275** 115530
[181] Thuy U T D, Nam P V, Van Chi N, Duong T X and Van Hoa N 2023 *J. Sci.: Adv. Mater. Devices* **8** 100586
[182] Luo C, Du L, Wu W, Xu H, Zhang G, Li S, Wang C, Lu Z and Deng Y 2018 *ACS Sustain. Chem. Eng.* **6** 12621–9

Chapter 2

Materials used to optimize triboelectric nanogenerator performance

Gaurav Khandelwal, Rajagopalan Pandey, Ben B Xu and Daniel M Mulvihill

Since their inception, triboelectric nanogenerators (TENGs) have become a popular choice for converting mechanical energy into electricity due to their unique properties, including a wide available choice of materials and device designs, high output, low weight, and cost-effectiveness. TENGs have been explored for a multitude of applications such as physical, chemical, and biological sensors, drug delivery, cell stimulation, wound healing, water wave energy and so on. The performance of TENGs depends on structural and material figures of merit (FOMs). Thus, materials are a key governing factor that influences TENG performance. This chapter focuses on the materials used for TENGs and provides a detailed overview of several important TENG materials, including conventional polymers, 2D materials, natural materials, porous crystalline materials, ferroelectric materials, and textiles. It is concluded that 2D materials show promise in their ability to serve a dual function as both electrodes and tribolayers. Metal–organic frameworks (MOFs) could be the favored selection for the creation of self-powered chemical sensors, while ferroelectric materials offer the potential to enhance TENG performance through polarization.

2.1 Introduction

Amidst the growing challenges posed by greenhouse gas emissions and the energy crisis, the pursuit of novel and alternative technological approaches to harness energy from the environment is gaining increasing significance. Among the most underutilized sources of energy is low-frequency mechanical energy, which is universally abundant but often goes to waste in our surroundings.

A nanogenerator is a device that converts mechanical energy into electricity through a small-scale physical transformation within a system [1–5]. This technology

doi:10.1088/978-0-7503-5749-4ch2

holds considerable potential for practical applications. Among a variety of nanogenerators, triboelectric nanogenerators have emerged as a promising alternative energy source, particularly for electronic devices, due to their economic viability, scalability, ambient sources of energy, and safety. As a burgeoning sector within energy conversion technologies, the TENG was reintroduced/rebranded in 2012 to distinguish it from similar high-voltage-producing devices such as the van de Graaff generator. Up until now, the TENG has undergone a rapid and thriving phase of development, encompassing architectural design, materials selection and modification, performance enhancement, power management, and the exploration of applications. A TENG is realized by combining two common phenomena—contact electrification and electrostatic induction. Wang *et al* proposed four distinct operational modes, viz. the vertical contact–separation mode, the lateral sliding mode, the single-electrode mode and the freestanding triboelectric-layer mode. These have progressively evolved and been proposed for TENGs [6] (figure 2.1). In all scenarios, opposite triboelectric charges are induced on the surfaces of materials through the contact-electrification process. Subsequently, electrostatic induction provides the impetus for converting mechanical stimuli into electrical energy when relative motion is initiated. The four modes vary in terms of their design, the relative motion of the active layers, their electrodes, applied force, and performance [7].

Essentially, enhancing TENG performance hinges on a concerted effort to significantly increase the quantity of produced triboelectric charges. The performance figure of merit (FOM_P), comprising structural FOM (FOM_S) and material FOM (FOM_M), was introduced to compare TENG device performance. The FOM_M relies on the surface charge density (σ) of the material and significantly

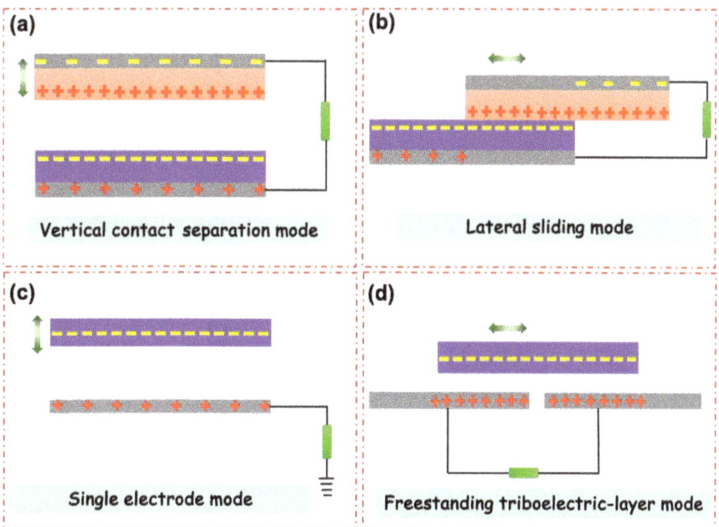

Figure 2.1. The working modes of TENGs: (a) vertical contact–separation, (b) lateral sliding, (c) single electrode, and (d) the freestanding triboelectric-layer mode (reprinted from [8], Copyright (2020), with permission from Elsevier.)

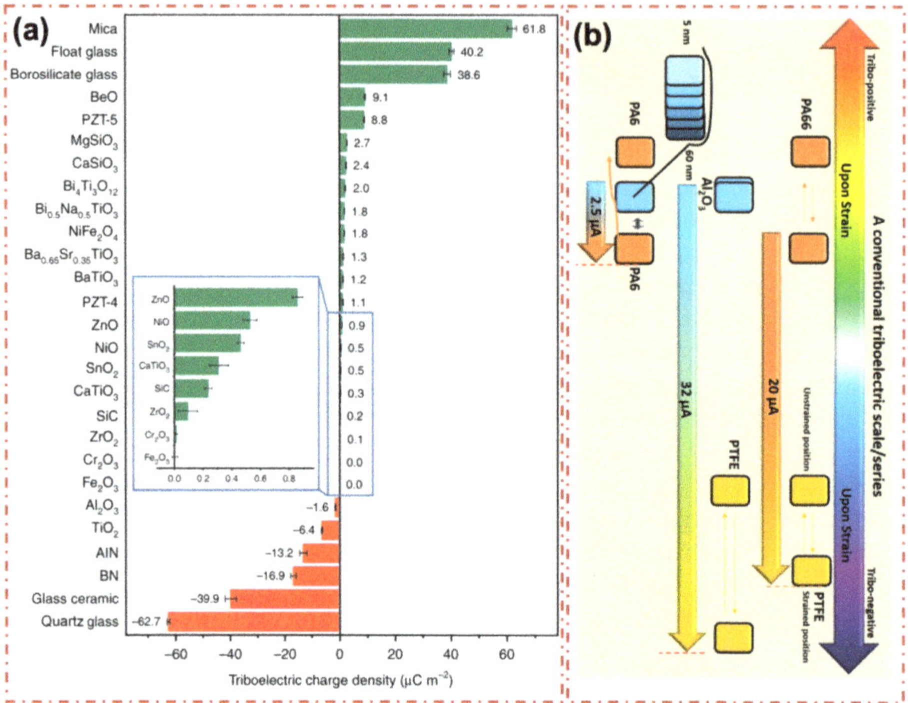

Figure 2.2. (a) A triboelectric series of inorganic nonmetallic materials (reproduced from [9], CC BY 4.0). (b) A schematic representation of the dynamic nature of materials in under-strained and unstrained conditions, demonstrating a dynamic triboelectric series concept (reprinted from [11], Copyright (2021), with permission from Elsevier).

influences TENG performance. Further, the larger the disparity in electron affinity between two triboelectric materials, the greater the potential for generating triboelectric charges, which can be qualitatively defined by the position of the material on the triboelectric scale.

The triboelectric series classifies materials based on their propensity to gain or lose electrons through contact and friction interactions with other materials. This series ranks materials in an order that indicates their relative tendency to become positively or negatively charged when they come into contact with each other and then separate. The triboelectric series has been quantified for the polymers and inorganic nonmetallic materials. Figure 2.2(a) shows a triboelectric series of inorganic nonmetallic materials, quantifying the charge density of 30 inorganic materials [9, 10]. Pandey *et al* proved for the first time that the dynamic nature of materials on the triboelectric scale is based on strain, as shown in figure 2.2(b) [11]. Materials at one end of the triboelectric series tend to become positively charged by donating electrons, while materials at the opposite end become negatively charged by obtaining electrons. The triboelectric series provides valuable insights into the behavior of materials in various electrical and energy-harvesting applications. This hierarchy can guide researchers seeking to optimize TENG performance from the perspective of materials selection. It is worth noting that the majority of materials in this list are polymers which possess an array of functional groups that are either electron-withdrawing components or electron-donating groups. These functional groups play a pivotal role in charge transfer and capture during the contact-electrification process, leveraging their distinct hybrid orbital configurations. In addition, polymers exhibit exceptional attributes including remarkable flexibility, machinability, stretchability, scalability, and low weight. As a result, polymers have inevitably emerged as the fundamental cornerstone of TENG technology.

This chapter presents an in-depth exploration of the different materials used for TENGs. It provides an insight into the importance of the materials used for TENG devices, a systematic comparison of different TENGs, and final conclusions.

2.2 Conventional polymer-based TENGs

Numerous materials have been utilized in the creation of TENGs since their inception. The performance of a TENG depends on the difference in electron affinity between the two materials. Fluorine-containing polymers such as polytetrafluoroethylene (PTFE), fluorinated ethylene propylene (FEP), and polyvinylidene fluoride (PVDF) are commonly used as the electron-withdrawing material. In contrast, polymers containing electron-donating groups such as nylon, silk, and wool are commonly used as tribopositive materials. PTFE is a synthetic polymer known for its exceptional non-stick and heat-resistant properties. Its molecular structure is composed of repeating units formed by the polymerization of tetrafluoroethylene (TFE) monomers. Its uniqueness lies in its strong carbon–fluorine bond, which imparts remarkable chemical resistance and thermal stability. PTFE stands out as the prevailing electronegative material at present, finding extensive utilization in the fabrication of TENGs [11–13]. Figure 2.3(a) illustrates novel approaches for creating advanced organic–inorganic hybrid PTFE–perovskite-based TENGs by employing techniques involving adjustments in chemical composition and the manipulation of ion migration within films through electric fields. Zhong *et al* presented a tilting-sensitive TENG adept at efficiently capturing swaying energy from unstable or fluctuating ship surfaces [14]. The device incorporated integrated blade arrangements on sliders, rendering it responsive to tilting actions and ensuring consistent power generation. The TENG integrated a slidable frame housing 14 PTFE blades and a stationary frame containing 15 aluminum (Al) blades. The intersection of these two blade structures gave rise to 28 fundamental TENG units within a single tilt-sensitive TENG (TS-TENG). In conditions involving low-frequency fluctuations, the TENG demonstrated a peak power density of 1.41 W m^{-3}. The TENG had the ability to power 30 LEDs and could be used to drive electronic devices. Zhao *et al* used an emulsion electrospinning technique to fabricate PTFE/poly(ethylene oxide) (PEO) nanofiber for TENG applications as shown in figure 2.3(b) [15]. To boost the resulting surface charge in the triboelectric setup, a straightforward procedure involving negative charge injection was applied to the electrospun PTFE. This treatment led to the creation of a TENG that exhibited a stable peak-to-peak open-circuit voltage (V_{oc}) of

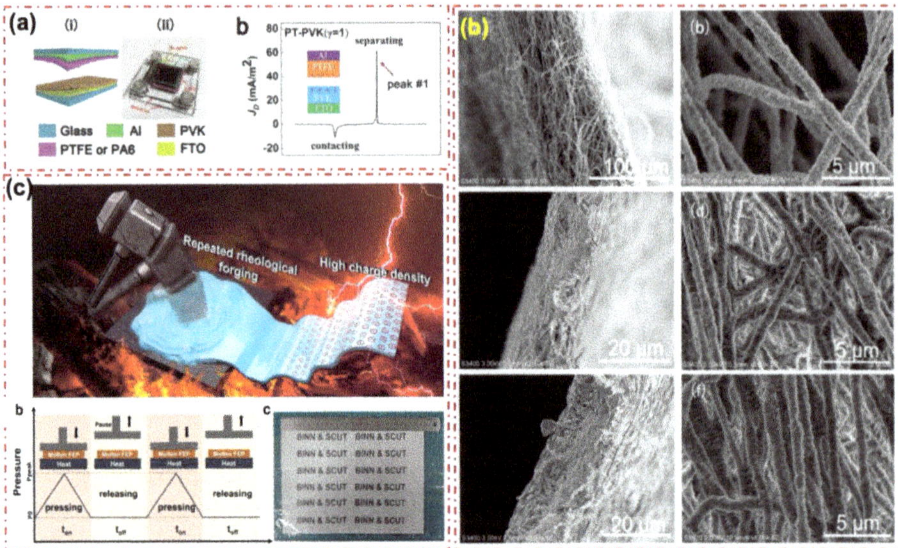

Figure 2.3. (a) A schematic structure and the corresponding perovskite-based TENG [12] John Wiley & Sons. © 2020 Wiley-VCH GmbH. (b) SEM images capture the fiber cross-section and surface morphologies of the PEO/PTFE composite nanofiber membrane, along with pristine electrospun PTFE/PEO samples before and after sintering at 350 °C and 370 °C (reprinted with permission from [15], Copyright (2018), American Chemical Society). (c) The figures illustrate the production of films with high triboelectric charge density using the rheological repeated forging technique. This technique enhances the properties of FEP films, resulting in improved performance (reproduced from [21], CC BY 4.0).

approximately 900 V, a short-circuit current density of around 20 mA m^{-2}, and a corresponding charge density of about 149 μC m^{-2}. Li *et al* showcased a buoy-like high-performance liquid–solid-interface TENG designed to harness low-frequency blue energy [16]. Composed of PTFE films and Ag electrodes, the liquid–solid TENG produced close to 50 times more energy than a conventional solid–solid TENG. Oscillating in the water, the device achieved a short-circuit current of 40 μA, a transferred charge of 1000 nC, and an open-circuit voltage of 400 V and had the ability to illuminate nearly 100 LEDs. In addition, the integration of multiple units into a network led to remarkable outputs: 300 V, 290 μA, and 16 725 nC were obtained from 18 units interconnected in parallel. This network configuration served as a direct power source for self-powered systems. Bhaskar *et al* formulated a cost-effective strategy for creating a micro-grooved architecture on PTFE using a simplified and economical thermal imprinting lithography technique, employing micro-pyramidal textured silicon as a master mold that was generated via a wet-chemical etching process [17]. The optimized TENG demonstrated maximum V_{OC}, I_{SC}, and charge density values of around 320 V, 15 μA, and 44 μC m^{-2}, respectively. Furthermore, enhancing the TENG's performance involved expanding the active area to 25 cm^2, resulting in values of about 625 V and 50.5 μA, respectively. This modification also showcased an energy conversion efficiency of 62.5% and an impressive output power density of roughly 252 W m^{-2}. Zhang *et al* utilized cotton film paired with PTFE to capture low-frequency mechanical energy [18]. The incorporation of conductive ink-coated cotton further enhanced the TENG's functionality. The TENG achieved values of 556 V and 26 μA and a maximum power density of 0.66 mW cm^{-2}. Concurrently, Lin *et al* embarked on a parallel pursuit by introducing a TENG concept inspired by pendulum dynamics, strategically designed to exploit water and wind energy while achieving heightened output at low frequencies [19]. Their intricate design features comprised an electrode layer (Cu), a triboelectric pendulum layer (Cu), and thin PTFE strips. Upon the application of an external stimulus, the triboelectric pendulum layer initiated oscillatory motion, rhythmically making delicate contact with the slender PTFE strips. With the goal of capturing wave energy on a larger scale, this device was seamlessly integrated into a 2×3 array formation and demonstrated the ability to charge a 100 μF capacitor from 0 to 3.1 V in just 780 s, effectively enabling self-powered temperature measurement. In addition, for the purpose of harnessing copious wind energy, multiple units were artfully suspended from tree branches, offering a platform for diverse applications powered by wind-driven energy. Recently, Ravi *et al* explored a liquid nanogenerator concept using hydrophobic surfaces in a silicone pipe to convert flow current into electrical energy using water reservoirs [20]. Commercial PTFE film creates a hydrophobic surface for mechanical energy conversion. The interaction between water and the embedded PTFE in the silicone tube generated power. Higher V_{oc} and I_{sc} values were observed for tap water, since it has a larger number of mobile ions than deionized (DI) water. The TENG setup achieved peak-to-peak V_{oc} and I_{sc} values of 29.5 V and 3.7 μA, respectively.

FEP is a notable polymer renowned for its exceptional combination of chemical resistance, thermal stability, and low friction properties. FEP belongs to the family of fluoropolymers and finds many applications across various industries due to its unique characteristics. This polymer's properties are derived from the incorporation of fluorine atoms within its molecular structure, boosting its electrophilic characteristics and making it a frequently employed electronegative material within triboelectric pairs. In a study by Xiao *et al*, a spherical TENG utilizing a spring-assisted multilayered design was introduced to effectively harness low-frequency, irregular, and random water wave energy [22]. The incorporation of a spring structure enhanced the spherical TENG's output by converting low-frequency water wave motions into higher-frequency vibrations. In addition, the multilayered structure enhanced space utilization, resulting in increased output. The distinct structure enabled a single spherical TENG unit to yield an output current of 120 μA, significantly surpassing previous rolling spherical TENGs. With water wave stimulation, these units achieved a peak output power of up to 7.96 mW. Integrating four units in the TENG array successfully powered LEDs and an electronic thermometer. Pan *et al* tried to understand the effects of liquids on TENG performance: a U-tube TENG was used with 11 liquids [23]. Factors such as polarity, dielectric constant, and affinity to FEP influenced the output. A pure water-based TENG yielded the best results of an V_{oc} of 81.7 V and a I_{sc} of 0.26 μA for shaking mode. It worked as a self-powered concentration sensor and achieved an accuracy of more than 92%. An enhanced water–FEP U-tube TENG efficiently harvested water-wave energy, producing an V_{oc} of 350 V, an I_{sc} of 1.75 μA, and a power density of 2.04 W m^{-3}, illuminating LEDs and powering a meter. The water–FEP U-tube TENG holds promise for electronics self-powered by water-wave energy. Zhang *et al* demonstrated an innovative adhesive-tape-peeling method to enhance the film's charge, significantly improving TENG performance [24]. This enhancement was demonstrated by an increase in the FEP film's surface potential from −360 to −2850 V, an increase in the TENG's output negative voltage peak from −83 to −363 V, and an elevation of the wind-driven TENG's average output power from 2.5 to 19.8 μW when charging a 10 μF capacitor. Furthermore, the TENG's ability to generate electricity from low-frequency human motion and directly power light-emitting diodes (LEDs) was demonstrated. Liu *et al* introduced a novel approach that produced FEP via repeated rheological forging [21]. This technique improved its mechanical attributes, optical clarity, and ability to maintain an ultrahigh tribocharge density as illustrated in figure 2.3(c). For a film 30 μm thick, the output charge density produced by a contact–separation nanogenerator reached 352 μC m^{-2}, which further increased to 510 μC m^{-2} in an air-breakdown mode. The repeated forging method proved effective in regulating the surface functional group composition, crystallinity, and dielectric constants of the fluorinated ethylene propylene, thereby enhancing its triboelectrification capacity.

PVDF is a versatile polymer with unique piezoelectric and pyroelectric characteristics that allow it to convert mechanical vibrations and temperature changes into electrical signals. Its excellent chemical resistance, durability, and flexibility also contribute to its use in sensors, actuators, and energy harvesting devices. It possesses a semicrystalline structure composed of repeating CH_2–CF_2 units. This arrangement contributes to its unique properties, making PVDF a valuable material in tribo- and piezoelectric energy harvesting and smart sensing. Shi *et al* introduced an effective approach that combined PVDF and graphene nanosheets through electrospinning to enhance TENG performance [25]. The resulting TENG exhibited impressive performance: an output voltage of ∼1511 V, a short-circuit current density of ∼189 mA m^{-2}, and a peak power density of ∼130.2 W m^{-2}—nearly eight times higher than that of a PVDF–polyamide-6 (PA6) TENG combination. Under impedance matching, it harvested ∼74.13 μJ of energy per cycle and achieved an average power density of 926.65 mW m^{-2}. Lee *et al* introduced a radical atom-transfer polymerization technique to control the dielectric constant of poly(tert-butyl acrylate) (PtBA)-grafted PVDF copolymers, which was shown to significantly boost output power [26]. The copolymer mainly comprised α phases with heightened dipole moments, which were attributed to π-bonding and the polar nature of ester functional groups in PtBA. This led to a nearly twofold increase in the dielectric constant, as verified by Kelvin probe force microscopy (KPFM). This augmentation in the dielectric constant substantially raised charge accumulation on the copolymer during contact, resulting in an output signal of 105 V and 25 μA cm^{-2}—a 20-fold power enhancement compared to the outputs of pure PVDF-based nanogenerators, achieved by surface potential tuning using a poling technique. Khandelwal *et al* introduced a facile phase inversion process that yielded PVDF films with a porous structure and high crystallinity (figure 2.4(a)) [27]. They introduced non-piezoelectric, semiconductive TiO_2 microparticles that

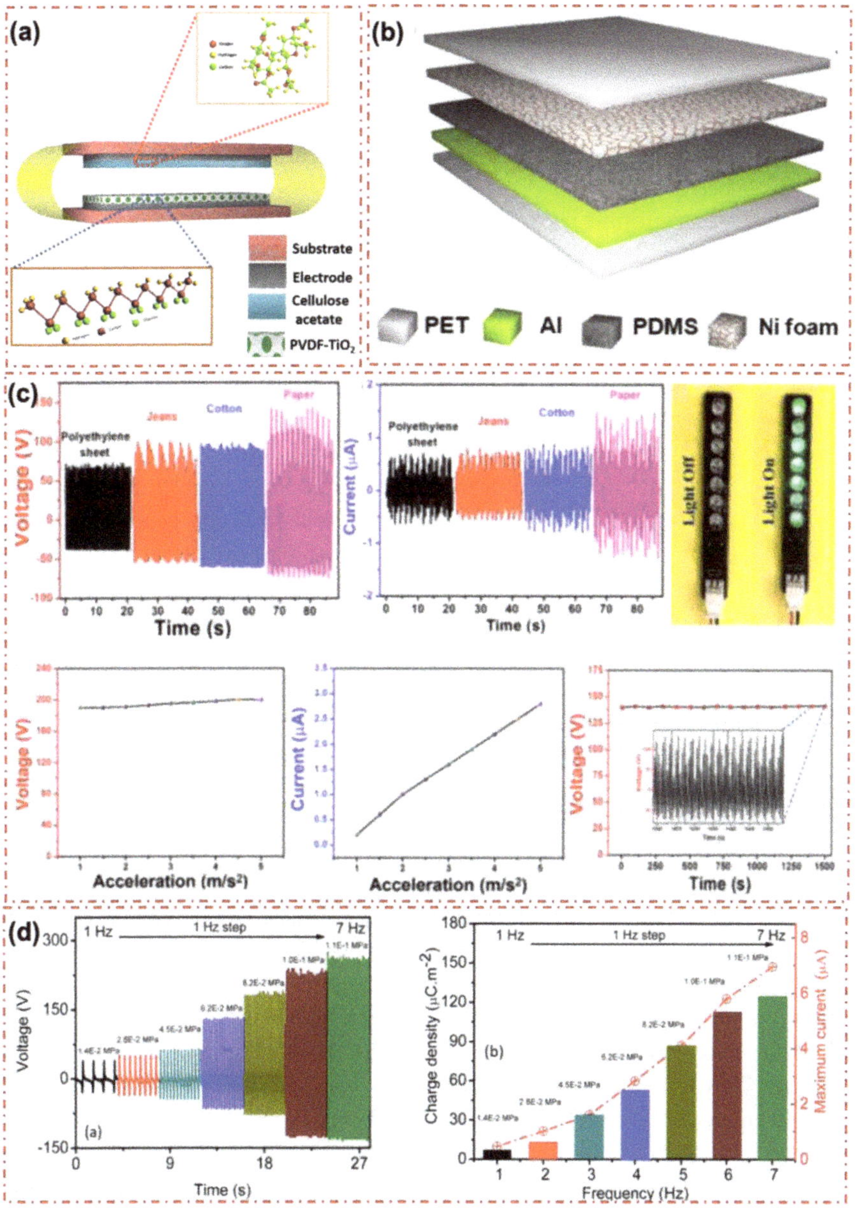

Figure 2.4. (a) Device design of a phase-inverted TENG (reprinted from [27], Copyright (2019), with permission from Elsevier). (b) The figure illustrates the structural design of a waterproof wave-energy TENG (reprinted from [28], Copyright (2019), with permission from Elsevier). (c) The electrical performance of a single-electrode textile-based triboelectric nanogenerator (ST-TENG) with different opposite contact materials, including an LED glow produced by the generated output and the output of the ST-TENG at different accelerations (reprinted from [29], Copyright (2020), with permission from Elsevier). (d) The V_{oc} measurements for PVDF-$ZnSnO_3$/PA6 TENGs at frequencies ranging from 1 to 7 Hz with a fixed 10 mm spacing. The figure tracks the variations in maximum short circuit current and the corresponding charge density within the 1–7 Hz frequency range (reprinted from [30], Copyright (2016), with permission from Elsevier).

acted as charge trapping sites to further augment the output. Furthermore, the device functioned as a versatile energy harvester for biomechanical and wind energy while also serving as a volatile organic compound sensor. The sensor showed excellent sensitivity to different benzene concentrations. As discussed in a later section, several ferroelectric materials have also been incorporated into PVDF to fabricate TENGs with enhanced performance.

Some other commonly used negative materials in TENG applications are polydimethylsiloxane (PDMS), Kapton, rubber, and polyvinyl siloxane (PVS). In this section, we briefly discuss other such polymers. PDMS is a versatile and widely used silicone-based polymer known for its unique properties and structural characteristics. PDMS is composed of repeating units of dimethylsiloxane, which consists of alternating silicon and oxygen atoms, forming a flexible and linear polymer chain. Arun *et al* presented a cellulose/PDMS composite TENG (C-TENG) with an Al electrode [31]. A 5 wt.% composite film generated a V_{oc} of 28 V and an I_{sc} of 2.8 μA, achieving an instantaneous peak power of 576 μW at a mechanical force of 32.16 N. The C-TENG powered LEDs and served as a self-powered motion detector for security applications, thus establishing its usability as an operational power source. Furthermore, it was integrated into an internal lithium-ion battery charging circuit to harness energy during a range of human movements. The same team subsequently demonstrated a fully encapsulated waterproof TENG that used PDMS and nickel foam [28]. In an innovative application, alphanumeric characters were encoded using Morse code to send a message over a considerable distance.

Polyimide (PI) (Kapton) is a versatile and high-performance polymer that finds extensive use in various industries due to its exceptional thermal stability, mechanical strength, and electrical insulating properties. It is composed of repeating imide monomer units which are characterized by a rigid and aromatic backbone. Kim *et al* focused on enhancing the TENG characteristic of PI through surface modification using the electrospinning technique [32]. PI has inherent solvent resistance, posing challenges for its conversion into nanofibers via electrospinning. The study addressed this by directly electrospinning using PI ink, in a simple one-step process. The impact of PI concentration on electrospinnability was examined, and uniform and continuous nanofibrous structures were obtained at 15 and 20 wt.% concentrations. A TENG utilizing the 20 wt.% concentration exhibited notable performance improvements, boasting an open-circuit voltage of 753V and a short-circuit current of 10.79 μA. Its power density reached 2.61 W m^{-2} at a 100 MΩ load resistance. The TENG remained stable during a taping experiment that lasted for 10 000 cycles, successfully harnessing electrical energy. The energy harvested from this TENG was sufficient to power more than 55 LEDs and drive small electronic devices, showcasing its potential as an effective wearable energy harvester. Arun *et al* employed an ST-TENG to enhance table functionality, creating a smart table that triggered an alarm when unauthorized access was detected [29]. The TENG harvested energy from various clothing materials as opposed to triboelectric materials. Extensive investigations of electrical performance using materials like jeans, cotton, and paper revealed that the optimal performance was obtained using paper, which yielded a peak output voltage of ~200 V and a current of 2.2 μA (figure 2.4(c)). Ultimately, the TENG-enhanced smart table was integrated with an Arduino board to effectively activate an alarm upon desk access, showing the potential of TENGs for advancing smart furniture applications.

In response to the global demand for face masks during the SARS-CoV-2 pandemic, Ghatak *et al* introduced an innovative approach to face mask design [33]. The proposed mask integrated three layers that functioned as a triboelectric filter while the outer layer served as an electrocution layer to enhance filtration and counteract the spread of the virus. Four pairs of triboelectric-series fabrics, including PVC–nylon, polypropylene (PP)–polyurethane (PU), latex rubber-PU, and PI-nylon, were studied to validate the mask's efficiency.

To the best of the authors' knowledge, Kumar *et al* were the first to use PVS in a TENG [34]. They developed a contact–separation-mode TENG using PVS in contact with mica. PVS is a tribonegative silicone-based viscoelastic polymer. It is a well-known biomaterial that is widely used in dentistry to make tooth imprints. The key characteristics of the PVS polymer include low cost, easy commercial availability, simple processing, low modulus, high flexibility, and nontoxicity. PVS was used partly because it is an ideal choice of material for accurately molding surface topographies (the study by Kumar *et al* focused on the effect of surface roughness on TENG output). The sample that produced the highest output (the PVS sample with the lowest roughness) produced a voltage of 222.8 V, a short-circuit current density of 53 mA m^{-2}, and a peak power density of 4.3 W m^{-2}.

Similarly, tribopositive materials are substances that tend to lose charges when they come into contact with other materials through friction. These materials typically have a lower electron affinity and are more likely to release electrons when exposed to friction. As discussed in the previous section, extensive research efforts have been devoted to tribonegative materials. In contrast, the selection of tribopositive materials has received comparatively less attention, despite their equal significance to tribonegative materials in enhancing surface charge density and consequent power density. The limited options available for positive insulator tribomaterials, mostly metals such as aluminum (Al) and copper (Cu), are evident from the triboelectric series.

Soin *et al* employed a phase-inverted PA6 membrane alongside a $ZnSnO_3$–PVDF composite to form a TENG [30]. Under cyclic excitation of 490 N at 3 Hz, the TENG exhibited notably elevated performance, delivering a voltage of 520 V and a charge density of 62.0 μC m^{-2}, roughly twice that of the pristine material. The same group later introduced PEO as a new tribopositive material, highlighting its superior positive tribopolarity in comparison to that of PA6 [35]. The PEO-based TENG generated up to 970 V, a current density of 85 mA m^{-2}, and a power output of approximately 40 W m^{-2} under a 50 N contact force. In contrast, the PA6/PDMS TENG yielded 630 V, 30 mA m^{-2}, and a power output of around 18 W m^{-2}, emphasizing the noteworthy attributes of PEO as a positive tribomaterial. These findings were substantiated by KPFM, which revealed a contact potential difference of 1.26 for PEO and 0.87 V for PA6. The group also introduced aniline formaldehyde resin (AFR) as a tribopositive material [36]. A straightforward reaction between aniline and formaldehyde under acidic conditions leads to the formation of aniline–formaldehyde condensate. Smooth AFR films with abundant surface functional groups rich in nitrogen and oxygen can be obtained using a subsequent high-temperature curing step under vacuum conditions. The surface functional groups exhibit a tribopositive charge, providing AFR with a notably higher positive tribopolarity than that of PA6. A TENG incorporating optimized thin AFR layers in conjunction with a PTFE film achieved remarkable performance metrics: a peak-to-peak voltage of around ~1000 V, a current density of approximately ~65 mA m^{-2}, a transferred charge density of about ~200 μC m^{-2}, and an instantaneous power output (energy pulse) of roughly ~11 W m^{-2}. In addition, KPFM measurements validated AFR's suitability, revealing higher surface potential values for AFR than those of PA6. Acharaya *et al* introduced a novel thermoplastic starch (TPS) as a positive triboelectric material. Their developed bioplastic exhibited exceptional durability, producing an open-circuit peak-to-peak output voltage of approximately 560 V, coupled with an output current density of around 120 mA m^{-2} and an instantaneous power density of 17 W m^{-2}. The integration of environmentally friendly TPS as a positive triboelectric component added the benefits of abundance, biodegradability, cost-effectiveness, and simplified fabrication, thereby enhancing its appeal for biomedical uses.

2.3 Two-dimensional (2D) materials for TENGs

Two-dimensional materials such as graphene, transition-metal dichalcogenides (TMDs), MXenes, etc. exhibit properties which are significantly different from those of their bulk counterparts [37]. Graphene exhibits high electrical conductivity (2000 S m^{-1}), a high surface area (2630 m^2 g^{-1}), excellent mechanical strength (1 TPa), high optical transparency (~97%), and high electron mobility (2 00 000 cm^2 V^{-1} S^{-1}). Similarly, TMDs such as MoS_2, $MoSe_2$, WSe_2, WS_2, etc. are atomically thick materials with a direct bandgap. Two-dimensional materials such as MoS_2 exhibit different phases; for example, the 1T metallic phase, which has high conductivity, can be used as an electrode material, while 2H material can be used as charge trapping material to improve the performance of TENGs [38–40]. Recently, a new class of 2D materials known as MXenes has been introduced. These materials comprise transition-metal carbides/nitrides or carbonitrides. MXenes are typically denoted by $M_{n+1}X_nT_x$, where M is a transition metal, X is carbon (C) or nitrogen (N), T = a functional group (–F, –O, or –OH), and n can be between one and three [41, 42]. The highly conductive behavior of MXenes makes them suitable for use as electrodes, and the presence of surface functional groups is responsible for their triboelectric behavior, which is close to that of PTFE [43].

Graphene was used as the transparent electrode in a flexible and transparent triboelectric generator (FTTG) with different electrode combinations i.e. graphene was used as both the bottom and top electrode (GG), indium tin oxide (ITO) was used as both the top and bottom electrode (II) [44], graphene was used as the bottom electrode and ITO as the top electrode (IG), and graphene was used on the top and ITO as the bottom electrode (GI). The FTTG with the GI electrode combination produced the best output (56 V). The different electrode combinations followed an output trend of GI > II > GG > IG. The variation in the output was attributed to the strain-responsive resistive behavior of the top layer. Similarly, graphene (1 nm) was used as the electrode in single-electrode TENGs which could conformably attach to human skin [45]. The transparency of the chemical vapor deposition (CVD)-grown graphene was subsequently improved using poly(3,4-ethylenedioxythiophene) polystyrene sulfonate (PEDOT:PSS) [46]. The surface engineering not only improved the transmission, but also enhanced the TENG's performance by 140%. Liquid exfoliated and laser-induced graphene (LIG) were also demonstrated for the fabrication of TENGs [47–51]. The LIG was obtained by laser irradiation of carbon sources such as PI or polyetherimide (figure 2.5(a)). A PI-based LIG/PI TENG produced 3100 V in single-electrode mode. In contact–separation mode, PI

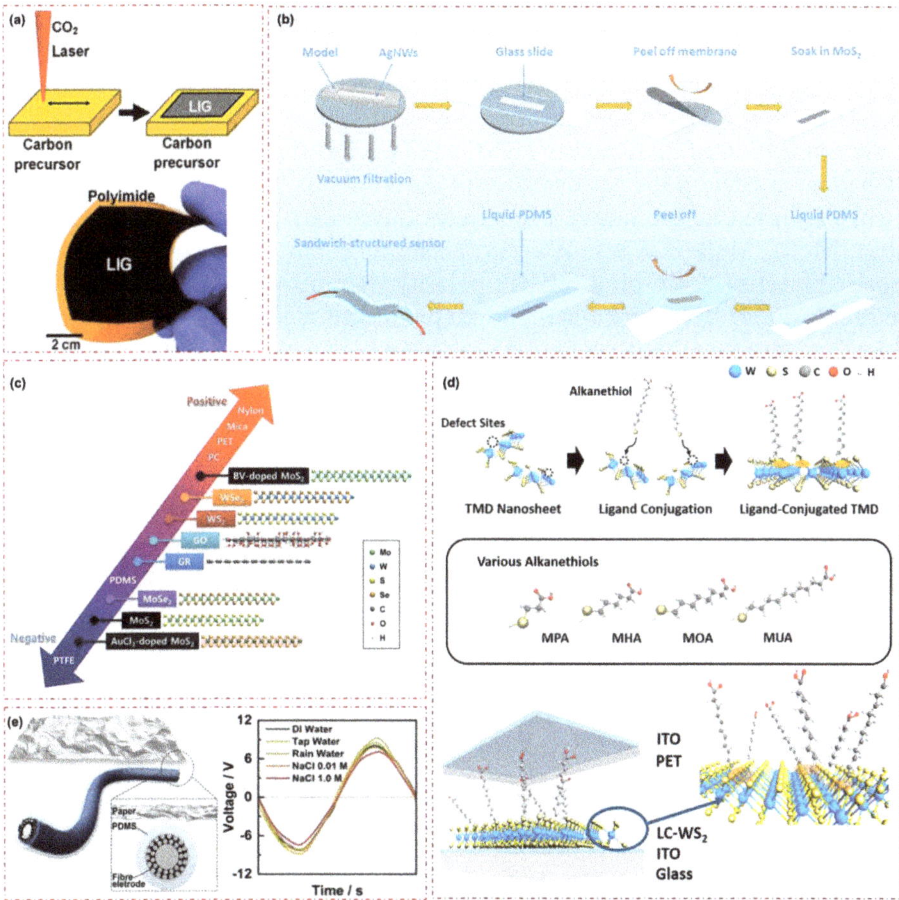

Figure 2.5. (a) The process used to produce the LIG electrode (reprinted with permission from [47], Copyright (2019), American Chemical Society). (b) A schematic illustration of the fabrication of a stretchable TENG (reprinted from [53], Copyright (2019), with permission from Elsevier). (c) A triboelectric series of 2D layered materials [54] John Wiley & Sons. © 2018 WILEY-VCH Verlag GmbH & Co. KGaA, Weinheim. (d) The process used to modify WS_2 with various alkanethiols and a TENG device design (reprinted with permission from [55], Copyright (2021), American Chemical Society). (e) The design of an MXene-coated yarn-based TENG and its outputs for different water and salt concentrations (reprinted from [56], Copyright (2020), with permission from Elsevier).

with PU as the active triboelectric layer produced an output of ~1 kV [47]. In other work, LIG electrodes have also been used for the fabrication of a 16 × 16 flexible triboelectric sensing array (TSA) for tactile sensing [52].

In terms of TENG active layers, a graphene oxide TENG (GONG) composed of Al/PI on one side and ITO/PI/graphene oxide (GO) on the other side produced an output of 2 V at 30 nA [52]. In other studies, reduced graphene oxide (rGO) was used as a charge-trapping material, and fluorinated graphene was used to impart self-cleaning and superhydrophobic properties to TENGs [57, 58]. Aligned graphene sheets (AGSs) were used as a filler in PDMS to improve TENG performance by enhancing overall device capacitance. The AGS led to a current enhancement of 1.44 times compared to the current produced by disoriented or disordered sheets [59].

TMDs have not been greatly explored for use in TENG electrodes. In this regard, metallic MOS_2 was mixed with silver nanowires to enhance its electrical conductivity for the fabrication of stretchable TENGs (STENGs) (figure 2.5(b)). Stretching by 50% had no impact on the performance of the STENGs [53]. However, TMDs have been widely used as the active triboelectric layer and as fillers in the polymeric matrix. MoS_2, $MoSe_2$, WSe_2, WS_2, graphene (G), and GO were arranged in the triboelectric series based on a relative polarity test and work function calculations using KPFM [54]. The work function measurement followed the sequence of MoS_2 (4.85 eV) > $MoSe_2$ (4.70 eV) > G (4.65 eV) > GO (4.56 eV) > WS_2 (4.54 eV)> WSe_2 (4.45 eV). Figure 2.5(c) depicts the 2D materials arranged in a triboelectric series with conventional polymers.

Chemical doping was used to tune the work function of MoS_2 [54]. MoS_2 was also used as a solid lubricant to improve wear resistance while simultaneously enhancing the charge density of a TENG [60]. Defect engineering involving the conjugation of thiol ligands to WS_2 using mercaptoundecanoic acid (MUA), mercaptooctanoic acid (MOA), mercaptopropionic acid (MPA), and mercaptohexanoic acid (MHA) (figure 2.5(d)) was later shown to tune TENG performance [55]. Figure 2.5(d) also shows the design of the WS_2-TENG with different conjugated ligands. The MUA-conjugation-based TENG showed a tenfold improvement in the electrical performance compared to that of the pristine WS_2 TENG [55].

Turning to MXenes, $Ti_3C_2T_x$ ink was coated onto a yarn to design a yarn-based TENG (figure 2.5(e)) [56]. The conductivity of the yarn was improved by decorating the $Ti_3C_2T_x$-coated yarn with Ag nanoparticles. Figure 2.5(e) depicts the outputs of the yarn-based TENG obtained by harvesting rainwater, tap water, DI water, and NaCl-containing water [56]. The $Ti_3C_2T_x$/cellulose nanofiber (CNF) composite was used as a liquid electrode in a shape-adaptive CNF/MXene CM-TENG in another study [61]. The CNF improved the $Ti_3C_2T_x$ interconnections by acting as an interlocking agent. The CM-TENG generated an output of ~300 V at 5.5 μA [61].

$Ti_3C_2T_x$ was also used as a filler in porous PDMS to enhance the performance of a TENG by increasing the surface potential of PDMS from −95 to −301 V [62]. The PDMS/MXene TENG showed a seven-fold (119 V and 11 μA) higher output compared to a pristine PDMS TENG [62]. Similarly, in another study, freeze-drying

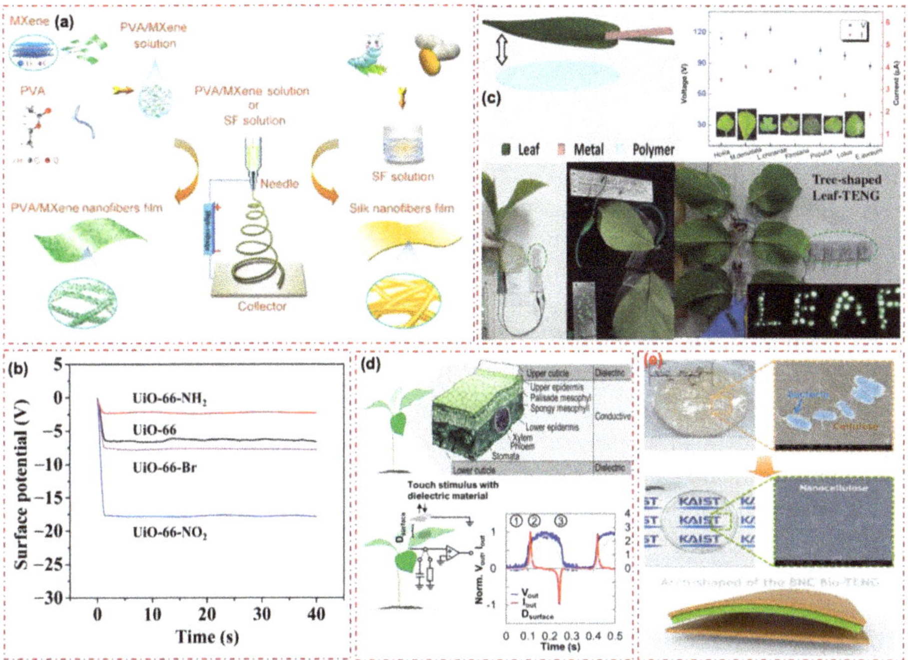

Figure 2.6. (a) The electrospinning process used to produce PVA/MXene or SF fibers (reprinted from [64], Copyright (2019), with permission from Elsevier). (b) The surface potentials of UIO-X with different X groups (reprinted from [67], Copyright (2022), with permission from Springer Nature). (c) Hosta leaf used for the fabrication of leaf-TENG, the outputs produced by different leaves, and the tree-shaped TENG [68] John Wiley & Sons. © 2018 WILEY-VCH Verlag GmbH & Co. KGaA, Weinheim. (d) A charge-generating mechanism in a rhododendron leaf [69] John Wiley & Sons. © 2018 WILEY-VCH Verlag GmbH & Co. KGaA, Weinheim. (e) Bacterial nanocellulose (BNC) scanning electron microscopy (SEM) image and TENG device design (reprinted from [70], Copyright (2017), with permission from Elsevier).

and vacuum impregnation were used to prepare PDMS/MXene composites [63]. However, there was no significant enhancement to either voltage or current output, despite the complexity of the preparation method. In several studies, MXene $Ti_3C_2T_x$ has been used as a filler in different polymers such as polyvinyl alcohol (PVA), poly(vinylidene fluoride-co-trifluoroethylene) (P(VDF-TrFE)), and PVDF to prepare electrospun fibers as a TENG active layer [64–66]. One example of the resulting TENGs was an all-fiber active-layer-based TENG with PVA/MXene nanofibers as the negative and silk fibroin (SF) nanofibers as the positive triboelectric layer [64]. The electrospun fiber preparation is shown in figure 2.6(a). The fabricated device was used to power an electrowetting-on-dielectric (EWOD) chip and to monitor body motions [64]. Two-dimensional materials are an excellent choice to fulfill the demand of flexible electronics. MXenes hold tremendous potential for TENG device fabrication, as they can play a dual role; so far, only $Ti_3C_2T_x$ has been explored in detail.

2.4 Metal–organic-framework- and covalent–organic-framework-based TENGs

Metal–organic frameworks (MOFs) are porous crystalline materials discovered in the 1990s that comprise a metal ion linked or bound to a ligand. Metal ions are considered to be secondary building units (SBUs). MOFs offers several unique advantages, such as high surface area, structural and functional tunability, high porosity, and ease of post-synthetic modification [71–74]. Since their discovery, MOFs have been shown to be applicable in catalysis, sensing, energy storage, energy harvesting, gas storage, gas separation, and so on [71–78]. Similarly, covalent–organic frameworks (COFs) are porous crystalline materials, except that they consist of lighter materials such as C, H, N, O, B, and Si, leading to low density [79]. In 2019, MOFs were used as an active layer in TENGs for the first time [80]. MOFs and COFs are brittle in nature and are hard to grow on substrates in such a way that they adhere to them strongly. For this reason, the majority of the reported MOFs/COFs were produced by attaching powder to a tape at high pressure, followed by the removal of unattached material using air or nitrogen [81, 82].

The first example of the use of MOFs in TENGs was the use of zeolitic imidazole framework 8 (ZIF-8) as a positive triboelectric layer. The ZIF-8 was grown from flexible ITO-coated polyethylene terephthalate (PET) for different numbers of cycles (15, 18, 20, 30, and 50 cycles). A TENG device made from Kapton and ZIF-8 grown for 20 cycles in vertical contact–separation mode produced an output voltage of 164 V at 7 μA. The fabricated ZIF-8 TENG was used for self-powered tetracycline sensing. The voltage output of the TENG decreased with an increase in tetracycline concentration, which was attributed to an interaction between tetracycline's benzene ring and the ZIF-8 imidazole ring. The sensor exhibited a sensitivity of 3.12 V μM^{-1} with and ability to be refreshed by a simple washing process using methanol [80]. Later, the same group demonstrated the use of various ZIF family members, namely ZIF-7, ZIF-9, ZIF-11, and ZIF-12, as an active layer in TENGs [83]. The isostructural MOF UiO-66-X (X = NH_2, NO_2, H, and Br) was subsequently grown on fluorine-doped tin oxide (FTO)-coated glass for use as a TENG active layer. The UiO-66 with X = NO_2 produced the highest output (23.79 V and 0.29 μA) among the variants tested [67]. Figure 2.6(b) shows the surface potentials of UiO-66 for different X, confirming that X=NO_2 had the highest surface potential when contact electrified with copper (Cu). The surface potential showed a trend of UiO-66-NH_2 < UiO-66 < UiO-66-Br < UiO-66-NO_2. This study demonstrated an attractive strategy for tuning the triboelectric properties of MOFs [67].

MOFs are closely related to metal–biomolecule frameworks (MBIOFs). One-dimensional copper aspartate (Cu-Asp) MBIOF nanofibers were demonstrated for TENG-based thioacetamide sensing [84]. The Cu-Asp nanofibers were coated on aluminum and PET by a simple tape-cast coating method. The coated Cu-Asp nanofibers remained stable even after the substrates were bent and dipped in water; no cracks or material removal was observed, confirming the easy scalability of this process. The Cu-Asp nanofibers were used as the active layer to fabricate free-standing triboelectric-layer mode and vertical contact–separation mode TENGs, achieving 200 V at 6 μA and 80 V at 1.2 μA, respectively. The vertical contact–separation mode TENG with reduced dimensions was used for thioacetamide sensing with a sensitivity of 0.76 V mM^{-1} [84].

MOFs have also been used as a filler in a polymer matrix to enhance the performance of TENGs. The first such example was the use of HKUST-1 as a filler in PDMS for the fabrication of a humidity-resistant TENG [85]. A 5 wt.% HKUST-1 in PDMS showed the best output performance, which was attributed to the MOF's electron-trapping abilities. Similarly, in another study, the MOF KAUST-8 was used as a filler in PDMS to improve active-layer hydrophobicity and enhance TENG performance [86]. A 0.5 wt.% KAUST-8-loaded PDMS-based TENG showed 11 times higher power density compared with that of a pristine PDMS-based TENG. The enhancement in TENG performance was due to the bifunctional properties of KAUST-8, i.e. improved charge-inducing and charge-trapping abilities [86].

COFs have also been demonstrated as active layers for TENGs. One such example is the use of a bromine-functionalized TPB-DBBA COF consisting of 1,3,5-tris (4-aminophenyl) benzene (TPB) and a 2,5-dibromobenzene-1,4-dicarboaldehyde (DBBA) linker as the active layer in a TENG [87]. The functionalized TPB-DBBA COF-based TENG showed enhancements of 1.3 and 2 times in the voltage and current, respectively, compared to the values obtained for a TPB-TP (TP: Terephthalaldehyde)-based TENG. The better output was ascribed to the high crystallinity, large surface area, high charge density and charge delocalization of TPB-DBBA [87].

Similarly, numerous MOFs and COFs have been demonstrated for TENGs. For further details, the reader is referred to [88]. The growth of MOFs/COFs on conductive substrates with strong adherence to the substrate is still challenging, thus limiting their applicability in TENGs. Overcoming these growth challenges could extend the use of MOF- and COF-based TENGs as multifunctional self-powered sensors capable of detecting multiple analytes with high sensitivity and selectivity in the near future.

2.5 Natural and biodegradable materials for TENGs

Increasing environmental pollution has led to the exploration of natural or biodegradable materials as active layers for TENGs. Several natural materials such as hosta leaf, seaweed (laver), egg white (EW), rice paper (RP), rhododendron plant leaf, and CNFs have been demonstrated for use as TENG active layers [68, 69, 89–91]. In addition, several synthetic biodegradable polymers, such as polylactic acid (PLA), polycaprolactone (PCL), poly(3-hydroxybutyrate-co-3-hydroxyvalerate) (PHB-V), PVA, and lipids, have also been used as TENG active layers [92, 93]. In this regard, figure 2.6(c) shows the use of hosta leaf in the construction of a TENG [68]. The hosta leaf acted as an electrode as well as an electrification layer, while polymethylmethacrylate (PMMA) served as the opposite triboelectric layer. The conductive properties of the leaf were due to the presence of water-based electrolytes in the leaf. The hosta leaf TENG produced an output voltage of 230 V and a current of 9.5 μA at a frequency of 2 Hz. However, the life of a leaf TENG is limited, as the leaf eventually dries out after several days. Figure 2.6(c) shows the performance of different leaves with PMMA as the opposite layer in the TENG device [68]. Other research groups subsequently studied the charge generation in a rhododendron plant leaf (figure 2.6(d)). The outer plant leaf acted as a partial dielectric, while the inner leaf tissues were capable of charge transfer due to ionic conduction. The leaf was used to design a single-electrode TENG with different opposite contact materials [69].

Cellulose from different sources, including wood pulp and bacteria, is widely used as the active layer in TENGs. The use of bacterial nanocellulose (BNC) for the fabrication of a bio-TENG was demonstrated in 2017 [70]. The BNC was produced from *Acetobacter xylinum* using the process shown in figure 2.6(e). The cellulose was then modified using thiol–ene click chemistry to improve the performance of the CNF-based TENG. Allicin was grafted onto CNF in different amounts [94]. The optimal allicin TENG, i.e. one grafted with 5 ml allicin, produced an output of 7.9 V at 5.13 μA. In another study, CNFs were modified with an amino group to fabricate an amino–CNF (A-CNF) TENG with FEP as the opposite layer. The A-CNF TENG produced an output voltage of 195 V, a current of 13.4 μA, and a charge of 65 nC. The amino modification also improved the device's performance when exposed to humidity [94].

A bioabsorbable natural-material-based TENG (BN-TENG) was reported that was made from a combination of bioresorbable materials including EW, cellulose, SF, RP, and chitin [90]. An SF encapsulation layer was used to tune the lifetime of the BN-TENG. A BN-TENG with SF and RP as the triboelectric layer produced the highest output (34 V, 0.32 μA, and 21.6 mW m^{-2}). Figure 2.7(a) shows the BN-TENG device and the arrangement of the natural biopolymers, which depended on their ability to gain or lose electrons. The encapsulated BN-TENG was implanted in a Sprague Dawley (SD) rat. The implanted device produced 4.5 V, which decreased to 1.5 V after 11 days of implantation. The device maintained excellent biocompatibility, and the generated output was used to regulate the beating of the cardiomyocyte cluster (C1, C2, and C3) [90]. Similarly, a biodegradable, implantable, multifunctional transient TENG (T^2ENG) was reported for epilepsy detection, monitoring, and treatment [95]. The T^2ENG was fabricated from silk and magnesium, in which the magnesium served as both the electrode and the negative triboelectric layer. The T^2ENG generated an output voltage of 60 V, a current of 1 μA, and a power density of 58.5 mW m^{-2}. The T^2ENG partially dissolved in 60 min when immersed in deionized (DI) water. The T^2TENG was implanted in an SD rat and was resorbed or degraded *in vivo*. A distinct signal produced by epilepsy was used to trigger the thermal unit of the medical system, as shown in figure 2.7(b) [95].

2.6 Ferroelectric TENG materials

Ferroelectric materials are dielectrics that have the ability to become permanently polarized after the removal of an electric field. The spontaneous polarization in ferroelectric materials changes with the electric field [96, 97]. Ferroelectric materials exhibit both piezoelectric and pyroelectric behavior. Ferroelectric materials are widely used to enhance the performance of TENGs.

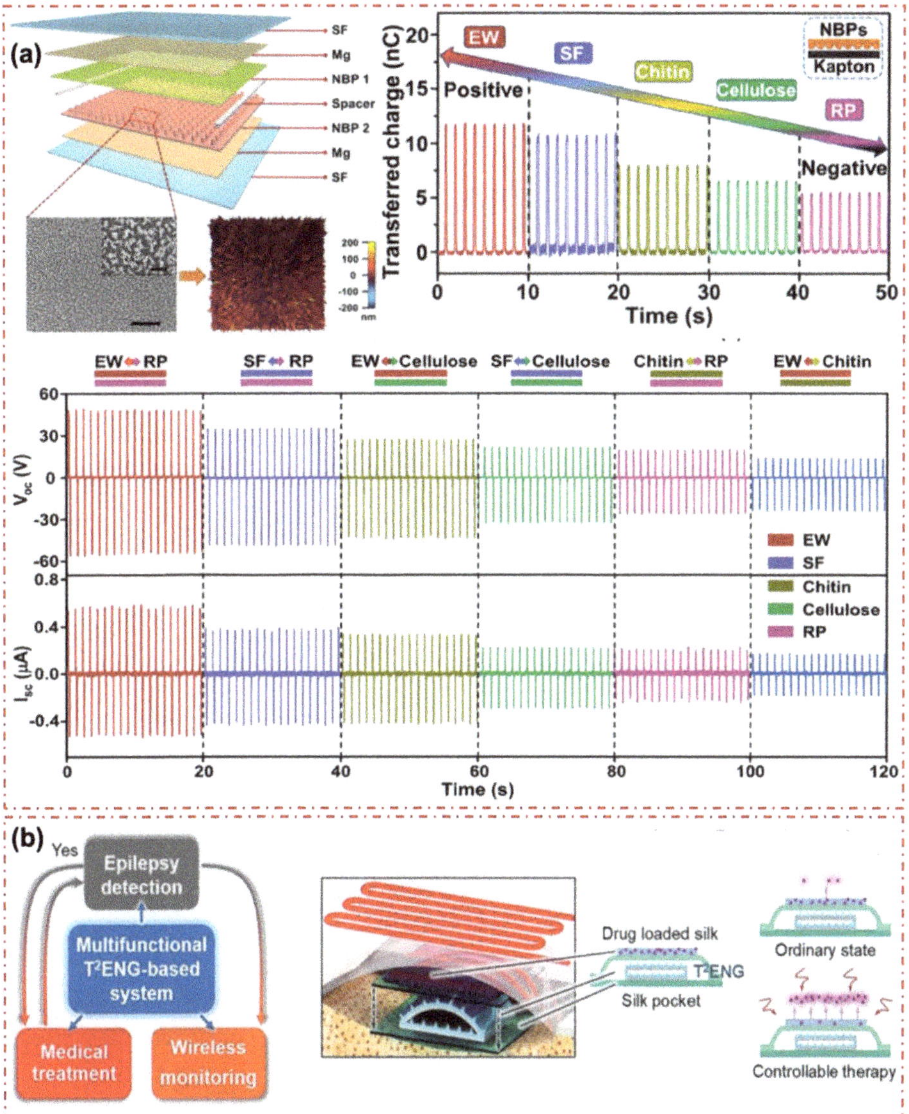

Figure 2.7. (a) The device design of the bioabsorbable TENG and the voltage outputs of different layer combinations [90] John Wiley & Sons. © 2018 WILEY-VCH Verlag GmbH & Co. KGaA, Weinheim. (b) The drug release process of the transient T^{2}ENG [95] John Wiley & Sons. © 2018 WILEY-VCH Verlag GmbH & Co. KGaA, Weinheim.

The triboelectricity on a ferroelectric polymer P(VDF-TrFE) film was studied using atomic force microscopy (AFM). KPFM was used to study the surface potential, and piezo force microscopy (PFM) was used to detect the polarization [98]. Figure 2.8(a) shows the post-poling polarization states detected by the conductive PFM tip. The brighter region represents upward polarization and the darker region represents downward polarization. The surface contrast decayed over

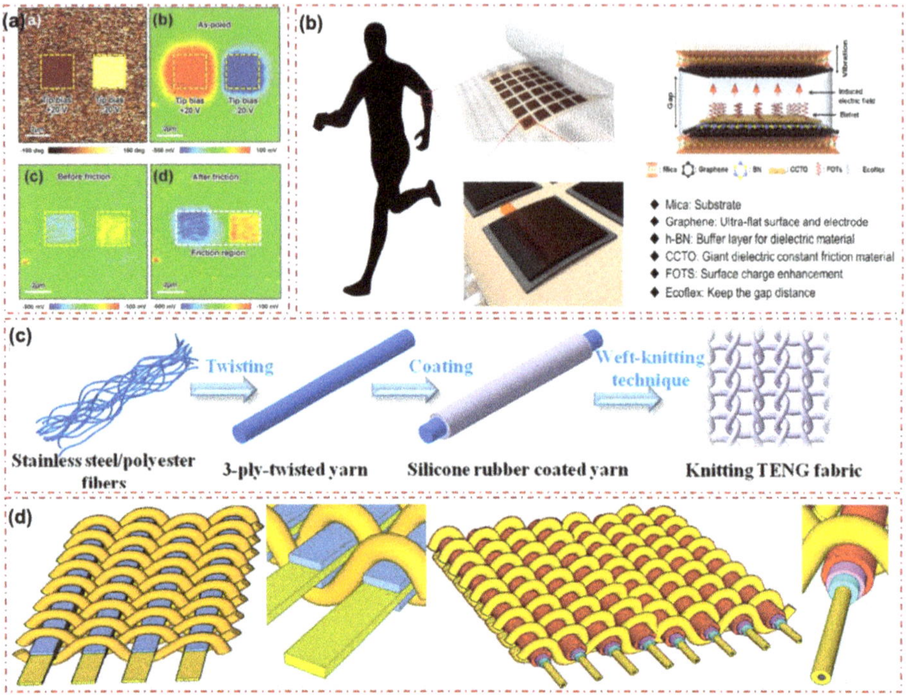

Figure 2.8. (a) PFM results showing the polarization states in P(VDF-TrFE) [98] John Wiley & Sons. © 2016 WILEY-VCH Verlag GmbH & Co. KGaA, Weinheim. (b) The design of a calcium copper titanate (CCTO)-based noncontact TENG comprising different layers (reprinted with permission from [99], Copyright (2021), American Chemical Society). (c) The weft knitting technique used for the fabrication of knitted yarn (reprinted with permission from [100], Copyright (2017), American Chemical Society). (d) The design of a hybrid energy harvester consisting of a TENG and a solar cell (reprinted from [101], Copyright (2016), with permission from Springer Nature).

time due to the discharge of extra injected charges. However, the sample maintained its surface contrast even after 24 h. The figure also shows the TENG device with unpoled and positively and negatively poled conditions, confirming the effect of polarization on the TENG output [98].

Ferroelectric materials have been used as fillers in suitable polymers to boost TENG performance. In this regard, strontium titanate ($SrTiO_3$) nanoparticles were mixed into PDMS at different concentrations. A 10 wt.% $SrTiO_3$ composite-based TENG showed a fivefold enhancement in output (338 V, 19 nC cm^{-2}) compared to that of pristine PDMS. The enhancement in output was attributed to an improved dielectric constant, better charge transfer, and improved surface charge density [102]. Similarly, barium titanate ($BaTiO_3$) was incorporated into P(VDF-TrFE) to improve the performance of a ferroelectric TENG (FE-TENG). The effect of poling on the device performance was also studied. The poled samples exhibited high surface potential due to the ferroelectric polarization. An FE-TENG based on 5 wt.% barium titanate produced an output voltage of 330 V at 0.3 mA, which was used to drive smart watches via a power management circuit (PMC) [103]. Min *et al* constructed similar ferroelectric TENGs, except that these were based on electrospun fibrous surfaces of P(VDF-TrFE) with $BaTiO_3$ nanofillers (in contact with a PET film countersurface) [104]. Regarding contact pressure, the maximum output ($V_{oc} = 315$ V and $J_{sc} = 6.7$ μA cm^{-2}) was significantly higher than those of TENGs with spin-coated P(VDF-TrFE)/$BaTiO_3$. It was hypothesized that electrospinning improved dipole alignment owing to the high applied voltages and aided the formation of a highly oriented crystalline β-phase via uniaxial stretching. P(VDF-TrFE) with tetragonal $BaTiO_3$ produced a higher output than that of P(VDF-TrFE) with cubic $BaTiO_3$ even though its permittivity was nearly identical. Thus, it was demonstrated that $BaTiO_3$ fillers boost output, not just by increasing permittivity but also by enhancing the crystallinity and amount of the β-phase (since tetragonal $BaTiO_3$ produced a more crystalline β-phase in greater amounts).

A rotating TENG was later fabricated by incorporating calcium copper titanate, $CaCu_3TiO_2$ (CCTO), into butylated melamine formaldehyde (BMF). CCTO is a strongly dielectric material with a value of 7500 [105]. An enhancement in device performance was attributed to the induction of strong polarization in the material by the electric field produced by the triboelectric charges. A device with a loading of 1 wt.% CCTO generated an output of 268 V [105]. The same research group also demonstrated the use of CCTO in noncontact-mode TENGs (nc-TENGs). Figure 2.8(b) depicts the device design, which comprised graphene electrodes, hexagonal boron nitride (h-BN) as a buffer layer to receive the deposited CCTO, and 1H,1H,2H,2H-perfluorooctyltrichlorosilane (FOTS)-coated CCTO for surface charge enhancement [99]. The optimized nc-TENG, which had a 0.5 mm gap, produced 15.1 V at 420 μA. Similarly, another strongly dielectric material $(Ba_{0.85}Ca_{0.15})(Ti_{0.90}Zr_{0.10})O_{3-x}BiHoO_3$ (BCZTBH) was used as the filler in PDMS to fabricate a multistack hybrid generator (MS-HG). The MS-HG generated an output voltage and current of 300 V and 6.6 μA, respectively [106].

2.7 Textile-based TENGs

Clothing is part of daily human life, and textiles are capable of undergoing multiple cycles of biomechanical deformation. Rapid advancement in technology has converted textiles into intelligent textiles capable of performing different functions, including sensing, energy harvesting, human–machine interfacing, etc. In terms of mechanical energy harvesting, TENGs are widely designed to take the form of textiles. Textile-based TENGs are designed as single fibers (yarns), fabrics based on 2D and 3D textile structures, or multilayer stacked fiber devices [107].

Fibers are the fundamental building blocks or smallest design units of textiles. Single-fiber or yarn-based TENGs are mostly designed using a core–shell or coaxial configuration. One such example is a coaxial wearable and stretchable fiber (coaxial triboelectric nanogenerator fiber, CTNF) TENG fabricated with aligned carbon nanotube (CNTs) as the outer and inner electrodes; PMMA microspheres and PDMS formed the frictional layers [108]. The gap between the PMMA and the PDMS was generated with the help of a sucrose-particle-based sacrificial layer. The removal of the sacrificial layer also created surface roughness on the PDMS. The CTNF with 200 nm PMMA microspheres produced an output of 1 V at 150 nA under a 10 N force [108]. Another study reported stretchable and washable yarn-based textiles capable of harvesting and storing biomechanical energy. For this purpose, silicone rubber was coated onto three-ply twisted stainless steel–polyester blended yarn. Figure 2.8(c) shows the fabrication process in which the yarns were knitted into fabrics using the weft knitting technique. The TENG produced an open-circuit voltage of 150 V, a short-circuit current of ~2.9 μA, and a charge of ~52 nC at a frequency of 5 Hz [100]. Bairagi *et al* focused on enhancing the performance of a textile TENG based on the commonly used fabrics silk and PET [109]. By depositing electrospun nylon 66 onto the silk and coating the PET with PVDF, they were able to obtain a significant boast in output: the output voltage and short-circuit current density increased by ~17 times (5.85–100 V) and ~16 times (1.6–24.5 mA m^{-2}), respectively, compared with the silk/PET baseline. The boost was likely to be the result of enhancing the tribopositivity/tribonegativity of the respective active layers and of an improved contact area facilitated by the nanofibers. In 2019, the scalable manufacture of triboelectric yarn was achieved by modifying the melt-spinning method. A spinning setup was utilized to coat silicone rubber onto a stainless steel yarn (single-electrode triboelectric yarn, SETEY). The SETEY was also capable of operating in different liquid environments, such as cyclohexane, diethyl ether, alcohol, water, ethyl acetate, etc [110].

Textile-based TENGs are often also integrated with other energy generation or storage. A flying-shuttle weaving process was used to weave solar cells using fiber-based TENGs [101]. Figure 2.8(d) illustrates the design of the hybrid energy-generating textile. The TENG comprised copper and PTFE frictional layers. The hybrid textile (4 cm × 5 cm) was successfully used to charge a commercial 2 mF capacitor to 2 V in 1 min [101]. In another study, a wearable and flexible TENG was integrated with a zinc-ion battery (ZIB) to drive electronic devices [111]. Figure 2.9(a) illustrates the fabrication process and 3D view of the spacer-fabric-based TENG with a ZIB. The TENG was fabricated by coating graphene onto nylon and PTFE onto nylon, as shown in the figure. The TENG produced an output voltage of ~15 V at a current of ~4 μA. The TENG was demonstrated to charge the ZIB to 0.90 V in 70 s, which further increased to 1.28 V in 29.65 min [111].

The output performance of 2D textile TENGs is poor due to limitations in the thickness direction and their small contact area. Three-dimensional textiles increase the possible contact area and can be useful and advantageous compared to 2D textiles. A commercial 3D spacer fabric was used for the fabrication of a TENG. Figure 2.9(b) shows a 3D illustration of a triboelectric textile (TET), which was produced by coating commercial PET fabric with PDMS followed by coating it in 20 layers of CNT sheets to form one electrode, while silver paste served as the opposite electrode [112]. A 5 × 5 cm^2 TET produced an output voltage, short circuit current, and power density of 500 V, 20 μA, and 153.8 mW m^{-2}, respectively. The fabricated TENG was used to power 49 LEDs and a liquid crystal display (LCD) [112]. In another study, computerized knitting technology was used to fabricate 3D all-fiber-based textile TENGs (all-textile triboelectric energy harvesters, A-TEHs). The A-TEHs comprised silver-coated nylon fibers as the electrodes and cotton and polyacrylonitrile (PAN) as dielectric layers. In single-electrode mode, an A-TEH

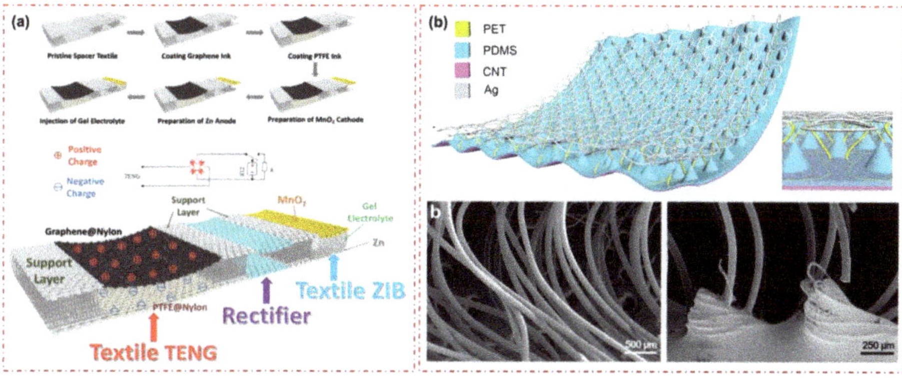

Figure 2.9. (a) The fabrication of the ZIB and the design of the TENG–ZIB hybrid device. Reproduced with permission [111] John Wiley & Sons. © 2018 WILEY-VCH Verlag GmbH & Co. KGaA, Weinheim. (b) A 3D illustration of the triboelectric textile produced from the 3D spacer fabric (reproduced from [112] with permission from the Royal Society of Chemistry).

exhibited a power density of 1768.2 mW m^{-2}. However, the power density was achieved at a load of 1200 N, which is too high for wearables [113].

A comparison of different textile-based device designs and their output performances can be found in [107]. Despite several advancements in textile TENGs, much attention is still required to fabricate a TENG with conventional textile properties, including breathability, wash durability, robustness, and scalability for industrial manufacturing technologies. Further, the performance of the majority of textile TENGs is still poor due to the low effective contact area between the active layers (owing to the structure of textiles).

2.8 Perspective and future directions

The performance of TENGs largely depends on their constituent materials. Rapid advancement in materials has improved the performance of TENGs and also extended their applications. Ferroelectric materials can greatly enhance the performance of TENGs when used as fillers, while materials such as MOFs can improve the sensitivity and selectivity of self-powered chemical sensors. Further TENG improvements can be made to enhance device performance and extend TENG applications by addressing the challenges listed below. Material pairs chosen to boost the contact area in TENGs also lead to higher output.

1. The choice of positive triboelectric materials is restricted. New eco-friendly materials with high surface charge density and ease of processing can be explored and tested for their performance and stability.
2. Materials that can replace fluorine-containing polymers such as Teflon while maintaining similar performance can reduce environmental pollution and lead to the fabrication of eco-friendly devices.
3. The high impedance of TENGs is the biggest challenge for real-time applications. Materials that produce high output while offering low device impedance should be a focal point of future research.

4. Humidity-resistant materials deserve more attention, since they can be used to fabricate devices which can maintain similar outputs under different environmental conditions.
5. MOFs are brittle and hard to grow on conductive substrates while achieving high adhesion. Methods such as hot pressing that can produce MOFs on conductive substrates also need more attention.

2.9 Conclusions

Materials have been rapidly advancing to meet the demand to enhance the output power of TENGs. Indeed, major increases in output have been realized since the first TENG paper was published in 2012. However, output is by no means the only requirement, and materials are also under development for purposes such as imparting functional properties, increasing durability, increasing flexibility, and increasing the selectivity and sensitivity of self-powered TENG-based sensors. Two-dimensional materials are promising as materials that can play a dual role as both electrode and active layers. MOFs may be the preferred choice for the development of self-powered chemical sensors, while ferroelectric materials can be used to boost TENG performance by permanent polarization. The new electrode materials can increase TENGs' flexibility and their ability to undergo multiple stretching or bending cycles. Materials selection is critical for TENG performance and durability, as mentioned in this chapter. However, the materials need to be tuned to obtain TENGs with high current performance and reduce the occurrence of wear and tear, especially in sliding-mode TENGs. It is hoped that the continuous development in materials will soon address these challenges and advance TENGs towards commercialization.

Acknowledgments

The authors acknowledge the support of the UK Engineering and Physical Sciences Research Council (EPSRC) in supporting this work through grant reference EP/V003380/1 ('Next Generation Energy Autonomous Textile Fabrics based on Triboelectric Nanogenerators').

References

[1] Pandey R, Khandelwal G, Palani I A, Singh V and Kim S-J 2019 A La-doped ZnO ultra-flexible flutter-piezoelectric nanogenerator for energy harvesting and sensing applications: a novel renewable source of energy *Nanoscale* **11** 14032–41

[2] Rajagopalan P, Singh V, Palani I A and Kim S-J 2019 Superior response in ZnO nanogenerator via interfaced heterojunction for novel smart gas purging system *Extreme Mech. Lett.* **26** 18–25

[3] Rajagopalan P, Jakhar P, Palani I A, Singh V and Kim S J 2020 Elucidations on the effect of Lanthanum doping in ZnO towards enhanced performance nanogenerators *Int. J. Precis. Eng. Manuf.-Green Technol.* **7** 77–87

[4] Rajagopalan P, Singh V and Palani I A 2018 Enhancement of ZnO-based flexible nano generators via a sol–gel technique for sensing and energy harvesting applications *Nanotechnology* **29** 105406
[5] Rajagopalan P, Singh V and Palani I A 2016 Investigations on the influence of substrate temperature in developing enhanced response ZnO nano generators on flexible polyimide using spray pyrolysis technique *Mater. Res. Bull.* **84** 340–5
[6] Khandelwal G, Maria Joseph Raj N P and Kim S-J 2021 Materials beyond conventional triboelectric series for fabrication and applications of triboelectric nanogenerators *Adv. Energy Mater.* **11** 2101170
[7] Khandelwal G and Dahiya R 2022 Self-powered active sensing based on triboelectric generators *Adv. Mater.* **34** 2200724
[8] Khandelwal G, Maria Joseph Raj N P and Kim S-J 2020 Triboelectric nanogenerator for healthcare and biomedical applications *Nano Today* **33** 100882
[9] Zou H *et al* 2020 Quantifying and understanding the triboelectric series of inorganic non-metallic materials *Nat. Commun.* **11** 2093
[10] Zou H *et al* 2019 Quantifying the triboelectric series *Nat. Commun.* **10** 1427
[11] Rajagopalan P, Huang S, Shi L, Kuang H, Jin H, Dong S, Shi W, Wang X and Luo J 2021 Novel insights from the ultra-thin film, strain-modulated dynamic triboelectric characterizations *Nano Energy* **80** 105560
[12] Huang S *et al* 2020 Controlling performance of organic–inorganic hybrid perovskite triboelectric nanogenerators via chemical composition modulation and electric field-induced ion migration *Adv. Energy Mater.* **10** 2002470
[13] Rajagopalan M M, Xu P, Palani S, Singh I A, Wang V, Wu X and W 2021 Enhancement of patterned triboelectric output performance by an interfacial polymer layer for energy harvesting application *Nanoscale* **13** 20615–24
[14] Zhong W, Xu L, Wang H, An J and Wang Z L 2019 Tilting-sensitive triboelectric nanogenerators for energy harvesting from unstable/fluctuating surfaces *Adv. Funct. Mater.* **29** 1905319
[15] Zhao P *et al* 2018 Emulsion electrospinning of polytetrafluoroethylene (ptfe) nanofibrous membranes for high-performance triboelectric nanogenerators *ACS Appl. Mater. Interfaces* **10** 5880–91
[16] Li X, Tao J, Wang X, Zhu J, Pan C and Wang Z L 2018 Networks of high performance triboelectric nanogenerators based on liquid–solid interface contact electrification for harvesting low-frequency blue energy *Adv. Energy Mater.* **8** 1800705
[17] Dudem B, Kim D H, Mule A R and Yu J S 2018 Enhanced performance of micro-architectured PTFE-based triboelectric nanogenerator via simple thermal imprinting lithography for self-powered electronics *ACS Appl. Mater. Interfaces* **10** 24181–92
[18] Zhang Z and Cai J 2021 High output triboelectric nanogenerator based on PTFE and cotton for energy harvester and human motion sensor *Curr. Appl Phys.* **22** 1–5
[19] Lin Z, Zhang B, Guo H, Wu Z, Zou H, Yang J and Wang Z L 2019 Super-robust and frequency-multiplied triboelectric nanogenerator for efficient harvesting water and wind energy *Nano Energy* **64** 103908
[20] Cheedarala R K and Song J I 2022 Harvesting of flow current through implanted hydrophobic PTFE surface within silicone-pipe as liquid nanogenerator *Sci. Rep.* **12** 3700
[21] Liu Z, Huang Y, Shi Y, Tao X, He H, Chen F, Huang Z-X, Wang Z L, Chen X and Qu J-P 2022 Fabrication of triboelectric polymer films via repeated rheological forging for ultrahigh surface charge density *Nat. Commun.* **13** 4083

[22] Xiao T X, Liang X, Jiang T, Xu L, Shao J J, Nie J H, Bai Y, Zhong W and Wang Z L 2018 Spherical triboelectric nanogenerators based on spring-assisted multilayered structure for efficient water wave energy harvesting *Adv. Funct. Mater.* **28** 1802634
[23] Pan L, Wang J, Wang P, Gao R, Wang Y-C, Zhang X, Zou J-J and Wang Z L 2018 Liquid-FEP-based U-tube triboelectric nanogenerator for harvesting water-wave energy *Nano Res.* **11** 4062–73
[24] Zhang H, Feng S, He D, Xu Y, Yang M and Bai J 2018 An electret film-based triboelectric nanogenerator with largely improved performance via a tape-peeling charging method *Nano Energy* **48** 256–65
[25] Shi L *et al* 2021 High-performance triboelectric nanogenerator based on electrospun PVDF-graphene nanosheet composite nanofibers for energy harvesting *Nano Energy* **80** 105599
[26] Lee J W *et al* 2017 Robust nanogenerators based on graft copolymers via control of dielectrics for remarkable output power enhancement *Sci. Adv.* **3** e1602902
[27] Khandelwal G, Chandrasekhar A, Pandey R, Maria Joseph Raj N P and Kim S-J 2019 Phase inversion enabled energy scavenger: a multifunctional triboelectric nanogenerator as benzene monitoring system *Sensors Actuators* B **282** 590–8
[28] Chandrasekhar A, Vivekananthan V, Khandelwal G and Kim S J 2019 A fully packed water-proof, humidity resistant triboelectric nanogenerator for transmitting Morse code *Nano Energy* **60** 850–6
[29] Chandrasekhar A, Vivekananthan V, Khandelwal G, Kim W J and Kim S J 2020 Green energy from working surfaces: a contact electrification–enabled data theft protection and monitoring smart table *Mater. Today Energy* **18** 100544
[30] Soin N *et al* 2016 High performance triboelectric nanogenerators based on phase-inversion piezoelectric membranes of poly(vinylidene fluoride)-zinc stannate (PVDF-$ZnSnO_3$) and polyamide-6 (PA6) *Nano Energy* **30** 470–80
[31] Chandrasekhar A, Alluri N R, Saravanakumar B, Selvarajan S and Kim S-J 2017 A microcrystalline cellulose ingrained polydimethylsiloxane triboelectric nanogenerator as a self-powered locomotion detector *J. Mater. Chem.* C **5** 1810–5
[32] Kim Y, Wu X and Oh J H 2020 Fabrication of triboelectric nanogenerators based on electrospun polyimide nanofibers membrane *Sci. Rep.* **10** 2742
[33] Ghatak B, Banerjee S, Ali S B, Bandyopadhyay R, Das N, Mandal D and Tudu B 2021 Design of a self-powered triboelectric face mask *Nano Energy* **79** 105387
[34] Kumar C, Perris J, Bairagi S, Min G, Xu Y, Gadegaard N and Mulvihill D M 2023 Multiscale *in situ* quantification of the role of surface roughness and contact area using a novel Mica-PVS triboelectric nanogenerator *Nano Energy* **107** 108122
[35] Ding P, Chen J, Farooq U, Zhao P, Soin N, Yu L, Jin H, Wang X, Dong S and Luo J 2018 Realizing the potential of polyethylene oxide as new positive tribo-material: over 40 W/m^2 high power flat surface triboelectric nanogenerators *Nano Energy* **46** 63–72
[36] Zhao P *et al* 2020 Expanding the portfolio of tribo-positive materials: aniline formaldehyde condensates for high charge density triboelectric nanogenerators *Nano Energy* **67** 104291
[37] Khandelwal G, Deswal S, Shakthivel D and Dahiya R 2023 Recent developments in 2D materials for energy harvesting applications *J. Phys.: Energy* **5** 032001
[38] Manzeli S, Ovchinnikov D, Pasquier D, Yazyev O V and Kis A 2017 2D transition metal dichalcogenides *Nat. Rev. Mater.* **2** 17033

[39] Wu C, Kim T W, Park J H, An H, Shao J, Chen X and Wang Z L 2017 Enhanced triboelectric nanogenerators based on MoS_2 monolayer nanocomposites acting as electron-acceptor layers *ACS Nano* **11** 8356–63

[40] Nardekar S S, Krishnamoorthy K, Manoharan S, Pazhamalai P and Kim S J 2022 Two faces under a hood: unravelling the energy harnessing and storage properties of 1T-MoS_2 quantum sheets for next-generation stand-alone energy systems *ACS Nano* **16** 3723–34

[41] Verger L, Natu V, Carey M and Barsoum M W 2019 MXenes: an introduction of their synthesis, select properties, and applications *Trends Chem.* **1** 656–69

[42] Gogotsi Y and Anasori B 2019 The rise of MXenes *ACS Nano* **13** 8491–4

[43] Dong Y, Mallineni S S K, Maleski K, Behlow H, Mochalin V N, Rao A M, Gogotsi Y and Podila R 2018 Metallic MXenes: a new family of materials for flexible triboelectric nanogenerators *Nano Energy* **44** 103–10

[44] Zhou J, Chen Y, Song X, Zheng B, Li P, Liu J, Qi F, Hao X and Zhang W 2017 Flexible transparent triboelectric nanogenerators with graphene and indium tin oxide electrode structures *Energy Technol.* **5** 599–603

[45] Chu H, Jang H, Lee Y, Chae Y and Ahn J-H 2016 Conformal, graphene-based triboelectric nanogenerator for self-powered wearable electronics *Nano Energy* **27** 298–305

[46] Yang J, Liu P, Wei X, Luo W, Yang J, Jiang H, Wei D, Shi R and Shi H 2017 Surface engineering of graphene composite transparent electrodes for high-performance flexible triboelectric nanogenerators and self-powered sensors *ACS Appl. Mater. Interfaces* **9** 36017–25

[47] Stanford M G, Li J T, Chyan Y, Wang Z, Wang W and Tour J M 2019 Laser-induced graphene triboelectric nanogenerators *ACS Nano* **13** 7166–74

[48] Shin D-W *et al* 2018 A new facile route to flexible and semi-transparent electrodes based on water exfoliated graphene and their single-electrode triboelectric nanogenerator *Adv. Mater.* **30** 1802953

[49] Zhao J, Pei S, Ren W, Gao L and Cheng H-M 2010 Efficient preparation of large-area graphene oxide sheets for transparent conductive films *ACS Nano* **4** 5245–52

[50] Lin J, Peng Z, Liu Y, Ruiz-Zepeda F, Ye R, Samuel E L G, Yacaman M J, Yakobson B I and Tour J M 2014 Laser-induced porous graphene films from commercial polymers *Nat. Commun.* **5** 5714

[51] Chyan Y, Ye R, Li Y, Singh S P, Arnusch C J and Tour J M 2018 Laser-induced graphene by multiple lasing: toward electronics on cloth, paper, and food *ACS Nano* **12** 2176–83

[52] Yan Z *et al* 2021 Flexible high-resolution triboelectric sensor array based on patterned laser-induced graphene for self-powered real-time tactile sensing *Adv. Funct. Mater.* **31** 2100709

[53] Lan L, Yin T, Jiang C, Li X, Yao Y, Wang Z, Qu S, Ye Z, Ping J and Ying Y 2019 Highly conductive 1D–2D composite film for skin-mountable strain sensor and stretchable triboelectric nanogenerator *Nano Energy* **62** 319–28

[54] Seol M, Kim S, Cho Y, Byun K-E, Kim H, Kim J, Kim S K, Kim S-W, Shin H-J and Park S 2018 Triboelectric series of 2D layered materials *Adv. Mater.* **30** 1801210

[55] Kim T I, Park I-J, Kang S, Kim T-S and Choi S-Y 2021 Enhanced triboelectric nanogenerator based on tungsten disulfide via thiolated ligand conjugation *ACS Appl. Mater. Interfaces* **13** 21299–309

[56] Jiang C, Li X, Ying Y and Ping J 2020 A multifunctional TENG yarn integrated into agrotextile for building intelligent agriculture *Nano Energy* **74** 104863

[57] Wu C, Kim T W and Choi H Y 2017 Reduced graphene-oxide acting as electron-trapping sites in the friction layer for giant triboelectric enhancement *Nano Energy* **32** 542–50
[58] Jiang C, Li X, Ying Y and Ping J 2021 Fluorinated graphene-enabled durable triboelectric coating for water energy harvesting *Small* **17** 2007805
[59] Xia X, Chen J, Liu G, Javed M S, Wang X and Hu C 2017 Aligning graphene sheets in PDMS for improving output performance of triboelectric nanogenerator *Carbon* **111** 569–76
[60] Zhao K *et al* 2022 High-performance and long-cycle life of triboelectric nanogenerator using PVC/MoS_2 composite membranes for wind energy scavenging application *Nano Energy* **91** 106649
[61] Cao W-T, Ouyang H, Xin W, Chao S, Ma C, Li Z, Chen F and Ma M-G 2020 A stretchable highoutput triboelectric nanogenerator improved by MXene liquid electrode with high electronegativity *Adv. Funct. Mater.* **30** 2004181
[62] Jiang C, Li X, Yao Y, Lan L, Shao Y, Zhao F, Ying Y and Ping J 2019 A multifunctional and highly flexible triboelectric nanogenerator based on MXene-enabled porous film integrated with laser-induced graphene electrode *Nano Energy* **66** 104121
[63] Wang D, Lin Y, Hu D, Jiang P and Huang X 2020 Multifunctional 3D-MXene/PDMS nanocomposites for electrical, thermal and triboelectric applications *Compos. Part A: Appl. Sci. Manuf.* **130** 105754
[64] Jiang C, Wu C, Li X, Yao Y, Lan L, Zhao F, Ye Z, Ying Y and Ping J 2019 All-electrospun flexible triboelectric nanogenerator based on metallic MXene nanosheets *Nano Energy* **59** 268–76
[65] Rana S M S, Rahman M T, Salauddin M, Sharma S, Maharjan P, Bhatta T, Cho H, Park C and Park J Y 2021 Electrospun PVDF-TrFE/MXene nanofiber mat-based triboelectric nanogenerator for smart home appliances *ACS Appl. Mater. Interfaces* **13** 4955–67
[66] Bhatta T, Maharjan P, Cho H, Park C, Yoon S H, Sharma S, Salauddin M, Rahman M T, Rana S M S and Park J Y 2021 High-performance triboelectric nanogenerator based on MXene functionalized polyvinylidene fluoride composite nanofibers *Nano Energy* **81** 105670
[67] Wen R, Feng R, Zhao B, Song J, Fan L and Zhai J 2022 Controllable design of high-efficiency triboelectric materials by functionalized metal–organic frameworks with a large electron-withdrawing functional group *Nano Res.* **15** 9386–91
[68] Jie Y, Jia X, Zou J, Chen Y, Wang N, Wang Z L and Cao X 2018 Natural leaf made triboelectric nanogenerator for harvesting environmental mechanical energy *Adv. Energy Mater.* **8** 1703133
[69] Meder F, Must I, Sadeghi A, Mondini A, Filippeschi C, Beccai L, Mattoli V, Pingue P and Mazzolai B 2018 Energy conversion at the cuticle of living plants *Adv. Funct. Mater.* **28** 1806689
[70] Kim H-J, Yim E-C, Kim J-H, Kim S-J, Park J-Y and Oh I-K 2017 Bacterial nano-cellulose triboelectric nanogenerator *Nano Energy* **33** 130–7
[71] Olorunyomi J F, Geh S T, Caruso R A and Doherty C M 2021 Metal–organic frameworks for chemical sensing devices *Mater. Horiz.* **8** 2387–419
[72] Wu J *et al* 2020 Metal–organic framework for transparent electronics *Adv. Sci.* **7** 1903003
[73] Zhou H-C J and Kitagawa S 2014 Metal–organic frameworks (MOFs) *Chem. Soc. Rev.* **43** 5415–8

[74] Furukawa H, Cordova K E, O'Keeffe M and Yaghi O M 2013 The chemistry and applications of metal-organic frameworks *Science* **341** 1230444

[75] Olorunyomi J F, Sadiq M M, Batten M, Konstas K, Chen D, Doherty C M and Caruso R A 2020 Advancing metal-organic frameworks toward smart sensing: enhanced fluorescence by a photonic metal-organic framework for organic vapor sensing *Adv. Opt. Mater.* **8** 2000961

[76] Okur S, Zhang Z, Sarheed M, Nick P, Lemmer U and Heinke L 2020 Towards a MOF e-nose: a SURMOF sensor array for detection and discrimination of plant oil scents and their mixtures *Sensors Actuators* B **306** 127502

[77] Li X, Li D, Zhang Y, Lv P, Feng Q and Wei Q 2020 Encapsulation of enzyme by metal–organic framework for single-enzymatic biofuel cell-based self-powered biosensor *Nano Energy* **68** 104308

[78] Kuppler R J, Timmons D J, Fang Q-R, Li J-R, Makal T A, Young M D, Yuan D, Zhao D, Zhuang W and Zhou H-C 2009 Potential applications of metal–organic frameworks *Coord. Chem. Rev.* **253** 3042–66

[79] Altundal O F, Altintas C and Keskin S 2020 Can COFs replace MOFs in flue gas separation? High-throughput computational screening of COFs for CO_2/N_2 separation *J. Mater. Chem.* A **8** 14609–23

[80] Khandelwal G, Chandrasekhar A, Maria Joseph Raj N P and Kim S-J 2019 Metal–organic framework: a novel material for triboelectric nanogenerator–based self-powered sensors and systems *Adv. Energy Mater.* **9** 1803581

[81] Hajra S, Sahu M, Padhan A M, Lee I S, Yi D K, Alagarsamy P, Nanda S S and Kim H J 2021 A green metal–organic framework-cyclodextrin MOF: a novel multifunctional material based triboelectric nanogenerator for highly efficient mechanical energy harvesting *Adv. Funct. Mater.* **31** 2101829

[82] Khandelwal G, Maria Joseph Raj N P and Kim S-J 2020 ZIF-62: a mixed linker metal–organic framework for triboelectric nanogenerators *J. Mater. Chem.* A **8** 17817–25

[83] Khandelwal G, Maria Joseph Raj N P and Kim S-J 2020 Zeolitic imidazole framework: metal–organic framework subfamily members for triboelectric nanogenerators *Adv. Funct. Mater.* **30** 1910162

[84] Khandelwal G, Ediriweera M K, Kumari N, Maria Joseph Raj N P, Cho S K and Kim S-J 2021 Metal–amino acid nanofibers based triboelectric nanogenerator for self-powered thioacetamide sensor *ACS Appl. Mater. Interfaces* **13** 18887–96

[85] Wen R, Guo J, Yu A, Zhai J and Wang Z l 2019 Humidity-resistive triboelectric nanogenerator fabricated using metal organic framework composite *Adv. Funct. Mater.* **29** 1807655

[86] Guo Y, Cao Y, Chen Z, Li R, Gong W, Yang W, Zhang Q and Wang H 2020 Fluorinated metal-organic framework as bifunctional filler toward highly improving output performance of triboelectric nanogenerators *Nano Energy* **70** 104517

[87] Zhai L, Cui S, Tong B, Chen W, Wu Z, Soutis C, Jiang D, Zhu G and Mi L 2020 Bromine-functionalized covalent organic frameworks for efficient triboelectric nanogenerator *Chem. A Eur. J.* **26** 5784–8

[88] Rajaboina R K, Khanapuram U K, Vivekananthan V, Khandelwal G, Potu S, Babu A, Madathil N, Velpula M and Kodali P 2023 Crystalline porous material-based nanogenerators: recent progress, applications, challenges, and opportunities *Small* **20** 2306209

[89] Khandelwal G, Minocha T, Yadav S K, Chandrasekhar A, Maria Joseph Raj N P, Gupta S C and Kim S-J 2019 All edible materials derived biocompatible and biodegradable triboelectric nanogenerator *Nano Energy* **65** 104016
[90] Jiang W *et al* 2018 Fully bioabsorbable natural-materials-based triboelectric nanogenerators *Adv. Mater.* **30** 1801895
[91] Kim I, Jeon H, Kim D, You J and Kim D 2018 All-in-one cellulose based triboelectric nanogenerator for electronic paper using simple filtration process *Nano Energy* **53** 975–81
[92] Khandelwal G, Min G, Karagiorgis X and Dahiya R 2023 Aligned PLLA electrospun fibres based biodegradable triboelectric nanogenerator *Nano Energy* **110** 108325
[93] Zheng Q, Zou Y, Zhang Y, Liu Z, Shi B, Wang X, Jin Y, Ouyang H, Li Z and Wang Z L 2016 Biodegradable triboelectric nanogenerator as a life-time designed implantable power source *Sci. Adv.* **2** e1501478
[94] Roy S, Ko H-U, Maji P K, Van Hai L and Kim J 2020 Large amplification of triboelectric property by allicin to develop high performance cellulosic triboelectric nanogenerator *Chem. Eng. J.* **385** 123723
[95] Zhang Y *et al* 2018 Self-powered multifunctional transient bioelectronics *Small* **14** 1802050
[96] Kim T Y, Kim S K and Kim S-W 2018 Application of ferroelectric materials for improving output power of energy harvesters *Nano Convergence* **5** 30
[97] Kim M P, Um D-S, Shin Y-E and Ko H 2021 High-performance triboelectric devices via dielectric polarization: a review *Nanoscale Res. Lett.* **16** 35
[98] Lee K Y, Kim S K, Lee J-H, Seol D, Gupta M K, Kim Y and Kim S-W 2016 Controllable charge transfer by ferroelectric polarization mediated triboelectricity *Adv. Funct. Mater.* **26** 3067–73
[99] Han S A, Seung W, Kim J H and Kim S-W 2021 Ultrathin noncontact-mode triboelectric nanogenerator triggered by giant dielectric material adaption *ACS Energy Lett.* **6** 1189–97
[100] Dong K, Wang Y-C, Deng J, Dai Y, Zhang S L, Zou H, Gu B, Sun B and Wang Z L 2017 A highly stretchable and washable all-yarn-based self-charging knitting power textile composed of fiber triboelectric nanogenerators and supercapacitors *ACS Nano* **11** 9490
[101] Chen J, Huang Y, Zhang N, Zou H, Liu R, Tao C, Fan X and Wang Z L 2016 Micro-cable structured textile for simultaneously harvesting solar and mechanical energy *Nat. Energy* **1** 16138
[102] Chen J, Guo H, He X, Liu G, Xi Y, Shi H and Hu C 2016 Enhancing performance of triboelectric nanogenerator by filling high dielectric nanoparticles into sponge PDMS film *ACS Appl. Mater. Interfaces* **8** 736–44
[103] Seung W *et al* 2017 Boosting power-generating performance of triboelectric nanogenerators via artificial control of ferroelectric polarization and dielectric properties *Adv. Energy Mater.* **7** 1600988
[104] Min G, Pullanchiyodan A, Dahiya A S, Hosseini E S, Xu Y, Mulvihill D M and Dahiya R 2021 Ferroelectric-assisted high-performance triboelectric nanogenerators based on electrospun P(VDF-TrFE) composite nanofibers with barium titanate nanofillers *Nano Energy* **90** 106600
[105] Kim J, Ryu H, Lee J H, Khan U, Kwak S S, Yoon H-J and Kim S-W 2020 High permittivity $CaCu_3Ti_4O_{12}$ particle-induced internal polarization amplification for high performance triboelectric nanogenerators *Adv. Energy Mater.* **10** 1903524
[106] Sahu M, Vivekananthan V, Hajra S, Abisegapriyan K S, Maria Joseph Raj N P and Kim S-J 2020 Synergetic enhancement of energy harvesting performance in triboelectric

nanogenerator using ferroelectric polarization for self-powered IR signaling and body activity monitoring *J. Mater. Chem.* A **8** 22257–68

[107] Dong K, Peng X and Wang Z L 2020 Fiber/fabric-based piezoelectric and triboelectric nanogenerators for flexible/stretchable and wearable electronics and artificial intelligence *Adv. Mater.* **32** 1902549

[108] Yu X, Pan J, Zhang J, Sun H, He S, Qiu L, Lou H, Sun X and Peng H 2017 A coaxial triboelectric nanogenerator fiber for energy harvesting and sensing under deformation *J. Mater. Chem.* A **5** 6032–7

[109] Bairagi S, Khandelwal G, Karagiorgis X, Gokhool S, Kumar C, Min G and Mulvihill D M 2022 High-performance triboelectric nanogenerators based on commercial textiles: electrospun Nylon 66 nanofibers on silk and PVDF on polyester *ACS Appl. Mater. Interfaces* **14** 44591–603

[110] Gong W, Hou C, Zhou J, Guo Y, Zhang W, Li Y, Zhang Q and Wang H 2019 Continuous and scalable manufacture of amphibious energy yarns and textiles *Nat. Commun.* **10** 868

[111] Wang Z, Ruan Z, Ng W S, Li H, Tang Z, Liu Z, Wang Y, Hu H and Zhi C 2018 Integrating a triboelectric nanogenerator and a zinc-ion battery on a designed flexible 3D spacer fabric *Small Methods* **2** 1800150

[112] Liu L, Pan J, Chen P, Zhang J, Yu X, Ding X, Wang B, Sun X and Peng H 2016 A triboelectric textile templated by a three-dimensionally penetrated fabric *J. Mater. Chem.* A **4** 6077–83

[113] Gong J, Xu B, Guan X, Chen Y, Li S and Feng J 2019 Towards truly wearable energy harvesters with full structural integrity of fiber materials *Nano Energy* **58** 365–74

IOP Publishing

Recent Advances in Materials for Energy Harvesting and Storage

Suresh C Pillai, Daniel M Mulvihill and Aswathy Babu

Chapter 3

Exploring the potential of 2D materials in energy harvesting via triboelectric nanogenerators

Mathew Sunil, E J Jelmy, Sherin Joseph and Honey John

The impending energy crisis due to the relentless growth in the consumption of fossil fuels, including oil and natural gas, is an increasing concern. The increased demand for nonrenewable energy sources has focussed research attention on the development of renewable and sustainable alternatives. Among numerous innovations, nanogenerators have emerged as an excellent solution for mitigating the energy crisis. Triboelectric nanogenerators have attracted significant interest, primarily because of their portability, stability, remarkable energy conversion efficiency, and ability to work effectively with various materials. This chapter provides insights into TENGs driven by 2D materials, highlighting fundamental materials properties, their involvement in TENG fabrication, and their diverse applications. This chapter summarizes the current state of the art and provides a roadmap for future research endeavors in this dynamic field. Despite notable progress in the field of 2D TENGs, several obstacles persist in both the advancement and real-world implementation of these systems. A central challenge revolves around the imperative to continually explore and unearth innovative ultrathin 2D nanomaterials. These materials constitute the core of TENGs, and the discovery of fresh variations with desirable attributes is indispensable for advancing the frontiers of TENG performance.

3.1 Introduction

Amidst the worries posed by global warming and energy shortage, the importance of exploring new renewable and environmentally friendly energy sources is a paramount concern [1]. In the effort to address this escalating energy issue, novel devices for energy harvesting are capturing significant interest [2]. Despite its accessibility, autonomy, continuity, and ubiquity, untapped and wasted mechanical energy is inherent in our daily lives. The notable and efficient strides in converting this mechanical energy into electrical energy have led to the emergence of triboelectric nanogenerators (TENGs) [3].

TENGs work by contact electrification and electrostatic induction [4]. The triboelectric series ranks materials according to their relative triboelectric properties; some materials are more likely to gain electrons (become negatively charged) and others are more likely to lose electrons (become positively charged) during contact and separation [5]. The selection of triboelectric material plays a significant role in determining a TENG's output performance. Various materials within the triboelectric series, such as organic polymers, metals, and inorganic substances, are currently used as potential candidates [6]. However, challenges exist in the use of polymer-based TENGs due to recombination with positive charges on the electrode. Metal-based TENGs face corrosion-related issues [7]. Therefore, the exploration of novel materials has become necessary. The use of two-dimensional (2D) materials presents a promising avenue for enhancing the choice of materials available for use in TENGs. Since the successful exfoliation of graphene from graphite in 2004, science has witnessed the intensive application of 2D nanomaterials in energy harvesting and storage [8, 9]. The enhanced characteristics and uses of 2D structures can be attributed to their larger surface area (which is specifically tailored) and their superior quantum confinement effect [10]. These fascinating characteristics and potential benefits inspired by their remarkable properties and applications have led to the discovery of other 2D materials, including graphene, graphene oxide (GO), MXenes, transition-metal dichalcogenides (TMDs), clay, borophene, silicene, hexagonal boron nitride (h-BN), etc [11]. Compared to other bulk materials, these 2D materials are attractive due to their extremely thin layers (at the atomic scale), extensive surface area, flexibility, and ability to withstand mechanical stress [12].

This chapter provides a comprehensive overview of the investigation and advancement in 2D materials within the domain of TENGs. It encompasses a detailed analysis of the merits and performance characteristics exhibited by various 2D materials when integrated into TENGs. Furthermore, this chapter provides insights into the practical applications of TENGs based on 2D materials, including energy harvesting, biomechanical energy harvesting, environmental monitoring, self-powered sensors, and other relevant areas, as depicted in figure 3.1. The discussion addresses the prevailing challenges associated with the use of 2D materials in TENGs, culminating in proposed future research directions.

3.2 The fundamentals of triboelectric nanogenerators

The triboelectric effect refers to the phenomenon in which two materials acquire opposite charges upon contact [4]. The triboelectric charges exist only on the material's surface and do not recombine or disappear but persist for a considerable period, although a few migrate. This effect can produce a high voltage due to the static and immobile triboelectric charges formed on the surface of the material. These triboelectric charges formed on the surface of the contact material induce charges on the conductive electrodes attached to the contact material's opposite side, generating an output current [13, 14]. When two surfaces carrying triboelectric

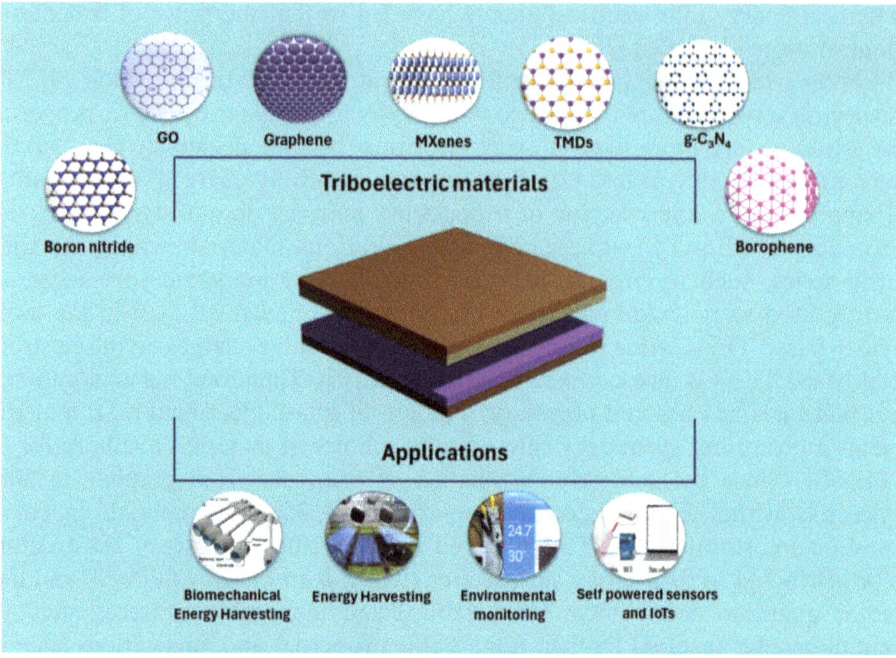

Figure 3.1. Various 2D materials and their practical application in TENGs.

charges experience relative motion due to an external force, an electric potential difference is created between the two electrodes, resulting in a transient movement of induced charges [13, 15]. Consequently, an electric current is produced, delivering effective output through a connected external load.

Four fundamental working modes have been identified to suit the practicality and feasibility of TENGs in different application domains. These include the contact–separation mode, the lateral sliding mode, the single-electrode mode, and the freestanding mode.

3.2.1 The vertical contact–separation mode

In this operational state, a stacked structure of two dielectric materials incorporates two electrodes on its upper and lower surfaces [16, 17]. A small space exists between the two dielectric materials, as shown in figure 3.2(a). The two materials come into contact with each other when subjected to an external mechanical force, inducing equal and opposite charges on the contact surfaces. When the force is removed, the two surfaces separate, and the contact-induced triboelectric charges produce a potential difference between the electrodes. Consequently, electrons move from one electrode to another, leading to a flow of electrons through the externally connected circuit that compensates for the potential difference. This movement generates an instantaneous charge Q on each electrode. When the two layers approach again, the reverse potential causes charge carriers to flow in the opposite direction. This periodic contact and separation produce an alternating current across the load.

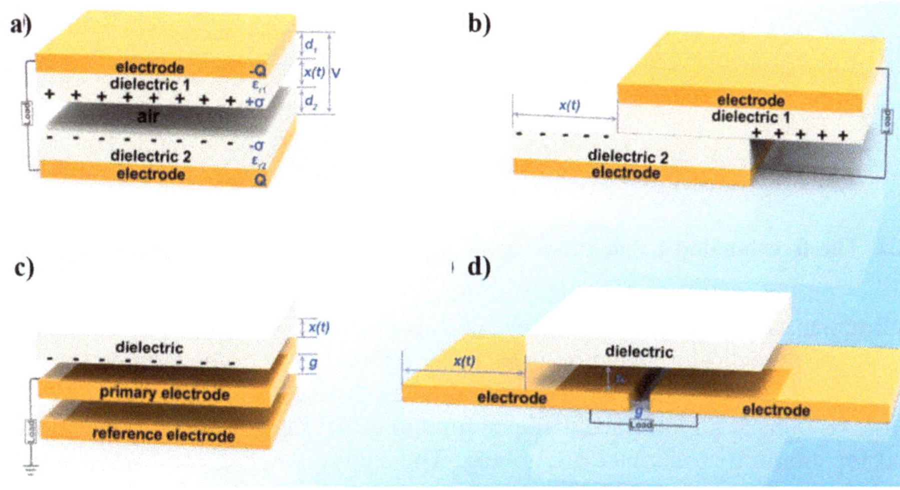

Figure 3.2. (a) Vertical contact–separation mode, (b) linear sliding mode, (c) single-electrode mode, and (d) freestanding triboelectric layer mode (reprinted from [21], Copyright (2022), with permission from Elsevier).

3.2.2 The lateral sliding mode

A TENG generates charges when subjected to periodic contact and sustained loading conditions. The introduction of a linear grating on the sliding surface is one of the methods introduced to greatly enhance the electrical output and energy-scavenging efficiency of TENGs. A sensing application based on the tangentially acting force has been recently investigated. In this mode, the friction caused by the relative movement of dielectric layers that operate through a sliding motion generates charges on the surface, as shown in figure 3.2(b) [18]. Compared to the contact–separation mode TENG, which requires a gap between the dielectric layers and relies on oscillation during contact and separation, the sliding-mode TENG generates electrical output solely through the sliding motion of the tribopair. This specific working mode eliminates the need for a specific working space between the dielectric layers and enables the device to function as an in-plane sensor and energy harvester [18–20].

3.2.3 The single-electrode mode

The use of electrodes and wires limits the extensive application of TENGs in energy harvesting and as self-powered sensors. This device setup restricts the practical utility of energy harvesting in stationary scenarios, particularly in contrast to situations involving freely moving objects [22]. In addition, certain materials or objects integrated into TENGs cannot be electrically linked to a closed circuit due to their mobility, such as in the case of a human walking or running [23]. To address these challenges, the single-electrode mode was introduced. In this method, when two materials of opposite polarity are made to come into contact, negative charges accumulate on the negative layer, and the positive tribolayer acquires positive charges as shown in figure 3.2(c). When the two layers are separated, a significant potential difference between the two layers triggers a flow of electrons from the ground to the attached electrode. This transient electron flow persists until the tribolayers return to their initial positions. As the two layers approach again, the electron flow reverses and flows from the electrode to the ground. The periodic contact and separation between the two layers generate an electrical output [24, 25].

3.2.4 The freestanding triboelectric mode

This is one of the four fundamental modes and is extensively used to harvest environmental vibrations. The absence of direct contact between the frictional layers advances its practical application in engineering. In this method, a region that slides against air or another medium acquires a charge, as shown in figure 3.2(d) [26]. This charge remains at an intermediate level for a considerable duration, eliminating the need for intermittent contact or rubbing. The charge thickness reaches saturation, which makes additional intermittent contact unnecessary. A pair of electrodes of uniform size is attached under the dielectric layer. The relative motion of the detached layer between these two electrodes creates an uneven charge distribution in the medium. This results in the flow of electrons between the attached two electrodes to balance the local charge distribution. This operating mode allows the fabrication of various device models, such as sliding, linear friction, and rotary loop structures. This design is suitable for harvesting energy from a moving object without requiring direct physical contact or even without contacting the object directly, such as when walking on the floor or being in a moving car [27].

Incorporating 2D materials into TENGs expands their horizons by enabling various design options and applications. These materials can be combined with polymers, nanomaterials, and other conventional triboelectric materials. The resulting hybrid structures can be configured for specific applications, including self-powered sensors, energy harvesters for IoT devices, and biomedical devices. The exclusive properties exhibited by 2D materials enable the scientific community to fabricate TENGs that can harvest energy from multiple sources simultaneously. Due to their small size and light weight, these materials provide insights that allow TENGs to be tailored for microelectronic and nanoelectronic devices.

3.2.5 Enhancement mechanisms

Electrodes and triboelectric materials constitute the integral components of a TENG. Within this context, integrating 2D nanomaterials substantially augments the operational efficacy of TENGs through two principal avenues. First, the overall output performance is bolstered by the use of electrically conductive 2D nanomaterials as electrodes, which elevates the production of induced charges and reduces internal resistance [28]. Second, these nanomaterials' exceptionally high specific surface area showcases them as optimal locales for capturing triboelectric charges within triboelectric layers [29]. This is attributed to the enhanced charge retention, which leads to a marked enhancement in output performance. In addition to these enhancements, due to the commendable optical and mechanical characteristics of the incorporated 2D nanomaterials, TENGs have found wider application in flexible and wearable electronics [30]. Furthermore, the facile enhancement of specific properties pertinent to 2D TENGs is facilitated by the proximity of these 2D materials to surface modifications. When 2D materials are used as electrodes or conductive layers in TENGs, they offer several advantages such as exceptional conductivity, a means of tailoring the surface chemistry, dielectric property enhancement, and improved mechanical characteristics.

3.2.5.1 Significant electrical conductivity

2D materials have garnered significant attention due to their remarkable electrical properties [31]. Graphene is an excellent conductor of electricity due to its high electron mobility and electrical conductivity [32]. Similarly, TMDs possess a tunable bandgap and high charge carrier mobility, making them a promising candidate for electronic and optoelectronic applications [30]. One of the critical factors that can alter the overall triboelectric output performance is the internal resistance of the fabricated TENG. The energy loss produced by high internal resistance diminishes the energy conversion efficiency. This issue can be mitigated by providing a low-resistance pathway for electron flow by introducing 2D materials into TENGs.

3.2.5.2 Surface properties and chemistry

The electrical properties of 2D materials are tailored using various methods, including chemical surface modifications and layer stacking, allowing TENGs to be customized for specific requirements [33]. Two-dimensional materials can establish different interactions with the polymer matrices via different bonds, thereby inducing polarization at the interface. These interactions formed between two materials can lead to an enhancement in charge transport during the triboelectrification process. In a study, Zhang *et al* harnessed the capabilities of GO to establish dipole–dipole interactions (specifically, between the hydrogen bond and the nylon filter membrane), thereby introducing charge separation at the contact surface [34]. The hydrogen bonding at the interface altered GO and nylon molecules' translational and rotational behaviors, which significantly modified the dielectric constant. The combined effect of charge transport and charge trapping in this composite material culminated in enhanced triboelectric performance.

3.2.5.3 Dielectric enhancement

The electrical output performance of a TENG is closely linked to its dielectric properties. Increasing the dielectric permittivity can enhance the system's ability to trap charges, reducing the rate at which surface charge dissipates [35]. This enhancement also leads to a noticeable increase in the surface charge density of polymer matrices that incorporate 2D materials. This combined impact is anticipated to yield improved triboelectric output performance. Bhavya *et al* explored this possibility in a TENG by employing liquid-phase-exfoliated 2D h-BN nanosheets (h-BNNSs) as coatings on biaxially oriented polyethylene terephthalate (BoPET) [36]. The TENG, which was designed to operate in contact–separation mode and used paper as the counter triboelectric material, demonstrated a remarkable output of 200 V.

The BNNS/BoPET layer functioned as an electron acceptor due to the presence of the embedded BNNs, which effectively captured electrons, resulting in heightened triboelectric negativity.

3.2.5.4 Exceptional mechanical characteristics

Two-dimensional materials are suitable for flexible TENGS and can be integrated into wearable devices and other applications due to their mechanical robustness and ability to withstand repeated bending and stretching [37]. The significant attributes of TENGs encompass the critical qualities of adaptability and resilience. A polydimethylsiloxane (PDMS) and MXene composite film possessed flexibility, robustness, and washability [38]. No obvious deterioration in the triboelectric response was observed when this prepared TENG, 1 cm × 1 cm in size, was subjected to 1000 cycles of full folding or 48 h of rolling. This exceptional electrical output and mechanical robustness was attributed to the secure adhesion of the MXene/PDMS composite film coating. This property makes it suitable for biomechanical motion sensing. Chen *et al* presented a stretchable TENG based on crumpled graphene with significantly improved output performance that was achieved through surface nanostructure modifications [39]. The output of flat TENG is 18.7 V and 6.6 μA which increased to 83 V and 25.78 μA as the crumple dregree reached 300 %. This TENG was able to endure substantial strain levels of up to 120% and exhibited remarkable stability and recoverability in its output performance.

3.3 Two-dimensional materials for triboelectric nanogenerators

3.3.1 Graphene and its derivatives

Following its discovery, graphene, the 2D form of carbon, has been the foremost and most explored significant 2D nanomaterial and is regarded as highly significant due to its unique electronic properties [40–43]. Graphene, which has a hexagonal arrangement of sp^2 hybridized carbon atoms, is mainly planar. It has a semiconducting nature and exhibits a zero bandgap, leading to exceptionally high charge mobility and tunable electronic properties. Its exemplary mechanochemical characteristics include outstanding strength and flexibility, excellent thermal conductivity, enhanced surface area, exceptional optical transparency, etc. It has therefore has emerged as a wonder material that is used in multifarious lines of work. Its current applications encompass prominent fields such as batteries, fuel cells, photovoltaics, supercapacitors, nanogenerators, automobiles, LEDs, and allied industries [44, 45]. Regarding nanogenerators, especially TENGs, given the large pool of material options available, there is great opportunity for development, increased output, and surpassing the current system's limitations. Graphene can be considered a potent triboelectric material, as it can store and retain electric charges for longer time periods due to its characteristic electrostatic properties [46]. By virtue of its attributes, graphene and its derivatives have been extensively utilized both as active components (individual layers or nanocomposites) and as the conductive electrode materials in TENGs [47, 48].

Initial studies of graphene-based TENGs were reported by Kim *et al* in 2014, in which the layer-by-layer deposition of graphene on copper foils was performed using the chemical vapor deposition (CVD) technique. The transparent layers were then transferred to a polyethylene terephthalate (PET) substrate, and the TENG was fabricated by pairing the graphene-covered PET with another PET substrate. Single layers of graphene deposited on the bottoms of both PET substrates served as electrodes. Readings were taken for monolayer (1L), bilayer (2L), trilayer (3L), and quad-layer (4L) randomly stacked graphene, but the output voltage and current were found to decrease as the number of layers increased. The reduction in output was due to the weak interlayer interactions between the randomly stacked graphene layers, which were transferred by a wet method. Also, the friction developed in 1L graphene was higher than the friction obtained in the randomly stacked layers. A similar TENG was also fabricated by the CVD technique; it had regularly grown graphene layers on nickel foils, which interestingly resulted in a high output performance of 9 V and 1.2 $\mu A\ cm^{-2}$ [49]. In another study, Chandrasekhar *et al* demonstrated a green method for the roll-to-roll transfer of CVD-deposited graphene on copper foil onto a PET/ethylene vinyl acetate (EVA) substrate using a hot water delamination process which allowed the reuse of the copper foils [50]. The film obtained was flexible and transparent and was intended for use in the fabrication of optoelectronic devices. An arch-shaped TENG was devised using the graphene/PET/EVA as the top friction layer and PDMS as the bottom layer. Both the active layers were provided with electrodes using similar graphene films to those of the active layers. The TENG recorded a maximum voltage and current output of 22 V and 0.9 μA, respectively, at a frequency of 4.3 Hz, and the device was also able to power 56 LEDs.

Later, Yin *et al* showed that voltage (a few millivolts) can be produced by moving a droplet of ionic liquid or seawater over a monolayer graphene film [51]. The triboelectricity thus produced resulted from the formation of a pseudocapacitor at the droplet/graphene interface; charging and discharging took place at the front and rear of the droplet, as shown in figure 3.3. The electric potential developed depended mainly on the number of drops, the droplet velocity, the ionic concentration, and the number of graphene layers. This mechanism can be effectively used to develop devices such as handwriting sensors, velocity sensors, mechanical energy harvesters, etc [52]. As a follow-up to the aforementioned work, Kwak *et al* showed that an electric power output of 1.9 μW was generated (i.e. almost 100 times greater than that of the earlier study) when a PTFE layer was added underneath the monolayer graphene and allowed to harvest energy by moving a water droplet/ionic liquid on the graphene layer [53]. During the process of graphene transfer onto PTFE, a triboelectric potential was observed, leading to the accumulation of positive and negative charges at the bottom and top surfaces of the graphene. A similar study was also reported by Zhao *et al* in 2018, in which self-powered monoelectrode graphene films were used to harvest rain energy. The mechanism involved charging and discharging cycles driven by π electrons (from graphene) and cations (from raindrops), resulting in the formation of pseudocapacitors at the graphene/raindrop

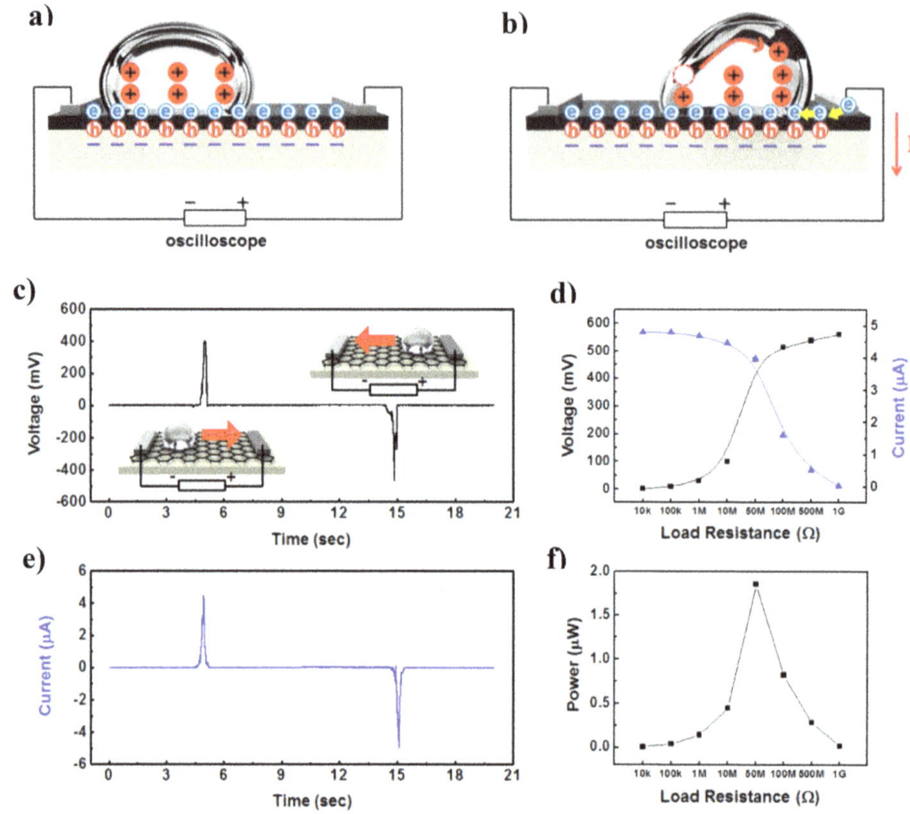

Figure 3.3. A mechanism for electric power generation. (a) The initial charge distribution of a water droplet and graphene/PTFE. (b) The redistribution of charge that occurs when the water droplet moves across the surface of graphene/PTFE. (c) The voltage generated by a single moving droplet on a graphene/PTFE structure. (d) The current produced by the graphene/PTFE structure during both forward and reverse motion of the water. (e) Variations in the output voltage and current based on the external load resistance. (f) Determination of the maximum power output from a single moving droplet as a function of the external load resistance (reprinted with permission from [51], Copyright (2016), American Chemical Society).

interface. A triboelectric output current of 2.15 μA/raindrop, a voltage of 129.83 μV/raindrop, and a power of 295.48 pW/raindrop were recorded.

Graphene has also been extensively studied as a nanofiller which can enrich the properties of polymer matrices, resulting in advanced and functional nanocomposites with augmented triboelectric performance. Aligned graphene sheets (AGS) were effectively embedded in PDMS using a repetitive spin coating approach by Xia *et al* [54]. This allowed for the efficient use of a fabricated AGS–PDMS TENG as a pressure sensor to determine the height and weight of objects. The introduction of AGS resulted in a threefold increase in the output performance of the PDMS TENG compared to its initial value. The enhanced performance was related to the formation of microcapacitors y the graphene sheets embedded in the PDMS matrix. Similarly, Qian *et al* also fabricated a PDMS/3D bilayer graphene/carbon cloth

(PDMS/3D BLGr/CC)-based flexible triboelectric nanogenerator (FTENG) with a maximum output voltage of 70 V, a current density of 9.3 μA cm^{-2}, and a power density of 0.65 mW cm^{-2} [55]. In addition, the FTENG retained 95.4% of its initial output performance for 1000 cycles of repeated bending tests at various angles. Recently, Yang *et al* employed the microinjection molding technique to fabricate films comprising a graphene–PTFE nanocomposite with the aim of enhancing their suitability for high-performance TENG applications [56]. The addition of a graphene nanofiller significantly improved the triboelectric output and resulted in high durability and stability of the TENG for more than 150 000 tapping cycles. The enhanced performance was attributed to the significant increase in the dielectric constant and the lubricating effect obtained by the addition of graphene nanosheets.

GO is an important 2D derivative comprising both an aromatic sp^2 and an aliphatic sp^3 hybrid carbon structure; it has lower electrical conductivity than that of the pure graphene structure. The lower electrical conductivity is mainly due to oxygen functional groups on the 2D surface. However, it has been widely utilized for the development of nanocomposite TENGs due to its appealing properties such as flexibility, ability to gain electrons due to its oxygen functional groups, and increased surface area, modulus, and charge effects. In 2017, Guo *et al* designed a GO-based single-electrode TENG (S-TENG) with only one triboelectric layer with higher energy harvesting efficiency, mechanical durability, and low weight [57]. The power density of the device was 3.13 W m^{-2}, and it could detect movements with an exceptional sensitivity of 388 μA MPa^{-1}. The device also showed prospects for wearable device applications due to its sensitive force detection and antimicrobial activity.

Later, in 2018, Harnchana tried a new approach, wherein a GO solution was employed to modify a PDMS matrix, resulting in a porous structure with an enhanced surface area to boost the triboelectric performance, as shown in figure 3.4 [58]. In addition, an anionic surfactant, sodium dodecyl sulfate (SDS), was also added to the GO/PDMS to enhance the power output of the TENG by accumulating charges and improving the charge transport properties. The PDMS@GO@SDS TENG achieved a maximum output performance of 438 V at 11 μA, which was three times greater than

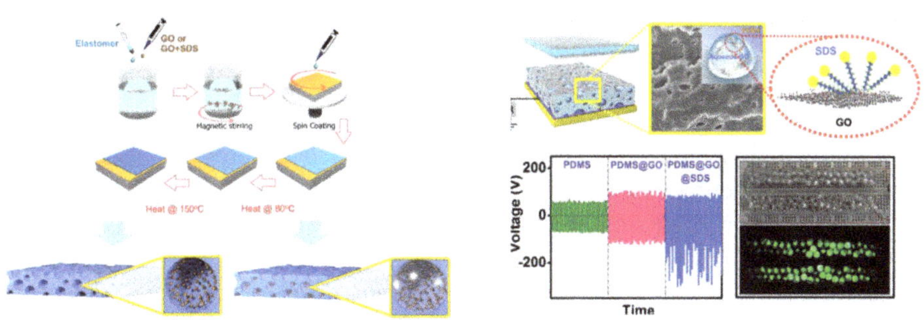

Figure 3.4. Schematics showing the fabrication of the PDMS@GO films that acted the as tribonegative materials and a voltage comparison of PDMS, PDMS@GO, and PDMS@GO@SDS films (reprinted with permission from [58], Copyright (2018), American Chemical Society).

that obtained for a pristine PDMS film. Parandeh *et al* developed eco-friendly triboelectric hybrid nanogenerators based on polycaprolactone (PCL)/GO and cellulose paper [59]. The PCL/GO layers were prepared using the electrospinning method and then paired with cellulose paper to fabricate the TENG, which produced an output of voltage 120 V, a current density of 2.5 mA m^{-2}, and a power density of 72.5 mW m^{-2} [60]. Similarly, Huang *et al* developed a TENG using electrospun nanofiber mats doped with GO, wherein a book-shaped TENG was fabricated using nanofiber mats of polyvinylidene fluoride (PVDF) and poly (3-hydroxybutyrate-co-3-hydroxyvalerate) (PHBV) as the two active friction layers. The GO was uniformly dispersed in the PVDF nanofiber matrix, providing charge-trapping sites and thus enhancing the charge storage properties of the TENG.

Reduced graphene oxide (rGO) is another derivative of graphene which has also been widely used in triboelectric nanogeneration applications. rGO is generally prepared via the chemical or thermal modification of GO, which is performed by reducing the content of oxygen functional groups. rGO is characterized by high chemical stability, a large surface area, and high electrical conductivity. Hence, rGO has been considered a useful tribonegative material for TENG applications. In 2016, Kaur *et al* reported the fabrication of an arch-shaped TENG comprising a thin film of rGO nanoribbons (rGONRs) with a PVDF polymer binder [61]. The presence of the rGO nanoribbons enhanced the surface roughness of the thin film, thus enhancing the charge storage and charge transfer capacity, as evidenced by cyclic voltammetry studies. Bhunia *et al* later demonstrated milliwatt power generation using a hybrid piezoelectric/triboelectric nanogenerator (HPTENG) fabricated using rGO/poly(vinylidene fluoride-co-trifluoroethylene) (P(VDF-TrFE)) nanocomposites [62]. It was found that the interactions that developed between the rGO and the P (VDF-TrFE) matrix broke the centrosymmetry of rGO, thus resulting in enhanced electrical output. A maximum voltage of 227 V and a power density of 0.28 W cm^{-3} were recorded for the HPTENG device.

Even though graphene has been extensively researched as an active friction layer for triboelectric applications, its applicability and demand as an electrode material have been equally significant. Being a flexible material, it also increases the functionality of graphene-based electrodes, since there is an increasing demand for flexible electrode materials due to the recent developments in flexible and wearable electronics. Consequently, in 2017, Yang *et al* attempted the surface engineering of graphene electrodes by spin coating the conductive polymer poly(3,4-ethylenedioxy-thiophene) polystyrene sulfonate (PEDOT:PSS, PH-1000) onto the graphene layers [63]. A poly(methyl methacrylate) (PMMA)-assisted transfer process was used to prepare the graphene on PET, and PH-1000 (dissolved in dimethyl sulfoxide, DMSO) was spin coated onto the films before they were further dried to create TENGs. For this TENG device, polyimides (PIs) were used as the triboelectric layer. The conductivity was increased after the surface engineering had taken place. It was found that the surface engineering of the graphene electrode enhanced the TENG output current density from 1.0 to 2.4 μA cm^{-2}, the output voltage from 26 to 52 V, and the output power from 5.5 to 12 μW.

In 2020, Pace *et al* compared the TENG performances of a gold electrode and graphene electrodes. The graphene electrodes were derived from few-layer graphene (FLG) produced by the wet-jet milling method [64]. A 26-fold increase in power density was obtained from the FLG-based electrodes. Hence, this confirmed that graphene is indeed an efficient charge collector with higher electrode capacitance than the gold electrode. Further, Pace *et al* reported a rise in capacitance in electrode structures composed of few layers of graphene. This increase was achieved through the integration of nitrogen-doped graphene (N-graphene), leading to a threefold improvement TENG output [65]. The effects of various N-graphene types, doping concentrations, and relative contents of N-pyridinic and N-graphitic sites on TENG performance were effectively studied.

3.3.2 MXenes

MXenes are an extensively researched category of two-dimensional nanosheets of transition-metal carbides, nitrides, and carbonitrides [66–68]. They encompass ~30 synthesized material compositions with millions of theoretically predicted variants [69]. They have gained prominence across various applications, including energy storage [66], electromagnetic interference shielding [70], and gas sensing [71]. These materials provide a diverse range of electrical and mechanical characteristics that can be tailored through composition adjustments and surface terminations. The overall chemical expression can be denoted by $M_{n+1}X_nT_z$, where M represents transition metals, X stands either for carbon (C) or nitrogen (N), and n can be 1, 2, or 3. The primary methods used to synthesize metallically conductive MXenes involve selectively removing the A element from the initial MAX phase, such as Ti_2AlC, Ti_3AlC_2, and Ti_4AlC_3. Due to their elevated electrical conductivity, MXenes have gained attention as electrode materials for TENG applications [69, 72]. The -OH, -O, or -F found on the surface provides a highly electronegative surface and is denoted by T_z.

In 2018, Dong *et al* pioneered the development of a flexible TENG utilizing $Ti_3C_2T_x$ as a triboelectric material based on the notion that it displays remarkable electrical conductivity and notable electronegativity [73]. A TENG was fabricated with a PET–indium tin oxide (PET-ITO)/MXene:PET-ITO configuration. The MXene that acted as an electrode was prepared by applying a homogeneous colloidal MXene solution to glass or PET-ITO films, as shown in figure 3.5(a). A significant open-circuit voltage was observed for both MXene/glass:PET-ITO and PTFE:PET-ITO configurations in a single-electrode setup. This similarity may be attributable to analogous electronegativity between MXene and PTFE due to the abundance of -F groups in MXene. The viability of MXene for use in TENG applications was confirmed in this study using a vertical contact–separation mode in which the MXene layer could effectively gather triboelectrically generated electrons from the PET surface. The flexible PET-ITO/MXene:PET-ITO TENG yielded an open-circuit voltage of ~500–650 V and a peak power of ~0.5–0.65 mW. This optimized configuration could power 60 LEDs and rapidly charge a 1 μF capacitor to 50 V. This TENG was able to function in two modes and thus achieved superior output performance. Due to the numerous functional groups present on the surface

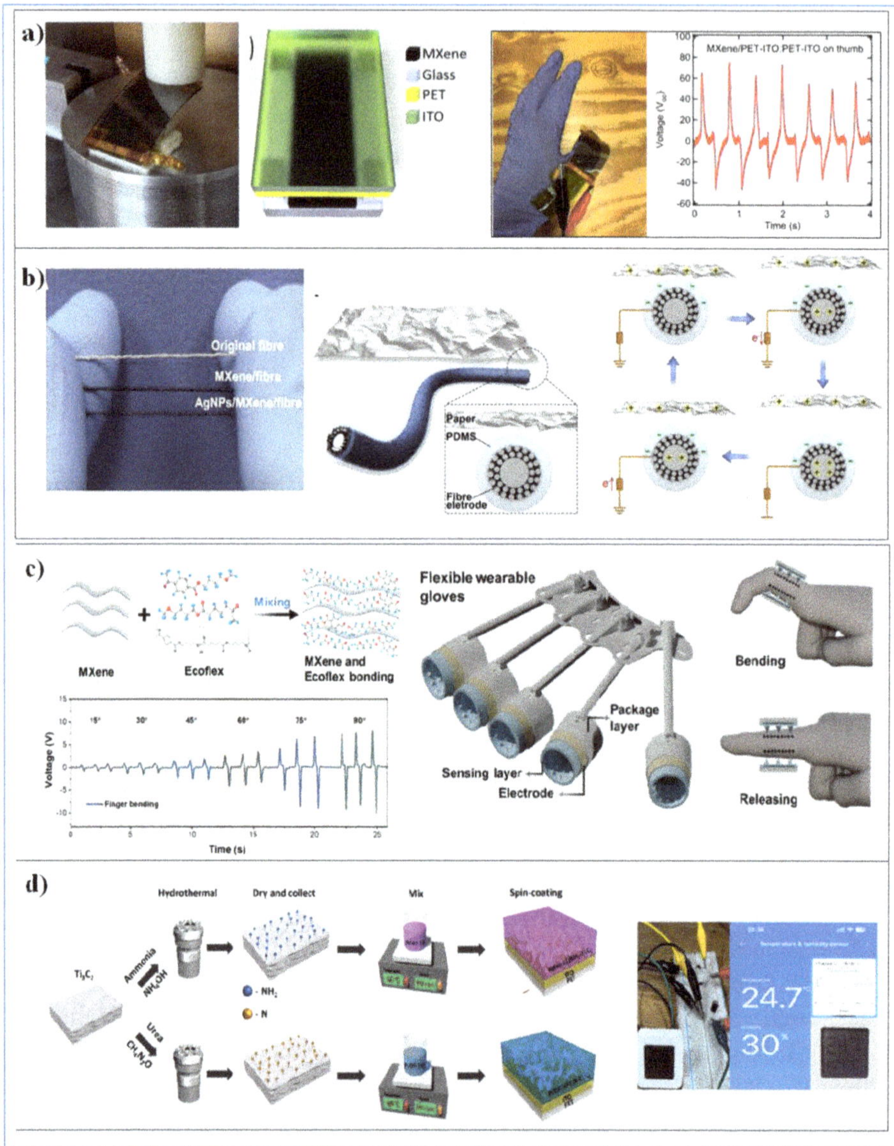

Figure 3.5. (a) A flexible MXene:PET/ITO TENG and its application as a biomechanical motion sensor (reprinted from [73], Copyright (2020), with permission from Elsevier). (b) An MXene-coated fiber used as an electrode for a TENG (reprinted from [74], Copyright (2020), with permission from Elsevier). (c) Three-dimensional printed gloves acting as a human–machine interface for various applications (reprinted from [76], Copyright (2023), with permission from Elsevier). (d) Ti_3C_2 functionalized with $-NH_2$ and –N for a humidity sensing application (reprinted from [77], Copyright (2023), with permission from Elsevier).

of MXenes and their high electronegativity, they are potential candidates for use as electrode materials in triboelectric applications. In view of the superior properties mentioned above, a stretchable fabric fiber was treated with a coating of MXenes and silver nanoparticles, making a suitable electrode [74]. In this study, the researchers examined the impact of varying the number of MXene ink coating cycles on the fiber resistance, as shown in figure 3.5(b). They found that increasing the cycles (up to six) significantly reduced the fiber resistance. The flexible, elastic, and mechanically durable single-fiber-based TENG utilized MXene-coated fiber as the electrode and PDMS as the triboelectric layer. Moreover, when a single fiber is woven into a textile, it can be used as agricultural textile, protecting crops while simultaneously harvesting energy from raindrops.

Hydrogels are novel matrices that can act as an alternative to conventional metal electrodes, as highlighted by their ability to stretch and self-heal. An MXene can act as an ion transport path and can thus improve the conductivity of a hydrogel by forming microchannels within it. The cross-linking of nanosheets can enhance their mechanical properties, including their stretchability. Using these ideas, Ti_3C_2Tx was incorporated into a PVA hydrogel matrix by Luo *et al* and the triboelectric response of the resulting material was studied in single-electrode mode using Kapton as the opposite contact material [75]. The MXene PVA hydrogel was encapsulated in Ecoflex silicone rubber. A streaming vibration potential model was observed for the MXene due to the contact between its surface and water in the PVA matrix in addition to the charges formed due to contact at the surface. This was the reason given for the enhanced triboelectric output of 230 V at 270 nA, which was four times higher than that of undoped PVA. The incorporation of MXene nanosheets endowed the TENG with excellent mechanical stability and stretchability, augmenting its overall output performance. The resultant TENG also found application as a system for handwriting recognition, utilizing the signals produced during compression and contact. An MXene–Ecoflex composite was prepared by Zhang *et al* to develop a toroidal self-powered triboelectric sensor featuring a pyramidal structure created by 3D printing technology [76]. A 3D-printed glove was made to ensure the sensing system was wearable; it integrated the pyramidal structure of the nanocomposite with flexible conductive cloth as an electrode. The adaptable pyramidal configuration demonstrated remarkable efficacy in upholding the intended separation between the positively charged skin on the fingers and the negatively charged layer of material. This eliminated the need for an additional spacer structure. A TENG 3 cm × 2 cm in size achieved a triboelectric response of 19.91 V at 1.15 μA and a high sensitivity of 0.088 V kPa^{-1} in the single-electrode mode. These characteristics allowed the generation of superior-quality signals, enabling the accurate identification of a wide range of finger movements. The applications of this device include gaming experiences, such as car rally games and control of balance tables, as well as interfaces for human–machine interactions to controlling appliances and robotic hands, as shown in figure 3.5(c).

A pivotal strategy for driving the use of TENGs is to create structural innovation, optimize the materials incorporated, and integrate them into various surveillance systems. Wu *et al* introduced an innovative polystyrene (PS)/MXene-based TENG with a unique drum-like design to monitor footballers [78]. Through comparative experimentation, they determined the optimal proportion of PS content by maintaining the MXene content at 2 wt%. A notable output of 141 V at 5.9 μA was observed for a TENG that paired PS/MXene with nylon in single-electrode mode.

The TENG was able to capture various dynamic postures such as walking, running, and long jumping when integrated into footballers' shoes. Due to its exceptional flexibility, this TENG can be deployed in various bodily locations, including the shoulders, neck, wrist, elbow, knee, and ankle, facilitating comprehensive motion monitoring of football players.

The surface alterations of MXenes that can be performed by functionalizing them with distinct surface functional groups are anticipated to enhance their physical attributes, including the bandgap and carrier mobility [79] [80]. In Cao *et al*'s study, chemically modified Ti_3C_2 nanoflakes (NFs) were used as either positive or negative additives within ferroelectric polymers as friction layers for a TENG [77]. Cao *et al* functionalized Ti_3C_2 with $-NH_2$ groups and –N, which exhibited positive and negative triboelectric properties, respectively, as shown in figure 3.5(d). The negative triboelectric friction layer employed in this study was formed by combining N-Ti_3C_2 with PVDF-TrFE, while the positive triboelectric layer was made by combining NH_2–Ti_3C_2 with nylon 11. This composite structure not only elevated the surface potential of each friction layer in the direction of the desired polarity but also intensified it through electrical polarization. The Ti_3C_2 possessed various functional groups, such as oxide (=O), hydroxyl (–OH), and fluorine (–F), which bonded with the methyl group present in the matrix polymers. The Ti_3C_2 nanosheets sandwiched the matrix polymer, leading to improved mechanical properties. When subjected to periodic contact and separation, the resulting TENG exhibited an outstanding performance of 250 V and 280 $\mu A\ cm^{-2}$, showcasing its potential for practical applications such as humidity sensing.

Capitalizing on the advantageous characteristics arising from the prominently electronegative surface and vigorous excitation of surface plasmons inherent to MXenes, several hybrid TENGs have been fabricated. A highly innovative research work proposed by Liu *et al* involved a PDMS/MXene-based TENG which could sense alterations in its surroundings and efficiently transform mechanical and light energies into electrical power [81]. Leveraging the electronegative characteristics of the PDMS/MXene composite, in vertical contact–separation mode the TENG exhibited a V_{oc} of 145 V and an I_{sc} of 27 μA. The output performance of the device underwent a 3.1-fold improvement under light stimulation compared to its performance in the absence of light stimulation. The fabricated TENG could effectively assess sound amplitude and wind velocities by capturing acoustic and kinetic energies.

3.3.3 Transition-metal dichalcogenides

TMDs have garnered considerable interest in recent times owing to their distinct electronic, optical, and mechanical properties. They have a layered structure in which a transition-metal atom (such as Mo or W) is positioned between two chalcogen atoms (such as S, Se, or Te). TMDs exhibit remarkable properties that make them favorable contenders for a wide array of applications, including electronic devices, optoelectronic components, catalytic processes, and energy storage systems [82–85]. TMDs possess a planar arrangement characterized by robust covalent bonds within planes and feeble van der Waals interaction between layers. In monolayer form, they have a direct bandgap, resulting in efficient light emission and absorption [86]. TMDs possess exceptional mechanical properties and have shown promising catalytic activity for various reactions, including hydrogen evolution, oxygen reduction, and water splitting [87]. TMDs have been investigated as viable electrode components in energy storage devices including batteries and supercapacitors [88].

Numerous research teams have investigated the TENG performance of systems based on MoS_2. Seol *et al* attempted a comprehensive exploration of the contact electrification characteristics of a range of 2D materials, including MoS_2, $MoSe_2$, WS_2, WSe_2, graphene, and graphene oxide. Their study aimed to elucidate the placement of these layered materials within the triboelectric series—a fundamental framework for understanding materials' charge generation upon contact and separation. Their research was employed to investigate the relative charging polarities of 2D materials when paired with conventional triboelectric materials. Their investigations involved the analysis of electric signals resulting from diverse material combinations to detect reliable contact charging polarities. As a result, a revised triboelectric series was formulated by integrating these 2D materials within the existing framework [89]. All the 2D materials employed in this investigation were situated in proximity to the tribonegative end of the triboelectric series. They were prepared using a consistent process that involved the chemical delamination of bulk flakes within a liquid medium, resulting in materials with identical thickness and surface roughness. To correlate the findings, the researchers employed two complementary approaches, namely Kelvin probe force microscopy (KPFM) and first-principles simulation [89]. The potential utility of monolayer MoS_2 as a functional substance for highly efficient energy harvesting was explored by Wu *et al.* In this study, Wu *et al* introduced monolayer MoS_2 as the triboelectric electron-acceptor layer within the friction layer of a TENG with the aim of significantly amplifying its output capabilities. A TENG that incorporated monolayer MoS_2 as the electron-acceptor layer showed an impressive peak power density of 25.7 W m^{-2} in the contact–separation mode. The mechanisms responsible for this substantial performance enhancement were linked to the remarkably efficient capture of triboelectric electrons within monolayer MoS_2 [90]. Figure 3.6(a) illustrates the TENG operating in vertical contact–separation mode with a monolayer MoS_2 film, showing the transfer of electrons from a PI layer to the MoS_2 monolayer. Vacuum-filtered films based on MoS_2 flakes were used in a TENG application by Cho *et al.* Output voltage measurements were taken when the filtered and annealed films came into contact with PET. The results demonstrated that the filtered film generated a higher voltage of 6.58 V compared to the annealed film's 3.48 V. Moreover, following the incorporation of MoS2 featuring an improved work function as an electron-acceptor layer on PI, a TENG with a PI/MoS_2:PI/PI configuration demonstrated a power density approximately 3.7 times higher than that of a TENG that used MoS_2 with an unenhanced work function [91].

Research conducted by Kim *et al* enhanced the electrical output of a TENG by carefully engineering defects using thiol-containing ligands on chemically

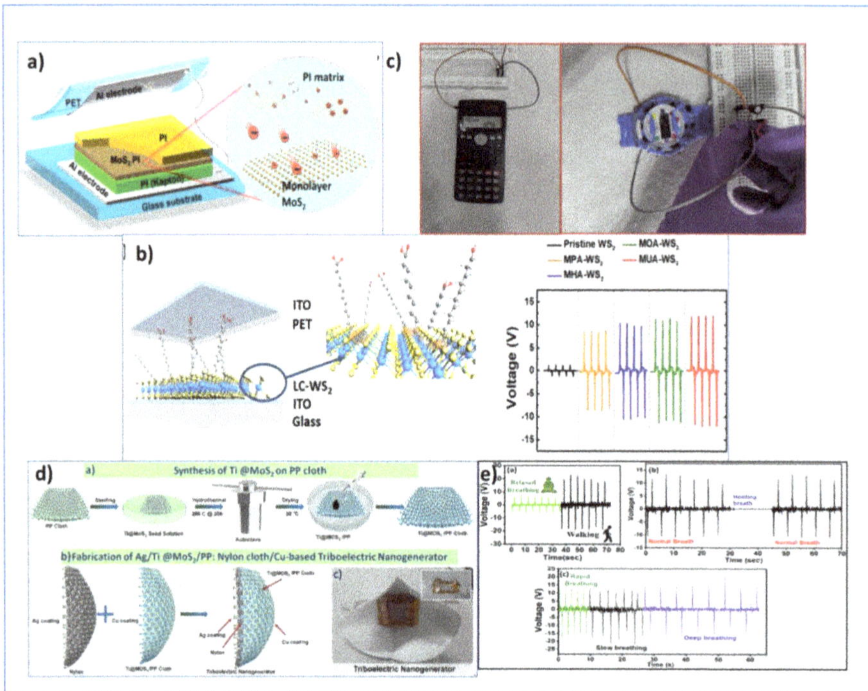

Figure 3.6. (a) A TENG with a MoS_2 charge-trapping layer in a PI polymer (reprinted with permission from [90], Copyright (2017), American Chemical Society); (b) a ligand-conjugated WS_2 TENG and the output voltages of various WS_2-based TENG devices (Reprinted with permission from [92], Copyright (2021), American Chemical Society); (c) an energy harvesting demonstration using MoS_2-incorporating PVDF films (reprinted from [95], Copyright (2023), with permission from Elsevier); (d) a smart mask made using Ti@MoS_2/PP; and (e) breathing patterns detected using a Ti@MoS_2/PP TENG. ((d) and (e) reprinted from [99], Copyright (2023), with permission from Elsevier.)

delaminated WS_2 nanosheets. The modified WS_2 nanosheets with ligands (ligand-conjugated WS_2, LC-WS_2) exhibited a remarkable tenfold enhancement in TENG properties compared to those of the original WS_2-based TENG. A comprehensive analysis was carried out to examine the underlying mechanisms by which thiolation affected the triboelectrification of WS_2, highlighting this modified system's exceptional durability in chemical and electrical aspects over an extended period [92]. A diagrammatic representation of the ligand-conjugated WS_2 TENG and the output voltages of the pristine WS_2 and LC-WS_2 TENG devices are shown in figure 3.6(b). In another study, Kim *et al* developed a large monolayer of CVD-grown MoS_2 for use in a TENG. To explore the effect of the formation of a depletion layer across a Schottky or p–n junction on the output performance of MoS_2 TENGs, they examined three TENGs with varying contacts: one featuring an ohmic contact, another with a Schottky contact, and a third with a p–n junction. TENGs employing ITO/MoS_2 with an ohmic contact exhibited nearly symmetrical outputs, consistent with conventional TENG behavior. In contrast, TENGs that utilized Au/MoS_2 with a Schottky contact and PPy/MoS_2 with a p–n junction displayed rectified behavior with enhanced voltage outputs compared to TENGs constructed from ITO/MoS_2. In this study, the voltage output (V_O) and short-circuit current (J_{SC}) generated by TENGs with Schottky and p–n junctions stemmed from charge diffusion processes that aimed to achieve thermal equilibrium with pressing [93]. Recently, a cutting-edge electrochemical triboelectric nanogenerator (e-TENG) was developed by applying techniques commonly employed in electrochemical capacitors [94]. This innovative TENG utilized FLG electrodes in combination with stable gel electrolytes containing liquid-phase exfoliated 2D-TMDs and polyvinyl alcohol. TENGs incorporating FLG and gel composites exhibited remarkable performance characteristics, including an open-circuit voltage of approximately 300 V, an instantaneous peak power of 530 mW m^{-2}, and high durability. The significant TENG property enhancement was attributed to the high electrical double-layer capacitance (EDLC) achieved by functionalizing the FLG electrodes with the gel composites. In a trial of various TMDs, TENGs based on W (WS_2 and WSe_2) exhibit notably improved lifetime stability, maintaining their electrical output for a minimum of 11 months after fabrication. In contrast, TENGs based on Mo (MoS_2 and $MoSe_2$) demonstrated a shorter lifespan than those of other 2D-TMDs. When they reached the voltage output of e-TENGs after one month and two months from the time of fabrication, a significant decrease (~60%) in TENG performance was evident for Mo-based TENGs, whereas the decrease was less for W-based TENGs (~15% to 4% decrease). This observation underscores the superior enhanced durability of TENGs based on W compared to their Mo-based counterparts, which is likely indicative of the greater resistance of W-based TMDs to oxidation [94].

The utilization of MoS_2 as an additive in various triboactive polymer matrices has sparked interest. Among the myriad triboelectric materials, polyvinylidene fluoride (PVDF) has garnered significant interest owing to its heightened tribonegativity, which is attributed to its fluorine content. In a research study, a high-performance TENG was developed that employed PVDF filled with MoS_2 to capture mechanical energy [95]. The influence of the MoS_2 concentration in the PVDF polymer was systematically investigated for different weight percentages of MoS_2 (0%, 3%, 5%, 7%, and 10%). Notably, the incorporation of MoS_2 led to enhancements in both the crystalline β-phase fraction and the dielectric permittivity of PVDF. Furthermore, the surface irregularity on the MoS_2-loaded PVDF samples increased, thereby contributing to an improved triboelectric output. Among the different setups examined, a TENG module which featured a 7% MoS_2 within the PVDF matrix for one layer, PDMS for the second layer, and a vertical contact–separation design demonstrated the most significant triboelectric output. This particular configuration achieved an output voltage of 189 V at a current of 1.61 μA. In contrast, a TENG produced using pure PVDF produced an output voltage and a current of 107 V and 0.88 μA, respectively. This unequivocally illustrates the impact of the MoS_2 filler in enhancing the TENG output of PVDF [95]. Energy harvesting studies using MoS_2-filled PVDF are demonstrated in figure 3.6. Similarly, MoS_2 acted as filler in PS that was used to develop a novel TENG with exceptionally high outputs. A PS/MoS_2 nanocomposite was used in a dual role as both the opposite contact material and an electron-acceptor layer, forming a bilayer structure with PS in conjunction with PS/MoS_2. A novel positive contact layer composed of the natural polymer *Alyssum homolocarpum* seed gum (AHSG) was introduced to optimize the negative layer's structure and thickness. This resulted in the creation of a PS+PS/MoS_2-AHSG (PPMA) TENG which produced an output voltage of approximately 1200 V, a current of approximately 0.74 mA, and a peak power output of 11.27 mW. This TENG was employed to weld silver nanorods (Ag NRs) [96]. The function of MoS_2 as an electrode component has been analyzed by researchers. Lan *et al* demonstrated the harvesting of energy from wind and the working of a self-powered wind speed sensor attached to plant leaves using a stretchable TENG composed of PDMS and a network of silver nanowires (AgNWs) enveloped by 2D metallic MoS_2 nanosheets. A simple technique was employed in the production of the composite film. The process involved immersing the AgNWs in a solution containing MoS_2 nanosheets. When the MoS_2 nanosheets were integrated with the AgNWs, they effectively reduced the resistance of the composite films by introducing a greater number of conductive pathways and expanding the junction area. This improvement was primarily attributed to the significant surface area provided by the MoS_2 nanosheets. The experimental results demonstrate that a stretchable TENG with a surface area of $1 \times 2.5\,cm^2$ can generate 0.16 W m^{-2} of power. This research opens up new avenues for stretchable materials for energy solutions and demonstrates promising possibilities in the realm of wearable electronics [97]. Recently, composite films consisting of MoS_2 NFs added to PVDF were developed for a TENG. Triboelectric energy harvesting was demonstrated using 1 wt% MoS_2 filler added to a PVDF composite film; this material yielded an impressive output power of ~18 μW cm^{-2}. The property enhancement was directly correlated to the enhancement in dielectric properties of the neat PVDF due to the addition of the MoS_2 filler [98].

Sensing applications using TENGs that incorporate MoS_2 are also described in the literature. A TENG was constructed using MoS_2 sheets modified with titanium via a hydrothermal method. The material thus produced was integrated into polypropylene (PP) cloth. A nylon cloth was layered on both sides, which served as a highly sensitive sensor for monitoring respiration and detecting ammonia gas [99]. The nanogenerator, designed with the configuration Cu/Ti@MoS_2/PP:nylon/Ag, was affixed to a mask used for respiration. During respiration cycles, it produced an open-circuit voltage (V_{oc}) of 29.3 V and a short-circuit current density (J_{SC}) of 42.7 μA cm^{-2}. In addition, the fabrication of a self-sustaining ammonia gas sensor was accomplished by amalgamating a TENG with the Ti@MoS_2/PP-based ammonia sensor. The self-sustaining ammonia gas sensor demonstrated remarkable performance across a wide sensing spectrum for ammonia gas, spanning from 200 to 2600 ppb at ambient room temperature. The proposed TENG mask and the breathing patterns are shown in figures 3.6(d) and (e) [99]. In another study, a flexible and wearable Au-decorated WS_2-based TENG was developed for sensing applications. Paper and PVA substrates were adopted for flexibility. Out of the different combinations tested, an Au–WS_2–PTFE counter-layered TENG generated the highest output V_{oc}. A photovoltaic investigation was conducted on the optimized TENG while exposed to illumination. The impact of light on the device's performance became evident in the increased V_{oc} value when the device was in ON state, as opposed to when it was in the OFF state.

The variation observed was ascribed to the generation of charge carriers induced by light, underscoring the material's capacity for detecting photons [100]. In another study, a flexible TENG was created to support self-powered temperature- and weight-sensing applications. This TENG incorporated key components such as MoS_2, nonconductive adhesive, graphite powder, and paper. To assess its performance, scientists placed a 100 W incandescent lamp 5 cm away from the MoS_2-based TENG and measured the V_{oc} across a specific temperature range and weight. Within a specific temperature (293–323 K) and weight range (50–72 kg), the dV_{oc}/dT and dV_{oc}/dW of the TENG were measured to be 0.093 V K^{-1} and 0.2 V kg^{-1}. This linear rise in V_{oc} can be ascribed to the limited thermal diffusion of charge carriers localized on the electrode surface. This phenomenon leads to an increased surface charge density as temperature rises, providing an explanation for the observed linear growth in V_{oc} [101]. Recently, a flexible textile-based single-electrode TENG utilizing a composite film of MoS_2–PDMS was demonstrated for the purpose of human motion sensing. A TENG containing 6 wt% MoS_2 displayed the most impressive output performance, featuring an V_{oc} of approximately 320 V, a J_{SC} density of around 30 μA cm^{-2}, and a maximum instantaneous output power density of roughly 3.2 mW cm^{-2}. The improvement in the output features of the TENG can be ascribed to the formation of a nanocapacitor network within the system. The TENG's ability to provide power was evaluated by charging commercial capacitors and illuminating 100 commercially available LEDs. The sensing capabilities of the TENG were tested by detecting various human movements, including wrist bending, arm bending, leg bending, and more [102]. Similarly, a WS_2-based motion sensor was developed by Chekke *et al.* They fabricated flexible and wearable single-electrode TENG devices using chemically exfoliated WS_2 nanosheet-coated cellulose paper and silk textile as the active layers. A TENG device paired with WS_2 and PTFE exhibited the highest V_{oc} (11.02 V) and was used for motion sensing applications. The sensor was capable of producing distinct output voltage signals and signal shapes based on different biomechanical actions [103].

3.3.4 Hexagonal boron nitride

h-BN stands out as an exceptional and adaptable material renowned for its impressive attributes and its diverse array of practical uses. h-BN boasts a crystal lattice structure with a hexagonal configuration akin to that of graphene, featuring alternating boron (B) and nitrogen (N) atoms meticulously arranged in a honeycomb pattern [104]. In this structure, each boron atom forms bonds with three nitrogen atoms, while each nitrogen atom establishes bonds with three boron atoms [105]. Unlike graphene, h-BN possesses remarkable electrical insulation properties, rendering it a suitable choice for insulation purposes in various electronic devices. It can also function as a lubricant when prepared in the form of ultrathin, atomically smooth layers, effectively reducing friction and wear in mechanical systems. Its chemical inertness and high resistance to attack from acids, bases, and most organic solvents greatly bolster its durability when exposed to harsh environments [105, 106]. In the realm of nanotechnology, h-BN plays a pivotal role as a substrate for the cultivation of high-quality graphene and other 2D materials. In addition, h-BN finds utility in optical devices and serves as a dielectric material in transistors and capacitors. Its biocompatibility and stability open up promising avenues for applications in drug delivery and medical imaging [106].

In early 2015, a study highlighted the significance of h-BN in the context of TENG applications. Han *et al* conducted research focused on achieving the defect-free, superior, and highly consistent production of Al_2O_3 with a precisely balanced chemical composition on a CVD-fabricated layered h-BN/pristine graphene system tailored for TENG applications. The primary role of the h-BN was to establish a defect-free graphene system, benefiting from its flat nature and its absence of dangling bonds while possessing lattice properties like those of graphene. Consequently, this allowed for the deposition of a high-quality oxide layer of Al_2O_3 on h-BN using atomic layer deposition. This was possible due to the partial ionic–covalent B–N bonding characteristic of h-BN, which differs from the properties of graphene. The distinct electrical power output variations observed in TENGs when Al_2O_3 interacted with graphene, both with and without h-BN, unquestionably underscore the crucial role played by interlayered h-BN [107]. Another study investigated the way in which the combination of two 2D nanomaterials affects the performance of TENGs. To do this, Parmar *et al* grew MoS_2–h-BN 2D/2D composite thin films on various substrates using pulsed laser deposition [108]. Comparative analyses were conducted with pristine MoS_2 and h-BN films. The TENG device featuring the MoS_2–h-BN composite film as an electron acceptor showed a peak-to-peak output voltage that was more than twice as high as that of the pristine MoS_2 or h-BN films. The TENG device and the output of the TENG are shown in figure 3.7(A) [108]. Studies of the use of h-BN as a filler in polymeric matrices are also available in the research literature [109, 110]. A research study was primarily concerned with creating a flexible TENG by incorporating 2D h-BN into PVDF nanofibers (NFs), combined with a positively charged nylon 11 nanofiber, for potential applications in wearable electronics. The study's results revealed a significant enhancement in the TENG's performance when 0.5 wt.% h-BNNSs were introduced. This improvement was attributed to the exceptional piezoelectric properties and the large specific surface area of the h-BNNSs, which facilitated the conversion of mechanical energy into electrical energy. Consequently, the TENG achieved remarkable open-circuit voltage and short-circuit current values of 500 V and 25.2 μA, respectively. The study also demonstrated the practical applications of this TENG, such as real-time monitoring of human body motion, and its use as a flexible wearable sensor (figure 3.7(B)) [109]. Very recently, the significance of incorporating h-BNNSs as a filler in a PDMS matrix to enhance its TENG response was reported. PDMS that incorporated h-BNNSs as a filler showed a threefold enhancement in TENG output (15 μA and 150 V) compared to the output of pristine PDMS at 10 N and 10 Hz [110].

As in the case of other 2D nanomaterials, h-BN can serve as a dielectric material within polymeric systems, enhancing the output performance of TENGs. To illustrate this phenomenon, scientists engineered a TENG by coating liquid-phase-exfoliated

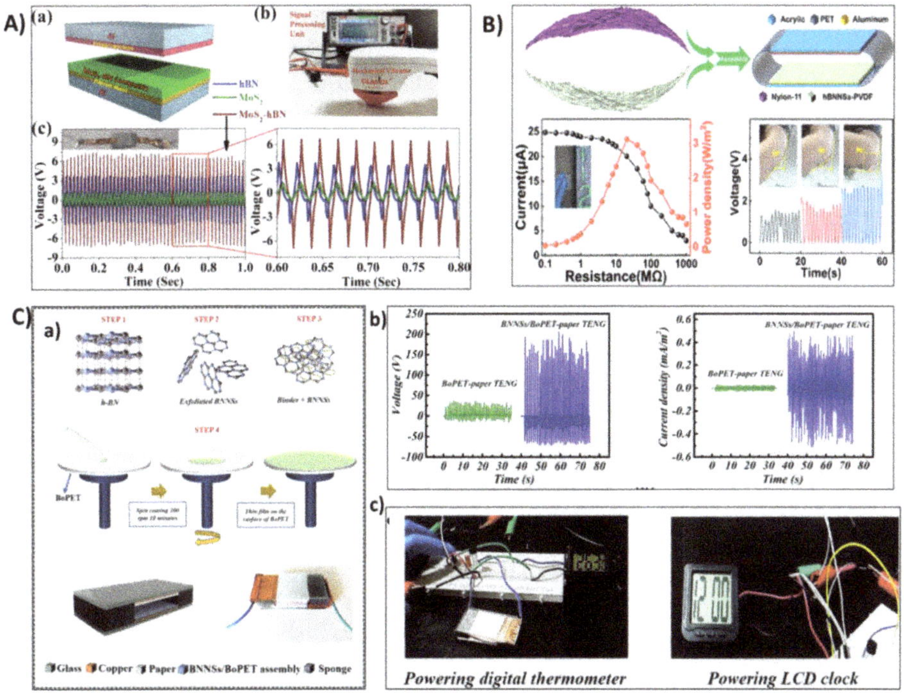

Figure 3.7. (A) 2D/2D composite thin films suggested for TENG applications. (a) A schematic of the TENG device. (b) The TENG measurement setup. (c) The measured TENG responses (reprinted with permission from [108], Copyright (2019), American Physical Society). (B) PVDF films with incorporated h-BNNSs paired with nylon nanofibers for human motion monitoring applications (reprinted with permission from [109], Copyright (2022), American Chemical Society). (C) Schematics of the h-BNNSs/BoPET-based TENG and its applications. (a) The exfoliation of the h-BNNSs and the various steps of TENG device fabrication. (b) TENG performance analysis. (c) Photos of energy harvesting studies carried out using the proposed TENG. ((a)–(c) reprinted from [111], Copyright (2021), with permission from Elsevier.)

2D h-BNNSs onto BoPET. This TENG device, featuring h-BNNSs-BoPET in conjunction with paper as the counterpart tribomaterial, exhibited substantially improved TENG output compared to the output of a TENG lacking the h-BNNS interlayer. Although it only experienced the moderate force of finger tapping (~3 N), the TENG constructed with h-BNNSs/BoPET–paper was capable of generating an output voltage of approximately ~200 V and a short-circuit current density of around ~0.48 mA m^{-2}. During load testing, the h-BNNSs/BoPET–paper TENG device achieved a peak electric power density of ~0.14 W m^{-2} at a resistive load of 200 MΩ. The integration of h-BNNSs significantly improved the electron-acceptance characteristics of the BoPET film, leading to heightened dielectric permittivity and improved TENG performance in the h-BNNSs/BoPET assembly. Furthermore, the fabricated h-BNNSs TENG effectively showcased its ability to charge electronic devices, including an LCD clock, a digital thermometer, and LEDs. The exfoliation of the h-BN, the proposed TENG device, and the TENG output generated are depicted in figure 3.7(C) [111]. Similarly, Pang *et al* introduced a TENG that utilized a PI/BNNS combination. In this setup, BNNSs were incorporated into a PI film, creating a sandwich-like structure known as a PI/BNNS/PI nanocomposite film, or PBP film. A PBP TENG demonstrated an enhanced triboelectric output of 65.9 V and 4.5 μA and a power density of 21.4 μW cm^{-2} at a load resistance of 10 MΩ, which was higher than the TENG outputs obtained without BNNS interlayers. This improvement was attributed to the elevated dielectric permittivity of BNNS, which played a vital role in augmenting the output performance of the TENG [112].

3.3.5 Graphitic carbon nitride (g-C_3N_4)

Graphitic carbon nitride (g-C_3N_4) is a 2D material composed of carbon and nitrogen atoms arranged in a hexagonal lattice. It exhibits excellent electrical conductivity and mechanical flexibility, making it suitable for use in triboelectric devices. Further, it possesses high sensitivity to mechanical deformations and frictional forces, which are essential characteristics for efficient triboelectric energy harvesting and sensing applications. One of the advantages of using g-C_3N_4 in triboelectric devices is its environmental friendliness. It is a nontoxic and sustainable material, which aligns with the growing emphasis on eco-friendly energy harvesting technologies.

Recent advancements in the use of g-C_3N_4 in the field of TENGs are noteworthy. For instance, researchers successfully incorporated g-C_3N_4 into nylon 66 via an electrospinning process and used the resulting material to construct a TENG for energy harvesting and sensing applications. This TENG device, crafted from lightweight, soft, highly porous materials with an effective surface area of 4 cm^2 exhibited an open-circuit voltage of 80 V, a short current of approximately 3 μA, and a charge generation capability of 50 nC at an impact force of 40 N and a frequency of 3 Hz. In addition, when connected to a load resistance of 500 MΩ, it achieved a remarkable maximum power density of 45 mW m^{-2}. This level of energy generation was used to power small portable electronic devices such as LEDs and the displays of calculators and watches. Furthermore, the TENG system incorporating g-C_3N_4 was evaluated for its ability to harvest wind energy and detect human motion (figure 3.8) [113]. Research has led to the development of a novel TENG based on g-C_3N_4 nanosheets for photosensing applications. The performance of the TENG device was investigated using different frictional surfaces. In a study, the behavior of triboelectric charge transfer was explored in both pristine and doped g-C_3N_4 materials. When subjected to continuous finger tapping, the TENG device generated an output voltage of 55 V in approximately 50 s. However, when exposed to UV illumination, the response was notably faster, taking only around 14 s. This enhanced performance under UV illumination was attributed to the photoinduced generation of carriers within the g-C_3N_4 material. The dual functionality exhibited by the TENG device, which served as both a nanogenerator and a flexible photosensor, is a highly commendable feature, demonstrating the versatility and potential of TENGs for use in various applications [114]. The role of g-C_3N_4 as a filler in polymeric matrices has also been reported [115]. A TENG device comprising

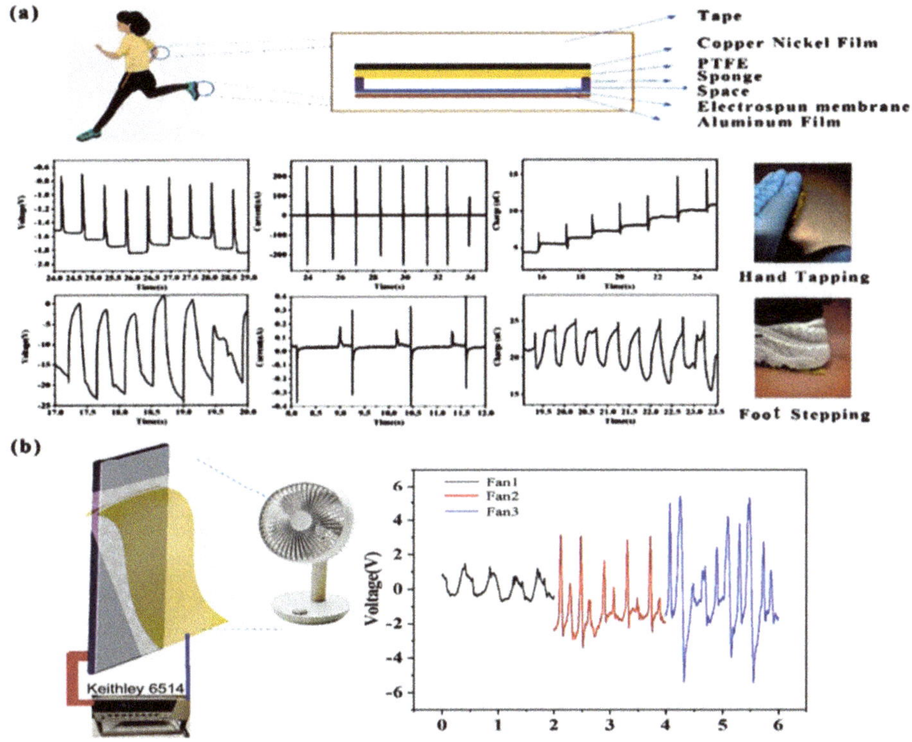

Figure 3.8. (a) A TENG device for energy harvesting and human motion sensing made from an electrospun membrane. (b) A demonstration of wind energy harvesting. Reproduced from [113]. CC BY 4.0.

2D g-C_3N_4/silicone paired with Kapton exhibited a high output voltage of ~550 V, a current of ~110 μA, and a power density of about 4.19 W m^{-2} [115]. The physical and chemical modification of g-C_3N_4 has been carried out to enhance TENG performance. For instance, a textile-based TENG (T-TENG) was developed that included an active layer of g-C_3N_4 nanosheets loaded with silver (AgCN) nanoparticles. When paired with Teflon as the counter triboelectric material and subjected to mechanical agitation, this AgCN/nylon bilayer T-TENG produced an open-circuit voltage reaching approximately 200 V. In just 30 s, it charged a commercial capacitor to approximately 85 V. The enhanced triboelectric performance was due to the modifications in surface area, dielectric properties, and the intrinsic resistivity of the host material [116]. In another study, researchers developed sustainable and biocompatible TENGs that employed corn husk and coconut coir fibers as the positive layers and 2D g-C_3N_4 nanosheets for the negative layers. These TENGs also functioned as photosensors. The TENG device based on corn husk generated a maximum output voltage, current, and power of 630 V, 0.79 mA, and 131 mW, respectively, when subjected to simple biomechanical forces. On the other hand, the coconut fiber-based TENG device produced an outputs of 581 V, 11.47 mA, and 1980 mW (for an area of 2×2 cm^2) [117].

3.3.6 Other 2D nanomaterials in TENG applications

Nanoclay, borophene, layered double hydroxides (LDHs), and metal–organic frameworks (MOFs) have found utility in the advancement of TENGs. The following sections provide a brief overview of recent advancements in TENGs utilizing these materials.

3.3.6.1 A 2D nanoclay-based TENG

Two-dimensional nanoclay, when incorporated into TENG materials, can enhance the lifespan of their mechanical strength, which is crucial for device performance. The tuneable surface properties and surface chemistry of 2D nanoclay can influence the charge transfer and related triboelectric properties of TENGs. Two-dimensional nanoclay can also act as the charge-trapping layer within a system. A TENG was reported that was composed of raw mica as the tribopositive layer and PTFE as the tribonegative frictional layer; the TENG had a maximum output power density of 62.82 mW m^{-2}. A TENG-attached shoe was constructed to demonstrate mechanical energy harvesting during walking [118]. Like other 2D materials, 2D nanoclay can be seamlessly integrated into polymer matrices to create nanocomposites. These nano-composites can be tailored to display specific electrical and mechanical characteristics, thus making them suitable for TENG applications. Furthermore, these TENGs offer the added benefit of being environmentally friendly fillers. For instance, 2D smectite clay (SC) has been introduced into PVDF to enhance the PVDF's tribonegativity. This incorporation significantly enhanced the power output of a SC TENG, increasing it from a mere 15 mW m^{-2} in the pristine TENG to an impressive 1450 mW m^{-2} in the composite [119]. Similarly, 2D mica was incorporated into flexible and stretchable thermoplastic polyurethane (TPU) nanofibers to enhance the performance of TENGs made from neat TPU. When combined with PVDF/MXene nanofibers, the TPU/mica nanofibers showed substantially increased TENG outputs. This improvement compared to neat TPU was attributed to the enhanced positive electrostatic surface potential resulting from the presence of mica, which enhanced the TENG's tribopositivity [120]. Environmentally friendly and biocompatible TENGs were developed by incorporating natural clays as fillers into a chitosan biopolymer matrix. Specifically, sepiolite, bentonite, and kaolin clays were chosen as fillers to create composites with chitosan. These chitosan-based TENGs, containing 3 wt% sepiolite, 1 wt% bentonite, and 1 wt% kaolin, demonstrated open-circuit voltages of 863, 996, and 963 V and power densities of 20.4 W m^{-2}, 26.5 W m^{-2}, and 22.8 W m^{-2}, respectively. These environmentally friendly TENGs constructed from biocompatible materials offer a cost-effective solution and hold promise for developing novel technologies utilizing natural resources [121].

3.3.6.2 The role of 2D metal–organic frameworks in TENGs

Recently, TENGs have found exciting applications when coupled with MOFs, creating a new class of energy harvesting devices. MOFs comprise metal ions interconnected by organic linkers and are characterized by their porosity. Renowned for their adjustable properties and substantial surface area, MOFs stand out as materials with versatile and customizable characteristics. Due to the porous nature of MOFs, TENGs based on MOFs can be tried for various applications including gas sensing and water wave energy harvesting. Energy harvesting from human motion has been demonstrated using a TENG based on silk fibroin (SF) composite films that incorporated a 2D MOF. The highest instantaneous power density (263 μW cm^{-2}) was achieved by aligning MOF NFs within the SF matrix in an in-plane fashion at a mass ratio of 0.2 wt%. This specific arrangement of NF MOFs embedded in the SF matrix created a network of numerous nanoscale capacitors, providing the composite film with an outstanding ability to store electric charge [122]. In another study, researchers utilized zeolitic imidazole framework-8 (ZIF-8) and Kapton as the key materials in constructing a MOF TENG. This MOF TENG produced a consistent output of 164 V at 7 μA in a vertical contact–separation configuration. To illustrate the practical utility of the MOF TENG in powering low-energy electronic devices, it was employed to create a self-powered UV counterfeit detection system and a tetracycline sensor (figure 3.9(A)). Notably, the sensor demonstrated exceptional selectivity and could be easily reused by means of a simple washing process [123]. Figure 3.9(A) (a) shows the activation of a temperature sensor for 10 s, driven by a 100 μF capacitor. After 10 s, the temperature sensor deactivated as the capacitor gradually discharged. Figure 3.9(A) (b) shows a straightforward and portable self-powered UV counterfeit detection system. This system effectively highlights the three concealed security features on Korean Won currency that are visible when exposed to UV light, features that are otherwise invisible to the naked eye [123]. Reports evaluating the performances of MOFs as fillers in polymer matrices are also available. For instance, a fluorinated MOF with charge-inducing and charge-trapping capabilities was incorporated into a PDMS matrix to improve PDMS composite films' negative triboelectricity and charge-trapping property. A TENG constructed using the composite film showed an

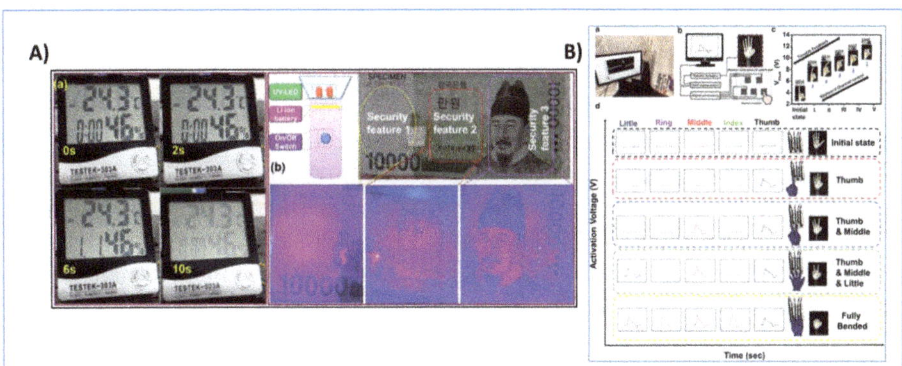

Figure 3.9. (A) A temperature sensor and counterfeit system. (a) The operation of the temperature sensor. (b) A UV counterfeit system powered by the proposed TENG. ((a) and (b) reproduced from [123], CC BY 4.0.) (B) Smart robotic assistance powered by a borophene-based TENG. (a) Images of the robotic system. (b) Setup and signal processing network in the system. (c) TENG output responses. (d) Photos of proposed human–robotic interaction using the TENG. ((a)–(d) reproduced from [126], CC BY 4.0.)

elevenfold increase in output power density as compared to the output of a TENG constructed using neat PDMS when paired with an Al foil as the triboelectric pair [124]. The chemical modification and structural engineering of MOFs have also been carried out to enhance TENG performance. For instance, a comparison was made between the TENG performance of PVDF modified with Zn-MOF and that of PVDF modified with Zn/Co-MOF. The PVDF composite modified with Zn/Co-MOF exhibited improved electron transfer and superior TENG output performance compared to unmodified MOF-based TENGs. In a practical application, the improved output generated by the Zn/Co-MOF@PVDF-TENG was employed to illuminate an ultraviolet lamp plate for the [2 + 2] photochemical cycloaddition of organometallic macrocycles [125].

3.3.6.3 Borophene-based TENGs

Two-dimensional borophene is a crystalline allotrope of boron that consists of a single layer of boron atoms arranged in a honeycomb lattice similar to that of graphene. It is an emerging nanomaterial with unique properties and potential applications. It is a promising material in TENG applications due to its exceptional electrical conductivity and mechanical flexibility. Although research into borophene for various applications is in its early stages, it has already proved its applicability for energy harvesting and sensing devices. A flexible borophene TENG (B-TENG), compatible with fabrics and constructed from a nanocomposite of borophene and Ecoflex, was introduced for energy harvesting, medical assistance, and wound healing purposes. The B-TENG was incorporated into a smart keyboard setup and integrated with a robotic system, creating an upper-limb medical assistance interface for individuals with disabilities. In addition, the B-TENG functioned as an active sensing component in the design of a lower-limb gait phase visualization platform. Furthermore, it has demonstrated its potential in delivering electrical stimulation (ES) for wound therapy, as evidenced by *in vitro* cellular behavior studies and corresponding animal model experiments. Photos of the smart robotic assistive design proposal are shown in figure 3.9(B) [126]. Hou *et al* created a wearable electronic skin using a TENG sensor based on borophene. This innovative sensor served multiple purposes, including health monitoring, voice recognition, wireless detection of human motion, and integration into smart robots for human–machine interaction. The sensor exhibited exceptional characteristics, such as high sensitivity, a wide pressure range, a low detection threshold, low power consumption, and excellent reproducibility [127]. Borophene research is currently in the early stages of development, and we can expect to see a growing body of literature supporting it in the near future.

3.4 Future perspectives

While considerable progress has been achieved in the domain of 2D TENGs, characterized by advancements in theoretical exploration and a variety of practical applications demonstrated, numerous hurdles persist in both the creation and effective utilization of these systems. One key challenge is the need for the continued exploration and discovery of novel ultrathin 2D nanomaterials. These materials serve as the foundation for TENGs, and discovering new variants with desirable properties is essential for pushing the boundaries of TENG performance. Surface modification of the triboelectric layers is a promising avenue for amplifying the charge density generated by TENGs. Investigating and optimizing these modifications can significantly enhance the overall performance of 2D TENGs. Some 2D nanomaterials exhibit sensitivity to external environmental factors, such as temperature and humidity. Controlling and mitigating these sensitivities is critical for the reliability and stability of 2D TENGs under various operating conditions. Another hurdle for the development of TENG technology lies in its measurements. Researchers used to characterize the electrical output performance of TENGs using metrics such as the open-circuit voltage, short-circuit current, and energy conversion efficiency; however, these suffer from inconsistent test conditions. Variables such as the working environment, contact area, applied load, material modifications, etc. can introduce discrepancies. The development of well-defined evaluation standards would facilitate more accurate comparisons of 2D TENG performance and the refinement of fabrication techniques [128, 129]. The major concerns in developing devices and products for use in various TENG fields need to be rectified. In the realm of wearable technology, the seamless integration of TENG modules into clothing and wearable devices aims to harness energy from body movements, demanding expertise in design to ensure sustained TENG performance even after multiple uses and cleaning cycles. It is crucial to guarantee the washability of these integrated modules. In addition, incorporating TENG modules into electronic gadgets can facilitate self-powered operation, reducing reliance on traditional batteries. In the context of IoT devices and sensors, utilizing TENG technology to power such devices without depending on external power sources is a key goal, although many current studies employ external batteries to energize microcontrollers linked to TENG modules. Efforts to integrate TENG technology into environmental sensors, particularly in remote or harsh locations where conventional power sources are impractical, hold significant potential for future exploration. Moreover, the prospect of powering implantable medical devices, sensors, and healthcare gadgets using TENGs is compelling, yet the clinical trials necessary for testing remain challenging for many research groups to attain. While 2D TENGs have shown immense promise, addressing these challenges will be instrumental in harnessing their full potential and achieving more reliable and efficient energy conversion systems. As advancements continue in materials science, nanotechnology, and energy harvesting, the scope and efficiency of 2D TENGs are poised to expand, unlocking novel avenues in energy harvesting and self-powered systems.

3.5 Summary

The chapter provided a concise overview of 2D nanomaterials employed in TENG applications. These 2D nanomaterials, including graphene, GO, MXenes, TMDs, h-BN, g-C_3N_4, nanoclays, MOFs, and borophenes, exhibit exceptional properties such as excellent electrical conductivity, exceptionally high specific surface area, impressive mechanical and optical characteristics, and adaptability for customization. Consequently, they are well suited for use as electrodes and triboelectric materials in TENGs. Furthermore, this chapter explored the various applications of TENGs that leverage these 2D materials, such as energy harvesting, self-powered sensing, and human–computer interaction. The substantial promise of 2D materials lies in their ability to function as modifiers seamlessly incorporated into foundational materials, thereby forming composite materials. Such materials possess the capability to amplify the dielectric characteristics of substances, regulate work functions, and facilitate the entrapment of charges. Furthermore, this chapter identified the hurdles facing TENGs and the potential pathways for their advancement. Overcoming key challenges in 2D materials—such as stability, scalability, integration with TENG modules, and durability—while capitalizing on their advantages, such as high surface area, remarkable electrical properties, customizability, mechanical strength, and charge-trapping capabilities, would enable the development of more efficient, robust, and versatile TENG devices, paving the way for their integration into various fields, including wearable technology, IoT devices, and energy harvesting systems.

References

[1] Pomerantseva E, Bonaccorso F, Feng X, Cui Y and Gogotsi Y 2019 Energy storage: the future enabled by nanomaterials *Science* **366** eaan8285

[2] Chen J and Wang Z L 2017 Reviving vibration energy harvesting and self-powered sensing by a triboelectric nanogenerator *Joule* **1** 480–521

[3] Liu Y, Wang L, Zhao L, Yu X and Zi Y 2020 Recent progress on flexible nanogenerators toward self-powered systems *InfoMat* **2** 318–40

[4] Wu C, Wang A C, Ding W, Guo H and Wang Z L 2019 Triboelectric nanogenerator: a foundation of the energy for the new era *Adv. Energy Mater.* **9** 1802906

[5] Zhang R and Olin H 2020 Material choices for triboelectric nanogenerators: a critical review *EcoMat* **2** e12062

[6] Khandelwal G, Raj N P M J and Kim S-J 2021 Materials beyond conventional triboelectric series for fabrication and applications of triboelectric nanogenerators *Adv. Energy Mater.* **11** 2101170

[7] Wen R, Guo J, Yu A, Zhai J and Wang Z L 2019 Humidity-resistive triboelectric nanogenerator fabricated using metal organic framework composite *Adv. Funct. Mater.* **29** 1807655

[8] Zhang H 2015 Ultrathin two-dimensional nanomaterials *ACS Nano* **9** 9451–69

[9] Novoselov K S, Geim A K, Morozov S V, Jiang D-E, Zhang Y, Dubonos S V *et al* 2004 Electric field effect in atomically thin carbon films *Science* **306** 666–9

[10] Baig N 2022 Two-dimensional nanomaterials: a critical review of recent progress, properties, applications, and future directions *Composites* A **165** 107362

[11] Zhou Y, Zhang J-H, Li S, Qiu H, Shi Y and Pan L 2023 Triboelectric nanogenerators based on 2D materials: from materials and devices to applications *Micromachines* **14** 1043

[12] Shanmugam V, Mensah R A, Babu K, Gawusu S, Chanda A, Tu Y *et al* 2022 A review of the synthesis, properties, and applications of 2D materials *Part. Part. Syst. Charact.* **39** 2200031

[13] Luo J, Gao W and Wang Z L 2021 The triboelectric nanogenerator as an innovative technology toward intelligent sports *Adv. Mater.* **33** 2004178
[14] Sun P, Jiang S and Huang Y 2021 Nanogenerator as self-powered sensing microsystems for safety monitoring *Nano Energy* **81** 105646
[15] Dong Y, Mallineni S S K, Maleski K, Behlow H, Mochalin V N, Rao A M *et al* 2018 Metallic MXenes: a new family of materials for flexible triboelectric nanogenerators *Nano Energy* **44** 103–10
[16] Fan F-R, Tian Z-Q and Wang Z L 2012 Flexible triboelectric generator *Nano Energy* **1** 328–34
[17] Zhu G, Pan C, Guo W, Chen C-Y, Zhou Y, Yu R *et al* 2012 Triboelectric-generator-driven pulse electrodeposition for micropatterning *Nano Lett.* **12** 4960–5
[18] He W, Liu W, Chen J, Wang Z, Liu Y, Pu X *et al* 2020 Boosting output performance of sliding mode triboelectric nanogenerator by charge space-accumulation effect *Nat. Commun.* **11** 4277
[19] Jing Q, Zhu G, Bai P, Xie Y, Chen J, Han R P *et al* 2014 Case-encapsulated triboelectric nanogenerator for harvesting energy from reciprocating sliding motion *ACS Nano* **8** 3836–42
[20] Bai P, Zhu G, Liu Y, Chen J, Jing Q, Yang W *et al* 2013 Cylindrical rotating triboelectric nanogenerator *ACS Nano* **7** 6361–6
[21] Tian Y, An Y and Xu B 2022 MXene-based materials for advanced nanogenerators *Nano Energy* **101** 107556
[22] Yang Y, Zhou Y S, Zhang H, Liu Y, Lee S and Wang Z L 2013 A single-electrode based triboelectric nanogenerator as self-powered tracking system *Adv. Mater.* **25** 6594–601
[23] Oh H J, Bae J H, Park Y K, Song J, Kim D K, Lee W *et al* 2020 A highly porous nonwoven thermoplastic polyurethane/polypropylene-based triboelectric nanogenerator for energy harvesting by human walking *Polymers* **12** 1044
[24] Niu S, Liu Y, Wang S, Lin L, Zhou Y S, Hu Y *et al* 2014 Theoretical investigation and structural optimization of single-electrode triboelectric nanogenerators *Adv. Funct. Mater.* **24** 3332–40
[25] Su Y, Zhu G, Yang W, Yang J, Chen J, Jing Q *et al* 2014 Triboelectric sensor for self-powered tracking of object motion inside tubing *ACS Nano* **8** 3843–50
[26] Wang S, Xie Y, Niu S, Lin L and Wang Z L 2014 Freestanding triboelectric-layer-based nanogenerators for harvesting energy from a moving object or human motion in contact and non-contact modes *Adv. Mater.* **26** 2818–24
[27] Son J-H, Kim W-G, Yun S-Y, Kim D-W and Choi Y-K 2023 Wearable bead-based triboelectric nanogenerator with dual-mode operation for monitoring abnormal behavior in dementia patients *Nano Energy* **114** 108642
[28] Han S A, Lee J, Lin J, Kim S-W and Kim J H 2019 Piezo/triboelectric nanogenerators based on 2-dimensional layered structure materials *Nano Energy* **57** 680–91
[29] Jiang Q, Wu C, Wang Z, Wang A C, He J-H, Wang Z L *et al* 2018 MXene electrochemical microsupercapacitor integrated with triboelectric nanogenerator as a wearable self-charging power unit *Nano Energy* **45** 266–72
[30] Jiang C, Wu C, Li X, Yao Y, Lan L, Zhao F *et al* 2019 All-electrospun flexible triboelectric nanogenerator based on metallic MXene nanosheets *Nano Energy* **59** 268–76
[31] Tan C, Cao X, Wu X-J, He Q, Yang J, Zhang X *et al* 2017 Recent advances in ultrathin two-dimensional nanomaterials *Chem. Rev.* **117** 6225–331

[32] Neto A C, Guinea F, Peres N M, Novoselov K S and Geim A K 2009 The electronic properties of graphene *Rev. Mod. Phys.* **81** 109
[33] Akinwande D, Petrone N and Hone J 2014 Two-dimensional flexible nanoelectronics *Nat. Commun.* **5** 5678
[34] Zhang R, Hummelgård M, Örtegren J, Andersson H, Olsen M, Chen D *et al* 2023 Triboelectric nanogenerators with ultrahigh current density enhanced by hydrogen bonding between nylon and graphene oxide *Nano Energy* **115** 108737
[35] Lee J W, Cho H J, Chun J, Kim K N, Kim S, Ahn C W *et al* 2017 Robust nanogenerators based on graft copolymers via control of dielectrics for remarkable output power enhancement *Sci. Adv.* **3** e1602902
[36] Bhavya A S, Varghese H, Chandran A and Surendran K P 2021 Massive enhancement in power output of BoPET-paper triboelectric nanogenerator using 2D-hexagonal boron nitride nanosheets *Nano Energy* **90** 106628
[37] Xu S, Wei G, Li J, Han W and Gogotsi Y 2017 Flexible MXene–graphene electrodes with high volumetric capacitance for integrated co-cathode energy conversion/storage devices *J. Mater. Chem.* A*5 17442–51*
[38] He W, Sohn M, Ma R and Kang D J 2020 Flexible single-electrode triboelectric nanogenerators with MXene/PDMS composite film for biomechanical motion sensors *Nano Energy* **78** 105383
[39] Chen H, Xu Y, Zhang J, Wu W and Song G 2019 Enhanced stretchable graphene-based triboelectric nanogenerator via control of surface nanostructure *Nano Energy* **58** 304–11
[40] Castro Neto A H, Guinea F, Peres N M R, Novoselov K S and Geim A K 2009 The electronic properties of graphene *Rev. Mod. Phys.* **81** 109–62
[41] Novoselov K S, Geim A K, Morozov S V, Jiang D, Zhang Y, Dubonos S V *et al* 2004 Electric field effect in atomically thin carbon films *Science* **306** 666–9
[42] Novoselov K S, Jiang D, Schedin F, Booth T J, Khotkevich V V, Morozov S V *et al* 2005 Two-dimensional atomic crystals *Proc. Natl. Acad. Sci. USA* **102** 10451–3
[43] Novoselov K S, Morozov S V, Mohinddin T M G, Ponomarenko L A, Elias D C, Yang R *et al* 2007 Electronic properties of graphene *Phys. Stat. Sol.* B*244 4106*
[44] Geim A K and Novoselov K S 2007 The rise of graphene *Nat. Mater.* **6** 183–91
[45] Han S A, Lee J, Lin J, Kim S-W and Kim J H 2019 Piezo/triboelectric nanogenerators based on 2-dimensional layered structure materials *Nano Energy* **57** 680–91
[46] Wang Z and Scharstein R 2009 Electrostatics of graphene: charge distribution and capacitance *Chem. Phys. Lett.* **489** 229–36
[47] Farah F, Haniff M and Mohamed M 2021 A review on applications of graphene in triboelectric nanogenerators *Int. J. Energy Res.* **46** 544–76
[48] Zhou Y, Zhang J-H, Li S, Qiu H, Shi Y and Pan L 2023 Triboelectric nanogenerators based on 2D materials: from materials and devices to applications *Micromachines* **14** 1043
[49] Kim S, Gupta M K, Lee K Y, Sohn A, Kim T Y, Shin K-S *et al* 2014 Transparent flexible graphene triboelectric nanogenerators *Adv. Mater.* **26** 3918–25
[50] Chandrashekar B N, Deng B, Smitha A S, Chen Y, Tan C, Zhang H *et al* 2015 Roll-to-roll green transfer of CVD graphene onto plastic for a transparent and flexible triboelectric nanogenerator *Adv. Mater.* **27** 5210–6
[51] Kwak S S, Lin S, Lee J H, Ryu H, Kim T Y, Zhong H *et al* 2016 Triboelectrification-induced large electric power generation from a single moving droplet on graphene/polytetrafluoroethylene *ACS Nano* **10** 7297–302

[52] Yin J, Li X, Yu J, Zhang Z, Zhou J and Guo W 2014 Generating electricity by moving a droplet of ionic liquid along graphene *Nat. Nanotechnol.* **9** 378–83

[53] Zhao Y, Pang Z, Duan J, Duan Y, Jiao Z and Tang Q 2018 Self-powered monoelectrodes made from graphene composite films to harvest rain energy *Energy* **158** 555–63

[54] Xia X, Chen J, Liu G, Javed M S, Wang X and Hu C 2017 Aligning graphene sheets in PDMS for improving output performance of triboelectric nanogenerator *Carbon* **111** 569–76

[55] Qian Y, Sohn M, He W, Park H, Subramanian K R V and Kang D J 2020 A high-output flexible triboelectric nanogenerator based on polydimethylsiloxane/three-dimensional bilayer graphene/carbon cloth composites *J. Mater. Chem.* A **8** 17150–5

[56] Yang P, Wang P and Diao D 2022 Graphene nanosheets enhanced triboelectric output performances of PTFE films *ACS Appl. Electron. Mater.* **4** 2839–50

[57] Guo H, Li T, Cao X, Xiong J, Jie Y, Willander M *et al* 2017 Self-sterilized flexible single-electrode triboelectric nanogenerator for energy harvesting and dynamic force sensing *ACS Nano* **11** 856–64

[58] Harnchana V, Ngoc H V, He W, Rasheed A, Park H, Amornkitbamrung V *et al* 2018 Enhanced power output of a triboelectric nanogenerator using poly(dimethylsiloxane) modified with graphene oxide and sodium dodecyl sulfate *ACS Appl. Mater. Interfaces* **10** 25263–72

[59] Huang T, Lu M, Yu H, Zhang Q, Wang H and Zhu M 2015 Enhanced power output of a triboelectric nanogenerator composed of electrospun nanofiber mats doped with graphene oxide *Sci. Rep.* **5** 13942

[60] Parandeh S, Kharaziha M and Karimzadeh F 2019 An eco-friendly triboelectric hybrid nanogenerators based on graphene oxide incorporated polycaprolactone fibers and cellulose paper *Nano Energy* **59** 412–21

[61] Kaur N, Bahadur J, Panwar V, Singh P, Rathi K and Pal K 2016 Effective energy harvesting from a single electrode based triboelectric nanogenerator *Sci. Rep.* **6** 38835

[62] Bhunia R, Gupta S, Fatma B, Prateek , Gupta R K and Garg A 2019 Milli-watt power harvesting from dual triboelectric and piezoelectric effects of multifunctional green and robust reduced graphene oxide/P(VDF-TrFE) composite flexible films *ACS Appl. Mater. Interfaces* **11** 38177–89

[63] Yang J, Liu P, Wei X, Luo W, Yang J, Jiang H *et al* 2017 Surface engineering of graphene composite transparent electrodes for high-performance flexible triboelectric nanogenerators and self-powered sensors *ACS Appl. Mater. Interfaces* **9** 36017–25

[64] Pace G, Ansaldo A, Serri M, Lauciello S and Bonaccorso F 2020 Electrode selection rules for enhancing the performance of triboelectric nanogenerators and the role of few-layers graphene *Nano Energy* **76** 104989

[65] Pace G, Serri M, Castillo A E R, Ansaldo A, Lauciello S, Prato M *et al* 2021 Nitrogen-doped graphene based triboelectric nanogenerators *Nano Energy* **87** 106173

[66] Anasori B, Lukatskaya M R and Gogotsi Y 2017 2D metal carbides and nitrides (MXenes) for energy storage *Nat. Rev. Mater.* **2** 1–17

[67] Naguib M, Mochalin V N, Barsoum M W and Gogotsi Y 2014 25th anniversary article: MXenes: a new family of two-dimensional materials *Adv. Mater,* **26** 992–1005

[68] Anasori B, Xie Y, Beidaghi M, Lu J, Hosler B C, Hultman L *et al* 2015 Two-dimensional, ordered, double transition metals carbides (MXenes) *ACS Nano* **9** 9507–16

[69] Tan T L, Jin H M, Sullivan M B, Anasori B and Gogotsi Y 2017 High-throughput survey of ordering configurations in MXene alloys across compositions and temperatures *ACS Nano* **11** 4407–18

[70] Shahzad F, Alhabeb M, Hatter C B, Anasori B, Man Hong S, Koo C M *et al* 2016 Electromagnetic interference shielding with 2D transition metal carbides (MXenes) *Science* **353** 1137–40

[71] Xiao B, Li Y-C, Yu X-F and Cheng J-B 2016 MXenes: reusable materials for NH_3 sensor or capturer by controlling the charge injection *Sens. Actuators* B **235** 103–9

[72] Zhang C, Anasori B, Seral-Ascaso A, Park S H, McEvoy N, Shmeliov A *et al* 2017 Transparent, flexible, and conductive 2D titanium carbide (MXene) films with high volumetric capacitance *Adv. Mater.* **29** 1702678

[73] Dong Y, Mallineni S S K, Maleski K, Behlow H, Mochalin V N, Rao A M *et al* 2018 Metallic MXenes: a new family of materials for flexible triboelectric nanogenerators *Nano Energy* **44** 103–10

[74] Jiang C, Li X, Ying Y and Ping J 2020 A multifunctional TENG yarn integrated into agrotextile for building intelligent agriculture *Nano Energy* **74** 104863

[75] Luo X, Zhu L, Wang Y-C, Li J, Nie J and Wang Z L 2021 A flexible multifunctional triboelectric nanogenerator based on MXene/PVA hydrogel *Adv. Funct. Mater.* **31** 2104928

[76] Zhang S, Rana S M S, Bhatta T, Pradhan G B, Sharma S, Song H *et al* 2023 3D printed smart glove with pyramidal MXene/Ecoflex composite-based toroidal triboelectric nanogenerators for wearable human–machine interaction applications *Nano Energy* **106** 108110

[77] Cao V A, Kim M, Lee S, Van P C, Jeong J-R, Park P *et al* 2023 Chemically modified MXene nanoflakes for enhancing the output performance of triboelectric nanogenerators *Nano Energy* **107** 108128

[78] Wu M 2023 A drum structure triboelectric nanogenerator based on PS/MXene for football training monitoring *AIP Adv.* **13** 085012

[79] Khazaei M, Arai M, Sasaki T, Chung C-Y, Venkataramanan N S, Estili M *et al* 2013 Novel electronic and magnetic properties of two-dimensional transition metal carbides and nitrides *Adv. Funct. Mater.* **23** 2185–92

[80] Han M, Yin X, Wu H, Hou Z, Song C, Li X *et al* 2016 Ti_3C_2 MXenes with modified surface for high-performance electromagnetic absorption and shielding in the x-band *ACS Appl. Mater. Interfaces* **8** 21011–9

[81] Liu Y, Li E, Yan Y, Lin Z, Chen Q, Wang X *et al* 2021 A one-structure-layer PDMS/MXenes based stretchable triboelectric nanogenerator for simultaneously harvesting mechanical and light energy *Nano Energy* **86** 106118

[82] Han S A, Bhatia R and Kim S-W 2015 Synthesis, properties and potential applications of two-dimensional transition metal dichalcogenides *Nano Convergence* **2** 17

[83] Zhang G and Zhang Y-W 2017 Thermoelectric properties of two-dimensional transition metal dichalcogenides *J. Mater. Chem.* C **5** 7684–98

[84] Upadhyay S N, Satrughna J A K and Pakhira S 2021 Recent advancements of two-dimensional transition metal dichalcogenides and their applications in electrocatalysis and energy storage *Emerg. Mater.* **4** 951–70

[85] Xie L M 2015 Two-dimensional transition metal dichalcogenide alloys: preparation, characterization and applications *Nanoscale* **7** 18392–401

[86] Qian Z, Jiao L and Xie L 2020 Phase engineering of two-dimensional transition metal dichalcogenides *Chin. J. Chem.* **38** 753–60

[87] Su H, Pan X, Li S, Zhang H and Zou R 2023 Defect-engineered two-dimensional transition metal dichalcogenides towards electrocatalytic hydrogen evolution reaction *Carbon Energy* **5** e296
[88] Kumar N, Ghosh S, Thakur D, Lee C-P and Sahoo P K 2023 Recent advancements in zero- to three-dimensional carbon networks with a two-dimensional electrode material for high-performance supercapacitors *Nanoscale Adv.* **5** 3146–76
[89] Seol M, Kim S, Cho Y, Byun K-E, Kim H, Kim J *et al* 2018 Triboelectric series of 2D layered materials *Adv. Mater.* **30** 1801210
[90] Wu C, Kim T W, Park J H, An H, Shao J, Chen X *et al* 2017 Enhanced triboelectric nanogenerators based on MoS_2 monolayer nanocomposites acting as electron-acceptor layers *ACS Nano* **11** 8356–63
[91] Cho D-H, Park S, Im B, Kim Y, Kim S-W, Lee S-K *et al* 2023 Eco-friendly mass production of MoS_2 flakes in pure water for performance enhancement of triboelectric nanogenerator *Appl. Surf. Sci.* **625** 157235
[92] Kim T I, Park I-J, Kang S, Kim T-S and Choi S-Y 2021 Enhanced triboelectric nanogenerator based on tungsten disulfide via thiolated ligand conjugation *ACS Appl. Mater. Interfaces* **13** 21299–309
[93] Kim M, Kim S H, Park M U, Lee C, Kim M, Yi Y *et al* 2019 MoS2 triboelectric nanogenerators based on depletion layers *Nano Energy* **65** 104079
[94] Pace G, del Rio Castillo A E, Lamperti A, Lauciello S and Bonaccorso F 2023 2D materials-based electrochemical triboelectric nanogenerators *Adv. Mater.* **35** 2211037
[95] Singh V and Singh B 2023 PDMS/PVDF-MoS_2 based flexible triboelectric nanogenerator for mechanical energy harvesting *Polymer* **274** 125910
[96] Jalili M A, Karimzadeh F, Enayati M H, Kalali E N and Kheirabadi N R 2023 Development of a triboelectric nanogenerator for joining of silver nanorods *Adv. Electron. Mater.* **9** 2201348
[97] Lan L, Yin T, Jiang C, Li X, Yao Y, Wang Z *et al* 2019 Highly conductive 1D–2D composite film for skin-mountable strain sensor and stretchable triboelectric nanogenerator *Nano Energy* **62** 319–28
[98] Jangra M, Thakur A, Dam S, Chatterjee S and Hussain S 2023 Enhanced dielectric properties of MoS_2/PVDF free-standing, flexible films for energy harvesting applications *Mater. Today Commun.* **34** 105109
[99] Veeralingam S and Badhulika S 2023 Ti@MoS_2 incorporated polypropylene/nylon fabric-based porous, breathable triboelectric nanogenerator as respiration sensor and ammonia gas sensor applications *Sens. Actuators* B **380** 133346
[100] Chekke T, Narzary R, Ngadong S, Satpati B, Bayan S and Das U 2023 Au decorated ultrathin WS_2-based single-electrode triboelectric nanogenerator for flexible self-powered photodetector *Sens. Actuators* A **349** 114076
[101] Karmakar S, Kumbhakar P, Maity K, Mandal D and Kumbhakar P 2019 Development of flexible self-charging triboelectric power cell on paper for temperature and weight sensing *Nano Energy* **63** 103831
[102] Mahapatra , Ajimsha A, Deepak R S, Sumit D, Aggarwal R, Kumar S *et al* 2023 Textile-integrated MoS2-PDMS single electrode triboelectric nanogenerator for vibrational energy harvesting and biomechanical motion sensing *Nano Energy* **116** 108829

[103] Chekke T, Narzary R, Ngadong S, Satpati B, Bayan S and Das U 2023 2D WS_2-based single-electrode triboelectric nanogenerator for power generation and motion sensing *J. Electron. Mater.* **52** 2685–94
[104] Wickramaratne D, Weston L and Van de Walle C G 2018 Monolayer to bulk properties of hexagonal boron nitride *J. Phys. Chem.* C **122** 25524–9
[105] Bhimanapati G R, Glavin N R and Robinson J A 2016 2D boron nitride: synthesis and applications *Semiconductors and Semimetals* ed F Iacopi, J J Boeckl and C Jagadish (Amsterdam: Elsevier) vol 95 3 101–47
[106] Roy S, Zhang X, Puthirath A B, Meiyazhagan A, Bhattacharyya S, Rahman M M *et al* 2021 Structure, properties and applications of two-dimensional hexagonal boron nitride *Adv. Mater.* **33** 2101589
[107] Han S A, Lee K H, Kim T-H, Seung W, Lee S K, Choi S *et al* 2015 Hexagonal boron nitride assisted growth of stoichiometric Al_2O_3 dielectric on graphene for triboelectric nanogenerators *Nano Energy* **12** 556–66
[108] Parmar S, Biswas A, Kumar Singh S, Ray B, Parmar S, Gosavi S *et al* 2019 Coexisting 1T/2H polymorphs, reentrant resistivity behavior, and charge distribution in MoS_2-hBN 2D/2D composite thin films *Phys. Rev. Mater.* **3** 074007
[109] Yang Z, Zhang X and Xiang G 2022 2D boron nitride nanosheets in polymer nanofibers for triboelectric nanogenerators with enhanced performance and flexibility *ACS Appl. Nano Mater.* **5** 16906
[110] Vijoy K V, Anlin Lazar K, John H and Saji K J 2023 Enhancing the triboelectric performance of flexible PDMS/boron nitride composite nanogenerators *AIP Conf. Proc.* **2783** 040001
[111] Bhavya A S, Varghese H, Chandran A and Surendran K P 2021 Massive enhancement in power output of BoPET-paper triboelectric nanogenerator using 2D-hexagonal boron nitride nanosheets *Nano Energy* **90** 106628
[112] Pang L, Li Z, Zhao Y, Zhang X, Du W, Chen L *et al* 2022 Triboelectric nanogenerator based on polyimide/boron nitride nanosheets/polyimide nanocomposite film with enhanced electrical performance *ACS Appl. Electron. Mater.* **4** 3027–35
[113] Xiao Y, Xu B, Bao Q and Lam Y 2022 Wearable triboelectric nanogenerators based on polyamide composites doped with 2D graphitic carbon nitride *Polymers* **14** 3029
[114] Bayan S, Bhattacharya D, Mitra R K and Ray S K 2020 Two-dimensional graphitic carbon nitride nanosheets: a novel platform for flexible, robust and optically active triboelectric nanogenerators *Nanoscale.* **12** 21334–43
[115] Ruthvik K, Babu A, Supraja P, Navaneeth M, Mahesh V, Uday Kumar K *et al* 2023 High-performance triboelectric nanogenerator based on 2D graphitic carbon nitride for self-powered electronic devices *Mater. Lett.* **350** 134947
[116] Bayan S, Pal S and Ray S K 2022 Interface engineered silver nanoparticles decorated g-C_3N_4 nanosheets for textile based triboelectric nanogenerators as wearable power sources *Nano Energy* **94** 106928
[117] Kheirabadi N R, Karimzadeh F, Enayati M H and Kalali E N 2023 Sustainable and photoresponse triboelectric nanogenerators based on 2D-gC_3N_4 and agricultural wastes *J Mater. Sci.: Mater. Electron.* **34** 1571
[118] Wang X, Tong W, Li Y, Wang Z, Chen Y, Zhang X *et al* 2021 Mica-based triboelectric nanogenerators for energy harvesting *Appl. Clay Sci.* **215** 106330

[119] Li W, Yan F, Xiang Y, Zhang W, Loos K and Pei Y 2023 Enhanced triboelectric nanogenerators based on 2D smectite clay nanosheets with a strong intrinsic negative surface charge *Nano Energy* **112** 108487

[120] Li W, Lu L, Yan F, Palasantzas G, Loos K and Pei Y 2023 High-performance triboelectric nanogenerators based on TPU/mica nanofiber with enhanced tribo-positivity *Nano Energy* **114** 108629

[121] Yar A, Okbaz A and Parlayıcı Ş 2023 A biocompatible, eco-friendly, and high-performance triboelectric nanogenerator based on sepiolite, bentonite, and kaolin decorated chitosan composite film *Nano Energy* **110** 108354

[122] Chen Z, Cao Y, Yang W, An L, Fan H and Guo Y 2022 Embedding in-plane aligned MOF nanoflakes in silk fibroin for highly enhanced output performance of triboelectric nanogenerators *J. Mater. Chem.* A **10** 799–807

[123] Khandelwal G, Chandrasekhar A, Maria Joseph Raj N P and Kim S-J 2019 Metal–organic framework: a novel material for triboelectric nanogenerator–based self-powered sensors and systems *Adv. Energy Mater.* **9** 1803581

[124] Guo Y, Cao Y, Chen Z, Li R, Gong W, Yang W *et al* 2020 Fluorinated metal–organic framework as bifunctional filler toward highly improving output performance of triboelectric nanogenerators *Nano Energy* **70** 104517

[125] Huang C, Lu G, Qin N, Shao Z, Zhang D, Soutis C *et al* 2022 Enhancement of output performance of triboelectric nanogenerator by switchable stimuli in metal–organic frameworks for photocatalysis *ACS Appl. Mater. Interfaces* **14** 16424–34

[126] Chen S-W, Huang S-M, Wu H-S, Pan W-P, Wei S-M, Peng C-W *et al* 2022 A facile, fabric compatible, and flexible borophene nanocomposites for self-powered smart assistive and wound healing applications *Adv. Sci.* **9** 2201507

[127] Hou C, Tai G, Liu Y, Liu R, Liang X, Wu Z *et al* 2022 Borophene pressure sensing for electronic skin and human–machine interface *Nano Energy* **97** 107189

[128] Liu Y, Ping J and Ying Y 2021 Recent progress in 2D-nanomaterial-based triboelectric nanogenerators *Adv. Funct. Mater.* **31** 2009994

[129] Zhou Y, Zhang J-H, Li S, Qiu H, Shi Y and Pan L 2023 Triboelectric nanogenerators based on 2D materials: from materials and devices to applications *Micromachines* **14** 1043

IOP Publishing

Recent Advances in Materials for Energy Harvesting and Storage

Suresh C Pillai, Daniel M Mulvihill and Aswathy Babu

Chapter 4

Piezoelectric energy transduction: materials, diverse applications, and challenges

Swagata Banerjee, Chirantan Shee, Satyaranjan Bairagi, Ben B Xu, Daniel M Mulvihill and S Wazed Ali

The scarcity of fossil fuels and increasing power requirements demand a search for renewable energy sources. In this context, the piezoelectric effect has drawn the attention of researchers due to its reliability and reproducibility. Mechanical energy is abundant in our surroundings and wasted in various forms. Piezoelectric energy harvesting can use wasted mechanical energy to power small electronics. Piezoelectric materials are a class of materials which can generate electrical output from an input mechanical stress and vice versa. The diverse range of piezoelectric materials, offering varying levels of adaptability and efficiency, presents numerous possibilities for electronic applications. Various emerging research in this field is aiming to improve the energy conversion ability of this particular class of materials and increase its flexibility in real-life applications. This chapter first discusses the basic concept of piezoelectricity and its mechanism. Next, different classes of piezoelectric material are addressed. Finally, several case studies are systematically summarized to capture research progress in this field and the possible application range of piezoelectric materials. Ongoing efforts involve alterations in the synthesis and composition of these materials as researchers strive to create environmentally friendly alternatives to their toxic lead-based counterparts. It is concluded that there are persistent challenges related to synthesis, phase stabilization, flexibility, and miniaturization that must be addressed to expand the horizons of piezoelectric technology.

4.1 Introduction

Rapid technological progress has led to novel scientific innovations that have influenced our lives in innumerable ways. Smart sensors are an outcome of technological advancements that have been designed to make lives easier and less

doi:10.1088/978-0-7503-5749-4ch4

challenging [1]. The constant evolution of these sensors with enhanced functionality raises the bar for energy requirements. Further, they require a constant source of power to enable uninterrupted performance. Batteries have been the dominant power source since their discovery. The size of batteries is proportional to their capacity to supply power. This leads to problems of miniaturization and integration. These power sources have a specific charging and discharging cycle. They require frequent charging cycles to keep them functioning continuously. The disposal of batteries after their exhaustion is also more complex. This highlights the need for continuous sources of power that are flexible in terms of miniaturization and integration.

Fossil fuels, a nonrenewable resource, are still the primary source of energy today. The limited reserve of this valuable commodity is inching towards extinction owing to the over-exploitation of fossil fuels to meet the huge demand for energy. Despite their ability to produce high power, fossil fuels are polluting in nature. Their combustion releases gases that promote the greenhouse effect, leading to serious environmental issues such as global warming [2]. Thus, bearing in mind the limited availability and the non-eco-friendly nature of fossil fuels, there is a need to switch to smart, clean, and sustainable energy sources. The use of ambient resources such as solar, wind, and water power as effective energy sources has extensively explored in the last few decades. However, their performance is lacking due to their seasonal and scattered availability [3]. Interestingly, there is a huge abundance of mechanical energy in our surroundings, which, if harvested efficiently, could act as a potential solution to the impending energy crisis. Piezoelectricity, a phenomenon discovered in the 19th century, is one such technique that converts mechanical energy into electrical power, and vice versa. This pressure-driven phenomenon is observed in piezoelectric materials, which are categorized into different classes with distinct properties. Piezoelectricity is gaining ground in the domain of energy research due to its nonpolluting nature, power supply capacity, and wide scope for materials development [4]. A wide variety of novel piezoelectric materials have been developed through different synthetic techniques to improve upon existing materials. Chemical modification through doping [5, 6] and surface treatment of nanomaterials [7] have resulted in new materials with enhanced piezoelectric properties. Improvements to existing techniques, for example replacing sintering with microwave sintering, have improved the quality of piezoelectric ceramics [8]. Further improvements in piezoelectric properties can be achieved through the process of poling. Novel combinations of filler and polymer materials have created piezoelectric composites with multifunctional properties [9, 10]. This chapter focuses on piezoelectricity, illustrating its mechanism and describing the different classes of piezoelectric materials, concentrating mainly on ceramics, polymers, and composites. The challenges and future scope of this field are also included in the concluding part of the chapter.

4.2 The history of piezoelectricity

The discovery of piezoelectricity dates back to the 19th century. In 1880, Pierre and Jacques Curie observed the interesting phenomenon of the charge creation in crystals as a result of the application of pressure. W Hankel was the first to name this phenomenon as piezoelectricity. The reversible nature of this effect was affirmed thermodynamically by Lipmann. A few years of research in this domain led to the unveiling of 20 natural crystal classes that were capable of generating charges in response to the application of pressure. The complex science behind this phenomenon restricted its wider application, and thus it remained confined to laboratory research. Research progress in the early years of the 20th century witnessed the application of piezoelectricity in ultrasonic transducers, microphones, accelerometers, etc.

Novel materials with piezoelectric properties were gradually unearthed in the mid 20th century. The potassium dihydrogen phosphate family was one of the first groups of piezoelectric materials to be discovered [11]. Barium titanate, a ferroelectric material with a very high dielectric constant, was discovered in the 1940s. This material was chemically modified to improve its thermal stability and piezoelectric constants. The discovery of barium titanate led to a new class of ABO_3-type perovskite structures. A few years later, the lead-based piezoelectric ceramic lead zirconate titanate (PZT) was discovered by Jaffe and co-workers. PZT-based systems showed exceptionally high piezoelectric coefficients that encouraged researchers to explore this material. Chemical modifications of PZT performed through doping produced a ternary system of materials. This system of ternary compounds helped to enhance the piezoelectric properties of the pristine material and eased the processability of such chemical compounds. Further, these materials were successfully applied in the fields of medical science and communications and automotive and military applications [12].

4.3 The mechanism of piezoelectricity

In the year after Pierre and Jacques Curie's discovery of the piezoelectric effect, Hankel proposed the name 'piezoelectricity' for this unique phenomenon. The Greek word 'piezo' means pressure. Piezoelectricity is essentially electricity that originates from pressure. When an external force is applied to a piezoelectric material, an electrical potential difference is created between the two ends of the material. This phenomenon is known as the direct piezoelectric effect. A piezoelectric material can be mechanically strained by applying an external electric field to it. This reverse phenomenon is known as the converse piezoelectric effect. A crystalline material that has non-centrosymmetry in its crystal structure can only exhibit a piezoelectric effect. That means in a piezoelectric material, the center of the positive charges does not coincide with that of the negative charges during the application of external force to the material. Figure 4.1 shows a pictorial representation of the piezoelectric effect.

Within a non-centrosymmetric material, the internal gravity center of positive charges may coincide with that of the negative charges in the absence of an external force, which ultimately leads to a neutral molecule. During the application of external stress, the centers of the positive and negative charges separate due to deformation in the molecular structure which ultimately produces tiny dipoles.

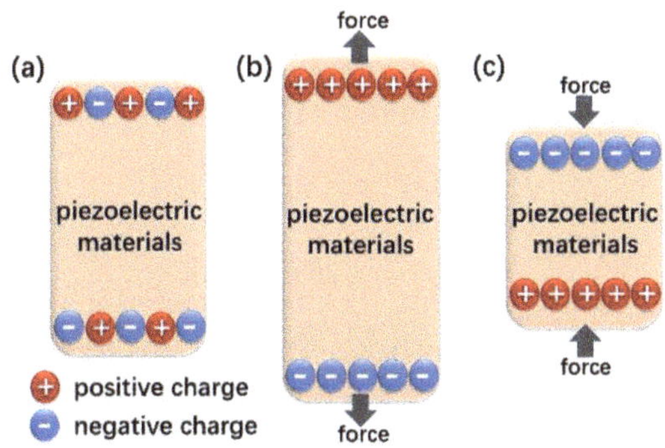

Figure 4.1. The piezoelectric effect (reprinted from [13], Copyright (2022), with permission from Elsevier).

Neighboring opposite charges may neutralize each other, and the linked charge is generated at the material's surface, thus producing a piezoelectric voltage. When a suitable conductive path is created between the two ends of the material, electrons flow through it due to the piezoelectric potential difference. When the material is elongated by an external force, the dipolar orientation increases in that direction, which simultaneously enhances the polarization. The linked charge at the surface also increases, and this ultimately raises the flow of electrons through the external conductor. On the other hand, if the material is compressed by an external force, the dipolar orientation is partially hampered, causing a decrease in the linked charge at the material's surface. This situation also requires a flow of electrons through the external conductor in order to reach an equilibrium. The direction of electron flow during stretching is obviously opposite to the direction in which the material is compressed. Thus, an alternating current can be produced by repeatedly squeezing and releasing the piezoelectric material. Piezoelectricity can efficiently convert wasted mechanical energy into its electrical counterpart. Piezoelectric energy harvesting efficiency can be improved by a process called poling, in which an inherent dipolar orientation is produced.

The polarization produced by piezoelectricity is directly proportional to the external stress applied. This can be mathematically expressed as follows (equation (4.1)):

$$P_{\mathrm{p}} = d\, T, \tag{4.1}$$

where P_{p} is the polarization vector (which is numerically the same as the surface charge density of the linked charges), d is the piezoelectric strain coefficient, and T is the stress applied to the material. For the reverse piezoelectric effect, the ratio between the strain produced and the applied electric field depends on the piezoelectric strain coefficient, i.e. $S_{\mathrm{p}} = d\, E$, where S_{p} is the produced strain and E is the applied electric field. Let us assume further that c is the elastic constant of the material, s is the compliance coefficient, and e is the piezoelectric stress constant. A few important mathematical equations for piezoelectricity are listed below (equations (4.2)–(4.5)).

$$T = c\,S \tag{4.2}$$

$$S = s\,T \tag{4.3}$$

$$P_{\mathrm{p}} = d\,T = d\,cs = e\,s \tag{4.4}$$

$$T_{\mathrm{p}} = c\,S_{\mathrm{p}} = c\,d\,E = e\,E \tag{4.5}$$

The subscript p denotes piezoelectric stress or piezoelectric strain. On the other hand, variables without subscripts refer to the externally applied stress and strain [14].

4.4 Piezoelectric materials

Various types of material can be piezoelectric in nature, depending on their chemical structures. Piezoelectric materials can be classified into four types, namely: piezoelectric crystals, piezoelectric ceramics, piezoelectric polymers, and piezoelectric composites. Each class has captured different application fields that depend on its own benefits and limitations. Although crystals have a higher mechanical quality factor, their processing techniques are costly and complex. Quartz crystal and ammonium dihydrogen phosphate are examples of piezoelectric crystals. Piezoelectric ceramics have higher dielectric constants, sensitivities, chemical stabilities, and coupling factors. On the other hand, the limitations associated with ceramics are: high density, brittleness, costly processing techniques, and lower maximum strains. Examples of piezoelectric ceramics include PZT, $BaTiO_3$ (barium titanate), $NaNbO_3$ (sodium niobate), $KNbO_3$ (potassium niobate), ZnO (zinc oxide), GaN (gallium nitride), etc. Piezoelectric polymers are flexible, economical, and lightweight and have easier processing techniques. However, the disadvantage of this class is its poor electromechanical coupling factor. Polyvinylidene difluoride (PVDF), polyacrylonitrile (PAN), and nylon 11 are some examples of this particular class. To utilize the advantages of both piezoelectric ceramics and polymers, these groups are often blended to form composite materials. Generally, the incorporation of fillers improves the electroactive-phase content of the polymer. Dipoles can be neutralized within a composite due to inconsistent polarization directions, which may be a limitation associated with this class [15].

4.4.1 Piezoelectric ceramic materials

Piezoelectric ceramic materials have gained prominence as energy-generating materials due to their commendable piezoelectric coefficient, which widens their scope of application. They are often applied in circuit components and used as transducers [16], actuators, and sensors. Recently, certain novel applications of these ceramics have been explored, which has further broadened the application domain of such materials.

PZT was one of the first lead-based piezoelectric ceramic materials to be discovered. It is basically a solid solution of two compounds, namely lead zirconate and lead titanate. It belongs to the class of ABO_3-type oxides, which have a perovskite structure. This structure is responsible for the various electromechanical properties of the material. The A site cations form a cuboid enclosing an oxygen octahedron. The B-site cation is usually placed inside the oxygen octahedron. In the ferroelectric state, the oxygen octahedron shifts off-center, accompanied by a shift of the B-site ions. This causes a charge imbalance in the crystal lattice, leading to spontaneous polarization between the cations and the anionic octahedron. In the case of PZT, the A site is occupied by lead ions, while the B site is occupied by randomly distributed zirconium or titanium ions. The high piezoelectric coefficients of PZT are also influenced by its phase structure. The term 'morphotropic phase boundary' (MPB) refers to a sudden alteration in the phase structure of PZT that is associated with a change in its composition. This boundary is unaffected by temperature changes. It is further influenced by the Zr/Ti ratio. The MPB generally occurs at a certain composition for which the energies of two different phases are quite similar; however, they differ in their elasticity. These compositions near the MPB are used for applications that demand extreme electromechanical properties or enhanced piezoelectric displacement. The cause of the improved piezoelectric properties at the MPB has been explained using different approaches. The MPB is characterized by a mixture of symmetries that allows easy poling of the polycrystalline ceramic. Further, it helps to lower the anisotropic energy, reducing the domain wall energy and hence facilitating domain mobility. This contributes to enhancing the ceramic's electromechanical properties to a great extent [11]. The large dielectric susceptibility of compositions near the MPB gives these compositions high piezoelectric constants. Further, their small elastic compliance causes large coupling coefficients. Some researchers are of the view that the presence of a monoclinic phase facilitates the switch between the rhombohedral and tetragonal phases when the ceramic is subjected to an electric field and that this ease of phase switching results in improved piezoelectric properties. The properties of PZT can be tailored for a given end use through different modification techniques. Zr-rich or Ti-rich PZT can be synthesized to produce rhombohedral or tetragonal phase PZT structures, respectively. The properties of tetragonal PZT are stable with respect to temperature, and it has a high Curie temperature. Tetragonal compositions are hence used in the design of communication circuits. PZT may further be modified by doping it with external elements, which is called donor doping or acceptor doping. Doping it with cationic elements that have lower valencies results in the creation of oxygen vacancies. These vacancies have a positive charge that couples with the acceptor ions to form electric dipoles. These electric dipoles usually occupy the domain boundary and constrict rotation to the non-180 degree domain. This raises the energy required for the rotation of the electric dipole, in turn increasing the coercive field, remanent polarization, and remanent strain of these acceptor-doped PZT ceramics. On the other hand, doping with higher-valence elements leads to the creation of cationic vacancies. The doped ions impart a semiconducting nature to the doped ceramic that reduces its resistivity. Parameters such as the remanent polarization, coercive field, and remanent strain are lower compared to those of acceptor-doped PZT (figures 4.2(a) and (b)). Generally, due to their electrical and mechanical characteristics, acceptor- and donor-doped PZT are referred to as hard and soft PZT, respectively [12].

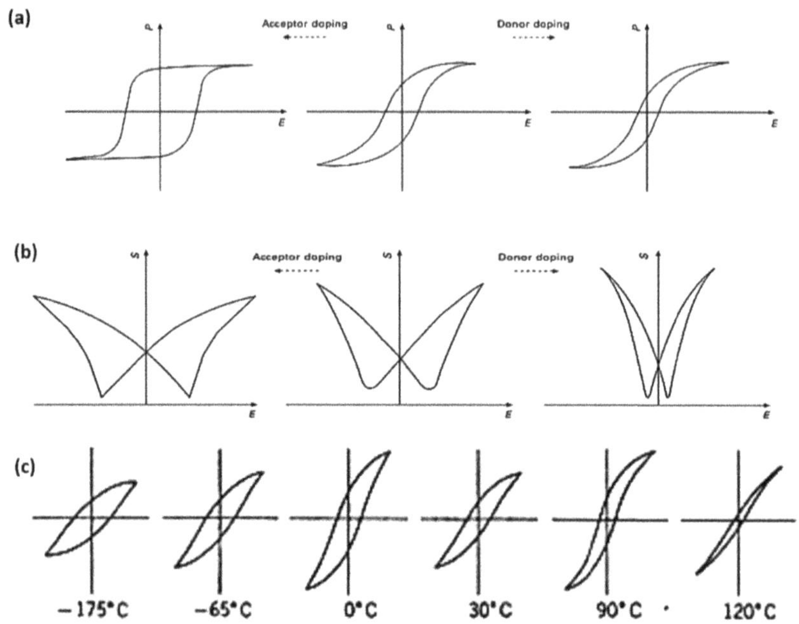

Figure 4.2. (a) The changes in the *P–E* hysteresis loop associated with acceptor and donor doping of PZT. (b) The changes in the butterfly curve due to the acceptor and donor doping of PZT. ((a) and (b) reprinted from [12], Copyright (2010), with permission from Elsevier. Copyright 2010 Woodhead Publishing Limited. All rights reserved.) (c) The changes in the *P–E* hysteresis loop due to increasing temperature (reproduced from [17], CC BY 4.0).

The toxicity concerns raised by the use of lead have shifted the research trend toward a search for lead-free alternative materials. In the category of lead-free piezoelectric ceramics, barium titanate was one of the first materials to be discovered. Barium titanate has very high piezoelectric coefficients; however, a comparatively low Curie temperature restricts its applications. Barium titanate shows a transition from a ferroelectric to a paraelectric nature at a temperature of 120 °C. This ceramic acquires an orthorhombic phase structure at 5 °C; at −90 °C, it has a trigonal structure. This ceramic has a perovskite structure. The smaller Ti ions are placed in the interstices of the cubic lattice created by the larger cation and the oxygen anions. Below the Curie temperature, this Ti ion keeps switching positions between energetically favorable states, relocating itself from the center of the octahedron. This causes changes in bond length and distorts the structure of the crystal lattice [18]. The electrical properties of barium titanate are influenced by the structure and motion of the domains formed during the transition from the paraelectric to the ferroelectric state below the Curie temperature. The configuration of domains is affected by the presence of additives and the microstructure formed during the process of manufacture. Often, the microstructure of the grains also influence the electrical properties of such ceramic materials [19]. Small grains that have homogeneous size distributions and a single domain structure result in appreciable ferroelectric characteristics. Coarse-grained structures result in inhomogeneous permittivity inside the crystal structure. Further, the microstructure affects the width and energy of the domains. Barium titanate is a ceramic that is known for its high dielectric constant and low dielectric loss factor. These properties has been widely explored for capacitor applications. The synthetic route of this ceramic affects the dielectric constant of the material. The microstructure of the material also influences the dielectric constant. The dielectric constant is inversely related to the grain size and its distribution. The dielectric constant is also affected by temperature, dopants, and frequency. The perovskite structure permits the inclusion of a variety of elements. Hence, elements of different valencies can be doped at the A- and B-sites of the perovskite structure to modify its dielectric properties [20]. This ceramic has commendable piezoelectric properties. It has high coupling coefficients and good thermal stability. Compared to those of its counterparts, the piezoelectric properties of this material are adequate at room temperature. Studies of the *P–E* hysteresis curve of this material show a significant dependence of these parameters on temperature. At higher temperatures, increased energy helps to reorient the domain walls, decreasing the coercive field. Less energy at lower temperatures renders it difficult to orient the domain walls, thus increasing the coercive field (figure 4.2(c)) [17].

Alkali niobates have also been explored as potential piezoelectric energy materials. Potassium sodium niobate (KNN), a member of this family of niobates, has shown great promise as an energy material. This piezoelectric ceramic is a chemical combination of potassium niobate and sodium niobate. Potassium niobate is ferroelectric in nature, while sodium niobate is antiferroelectric in nature. These two compounds are known to form solid solutions over a wide range of compositions. This gives rise to KNN with different stoichiometric ratios of potassium and sodium. An optimum K/Na ratio of 52/48 is known to show the highest piezoelectric performance. This composition-driven piezoelectric performance of KNN ceramics is referred to as the MPB [21]. In addition to the effect of the MPB, KNN also exhibits a change in its crystalline structure with a change in temperature. This phenomenon is called the polymorphic phase transition (PPT). KNN has a Curie temperature as high as 400 °C. It further shows a phase transition from the orthorhombic phase to the tetragonal phase at around 210 °C. The rhombohedral phase prevails at temperatures as low as −123 °C. Often, the piezoelectric properties of KNN peak at the phase boundaries, due to easier dipole orientation. However, there are also reports of the coexistence of multiple phases that help to improve the piezoelectric properties of the material [22]. As stated earlier, different synthetic routes for piezoelectric ceramics yield piezoelectric constants of varied magnitude. One process common to most synthetic routes is the process of sintering. Sintering is a high-temperature process that helps in the densification of ceramic materials. Often, such high temperatures result in the volatilization of alkali metal elements. Hence, modifications of the sintering process have been carried out to address such issues [23, 24]. Despite the advantageous properties of KNN, it still lacks excellence in some aspects. The electrical properties of KNN are very sensitive to the temperature maintained during their process of synthesis. Further, KNN has low mechanical quality. Moreover, there are bottlenecks in the process of developing

Table 4.1. A comparative analysis of widely studied piezoelectric ceramics [29].

Material	d_{33} (pC N^{-1})	g_{33} (Vm N^{-1})	Electromechanical coupling factor	Curie temperature (°C)
PZT	360	0.025	0.6–0.7	350
KNN	80–160	0.0469	0.51	400
$BaTiO_3$	191	0.0114	0.49	130
BNT	75		~0.21	<290
KNN-$LiSbO_3$	283		0.46	237

and enhancing the properties of KNN that need to be addressed to realize the electrical properties of KNN to its full potential [25].

Bismuth (Bi)-based materials are also an interesting class of piezoelectric ceramics. Bismuth sodium titanate (BNT) was first discovered in 1960 [26] and found to have appreciable piezoelectric characteristics. Its depolarization temperature is around 290 °C [27]. However, its high coercive field poses a challenge for the poling of this material. Hence, several approaches to modifying this ceramic have been undertaken to address these issues. Bismuth potassium titanate (BKT) has a high Curie temperature of around 380 °C; however, it faces densification-related issues during its synthesis. Sintering aids are often necessary to produce dense structures of these ceramics [28]. Table 4.1 provides a brief summary of the different piezoelectric ceramics and their piezoelectric properties.

4.4.2 Polymeric piezoelectric materials

The disadvantages of single-crystal and ceramic piezoelectric materials can be partially mitigated using piezoelectric polymers. Piezoelectric polymers are associated with advantages such as low density, flexibility, easier processing techniques, cost-effectiveness, high impact resistance, toughness, and high strength. The piezoelectric stress constant (g_{31}) is greater for polymers than for ceramics. On the other hand, the piezoelectric strain constant (d_{31}) of ceramics is higher compared to that of polymers. Polymers, with higher stress constants, have lower dielectric constants, elastic stiffness, and density, making them good sensors (i.e. they have better voltage sensitivity). Electrical field strength is also higher for polymers than for ceramics. Within a semicrystalline polymer, the piezoelectric property comes from the crystalline polar phase. Polymers generally exhibit polymorphic crystalline phases, among which a few may be polar in nature. The content of the electroactive phase can be increased by transforming the nonpolar phases into electroactive ones through techniques such as mechanical stretching, thermal annealing, and electrical poling. A typical electric field of around 50 MV m^{-1} is desirable for poling polymeric materials. According to dielectric theory, polarization within an amorphous polymer is quasi-stable, in contrast to the thermal equilibrium conditions within inorganic and semicrystalline polymers. In amorphous piezoelectric

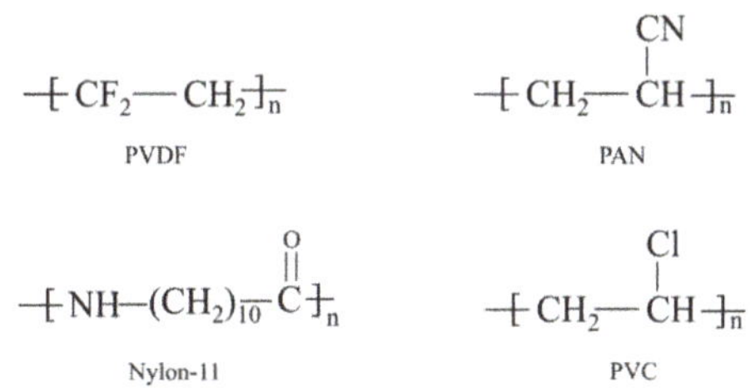

Figure 4.3. The chemical structures of a selection of common piezoelectric polymers.

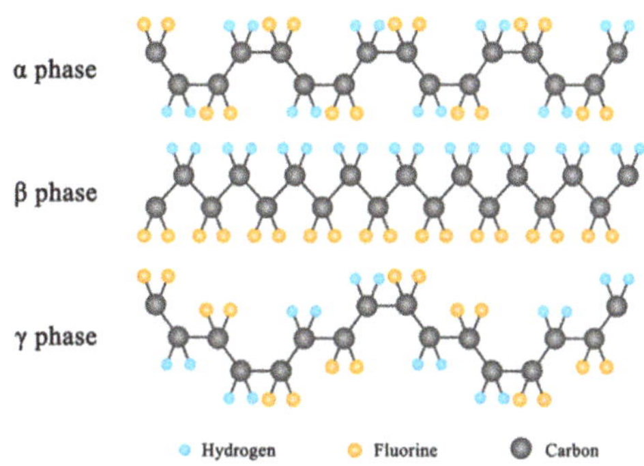

Figure 4.4. The crystalline phases of PVDF (reprinted from [32], Copyright (2020), with permission from Elsevier).

polymers, the glass transition temperature (T_g) decides the use temperature of the polymer as well as the poling conditions. When an amorphous polymer experiences an appropriate electric field at a temperature equal to or above its T_g, the dipoles are oriented in a preferential direction. At this stage, lowering the temperature below T_g and keeping the electric field constant freezes the molecular dipoles in a desirable orientation. However, in a semicrystalline polymer the polarization is somewhat locked by the crystalline orientation of the polymeric chains, improving its use temperature beyond the T_g. On the other hand, in an amorphous polymer, the use temperature is restricted by T_g itself. Time, temperature, and pressure determine the dipole relaxation for amorphous piezoelectric polymers [30]. Figure 4.3 shows the chemical structures of a selection of piezoelectric polymers.

PVDF, a polymer of difluoroethylene, exhibits superior piezoelectricity among piezoelectric polymers. PVDF is often explored as a copolymer with trifluoroethylene, which can offer a crystallinity of up to 90%, improving the piezoelectric property. Such a copolymer offers a working temperature of 100 °C due to improved thermal stability. PVDF exhibits different crystalline phases, namely α, β, and γ, depending on its molecular conformation (figure 4.4). Although the α-phase is a thermodynamically stable state, it does not exhibit the piezoelectric effect. This crystalline phase can be converted to the electroactive β-phase by several techniques such as mechanical stretching, electrical poling, and thermal annealing [31].

PAN is another productive piezoelectric polymer. PAN offers two types of conformations in its solid state. These are zigzag (sawtooth) and 3^1-helical conformations. The zigzag conformation is electroactive and piezoelectric in nature. This conformation, an all-trans conformation, offers a higher dipole moment compared to that of β-phase PVDF. For this reason, electrospun PAN membranes can offer higher piezoelectric outputs compared to those of electrospun PVDF membranes [33]. Table 4.2 depicts the properties of a few piezoelectric polymers.

4.4.3 Composite piezoelectric materials

Piezoelectric ceramic materials have appreciable piezoelectric constants. However, their rigid and fragile nature curbs their applications in remote locations. Piezoelectric polymers, on the other hand, have good flexibility and a moderate piezoelectric coefficient. Often, a combination of two or more materials with different properties results in the formation of a novel entity which has characteristics remarkably different from those of its constituent elements. This defines the concept of composites. Piezoelectric composites are made up of one or more piezoelectric materials and have properties different from those of their constituents. Composites help to mask drawbacks while utilizing the advantages of their constituent elements. Composites can be differentiated based on their connectivity pattern. They are depicted by the notation 'a–b'; where 'a' represents the connectivity of the dispersed phase and 'b' represents the connectivity of the continuous phase. The 0–3 composite is preferred over other combinations due to its shape-forming flexibility and scalability [29].

PVDF and PZT are two of the most promising piezoelectric materials discovered so far. Composites of PVDF/PZT have been explored to obtain novel materials with appreciable properties. PZT was synthesized through a solid-route method and incorporated into PVDF through the solution-casting method. Homogeneous solution-cast composite films were prepared by compression molding. The electromechanical response of the composite improved compared to that of pristine PVDF due to the addition of an electroactive filler material, PZT. However, filler agglomeration at high filler loadings adversely affected the electromechanical properties of the composite. The optimized composite of the nanocomposite produced an output voltage and power density of 55 V and 39 $\mu W\ cm^{-2}$, respectively—suitable for driving low-powered gadgets [34]. Similar studies have been carried out to explore the application of such films as acoustic sensors [35] and to study their mechanical properties [36]. Electrospun nanofibers of PVDF that incorporated PZT particles have also been studied. The nanofibers' average diameter increased with the PZT content owing to the increase in the viscosity of the electrospinning solution. Further, filler agglomeration was also observed at high filler loadings. The PZT particles also helped to induce a local electric field that aided in the transformation of the PVDF polymer chains from the α phase to the β phase. The electrospun nanocomposite's enhanced dielectric and piezoelectric properties were attributed to the presence of PZT, its interaction with the PVDF polymer chains, and the stabilization of the electroactive β phase of PVDF [37].

Table 4.2. The properties of piezoelectric polymers [29, 30].

Polymer	Polymer type	T_m (°C)	T_g (°C)	g_{33} (Vm N^{-1})	d_{33} (pC N^{-1})	d_{31} (pC N^{-1})	Maximum use temperature (°C)	Curie temperature (°C)
PVDF	Semicrystalline	175	−35	−0.33	−24 to −34	20–28	80	75–80
poly(vinylidene fluoride-co-trifluoroethylene) (P(VDF-TrFE))	Semicrystalline	—	—	—	−25 to −40	—	—	110
PAN	Amorphous	—	90	—	—	2	—	—
Nylon 11	Semicrystalline	195	68	—	—	3 @ 25 °C and 14 @ 107 °C	185	
Cellular polypropylene	Semicrystalline	—	—	30	130–2100	—	—	—
Poly(vinyl chloride) (PVC)	Amorphous	—	80	—	—	5	—	—

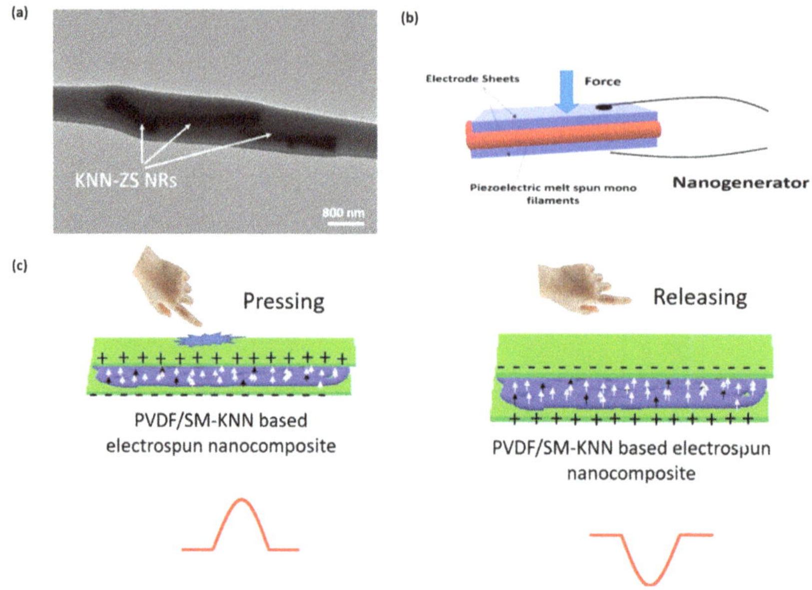

Figure 4.5. (a) Modified KNN nanorods incorporated into electrospun PVDF nanofiber (reprinted from [40], Copyright (2022), with permission from Elsevier), (b) the design of a piezoelectric nanogenerator (reprinted from [39], Copyright (2019), with permission from Elsevier), and (c) the working mechanism of the piezoelectric nanogenerator (reprinted from [44], Copyright (2020), with permission from Elsevier).

Solution-cast films of PVDF/PZT and PVDF/BT were prepared through the solution-casting method using dimethylacetamide as the solvent. The resultant films crystallized the polymer in its β phase. The ε' values of the composites were higher than that of the pristine polymer owing to the higher ε' value of the incorporated ceramic. The higher ε' values observed at higher frequencies were attributed to the loss caused by domain wall movements and dipolar orientation [38].

Composites of PVDF with lead-free piezoelectric ceramics have also been studied extensively. PVDF/KNN piezoelectric composites have been prepared and explored for their electrical properties [39–41] (figure 4.5). The PVDF/KNN composites were prepared by hot compression molding. This process helped to achieve composites with good densification. The thermal stability of the composites increased with the increase in KNN content. Further, the dielectric properties also increased as the concentration of KNN in the composite increased. There is a large difference in the dielectric properties of PVDF and KNN. The relative permittivity of the composite increased due to the interfacial polarization between KNN and PVDF caused by the difference in their dielectric properties. As the content of KNN increased, the interfaces in the composite increased, leading to higher dielectric loss. The prepared composite sample had better piezoelectric and dielectric properties than those of the pristine polymer. This nontoxic combination was found suitable for various energy harvesting and biological applications [42]. One of the major issues encountered during the fabrication of piezoelectric composites is the distribution of the filler in the polymer matrix. High filler loadings result in filler agglomeration, reducing the interaction between the polymer chains and the reinforcement, thus adversely affecting the piezoelectric performance of the composite. As a solution, fillers have been surface modified with chemicals such as 3-aminopropyltrimethoxysilane (APS), which reduce their surface energy and thus lower their agglomeration tendency. Incorporating such surface modified fillers into a PVDF polymer matrix resulted in better distribution of the filler inside the matrix and hence higher piezoelectric performance [43]. KNN has also been doped with various elements and incorporated into polymers to form electrospun nanocomposite fibers. The piezoelectric and triboelectric performances of these materials were found to be appreciable [6].

Copolymers of PVDF have also been investigated for their piezoelectric properties. PVDF-TrFE, a copolymer of PVDF, has higher electroactivity owing to the presence of additional fluorine in place of hydrogen. The main approach taken to enhance the piezoelectric property of polymeric composites is to increase the content of their electroactive phase. For this reason, dielectric fillers, metal, and metal-oxide nanoparticles have been added to polymeric composites. Often, higher loadings of such reinforcement materials compromise the flexibility of the polymeric nanocomposites. Fe-doped ZnO nanofillers were incorporated into a PVDF-TrFE polymer matrix through the solution-casting method. Analysis of the samples revealed a morphological transition of the filler from spherical to a coagulated structure upon doping with iron. Incorporating modified ZnO filler into the PVDF-TrFE matrix stabilized the electroactive phase of the polymer. The inclusion of modified filler in the matrix increased the number of dipoles in the composite films, which resulted in an improvement in their dielectric parameter [45].

4.5 Applications of piezoelectric materials

Piezoelectric materials find widespread applications in different domains of science and engineering. The most common application of piezoelectric materials is in the form of energy harvesters that utilize both low- and high-frequency vibrations. Apart from their use in energy harvesters, such materials also act as efficient sensors. As sensors, these materials are used to monitor the motion of different body parts in the form of electrical signals. Specific electrical signals are produced for specific kinds of movements. Any abnormalities in movement can be detected from anomalies in the electrical output. Piezoelectric sensors have been used for eyeball tracking [46], joint disorders [47], etc. Electrical stimulation has proven to be effective in regulating cell activity. Hence, there are a number of studies illustrating the use of piezoelectric devices in governing cell activity [48, 49]. Piezoelectric nanogenerators are also used for wound-healing applications. When a nanogenerator is attached to the wound site, the electric field between the electrodes penetrates the skin to mimic the endogenous electric field that facilitates wound healing [50, 51]. Piezoelectric devices have also been employed as implants. Such implants provide timely warning to medics in cases of emergency that occur prior to a scheduled appointment [52, 53]. Thus, piezoelectric nanogenerators are utilized in numerous ways in the field of bioscience.

Piezoelectric devices can be used as high-temperature sensors. The suitable temperature range of a device is governed by its Curie temperature. Parameters such as electromechanical coupling and piezoelectric coefficients under different temperature conditions need to be considered when deciding the suitability of a sensor for high-temperature applications. For instance, while aurivillius-structured ferroelectrics can be used at temperatures of up to 600 °C, langasite crystals are used below the region of 600 °C. Bismuth ferrite is a recent piezoelectric and ferroelectric material that can be used at temperatures beyond 600 °C and is gaining increasing attention. It is used in aerospace and automotive applications [54].

Apart from functioning as sensors, piezoelectric devices find wide acceptance in the field of structural health monitoring. Structural health monitoring involves the observation and evaluation of the structural integrity of composites. The timely detection of structural damage to a composite is critical with respect to its function [55]. Piezoelectric materials act as effective self-powered sensors for structural health monitoring and repair. The energy harvesting capability of these materials provides the power necessary to aid the uninterrupted sensing performance of piezoelectric sensors. Thus, they provide a two-way benefit. Piezoelectric sensors are used to locate the damage site in composites. Processing of the output signals received from the piezoelectric sensors helps to detect the position of damage. Further, such sensors also find application in detecting delamination in beams [56].

Nowadays, piezoelectric devices are being intertwined with wireless communication technology to widen their horizon of utilization. This has resulted in the development of wireless piezoelectric circuits that are capable of harvesting mechanical energy from remote locations. These technologies make the application more straightforward and user friendly. They provide realistic data, real-time monitoring, and control.

4.6 Challenges and future opportunities

Piezoelectric energy harvesting has emerged as a renewable and clean energy source. There are different types of piezoelectric material which have their own benefits and limitations. In order to partially mitigate the disadvantages and explore the respective advantages, piezoelectric composites were developed. To obtain desirable electrical and physical properties from a composite, connectivity between the filler and matrix is crucial. For this purpose, manufacturing conditions are vital. Suitable poling conditions should be chosen for synchronous polarization of the filler and matrix within a piezoelectric composite. The poling process involves a high electrical field and high temperatures, making it expensive. Moreover, the synthesis of piezoelectric ceramics is complex and costly due to their lower yield. Electrospinning has become very popular for the production of piezoelectric membranes, since this process does not demand a separate poling operation. However, this process involves the use of solvents and high electric fields. In addition, its lower production speed restricts the commercialization of this piezoelectric material manufacturing technique.

Although piezoelectric output is reliable and reproducible, the electrical output is comparatively small. The low amount of electrical energy can be stored in energy storage devices for further use. Alternatively, another effect such as triboelectricity can be combined with piezoelectricity in a single hybrid energy harvester to improve electrical output. Differentiating these piezoelectric and triboelectric outputs from such a hybrid device is still challenging for research purposes. In the context of wearable electronics, in which fabrics are used as piezoelectric devices, the right choice of electrode is difficult. The use of metallic wires during weaving or knitting processes obviously makes the fabric heavy and rigid. When metallic tapes are attached to the fabric to make electrodes, the flexibility and breathability of the fabrics are indeed reduced. For this purpose, the demand for flexible and porous electrodes is clear. The use of conductive coatings for electrodes is also a challenge for piezoelectric fabrics due to their porous nature. Despite several advances, suitable electrode configurations for wearable electronics still require significant research effort.

To conclude this section, we may recall the importance of piezoelectric energy harvesting in the context of the energy crisis. Although remarkable advancements have already occurred in this field, the previously mentioned challenges demand a lot of dedicated research. Along with energy harvesting, several applications such as sensing, structural health monitoring [55, 56], wound healing [50, 51], and sound absorption [57, 58] have already been explored. The design and development of efficient piezoelectric devices would obviously benefit all application arenas. It can be expected that further research will eliminate the identifiable obstacles, leading to the development of superior piezoelectric devices and enlarging the associated application realm.

4.7 Conclusions

The piezoelectric phenomenon was discovered a few centuries ago. However, its potential and scope of application have been realized relatively more recently. The wide variety of piezoelectric materials with respect to flexibility and piezoelectric efficiency opens varied avenues for implementation. However, there are challenges related to synthesis, phase stabilization, flexibility, and miniaturization that need to be resolved to broaden the scope of piezoelectric applications. With the growing concern about lead toxicity, the research trend has now shifted to less toxic and more eco-friendly materials. Modifications in the synthesis, composition, and structure of lead-free materials are constantly carried out to develop lead-free alternatives to lead-based materials. Efficient poling processes for the synchronous dipolar orientation of constituents in a piezoelectric composite are of crucial importance. The design of suitable electrodes for wearable electronics is still challenging, demanding further work in this field. Finally, it can be concluded that piezoelectric energy harvesting is a very efficient technique for the production of clean energy, opening a vast scope of possibilities. It has also been emphasized that the piezoelectric effect can be effectively combined with other energy harvesting approaches to develop high-performance hybrid devices (e.g. combined piezo/tribo devices). This chapter provided an overview of piezoelectricity, piezoelectric materials, and their suitable areas of application. Research progress in this field hints at the promising potential of piezoelectricity as part of the solution for small-scale local energy harvesting.

References and further reading

[1] Javaid M, Haleem A, Rab S, Pratap R and Suman R 2021 Sensors for daily life: a review *Sens. Int* **2** 100121

[2] Polina Maciejczyk G T, Chen L-C and Thurston G 2021 The role of fossil fuel combustion metals in $PM_{2.5}$ air pollution health associations *Atmosphere (Basel)* **12** 1086

[3] Cao S and Li J 2017 A survey on ambient energy sources and harvesting methods for structural health monitoring applications **9** 1–14

[4] Toprak A and Tigli O 2014 Piezoelectric energy harvesting: state-of-the-art and challenges *Appl. Phys. Rev. 1* **031104** 1–14

[5] Wang X, Huan Y, Ji S, Zhu Y, Wei T and Cheng Z 2022 Ultra-high piezoelectric performance by rational tuning of heterovalent-ion doping in lead-free piezoelectric ceramics *Nano Energy* **101** 107580

[6] Banerjee S, Bairagi S and Ali S W 2022 A lead-free flexible piezoelectric-triboelectric hybrid nanogenerator composed of uniquely designed PVDF/KNN-ZS nano fibrous web *Energy* **244** 123102

[7] Kim S R, Yoo J H, Kim J H, Cho Y S and Park J W 2021 Mechanical and piezoelectric properties of surface modified (Na,K)NbO_3-based nanoparticle-embedded piezoelectric polymer composite nanofibers for flexible piezoelectric nanogenerators *Nano Energy* **79** 105445

[8] Borrell A, D. Salvador M, Miranda M, L. Penaranda-Foix F and M. Catala-Civera J 2014 Microwave technique: a powerful tool for sintering ceramic materials *Curr. Nanosci.* **10** 32–5

[9] Cao X, Xiong Y, Sun J, Zhu X, Sun Q and Wang Z L 2021 Piezoelectric nanogenerators derived self-powered sensors for multifunctional applications and Artificial intelligence *Adv. Funct. Mater.* **31** 1–31

[10] Lin Y and Sodano H A 2008 Concept and model of a piezoelectric structural fiber for multifunctional composites *Compos. Sci. Technol.* **68** 1911–8

[11] Zhu X 2010 *Piezoelectric Ceramic Materials: Processing, Properties, Characterization, and Applications* (Hauppauge, NY: Nova Science Publishers)

[12] Kimura M, Ando A and Sakabe Y 2010 Lead zirconate titanate-based piezo-ceramics *Advanced Piezoelectric Materials: Science and Technology* (Cambridge: Woodhead Publishing Limited) pp 89–110

[13] Jiang Z, Tan X and Huang Y 2022 Piezoelectric effect enhanced photocatalysis in environmental remediation: state-of-the-art techniques and future scenarios *Sci. Total Environ.* **806** 150924

[14] Arnau A and Soares D 2008 *Fundamentals of Piezoelectricity* (Berlin: Springer)

[15] Zhang C, Fan W, Wang S, Wang Q, Zhang Y and Dong K 2021 Recent progress of wearable piezoelectric nanogenerators *ACS Appl. Electron. Mater.* **3** 2449–67

[16] Ogbonna V E, Popoola A P I and Popoola O M 2022 Piezoelectric ceramic materials on transducer technology for energy harvesting: a review *Front. Energy Res.* **10** 1–7

[17] Vijatović M M, Bobić J D and Stojanović B D 2008 History and challenges of barium titanate: part II *Sci. Sinter.* **40** 235–44

[18] Vijatovi M M, Bobi J D and Stojanovi B D 2008 History and challenges of barium titanate : part II**40** 235–44
[19] Tewatia K, Sharma A, Sharma M and Kumar A 2020 Factors affecting morphological and electrical properties of barium titanate: a brief review *Mater. Today Proc.* **44** 4548–56
[20] Hiruma Y, Nagata H and Takenaka T 2003 Dielectric, ferroelectric and piezoelectric properties of barium titanate and bismuth potassium titanate solid-solution ceramics *J. Ceram. Soc. Japan* **1128** 1125–8
[21] Dai Y J, Zhang X W and Chen K P 2009 Morphotropic phase boundary and electrical properties of $K_{1-x}Na_xNbO_3$ lead-free ceramics *Appl. Phys. Lett.* **94** 1–4
[22] Koruza J, Liu H, Höfling M, Zhang M H and Veber P 2020 K,Na)NbO_3-based piezoelectric single crystals: growth methods, properties, and applications *J. Mater. Res.* **35** 990–1016
[23] Li H *et al* 2022 K, Na)NbO_3 lead-free piezoceramics prepared by microwave sintering and solvothermal powder synthesis *Solid State Commun.* **353** 114871
[24] Dubernet M *et al* 2022 Synthesis routes for enhanced piezoelectric properties in spark plasma sintered Ta-doped KNN ceramics *J. Eur. Ceram. Soc.* **42** 2188–94
[25] Wu J 2020 Perovskite lead-free piezoelectric ceramics *J. Appl. Phys.* **127** 190901
[26] Zhou C and Liu X 2007 Dielectric properties and relaxation of $Bi_{0.5}Na_{0.5}TiO_3$–$BaNb_2O_6$ lead-free ceramics *Bull. Mater. Sci.* **30** 575–8
[27] Reichmann K, Feteira A and Li M 2015 Bismuth sodium titanate based materials for piezoelectric actuators *Materials* **8** 8467–95
[28] HIRUMA T T Y and NAGATA H 2007 Grain-size effect on electrical properties of $(Bi_{1/2}K_{1/2})TiO_3$ ceramics *Jpn. J. Appl. Phys.* **46** 1–5
[29] Habib M, Lantgios I and Hornbostel K 2022 A review of ceramic, polymer and composite piezoelectric materials *J. Phys. D: Appl. Phys.* **55** 423002
[30] Harrison J S and Ounaies Z 2001 *Piezoelectric Polymers* NASA NASA/CR-2001-211422 ICASE report no. 2001-43
[31] Holterman J and Groen P 2018 An Introduction to Piezoelectric Materials and Components www.applied-piezo.com
[32] Lu L, Ding W, Liu J and Yang 2020 Flexible PVDF based piezoelectric nanogenerators *Nano Energy* **78** 105251
[33] Wang W *et al* 2019 Unexpectedly high piezoelectricity of electrospun polyacrylonitrile nanofiber membranes *Nano Energy* **56** 588–94
[34] Wankhade S H, Tiwari S, Gaur A and Maiti P 2020 PVDF–PZT nanohybrid based nanogenerator for energy harvesting applications *Energy Rep.* **6** 358–64
[35] Jain A, Kumar S J, Kumar M R, Ganesh A S and Srikanth S 2014 PVDF-PZT composite films for transducer applications *Mech. Adv. Mater. Struct.* **21** 181–6
[36] Pradhan S K *et al* 2020 Piezoelectric and mechanical properties of PVDF-PZT composite *Ferroelectrics* **558** 59–66
[37] Chamankar N, Khajavi R, Yousefi A A, Rashidi A and Golestanifard F 2020 A flexible piezoelectric pressure sensor based on PVDF nanocomposite fibers doped with PZT particles for energy harvesting applications *Ceram. Int.* **46** 19669–81
[38] Gregorio R, Cestari M and Bernardino F E 1996 Dielectric behaviour of thin films of β-PVDF/PZT and β-PVDF/$BaTiO_3$ composites *J. Mater. Sci.* **31** 2925–30
[39] Bairagi S and Ali S W 2019 A unique piezoelectric nanogenerator composed of melt-spun PVDF/KNN nanorod-based nanocomposite fibre *Eur. Polym. J.* **116** 554–61

[40] Banerjee S, Bairagi S and Ali S W 2022 A lead-free flexible piezoelectric-triboelectric hybrid nanogenerator composed of uniquely designed PVDF/KNN-ZS nanofibrous web *Energy* **244** 123102
[41] Bairagi S and Ali S W 2020 A hybrid piezoelectric nanogenerator comprising of KNN/ZnO nanorods incorporated PVDF electrospun nanocomposite webs *Int. J. Energy Res.* **44** 5545–63
[42] Mirzazadeh Z, Sherafat Z and Bagherzadeh E 2021 Physical and mechanical properties of PVDF/KNN composite produced via hot compression molding *Ceram. Int.* **47** 6211–9
[43] Bairagi S and Ali S W 2020 Flexible lead-free PVDF/SM-KNN electrospun nanocomposite based piezoelectric materials : signi fi cant enhancement of energy harvesting efficiency of the nanogenerator *Energy* **198** 117385
[44] Bairagi S and Ali S W 2020 Flexible lead-free PVDF/SM-KNN electrospun nanocomposite based piezoelectric materials: significant enhancement of energy harvesting efficiency of the nanogenerator *Energy* **198** 117385
[45] Sahoo R, Mishra S, Unnikrishnan L and Mohanty S 2020 Enhanced dielectric and piezoelectric properties of Fe-doped ZnO/PVDF-TrFE composite films *Mater. Sci. Semicond. Process.* **117** 105173
[46] Lee S *et al* 2014 Ultrathin nanogenerators as self-powered/active skin sensors for tracking eye ball motion *Adv. Funct. Mater.* **24** 1163–8
[47] Hong Y *et al* 2021 Highly anisotropic and flexible piezoceramic kirigami for preventing joint disorders *Sci. Adv.* **7** 1–11
[48] Guo H *et al* 2023 Cell activity manipulation through optimizing piezoelectricity and polarization of diphenylalanine peptide nanotube-based nanocomposite *Chem. Eng. J.* **468** 143597
[49] Zhou L *et al* 2022 Advances in applications of piezoelectronic electrons in cell regulation and tissue regeneration *J. Mater. Chem.* B **2** 8797–823
[50] Wang A *et al* 2018 Piezoelectric nanofibrous scaffolds as *in vivo* energy harvesters for modifying fibroblast alignment and proliferation in wound healing *Nano Energy* **43** 63–71
[51] Bhang S H *et al* 2017 Zinc oxide nanorod-based piezoelectric dermal patch for Wound healing *Adv. Funct. Mater.* **27** 1603497
[52] Kwak J, Kingon A I and Kim S H 2012 Lead-free $(Na_{0.5},K_{0.5})NbO_3$ thin films for the implantable piezoelectric medical sensor applications *Mater. Lett.* **82** 130–2
[53] İlik B *et al* 2017 Thin film PZT acoustic sensor for fully implantable cochlear implants *Proceedings* **1** 366
[54] Banerjee S, Bairagi S and Wazed Ali S 2021 A critical review on lead-free hybrid materials for next generation piezoelectric energy harvesting and conversion *Ceram. Int.* **47** 16402–21
[55] Sapkal S, Kandasubramanian B and Panda H S 2022 A review of piezoelectric materials for nanogenerator applications *J. Mater. Sci., Mater. Electron.* **33** 26633–77
[56] Kumar R, Singh A and Tiwari M 2020 Investigation of crack repair in orthotropic composite by piezoelectric patching *Mater. Today Proc.* **21** 1303–12
[57] Wu C M and Chou M H 2016 Sound absorption of electrospun polyvinylidene fluoride/graphene membranes *Eur. Polym. J.* **82** 35–45
[58] Wu C M and Chou M H 2020 Acoustic–electric conversion and piezoelectric properties of electrospun polyvinylidene fluoride/silver nanofibrous membranes *Express Polym. Lett.* **14** 103–14
[59] Wang Y, Zhu L and Du C 2021 Progress in piezoelectric nanogenerators based on pvdf composite films *Micromachines* **12** 1278

IOP Publishing

Recent Advances in Materials for Energy Harvesting and Storage

Suresh C Pillai, Daniel M Mulvihill and Aswathy Babu

Chapter 5

Supercapacitors

Yong Ma, Zhanhu Guo, Yuan Wang, Yinzhu Jiang, Daniel M Mulvihill, Chunjiang Jia, Sheng Dai and Ben Bin Xu

Nowadays, supercapacitors (SCs) are attracting significant attention due to their large specific capacitance, extended cycle life, high power and energy densities, and wide operating temperature range. Electrode materials, a crucial component of SCs, directly influence their electrochemical performance. Layered bimetallic hydroxides (LBHs) have garnered considerable interest owing to their distinctive two-dimensional (2D) layered morphology, straightforward preparation method, extensive specific surface area, and high theoretical specific capacitance. In addition, other electrode materials, such as MXenes and conductive polymers, have also exhibited promise in the field of SCs. In recent years, researchers have successfully synthesized LBHs using various methods and demonstrated the diverse properties arising from these approaches. Furthermore, recent modifications to LBHs have been proposed, including techniques for adjusting their composition and structure to achieve distinct properties. Meanwhile, different variants of LBHs, such as bilayer hydroxides, MXenes, and conductive polymers, have found widespread application in SCs, showcasing their exceptional electrochemical properties. Furthermore, the energy storage mechanisms of these electrode materials have been extensively investigated. In summary, LBHs and other electrode materials such as MXenes and conductive polymers have demonstrated significant potential for use in the realm of SCs. In this chapter, we primarily review recent developments in the preparation methods and performance modulation strategies of MXenes, conductive polymers, and LBHs. We also introduce the electrochemical properties and energy storage mechanisms of different MXenes, conductive polymers, and LBH variants. We hope that this review will serve as a valuable reference for the ongoing advancement of SCs based on these materials.

doi:10.1088/978-0-7503-5749-4ch5

5.1 Introduction

The rapid advancement of science and technology has heightened humanity's energy requirements. Over an extended period, the persistent utilization of petroleum, coal, and other fossil fuels has led to a slight shortfall in nonrenewable energy resources. Consequently, the disparity between societal progress and energy provision has grown increasingly pronounced. [1–5]. To effectively address the energy supply challenge, it is imperative to mitigate the adverse environmental effects of fossil fuel combustion and promote the sustainable well-being of the ecological environment [6–9]. The development of efficient energy storage devices is regarded as a strategic technology essential for the advancement of renewable energy sources, including metal-ion batteries [10–12], fuel cells [13, 14], phase change energy, solar cells, metal–air batteries [15–17] and SCs [18–20]. SCs have garnered significant attention among these promising devices due to their advantages, which encompass a wide operating temperature range, high power density, extended cycle life, and excellent recoverability [21–25]. The electrochemical performance of SCs is primarily linked to the properties and structures of their electrode materials [26–29].

Depending on the energy storage mechanism, electrode materials are primarily categorized into two distinct groups [30]. One of these categories includes carbon materials, which offer excellent power density due to their rapid physical processes, based on the principles of electrochemical double-layer capacitors [31–33]. However, due to the absence of rapid reversible reactions, they exhibit lower energy densities [34, 35]. The other category comprises pseudocapacitive materials that undergo faradaic charge transfer reactions at electrochemically active sites. This group includes materials such as transition-metal oxides [36, 37]. Examples of such pseudocapacitive materials are MnO_2 and RuO_2, both of which serve as effective electrodes due to their high theoretical capacitances of 1380 and 1200–2200 F g^{-1}, respectively [38, 39]. Nonetheless, pseudocapacitive materials suffer from poor cyclic stability, a limitation that significantly restricts their practical applications [40, 41]. Therefore, the search for a suitable electrode material with robust cycling stability that retains the advantages of excellent rate capacity and high specific capacity has become even more crucial [42, 43].

Due to their remarkable physical and chemical properties, 2D materials offer not only electrical double-layer capacitance through ion adsorption/desorption but also pseudocapacitance through faradaic redox reactions that occur on their surfaces [44–46]. Certain 2D materials are capable of simultaneously providing both electrical double-layer capacitance and pseudocapacitance [47–49]. Due to their unique features, such as good elasticity and excellent mechanical properties, they offer significant advantages in terms of load-bearing capacity and quality [50, 51]. Two-dimensional materials, including graphene, have emerged as crucial electrochemically active materials in SCs [52], metal–organic frameworks (MOFs) [53, 54], LBHs [55], and MXenes [24, 56].

MXenes are a vast class of 2D transition-metal nitrides or carbides; examples include Mo_2C [57], Nb_2C [58], Ti_2C [59], V_2C [60], Ti_3C_2 [46, 61], etc. Figure 5.1(a) shows the distribution of the main selected elements of MXenes in the periodic table.

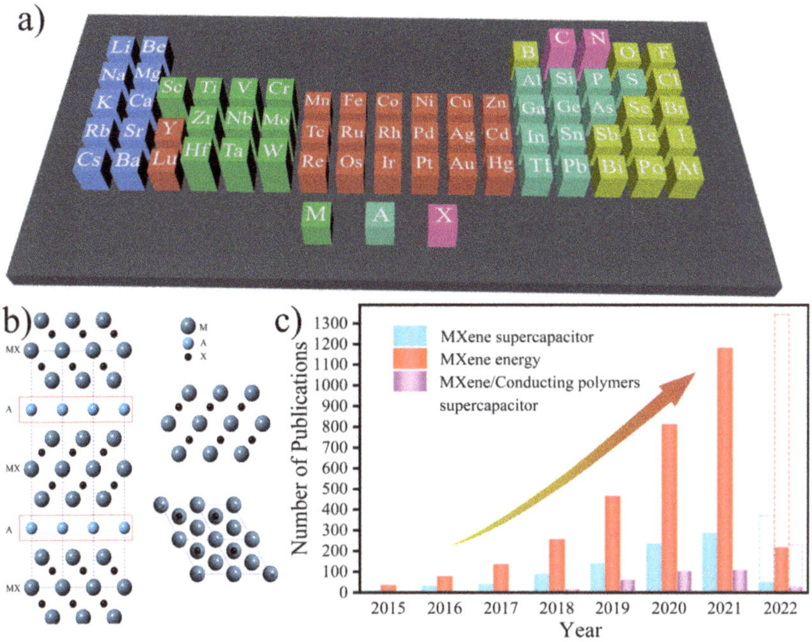

Figure 5.1. (a) A fragment of the periodic table showing the main elements of the MAX phase. (b) An illustration of the crystal structure of the MAX phase (reprinted with permission from [64], Copyright (2013), American Physical Society). (c) Annual publication volumes of MXene-related articles. (Source: Web of Science. Search index: MXene energy, MXene superconductor or MXene/conductive polymer superconductor.)

Their general molecular formula is $M_{n+1}AX_n$ (n=1, 2, 3), in which M, A, and X separately denote the transition metal, group III_A and IV_A elements, and nitrogen or carbon [62, 63]. In the densely packed M layer, X atoms occupy the central positions of octahedral structures, while the M and A layers are arranged alternately to constitute the MAX phase, as illustrated in figure 5.1(b) [64]. MXenes can be obtained by selectively etching the element A. This etching process is conducted in a solution, leading to the introduction of –O, –OH, or –F functional groups at the material's periphery [65]. Therefore, their general molecular formula can be expressed as $M_{n+1}X_nT_x$ (n=1, 2, 3) [66], in which M, X, and T denote the early transition metal, carbon or nitrogen, and surface terminating groups of –O, –OH, or –F, respectively [67].

Since the discovery of MXenes by Gogotsi *et al* [68] in 2011, MXene materials have garnered significant attention due to their substantial specific surface area and exceptional electrical conductivity. Figure 5.1(c) illustrates the number of papers reporting MXenes in the field of energy and SCs in recent years. The number of articles published annually exhibits a steady upward trend, indicating the increasing popularity of MXene research. Lukatskaya *et al* [69] stripped Ti_3C_2 with dimethyl sulfoxide and then adopted ultrasound to obtain few-layer Ti_3C_2 sheets. Following centrifugation, filtration, and drying processes, the self-supported 'Ti_3C_2 paper' electrodes had thicknesses ranging from 2 to 20 μm. In a KOH electrolyte, these 'paper' electrodes exhibited a volumetric capacitance of 450 F cm^{-3} at a scan rate of 2 mV s^{-1}, and they maintained a capacitance of 280 F cm^{-3} even at a higher scan rate of 100 mV s^{-1}. Ghidiu *et al* [70] etched Ti_3AlC_2 with hydrochloric acid and lithium fluoride and obtained clayey Ti_3C_2, then rolled the clayey Ti_3C_2 into thin sheets with a roller and directly made an electrode. In a 1 M H_2SO_4 electrolyte, the volumetric capacitance of these electrodes reached an impressive 900 F cm^{-3} at 2 mV s^{-1} and remained substantial at 730 F cm^{-3} even at a higher scan rate of 100 mV s^{-1}. The hydrophilicity, metallic conductivity, and surface redox reactions of MXenes play pivotal roles in the fabrication of high-performance electrode materials. However, due to the influence of van der Waals forces, adjacent MXene flakes tend to agglomerate or self-accumulate, resulting in a reduction in specific surface area and consequently impeding ion diffusion between the layers. This phenomenon significantly hampers the practical applicability of MXene-based electrodes [71, 72]. Fortunately, the integration of conductive polymers with MXene has the potential to enhance their electrochemical properties while also imparting superior mechanical characteristics [73–76].

The conductive polymers most commonly investigated for application in SC devices include polyaniline (PANI), polypyrrole (PPy), polythiophene (PTh), and poly(3,4-ethylenedioxythiophene) (PEDOT), which have good intrinsic conductivity [77–81]. Table 5.1 [82] shows the typical conductivity of different conductive polymers in the doped state. In contrast to conventional polymers, which have a bandgap of around 10 eV, conductive polymers exhibit a lower bandgap in the range of 1–3 eV. In addition, these polymers demonstrate specific doped/undoped behaviors, exhibit various morphologies, and possess notably rapid charge and discharge capabilities [83, 84].

Since 2015, the development of MXenes in the field of energy storage has experienced rapid growth, resulting in a substantial body of literature dedicated to SC research [85–91]. For example, Xu *et al* [87] discussed the use of MXenes for energy storage in sodium-ion batteries, comparing theoretical and experimental differences in electrochemical performance. Nan *et al* [85] summarized the use of MXenes in different energy storage devices and presented some challenges of MXenes for future energy storage. Wu *et al* [86] introduced the use of MXenes in potassium-ion batteries with some examples and explained the potassium storage mechanism of MXenes. Nasrin *et al* [92] elaborated the impact of improving 2D/2D MXene heterostructures on SC performance. The reviews of MXene applications in energy storage, such as SCs and batteries, are relatively numerous; however, there are relatively few reviews of the use of MXene/conductive polymer composites in SCs.

Table 5.1. The conductivities of typical conductive polymers [82].

Polymer	PANI	PPy	PTh	PEDOT
Conductivity (S cm^{-1})	0.1–5	10–50	300–400	300–500

Recently, there has been a surge of research interest, both domestic and international, in LBHs. This heightened attention is due to their adjustable composition, structure, and morphology [21, 93–98]. With their unique layered structure and anion exchange properties [74, 99–102], LBHs have been widely used in catalysis, adsorption, pharmacy, photochemistry, electrochemical energy storage equipment, and other fields [103–107]. LBHs are denoted by the general structural formula [108]: $M^{2+}_{1-X}M^{3+}_{X}(OH)_2A^{n-1}_{x/n}{\cdot}mH_2O$, where M^{2+} and M^{3+} represent divalent (e.g. M = Ni, Co, Cu, Mg, etc.) and trivalent (e.g. M = Al, Fe, Mn, Cr, etc.) metal ions, respectively [31, 109, 110], and A represents the interlayer anion separation (e.g. $CO_3{}^{2-}$, $NO_3{}^{-}$, Cl^-, etc.), which is used to balance the overall positive charge of LBH's main layer [111–114]. LBHs with different physicochemical properties are obtained by varying the molar ratio of M^{2+}/M^{3+} [115–117], the properties of the metal cations, the type of interlayer anions, and other conditions [118–123].

LBHs have garnered extensive and in-depth research attention within the field of electrochemistry [124–130]. Sarfraz *et al* [131] summarized and discussed the preparation method of LBHs and their latest uses in SCs. Zhao *et al* [132] focused on summarizing the relationship between various LBH composites, material structures, and electrochemical properties and the corresponding charge storage mechanism. Yan *et al* [133] briefly introduced the preparation method, structural design, chemical modification of SCs. They also discussed effective methods for improving SC performance and the future development of NiMn–LBH.

Pristine LBHs are also recognized as superior SC electrodes. This recognition is attributed to their layered structure and the ability to adjust interlayer spacing, resulting in a significant specific surface area and the facilitation of sufficient ion transport rates [93, 116, 134]. Nevertheless, pristine LBHs are afflicted by a substantial refilling effect, which can lead to a reduction in their overall exposed surface area and hinder the transport of electrolyte ions [135–137]. In addition, their low conductivity can impede the electron transfer process and limit the full utilization of active sites [138–140]. The ultrathin nanosheet structure of LBHs often results in rapid structural deterioration under demanding electrochemical conditions, ultimately impacting their performance and utility [141]. To address these limitations, researchers have developed hybrid LBH-based nanostructures, which hold promise as versatile nanomaterials. The key to enhancing their electrical conductivity and multifunctional properties lies in fabricating these materials with a substantial specific surface area and an increased number of active sites [119, 142–144]. The utilization of LBH materials as electrode materials for SCs primarily relies on their distinctive lamellar pore structure, which offers the requisite large specific surface area and a multitude of reactive sites essential for SCs. This development has evolved a new generation of environmentally friendly and efficient electrochemical functional materials.

To expedite access to recent research on MXenes, conductive polymers, and LBHs in the context of SCs, this chapter provides a comprehensive summary of recent studies, focusing on the utilization of these materials in SC applications. Specifically, this chapter centers on the application of 2D materials in SCs, namely

MXenes, LBHs, and composite materials that include conductive polymers. It encompasses discussions of the synthetic and preparatory methods of these materials as well as an exploration of advanced electrode materials. Furthermore, it delves into the applications of these materials in both symmetric SCs (SSCs) and asymmetric SCs (ASCs), providing detailed examples for each. In addition, it highlights current challenges in the development of MXenes, conductive polymers, and LBHs for SC applications. In conclusion, it offers insights into the future prospects of the use of MXenes, conductive polymers, and LBHs within the realm of SCs.

5.2 Synthetic methods

5.2.1 The preparation of MXenes

5.2.1.1 Hydrofluoric acid etching

There are dozens of known types of MXenes. Ti_3AlC_2 is taken as an example to explain the main preparation methods of MXene lamellas [63, 116]. As shown in figure 5.2(a) [145], the first reported method for preparing MXenes was to etch bulk Ti_3AlC_2 with hydrofluoric acid (HF). This preparation method requires cation intercalation to get layered Ti_3AlC_2 in a single layer or a few layers. However, there is often more than a single reaction in the etching process. The etching reaction of Ti_3AlC_2 in HF is expressed as follows [146]:

$$Ti_3AlC_2(s) + 3HF(aq) = AlF_3(a) + Ti_3C_2(s) + \frac{3}{2}H_2(g) \tag{5.1}$$

$$Ti_3C_2(s) + 2H_2O(aq) = Ti_3C_2(OH)_2(s) + H_2(g) \tag{5.2}$$

$$Ti_3C_2(s) + 2HF(aq) = Ti_3C_2F_2(s) + H_2(g). \tag{5.3}$$

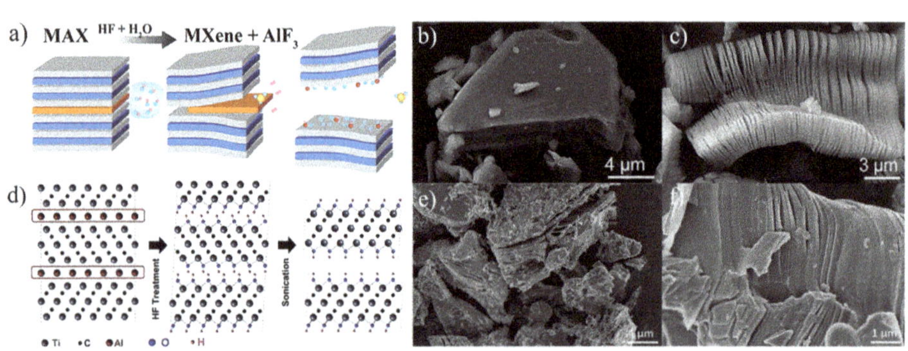

Figure 5.2. (a) An illustration of a layered Ti_3C_2 MXene synthesized by HF etching (reprinted with permission from [145], Copyright (2016), American Chemical Society). Scanning electron microscopy (SEM) images of (b) the MAX phase before treatment and (c) Ti_3AlC_2 after HF treatment. ((b) and (c) reprinted with permission from [147], Copyright (2012), American Chemical Society.) (d) An illustration of the Ti_3AlC_2 exfoliation process [148] John Wiley & Sons. Copyright © 2011 WILEY-VCH Verlag GmbH & Co. KGaA, Weinheim. (e),(f) SEM images of Ti_3AlC_2 after LiF–HCl treatment. ((e) and (f) Reprinted from [152], Copyright (2017), with permission from Elsevier.)

Equation (5.1) is an essential step in the formation of layered Ti_3C_2. Equations (5.2) and (5.5.3) are reactions that generate the corresponding surface terminals (–OH, –F) on the Ti_3C_2 lamellae, and the actual terminal is most likely a combination of –OH and –F. In the whole reaction, the solid dense Ti_3AlC_2 was treated with HF, and the Al atoms were then selectively stripped to form loosely stacked accordion-like $Ti_3C_2T_x$ (see figure 5.2(b) [147] and figure 5.2(c) [147]). The $Ti_3C_2T_x$ was then dispersed by ultrasound to form a single-layer or few-layer MXene. As shown in figure 5.2(d) [148], after the Al atoms were stripped, the exfoliated Ti_3C_2 layers had two naked Ti atoms per unit formula, and these were satisfied by suitable ligands. As this experiment was carried out in an aqueous solution, the system was rich in the ligands of fluorine ions, –OH and –F [148]. HF etching is a dynamic process. During the etching process, the etching conditions, including the HF dosage, reaction time, reaction temperature, etc. play roles that affect the resulting morphology and yield and the proportion of surface terminating groups of $Ti_3C_2T_x$ [149–151].

5.2.1.2 Fluoride-based salt etching

The synthesis of Ti_3C_2 through HF etching has gained widespread acceptance. However, this method involves the use of HF, a colorless, smoky, highly irritating, and highly corrosive liquid. Consequently, researchers have been actively exploring alternative, milder, and safer etching methods to replace the use of HF. In 2014, Ghidiu *et al* [70] were the first to etch Ti_3AlC_2 with a mixture consisting of LiF and HCl to obtain the MXene $Ti_3C_2T_x$. The resulting MXene had a clay-like plasticity after it had absorbed water, and could be rolled into a thin film with a thickness of tens of microns. The dried solid powder particles had high electrical conductivity. In addition, figures 5.2(e) and (f) [152] show the morphology of $Ti_3C_2T_x$ obtained after LiF–HCl etching the MAX phase. In the etching process, the use of a much gentler and environmentally friendly LiF–HCl etching agent to treat MAX facilitates the pre-intercalation of Li^+, which increases the MXene layer spacing and weakens the interlayer interaction [153, 154]. This approach facilitates MXene delamination during ultrasonic processing, resulting in the acquisition of few- or single-layer MXene flakes. It is worth emphasizing that LiF–HCl is among the most commonly employed *in situ* HF etching agents for the synthesis of high-quality MXenes [155]. Drawing inspiration from LiF–HCl etching, numerous other combinations of acids and fluoride salts, such as NaF, KF, and NH_4F, have been utilized. Notably, when H_2SO_4 is employed as a replacement for HF, it remains effective in etching $Ti_3C_2T_x$ to produce the desired outcome [70]. This phenomenon arises from the chemical reaction between the fluoride salt and the acid, leading to the generation of HF. Remarkably, within an aqueous solution of fluoride salt, a significant quantity of cations and H_2O molecules spontaneously infiltrates the interlayers of Ti_3C_2 sheets, thus producing an MXene that has surface modification and larger interlayer spacing [69, 156].

5.2.2 The preparation of conductive polymers

5.2.2.1 The hard template method

In the production of well-defined nanostructures, the hard template method stands out as a widely employed and effective approach [157]. The process unfolds as follows: (1) penetration or adsorption of monomers onto a template's surface or within its voids, (2) *in situ* reaction or curing of the dispersed monomers, and (3) subsequent removal of the template [158, 159]. Commonly utilized templates include silica nanomaterials, anodic aluminum oxide (AAO), orbital etching films, and various nonconductive polymer nanomaterials [160–163]. These templates typically manifest as porous cylinders or hollow structures. Consequently, the dimensions of the resultant structures are directly dependent on the pore size of the template. Their structure and length can be controlled by adjusting reaction conditions and varying monomer concentrations [164, 165]. This gives rise to the possibility of two distinct modes for nanostructured conductive polymers synthesized using the hard template method. One mode corresponds to a slow reaction rate with an ample supply of monomers, allowing for gradual monomer diffusion into the template, resulting in the formation of nanowire structures [159, 166]. Conversely, in the other scenario, under conditions of a rapid reaction rate and inadequate monomer supply, monomers swiftly migrate from the bulk solution to the pore wall due to interactions with the polymer. This leads to their polymerization and the formation of nanotube structures [167, 168]. Figure 5.3(a) [169] presents the preparation of conductive polymer nanostructures using porous membranes, nanofibers, and colloidal particles as hard templates.

Xue *et al* [170] discovered that PPy nanotube arrays could be grown inside an AAO template by lowering the template temperature and prepared samples using micro cold-wall vapor deposition technology. However, the traditional sacrificial template has some unavoidable disadvantages. For example, in sacrificial templates (such as AAO templates), the structure of the conductive polymers is affected by the morphology and shape of the electrode. Fortunately, in figure 5.3(b), Park *et al* [171] prepared PANI nanosheets by employing ice as a movable template [171]. Due to the low energy and orderly suspended –OH groups on the ice surface, the use of ice can effectively limit the growth of polymer films. In addition, the ice template removal process is simple, while ensuring the continuity and integrity of the film [172]. However, it is very difficult to prepare complex nanostructured conductive polymers by the hard template method due to the limited types of template structures available. Therefore, it is necessary to choose other preparation methods, such as the soft template method.

5.2.2.2 The soft template method

Compared to the hard template method, the soft template approach is relatively straightforward and is well suited for the large-scale synthesis of nanostructured conductive polymers. In the soft template method, colloidal materials, surfactants, and guided structural molecules are commonly employed to create micelles. These micelles, in turn, guide the growth of conductive polymers at the nanoscale, resulting in the formation of nanomaterials [157, 173, 174]. Figure 5.3(c) [169] illustrates the process of preparing nanotubes, nanospheres, and nanoarrays using micelles, nanowires, and monomer droplets as soft templates. It is important to note that the size and morphology of the nanostructures of conductive polymers produced through the soft template method are strongly influenced by various reaction

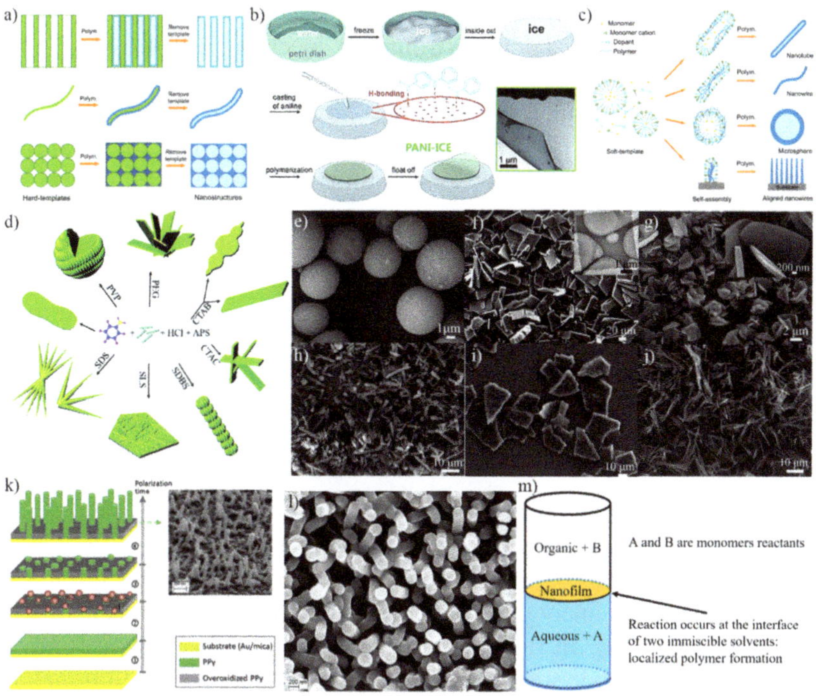

Figure 5.3. (a) An illustration of the preparation of conductive polymer nanostructures using porous membranes, nanofibers, and colloidal particles as hard templates (reproduced from [169], CC BY 4.0). (b) The preparation processes of 2D PANI nanosheets on ice surfaces (Inset: the transmission map of the complete 2D nanosheet) [171] John Wiley & Sons. © 2015 WILEY-VCH Verlag GmbH & Co. KGaA, Weinheim. (c) Illustration of the preparation of nanotubes, nanospheres, and nanoarrays using micelles, nanowires, and monomer droplets as soft templates (reproduced from [169], CC BY 4.0). (d) A schematic diagram of the preparation of PANI with different structures. SEM images of PANI with different structures, such as (e) spherical, (f) diamond plate-shaped, (g) flower-shaped, (h) cylindrical, (i) block-shaped, and (j) dendritic. ((d)–(j) Reproduced from [178] with permission from the Royal Society of Chemistry.) (k) A schematic diagram of the preparation of PPy nanowires by electrochemical polymerization (inset: SEM image of PPy nanowires). (l) An SEM image of a PPy nanowire from above. ((k) and (l) Reprinted from [186], Copyright (2018), with permission from Elsevier.) (m) A schematic of an interfacial polymerization reaction.

conditions. These conditions include the type and concentration of dopant, the concentrations of the oxidant and the monomer, the reaction temperature, etc [175–177]. These parameters can be used as a simple means to regulate the molecular structure and size [157, 158]. The most commonly employed surfactants feature hydrophilic heads and hydrophobic tails and are primarily composed of alkyl chains. Through a synergistic interaction between these molecules, these surfactants assemble into nanotube structures, which serve as templates that precisely regulate the growth of conductive polymers.

Soft templates with adjustable structures present an excellent opportunity for the fabrication of intricate three-dimensional (3D) nanostructures, a feat not easily achievable using hard template methods [158]. Ma *et al* [178] used this method to prepare different PANI nanostructures (figure 5.3(d)) [178]. In other words, at room temperature and in a weakly acidic environment, by changing the type of surfactant, PANI can be prepared with different structures such as spherical (figure 5.3(e)) [178], diamond plate-shaped (figure 5.3(f)) [57], flower-shaped (figure 5.3(g)) [178], cylindrical (figure 5.3(h)) [178], block-shaped (figure 5.3(i)) [178], branch-shaped (figure 5.3(j)) [178], cloud-shaped, rose-shaped structures, etc. These complex nanostructures are difficult to fabricate using hard templates.

Furthermore, the structures and properties of conductive polymers can be tailored by modifying the soft template materials and adjusting the reaction conditions. However, it is important to note that the inherent limitations of soft templates include their relatively poor stability and the challenge of achieving precise control over the final structure. In addition, some soft templates, such as tobacco mosaic virus, DNA, and sodium alginate, can be costly to produce when used to synthesize conductive polymers [158, 179, 180].

5.2.2.3 Electrochemical polymerization

Electrochemical polymerization is a dependable method for the fabrication of electrode-type conductive polymers. In this process, the thickness and morphology of the polymers can be finely controlled by adjusting the deposition rate and the applied charge [181–184]. In this method, doped conductive polymers are directly anchored onto the electrode surface without the requirement for additional catalysts [167, 185]. Nonetheless, conductive polymers synthesized solely through electrochemical polymerization often exhibit irregular structures [186]. To achieve the desired morphology, it is necessary to identify an appropriate template that complements the structure and size of the conductive polymer.

Numerous efforts have been undertaken to eliminate the use of templates. For instance, conductive polymer nanowires can be synthesized through electrochemical methods by applying bias electrodes in the presence of an aqueous monomer solution [187, 188]. This method involves the interaction of three components: the electrode, the conductor, and either another electrode or a target material [187]. As shown in figures 5.3(k) and (l) [186], Chouvy *et al* [186] used this method to prepare oriented PPy nanowires and found that as the solution contains a lower concentration of non-acidic anions and a higher concentration of weakly acidic anions, the size of the sample increases. Given that electrochemical reactions offer a high degree of control over the synthesis of conductive polymers and their nanostructures, the process of electrodeposition within the interelectrode channel has garnered significant interest for use in the fabrication of molecular devices [186, 187, 189].

5.2.2.4 Interfacial polymerization

In contrast to traditional polymerization in aqueous solutions, interfacial polymerization takes place within an organic and aqueous immiscible two-phase system. This method initiates the polymerization reaction with a reduced number of nucleation sites [190]. As shown in figure 5.3(m), interfacial polymerization is a templateless method in which dopant anions and a monomer are used at the liquid/liquid interface to facilitate the aggregation of monomer anions. This method can synthesize conductive polymers films with high specific surface area or particular electrochemical properties [190, 191]. Conductive polymer forms at the interface and rapidly diffuses into the aqueous phase, allowing the interface to react further. The polymerization rate is affected by the concentrations of the monomer and the oxidant, and the types of nanostructures obtained depend on the organic solvent, acid dopant, reaction time, and other factors [192]. For instance, the length of PANI nanofibers increases as the acid strength increases [158, 190]. The interfacial polymerization method displays many advantages: (1) synthetic and purification processes are carried out without any templates, (2) this method has high yield and is easily expanded and repeated, and (3) the obtained product is easily dispersed in the aqueous phase, facilitating biological applications and environmentally friendly processes in the environment [158, 190, 193].

5.2.3 The preparation of MXene/conductive polymer composites

MXene, as a high-performance electrode material, holds significant potential for applications in the field of SCs [46, 194]. Nevertheless, akin to other 2D materials, MXene thin films are susceptible to restacking, leading to a decrease in ion transport within the electrode. This phenomenon significantly impacts the specific capacitance and rate capability of the individual electrode [195, 196]. To address this challenge, extensive research efforts have concentrated on designing MXene electrode structures and enhancing their ion transport characteristics. These efforts encompass the development of microporous and hydrogel $Ti_3C_2T_x$ thin films as well as the design of composite electrodes featuring $Ti_3C_2T_x$ and carbon nanomaterials [197–199]. While certain strategies have proven effective in enhancing ion transport and the high-speed electrochemical performance of electrodes, it is worth noting that some of these approaches may lead to a reduction in electrode density. This can result in decreased electrode volume capacitance, rate capability, and energy density [200–202]. Recent studies have demonstrated that the deposition of conductive polymers, such as PANI, PPy, and PEDOT, onto the surface of MXene sheets can enhance the electrochemical properties of standalone MXene electrodes [203–206].

5.2.3.1 The preparation of MXene/PANI composites

Xu *et al* [207] came up with a new strategy for producing MXene/PANI composites by chemically oxidizing polymerizing aniline monomers on $Ti_3C_2T_x$ MXene nanosheets under acidic conditions. The surface of MXene nanosheets is rich in –O, –F, and –OH functional groups, which can provide nucleation sites for the deposition of aniline on the surface of the MXene [208]. Through the gradual addition of ammonium persulfate (APS), the aniline monomer underwent continuous polymerization on the functional groups present on the surface of MXene. This process ultimately resulted in the formation of irregular porous PANI structures on the MXene nanosheets, creating pathways for the transport of electrolyte ions. Consequently, the inclusion of these additional ion transport channels led to improved electron transfer efficiency and enhanced electrochemical activity in the MXene/PANI composite. In addition, Zhao *et al* [209] used the method of *in situ* polymerization at low temperature to modify PANI nanoparticles on the surface or between MXene nanosheets, as shown in figure 5.4(a) [209]. One can see from figure 5.4(b) [209] that the PANI nanoparticles in the MXene/PANI composite are relatively regular and distributed relatively uniformly. This method effectively reduced the damage to MXene nanosheets and ensured the polymerization quality of PANI as far as possible [210, 211].

Beidaghi *et al* [203] reported a method for preparing MXene/PANI composites by oxidant-free polymerization. Their method used aniline monomers to synthesize MXene/PANI composites on the periphery of MXene nanosheets through *in situ* polymerization without the use of additional oxidants, as shown in figure 5.4(c) [203]. The amount of polymer deposition on the periphery of MXene nanosheets is a key parameter affecting the electrochemical performance of the electrodes. A cross-sectional scanning electron microscopy (SEM) image of the nanosheets (figure 5.4(d)) [203] shows that MXene and PANI nanoparticles are closely combined to form a dense interlayer structure with pores. By minimizing the quantity of PANI deposited on the surface of the MXene layer, it is possible to create an MXene/PANI composite electrode that exhibits a mass capacitance of 503 F g^{-1} and a volume capacitance of 1682.3 F cm^{-3} while maintaining its electrochemical performance. This composite approach effectively manages the PANI content, resulting in an electrode with enhanced ion transport capacity, improved electrochemical performance, and exceptional cycling stability (with a capacitance retention rate of 98.3% after 10 000 cycles). Moreover, compared to previously reported $Ti_3C_2T_x$ electrodes, this method enables the preparation of high-quality, thicker MXene/PANI electrodes without compromising their electrochemical performance.

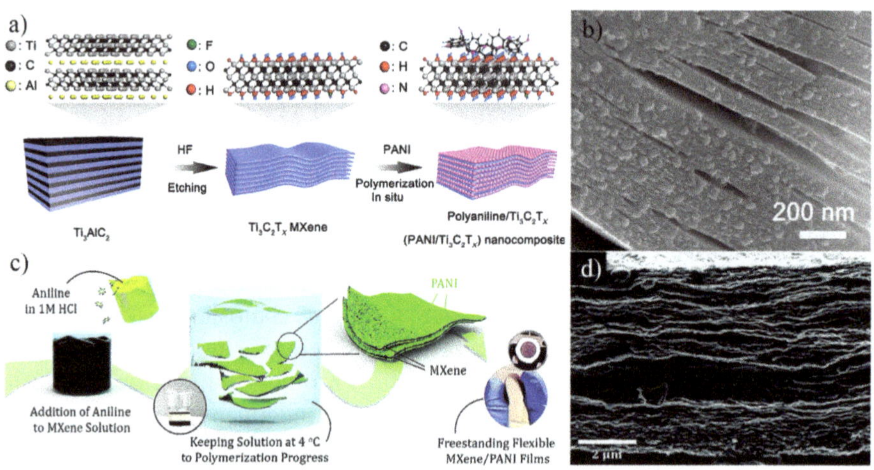

Figure 5.4. (a) A diagram of the preparation of PANI/$Ti_3C_2T_x$ by *in situ* polymerization at low temperature. (b) An SEM image of the MXene/PANI composite. ((a) and (b) reproduced from [209], CC BY 4.0.) (c) A diagram of the preparation of PANI/MXene by oxidant-free polymerization. (d) A cross-sectional SEM image of MXene/PANI films. ((c) and (d) Reproduced from [203] with permission from the Royal Society of Chemistry.)

5.2.3.2 The preparation of MXene/PPy composites

As shown in the mechanism diagram of figure 5.5(a) [212], Jian *et al* [212] used a one-step coelectrodeposition route to prepare MXene/PPy composite films on indium tin oxide (ITO) glass. With MXene nanosheets as the core for polymerization, pyrrole monomers gradually polymerize on the surface or between the MXene nanosheets to obtain a 3D carambola-like MXene/PPy composite film, as shown in figure 5.5(b) [212]. The incorporation of PPy, which exhibits pseudocapacitance, can yield a higher capacitance than that of the pure MXene. Simultaneously, the inclusion of $Ti_3C_2T_x$ mitigates the expansion issues typically associated with PPy during charging and discharging processes [213, 214]. It is worth

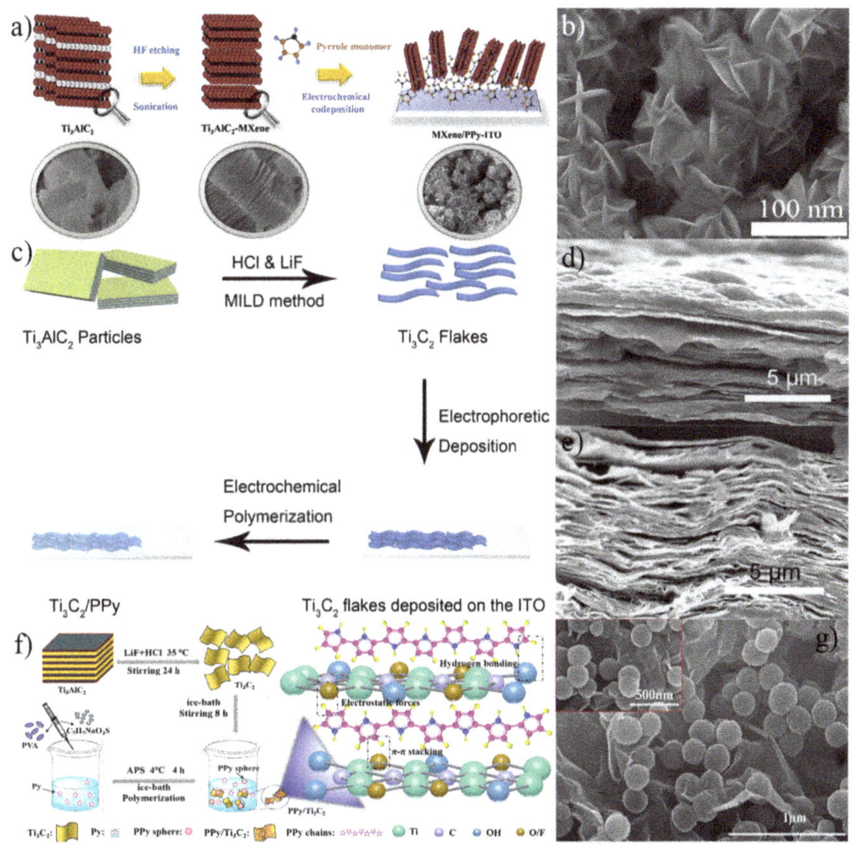

Figure 5.5. (a) A diagram of the mechanism of MXene/PPy composite preparation. Inset: SEM images of Ti_3AlC_2, Ti_3AlC_2–MXene, and MXene/Ppy. (b) An SEM image of the 3D starfruit-like MXene/PPy composite. ((a) and (b) reprinted from [212], Copyright (2019), with permission from Elsevier.) (c) A flowchart of the preparation of the $Ti_3C_2T_x$/PPy composite membrane by electrochemical polymerization. SEM images of the hybrid $Ti_3C_2T_x$/PPy membrane (d) and the pure $Ti_3C_2T_x$ film (e). ((c)–(e) reprinted from [215], Copyright (2020), with permission from Elsevier.) (f) A schematic diagram of the synthesis of the PPy/Ti_3C_2 nanocomposite. (g) An SEM image of the PPy/Ti_3C_2 composite. ((f) and (g) Reprinted from [216], Copyright (2020), with permission from Elsevier.)

noting that PPy intercalates the $Ti_3C_2T_x$ nanosheets, increasing the interlayer spacing of $Ti_3C_2T_x$ and further improving the electrochemical performance of the material. In addition, Zhang *et al* [215] fabricated Ti_3C_2/PPy film via HCl-LiF *in situ* etching, electrophoretic deposition, and electrochemical polymerization (figure 5.5(c)) [215]. From a comparison of figures 5.5(d) and (e) [215], it can be clearly seen that the cross-sections of the PPy film and the Ti_3C_2/PPy film are significantly different. At a current density of 1 mA cm^{-2}, the planar SCs fabricated using Ti_3C_2/PPy composite films exhibited areal capacitances of 109.4 mF cm^{-2} in a 2 M H_2SO_4 solution and 86.7 mF cm^{-2} when used with a polyvinyl alcohol (PVA)/H_2SO_4 solid electrolyte. These values were achieved due to the synergistic effects of Ti_3C_2 and PPy.

Wu *et al* [216] prepared PPy/Ti_3C_2 MXene heterostructure nanocomposites by one-step *in situ* polymerization, as shown in figure 5.5(f) [216]. Starting from a uniformly dispersed Ti_3C_2 nanosheet suspension, the pyrrole monomer was *in situ* polymerized to generate PPy nanospheres on the periphery of the MXene sheet layers to generate a PPy/Ti_3C_2 MXene heterostructure nanocomposite material. As can be seen in figure 5.5(g) [216], PPy and Ti_3C_2 formed a heterogeneous composite material with a regular particle size and a uniform distribution. The heterogeneous structure of PPy/Ti_3C_2 MXene composites provides many active sites for charge or ion transfer, accelerates the rapid accessibility and penetration of ions in the electrolyte, and reduces the inherent resistance and charge transfer resistance [217, 218], improving the electrochemical properties significantly.

Furthermore, MXene nanosheets demonstrate favorable hydrophilicity due to the presence of –F, –O, and –OH groups on their surface. These functional groups offer additional nucleophilic reaction sites for the polymerization of pyrrole monomers [219]. MXene nanosheets and the PPy framework form a strong bond through advantageous π–π conjugation and electrostatic forces. This effective fusion between the ultraflat Ti_3C_2 layers and PPy nanospheres prevents the undesired spontaneous agglomeration and accumulation of the Ti_3C_2 layers [216].

5.2.3.3 The preparation of MXene/PEDOT composites

Zhu *et al* [220] prepared a flexible MXene/PEDOT–poly(styrenesulfonate) (PEDOT:PSS) film using the vacuum-assisted filtration method, as shown in figure 5.6(a) [220]. In this method, the prepared MXene solution was mixed with a PEDOT:PSS dispersion in proportion, and an MXene/PEDOT:PSS composite film (figure 5.6(b)) [220] was obtained by vacuum filtration. To improve the conductivity of the composite film, the MXene/PEDOT:PSS composite film was treated with H_2SO_4 to remove the nonconductive PSS, and a flexible and independent high-performance MXene/PEDOT composite film (figure 5.6(c)) [220] was obtained. Upon comparing the two cross-sectional SEM images, it becomes evident that the regularity of the composite membrane was enhanced by treating it with concentrated H_2SO_4. This improved regularity contributed to an enhancement in its performance. As shown in figure 5.6(d) [220], it can be seen from the stress–strain curve that after the addition of a small quantity of PEDOT:PSS and H_2SO_4, the tensile strength of the film was significantly improved. In addition, compared with pure MXene film and other

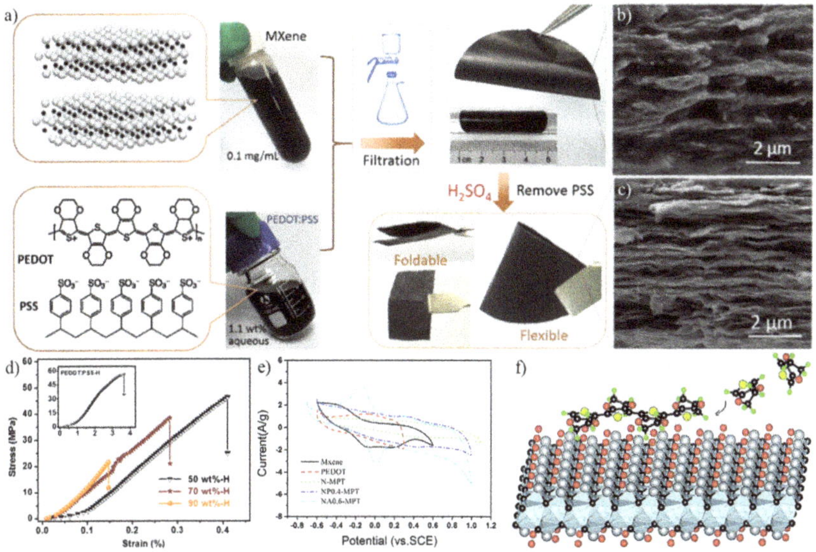

Figure 5.6. (a) A flowchart of the preparation of the MXene/PEDOT:PSS composite membrane by the vacuum-assisted filtration method. An SEM image of the MXene/PEDOT:PSS composite membrane (b) and an MXene/PEDOT composite membrane treated with concentrated H_2SO_4 (c). (d) The stress–strain curves of MXene/PEDOT:PSS composite films with different mass ratios. ((a)–(d) reprinted from [220], Copyright (2020), with permission from Elsevier.) (e) Cyclic voltammetry (CV) curves of the obtained samples at 10 mV s^{-1} (reprinted from [224], Copyright (2018), with permission from Elsevier). (f) An illustration of polymerizing 3,4-ethoxylene dioxy thiophene (EDOT) on the surface of an MXene (reproduced from [226] with permission from the Royal Society of Chemistry).

MXene/polymer composite films, the MXene/PEDOT:PSS composite film had greatly enhanced mechanical performance and maintained good electrical conductivity [221–223]. This method suggests ways to prepare lightweight, flexible, and high-performance MXene/PEDOT composites with high mechanical strength.

Chen *et al* [224] prepared 1,5-naphthalenedisulfonic acid-doped MXene/PEDOT (MPT) composites by *in situ* chemical oxidation. NP-MPT and NA-MPT composites were prepared by codoping MPT with phosphomolybdic acid and anthraquinone-2-sulfonate. Notably, NA-MPT exhibited a well-defined layered composite structure, boasting an impressive voltage window of up to 1.8 V and a specific capacitance of 323 F g^{-1}. It is worth highlighting that this specific capacitance value is more than twice those of both pure PEDOT and pure MXene, as illustrated in figure 5.6(e) [224]. This codoping method can be used as a basis for other MXene/polymer composite applications [21, 225].

Gogotsi *et al* [226] imitated the preparation of MXene/PPy composites by *in situ* polymerization without oxidants to prepare MXene/PEDOT composites, as shown in figure 5.6(f) [226]. In contrast to conventional PEDOT polymerization, while the degree of oxidation of EDOT may not be exceedingly high, it is nonetheless adequate to create the conditions necessary for PEDOT to undergo complete doping polymerization when combined with an MXene [227, 228]. This method eliminates the requirement for any oxidizing agent. It simply involves mixing the etched $Ti_3C_2T_x$ MXene solution with an aqueous EDOT solution to facilitate the polymerization of EDOT on the periphery of the MXene sheets [229, 230]. In comparison to MXene and pure PEDOT, the composite material demonstrates excellent cycle stability and rate performance. This effective synergy between $Ti_3C_2T_x$ and PEDOT enhances both ion storage capacity and charge storage capacity [231–234].

5.3 Electrodes based on MXene/conductive polymers

The performance of SCs is affected by many factors, such as the electrode material [235, 236], the choice of electrolyte [237], and the size of the potential window [238, 239], among which the electrode material is a key factor affecting the performance of the capacitor [240, 241]. Therefore, a lot of effort has been invested in the research and development of logical structural designs for electrode materials that promote effective electron transmission and ion diffusion in SCs [239, 242]. Advanced electrode materials must possess a range of key attributes, including temperature stability, high conductivity, a substantial specific surface area, robust chemical stability, corrosion resistance, and environmental friendliness. In addition, they should be cost-effective [91, 243, 244]. At present, the most researched electrode materials mainly include carbon-based materials [245], conductive polymers [246–248], and transition-metal oxides [249, 250]. This chapter focuses on composite electrodes made from MXenes and conductive polymers and their application in SCs.

5.3.1 MXene-based electrodes

Since their inception in 2011, MXenes have found extensive application in various studies, which has been accompanied by ongoing discoveries of new MXene variants. To date, a substantial portion of MXene research has centered on $Ti_3C_2T_x$, which is attributed to its outstanding electrical conductivity of 6000–10 000 S cm^{-1} and mechanical stability [149, 251, 252]. Yan *et al* [200] prepared a standalone and binderless MXene film electrode with excellent volume performance by vacuum-assisted filtration and the assembly of standalone membrane technology. This MXene film electrode is known for ultrahigh density and an excellent pseudocapacitance energy storage mechanism, showing a volume capacitance of 1222 F cm^{-3} [200]. Nonetheless, this approach has notable limitations and is only viable at low scan rates. The volumetric capacitance of MXene films experiences significant degradation at high scanning rates or under rapid charge and discharge conditions [65, 253, 254]. Li *et al* [255] prepared a flexible suspended $Ti_3C_2T_x$ MXene film with a wave structure (figure 5.7(a)) [255] using a mechanical compression method. The wavy structure mitigated the issue of self-stacking to some extent. In addition, this structure served as a stable conduit for the rapid and efficient transfer of ions and charges. The compact structure of the $Ti_3C_2T_x$ MXene film electrode delivered an exceptionally high volumetric capacitance, reaching 1277 F cm^{-3} at 10 mV s^{-1} in 3 M H_2SO_4. Impressively, even at a high scanning rate

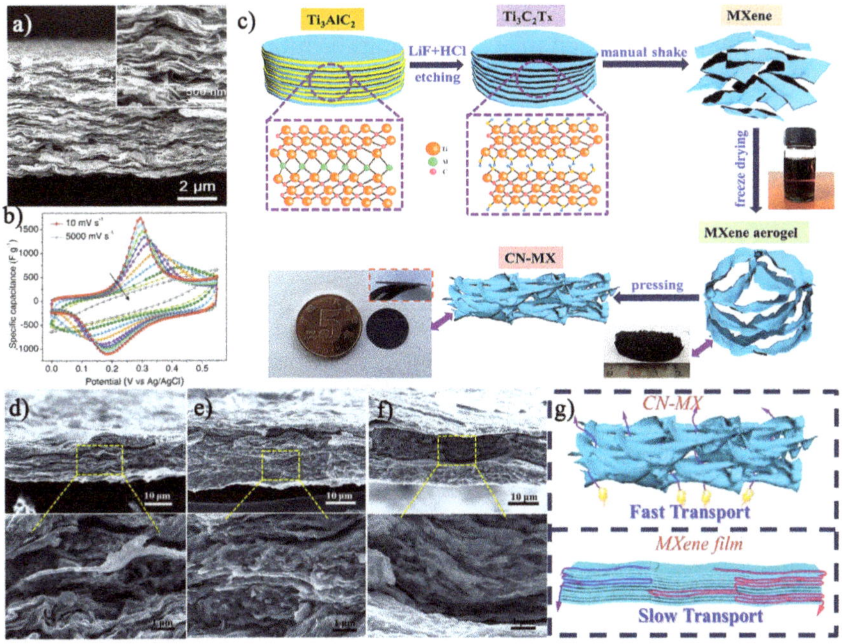

Figure 5.7. (a) An SEM image of the wavy MXene. (b) The CV curve of the $Ti_3C_2T_x$ MXene film electrode tested in the range of 10 to 5000 mV s^{-1}. ((a) and (b) reprinted from [255], Copyright (2020), with permission from Elsevier.) (c) A schematic diagram of the preparation process used for the MXene film. SEM images of cross-sections of films prepared under different mechanical pressures: (d) CN–MX_{20}, (e) CN–MX_{40}, and (f) CN–MX_{50}. (g) A schematic diagram of the ion diffusion of CN–MX and MXene films. ((c)–(g) Reprinted with permission from [256], Copyright (2020), American Chemical Society.)

of 5000 mV s^{-1}, it maintained a capacitance of 790 F cm^{-3}, as depicted in figure 5.7(b) [255].

Fan *et al* [256] were inspired by porous graphene film electrodes and processed 3D MXene aerogels into dense porous MXene films by mechanical pressing, as shown in figure 5.7(c) [256]. From the cross-sectional scans of figures 5.7(d)–(f) [256], it can be seen that when the mechanical pressure increased, the structure of the MXene film became denser and the sample layer arrangement became more regular, which was more conducive to improving the volumetric capacitance. According to the reports, at 500 mV s^{-1}, the MXene film obtained at a mechanical pressure of 40 MPa still had a volume capacitance of 462 F cm^{-3}, while the value of the traditional MXene film was reduced to 395 F cm^{-3} [256]. The folded structure of MXene nanosheets combined to form a tightly connected pore structure, which greatly increased the accessibility of active sites for electrolyte ions, thereby improving transport efficiency and the consequent electrochemical storage capacity, as shown in figure 5.7(g) [256].

Furthermore, this compact MXene film with a folded morphology is self-supporting and does not require adhesives, making it suitable for direct application as the working electrode. In contrast to conventional MXene films produced through vacuum-assisted filtration, it exhibits significantly enhanced rate performance and volumetric capacitance, rendering it more conducive to large-scale production [256]. Nonetheless, it is important to note that the film's abundant pore structure results in lower density and conductivity compared to those of traditional MXene films. Consequently, this contributes to a volumetric capacitance of only 932 F cm^{-3} at low scan rates [203]. MXene has natural advantages for SCs, but the application of independent membranes is very limited due to performance problems [257].

5.3.2 Binary composite electrodes made from MXenes and conductive polymers

5.3.2.1 MXene/PANI composite electrodes

Wu *et al* [258] successfully prepared organ-like amino-Ti_3C_2 (N-Ti_3C_2)/PANI composites using a two-step electrochemical route. As shown in figure 5.8(a) [258], N-Ti_3C_2 was first deposited or coated onto an FTO glass substrate through an electrochemical reaction. The ordered structure of the Ti_3C_2 MXene was then used as the carrier. Using a constant voltage, the PANI chain was connected to the FTO glass substrate through electrochemical polymerization. From the SEM image of figure 5.8(b) [258] and the energy-dispersive X-ray spectroscopy (EDS) spectra of figures 5.8(c)–(e) [258], it can be seen that the N-Ti_3C_2 and PANI were effectively combined. Figure 5.8(f) [258] shows the special binding used between the N-Ti_3C_2 and PANI, which is different from the usual direct binding of TI_3C_2 to PANI. The amine nitrogen on the PANI chain and the amino group carried by N-Ti_3C_2 are tightly

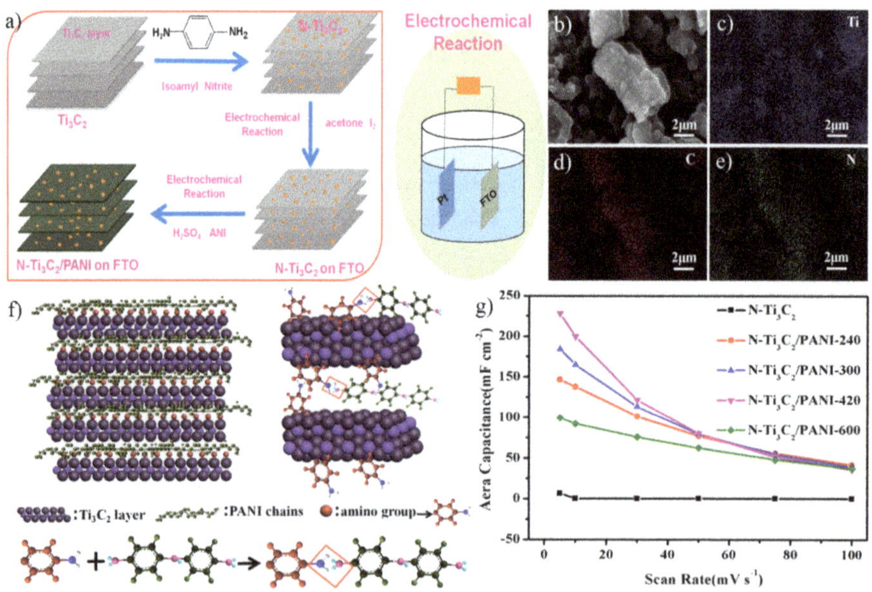

Figure 5.8. (a) A schematic diagram of the preparation of N-Ti_3C_2/PANI. (b) An SEM image of N-Ti_3C_2/PANI. EDS spectra: (c) Ti, (d) C, and (e) N. (f) An atomic schematic diagram of PANI intercalation in N-Ti_3C_2. (g) The specific capacitances of pure N-Ti_3C_2 and N-Ti_3C_2/PANI electrodes from 5 to 100 mV s^{-1}. ((a)–(g) Reprinted from [258], Copyright (2019), with permission from Elsevier.)

combined through chemical bonds, which increase the spacing and accessible surface area of the Ti_3C_2 MXene, effectively preventing the self-stacking of MXene sheets.

Furthermore, the organ-like N-Ti_3C_2/PANI composites formed through covalent grafting, established a precise and rapid channel for the transfer of charges and ions, significantly enhancing the charge transfer rate of the composites. This unique structure and bonding method contributed to the excellent electrochemical performance of N-Ti_3C_2/PANI. Specifically, in a 0.5 M H_2SO_4 electrolyte solution at 5 mV s^{-1}, N-Ti_3C_2/PANI exhibited exceptional performance, achieving a maximum surface capacitance of 228 mF g^{-1}. Notably, this capacitance was 32 times higher than that of the original Ti_3C_2 film, as depicted in figure 5.8(g) [258]. In addition, the N-Ti_3C_2/PANI composite electrode showed an 85% capacitance retention rate after 1000 cycles.

Li *et al* [259] fabricated an MXene/PANI material covered with PANI chains via the hydrothermal method. The MXene served as the supporting framework, while PANI served to link adjacent MXene layers. In the fabricated composites, the MXene enhanced flexibility and established a 3D network structure. The incorporation of PANI expanded the interlayer spacing of MXene, facilitating the rapid transfer of ions and charges [260]. In a 6 M KOH solution, the MXene/PANI composite electrode displayed a mass specific capacitance of 563 F g^{-1} at 0.5 A g^{-1}, which was almost 2.3 times that of pure MXene (figures 5.9(a) and (b) [259]). In addition, at 5 A g^{-1}, the capacitance retention rate after 10 000 cycles was as high as 95.15%.

Chen *et al* [261] used a simple chemical oxidation polymerization method to study a new type of MXene/PANI composite. This method was used to synthesize a high-performance SC electrode, as shown in figure 5.9(c) [261]. In this approach, the dopant DL tartaric acid (DLTA) was introduced into a $Ti_3C_2T_x$ MXene solution and thoroughly mixed. Subsequently, supramolecular self-assembly technology was employed to assemble DLTA around the periphery of the MXene. During the supramolecular self-assembly process, the abundant electronegative oxygen-containing functional groups promoted the orderly polymerization of aniline monomers, resulting in the formation of well-structured $Ti_3C_2T_x$ MXene–DLTA/PANI (TDP) self-assembled composites, as depicted in figure 5.9(d) [261].

The achievement of supramolecular self-assembly predominantly hinges on the chiral interaction of the MXene in conjunction with the flexible spatial distribution of DLTA, which possesses numerous negatively charged oxygen-containing functional groups such as –OH and –COOH. As can be seen from figure 5.9(e) [261], in 1 M of H_2SO_4 electrolyte solution, a TDP electrode material prepared with a mass ratio of MXene to aniline of 2:8 exhibited a specific capacitance of 452 F g^{-1} at 1 A g^{-1}, much higher than those of the pristine MXene (61 F g^{-1}) and PANI (263 F g^{-1}). In addition, the TDP electrode had a voltage window of up to 1.9 V at 4 A g^{-1}, and it still had a capacitance retention rate of 61% after 2000 cycles, as shown in figure 5.9(f) [261].

5.3.2.2 MXene/PPy composite electrodes

Wu *et al* [262] prepared organ-like $Ti_3C_2T_x$/PPy nanocomposites by low-temperature chemical oxidation. In this process, a pyrrole monomer was polymerized *in situ* on $Ti_3C_2T_x$ nanosheets at low temperature to form clear and uniformly dispersed

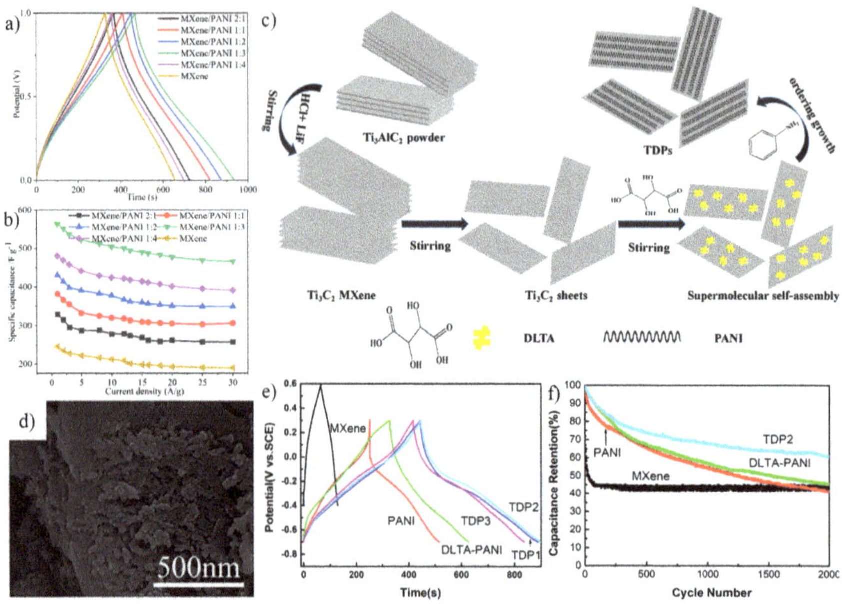

Figure 5.9. (a) Galvanostatic charge–discharge (GCD) curves and (b) the specific capacitances of pure MXene and MXene/PANI electrodes with different mass ratios at 0.5 A g^{-1}. ((a) and (b) reprinted from [259], Copyright (2021), with permission from Elsevier.) (c) A schematic diagram of the preparation procedure of $Ti_3C_2T_x$ MXene–DL tartaric acid/PANI (TDP). (d) An SEM image of TDP2 (when the mass ratio of MXene/aniline is 2:8, the obtained samples are respectively named TDP2). (e) GCD curves of MXene, PANI, DL tartaric acid–PANI and TDPs at 1 A g^{-1} and (f) cyclic curves of MXene, PANI, DL tartaric acid–PANI and TDPs at 4 A g^{-1}. ((c)–(f) Reprinted with permission from [261], Copyright (2020), American Chemical Society.)

PPy nanoparticles, as shown in figure 5.10(a) [262] and figure 5.10(b) [262]. The $Ti_3C_2T_x$ nanosheets, which served as a polymeric scaffold, effectively inhibited the growth of PPy, prevented PPy self-stacking, and enhanced the structural stability of the $Ti_3C_2T_x$/PPy composites. The composite material produced through this method relied primarily on a combination of hydrogen bonding and electrostatic forces between the $Ti_3C_2T_x$ nanosheets and the PPy chains. In addition, the intercalation effect of uniform PPy nanoparticles expanded the interlayer spacing of the $Ti_3C_2T_x$ nanosheets. Simultaneously, the highly oriented polymer molecular chains created more pathways for charge transfer and the diffusion of electrolyte ions, consequently increasing the specific capacitance and reducing the charge transfer resistance. Notably, as depicted in figure 10(c) [262], the $Ti_3C_2T_x$/PPy composite electrode with the best ratio showed a specific capacitance of 184.36 F g^{-1} at 2 mV s^{-1}, 37% higher than that of a pure $Ti_3C_2T_x$ MXene electrode (133.91 F g^{-1}). At 1 A g^{-1}, the capacitance of the $Ti_3C_2T_x$/PPy composite electrode remained 83.33% after 4000 charge–discharge cycles (figure 5.10(d)) [262]. This enhancement of the material's electrochemical performance and cycle stability can be attributed to the synergistic effects between $Ti_3C_2T_x$ nanosheets and PPy nanoparticles as well as the utilization of different energy storage mechanisms. Importantly, this method provides a

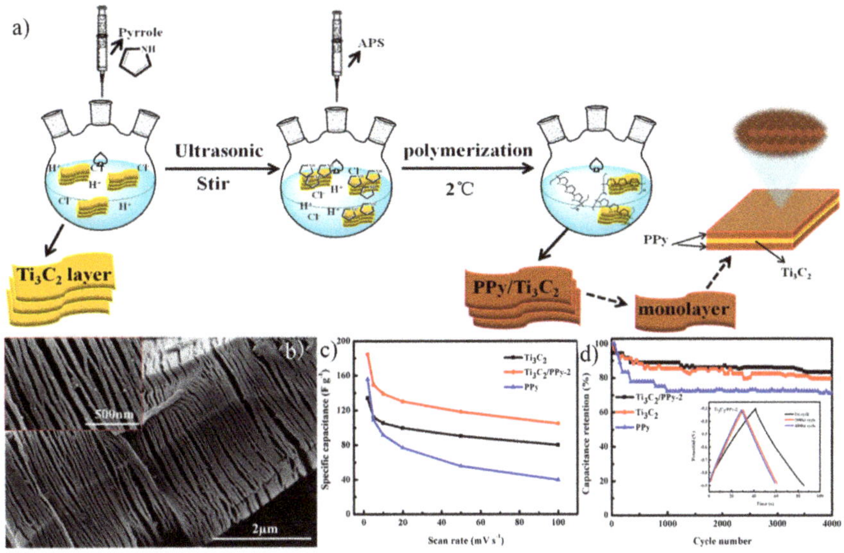

Figure 5.10. (a) A schematic illustration of the preparation of $Ti_3C_2T_x$/PPy composites through the low-temperature *in situ* polymerization of PPy on Ti_3C_2 nanosheets. (b) An SEM image of $Ti_3C_2T_x$/PPy (the inset shows a partially enlarged view). (c) The specific capacitances of PPy, Ti_3C_2, and Ti_3C_2/PPy at different scanning rates. (d) Cycle curves of PPy, Ti_3C_2, and Ti_3C_2/PPy at 1 A g^{-1}. ((a)–(d) Reprinted from [262], Copyright (2019), with permission from Elsevier.)

cost-effective and straightforward approach for the large-scale preparation of $Ti_3C_2T_x$/PPy composites.

Tong *et al* [263] successfully prepared an independent $Ti_3C_2T_x$/PPy composite film electrode by electrochemically depositing PPy on a $Ti_3C_2T_x$ film prepared using the vacuum-assisted filtration method (figure 5.11(a)) [263]. The $Ti_3C_2T_x$/PPy composite film had a special structure in which PPy was formed on the periphery of the $Ti_3C_2T_x$ film and inserted into the gap between the $Ti_3C_2T_x$ nanosheet layers, as shown in figures 5.11(b)–(e) [263]. This unique structure resulted in a robust and effective binding between the $Ti_3C_2T_x$ and the PPy. The favorable combination of $Ti_3C_2T_x$ and PPy enhanced the tensile strength of the $Ti_3C_2T_x$/PPy film to 48.2 MPa, significantly surpassing that of the original $Ti_3C_2T_x$ electrode (9.9 MPa). Furthermore, owing to this distinctive structure and the synergistic effects of its various components, the optimized $Ti_3C_2T_x$/PPy film electrode exhibited a specific capacitance of 420.2 F g^{-1} at 1 A g^{-1}, which was notably higher than those of Ti_3C_2 and PPy alone, as illustrated in figures 5.11(f) and (g) [263]. In addition, even at 20 A g^{-1}, the capacitance retention rate of the $Ti_3C_2T_x$/PPy film electrode still reached 86% after 10 000 cycles, showing good electrochemical stability, indicating that $Ti_3C_2T_x$/PPy film can be used as an SC electrode.

Lee *et al* [264] used *in situ* synthesis to prepare 3D $Ti_3C_2T_x$@PPy nanocomposites. A multilayer MXene nanosheet structure was wound in a 3D PPy nanowire matrix to form an MXene/PPy composite with an ideal 3D interconnected porous

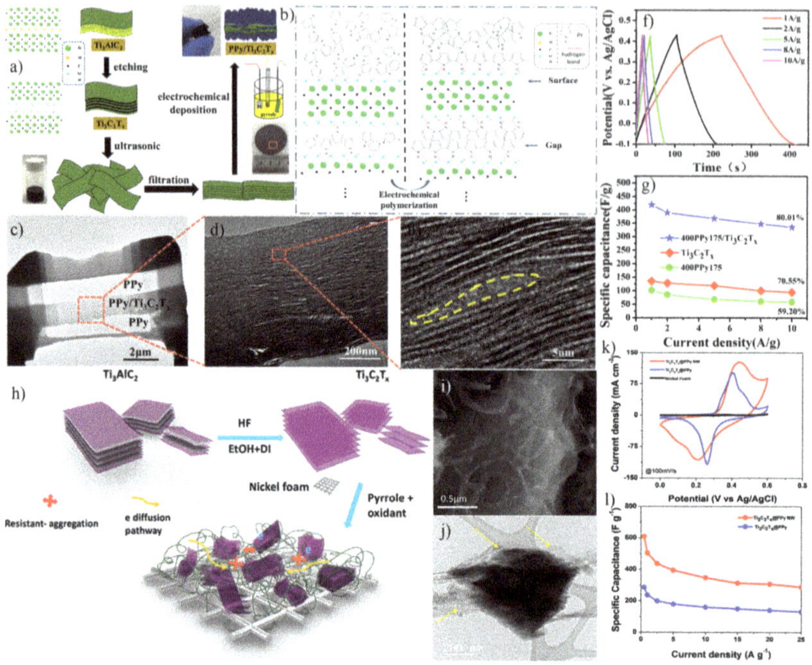

Figure 5.11. (a) A flow chart of the preparation of PPy/$Ti_3C_2T_x$ by the electrodeposition of PPy on a $Ti_3C_2T_x$ thin film [263]. (b) A schematic illustration of the electrochemical polymerization of PPy on the surface and gap of a Ti_3AlC_2 film. (c)–(e) Transmission electron microscopy (TEM) images of PPy/$Ti_3C_2T_x$ at different scales [263]. (f) GCD curves of the PPy/$Ti_3C_2T_x$ electrode at various current densities [263]. (g) The specific capacitances of PPy/$Ti_3C_2T_x$, pure $Ti_3C_2T_x$, and PPy electrodes at different current densities. ((a)–(g) Reprinted from [263], Copyright (2020), with permission from Elsevier.) (h) A schematic illustration of the preparation of the 3D $Ti_3C_2T_x$@PPy nanowire. An SEM image (i) and a TEM image (j) of the 3D $Ti_3C_2T_x$@PPy nanowire. (k) CV curves of different electrodes. (l) The specific capacitances of the PPy/$Ti_3C_2T_x$ nanowire and the PPy/$Ti_3C_2T_x$ electrode at different current densities. ((h)–(l) [264] John Wiley & Sons. © 2019 Wiley-VCH Verlag GmbH & Co. KGaA, Weinheim.)

structure and high conductivity, as shown in figures 5.11(h)–(j) [264]. The unique structure arising from the entanglement of MXene and PPy facilitated the swift diffusion of electrolyte ions and established a concise and uninterrupted charge transfer pathway, thereby maximizing the utilization of the active materials. Moreover, the porous interconnected 3D network interconnected the MXene sheets while preventing them from self-stacking. As can be seen from figures 5.11(k) and (l) [264], at 0.5 A g^{-1}, 3D $Ti_3C_2T_x$@PPy displayed a specific capacitance of 610 F g^{-1}, while the previously reported $Ti_3C_2T_x$@PPy and pure $Ti_3C_2T_x$ MXene separately exhibited 298 F g^{-1} and 138 F g^{-1}. Three-dimensional $Ti_3C_2T_x$@PPy has more than double the specific capacitance of the previously reported $Ti_3C_2T_x$@PPy. Significantly, at 4 A g^{-1}, after 14 000 cycles, a 3D $Ti_3C_2T_x$@PPy electrode maintained almost 100% stability, which was better than the stabilities of other advanced MXene-based SCs.

5.3.2.3 MXene/PEDOT composite electrodes

Tao *et al* [265] proposed a way of preparing vacant $Mo_{1.33}C$ MXene. The resulting vacant MXene film had a high volume capacitance of 1153 F cm^{-3} and a high conductivity of 29 674 S m^{-1}. Based on the vacant $Mo_{1.33}C$ MXene, Qin *et al* [266] prepared an $Mo_{1.33}C$ MXene/PEDOT:PSS film by a simple hydrothermal method and a vacuum-assisted filtration method. To further optimize its electrochemical properties, the $Mo_{1.33}C$ MXene/PEDOT:PSS film was soaked in concentrated H_2SO_4 for 24 h to remove the nonconductive PSS (figure 5.12(a)) [266]. Comparing figure 5.12(b) [266] and figure 5.12(c) [266], it is easy to see the difference between the MXene/PEDOT:PSS films before and after immersion in concentrated H_2SO_4. After soaking, the nonconductive PSS in the MXene/PEDOT film has a more obvious layered structure, which is beneficial to ion transport.

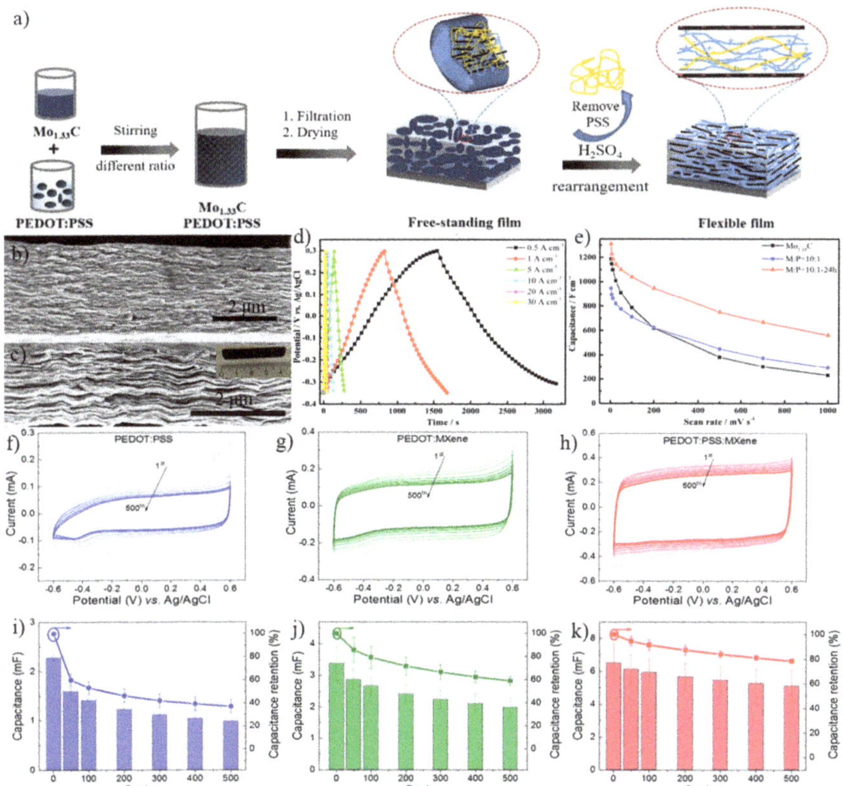

Figure 5.12. (a) A schematic diagram of the preparation of the MXene/PEDOT film. SEM images of (b) MXene/PEDOT:PSS and (c) MXene/PEDOT. (d) Different current density GCD curves of MXene/PEDOT: PSS after immersion in concentrated H_2SO_4 for 24 h. (e) The specific capacitances of $Mo_{1.33}C$, MXene/PEODT: PSS, and MXene/PEDOT electrodes at different scan rates. ((a)–(e) [266] John Wiley & Sons. © 2017 WILEY-VCH Verlag GmbH & Co. KGaA, Weinheim.) CV curves of PEDOT:PSS (f), PEDOT:MXene (g), and PEDOT:PSS:MXene (h) after 500 CV cycles at 100 mV s^{-1}. The cycling stabilities of PEDOT:PSS (i), PEDOT: MXene (j), and PEDOT:PSS:MXene (k). ((f)–(k) Reproduced from [221], CC BY 4.0.)

Furthermore, the resistivity of the composite film was significantly diminished, and its capacitance value was augmented (reaching a maximum volume capacitance of 1310 F cm^{-3}) following the treatment with concentrated H_2SO_4. The conductive PEDOT intercalation widened the interlayer spacing of the MXene sheets, while the oriented PEDOT nanofibers established a network structure between the MXene sheets. This network structure provided a rapid channel for ion transmission and facilitated a swift and reversible redox reaction. The galvanostatic charge–discharge (GCD) curves of figure 5.12(d) [266] and the capacitance curves of figure 5.12(e) [266] show that the $Mo_{1.33}C$ MXene/PEDOT:PSS film with a mass ratio of 10:1 had excellent specific capacitance (2 mV^{-1}, 1310 F cm^{-3} or 452 F g^{-1}) after concentrated H_2SO_4 treatment, which was superior to the specific capacitances of MXene-based electrode materials reported previously [70, 267, 268].

Inal *et al* [221] prepared PEDOT:PSS:MXene films through electrochemical polymerization and codoping. Compared to a thin film composed solely of a single dopant and PEDOT, the incorporation of PSS and MXene as co-dopants with PEDOT offered superior synergy, combining the attributes of MXene with those of PEDOT to yield a polymer composite with greater specific capacitance and energy density. This can be observed in the cyclic voltammetry (CV) curves depicted in figures 5.12(f)–(h) [221] and the cycle curve of figures 5.12(i)–(k) [221]. It can be seen that under the same conditions, the PEDOT:PSS:MXene film (607 ± 85.3 F cm^{-3}, the capacity retention rate after 500 cycles is 78%) had a higher capacity than the capacities of PEDOT:PSS (195.6 ± 1 F cm^{-3}, 37%) and PEDOT:MXene (358.9 ± 16.7 F cm^{-3}, 58%), higher volume capacitance, and higher stability. The codoping method significantly improved the electrochemical performance and stability of the PEDOT films, and electrochemical polymerization is a simple and easy-to-operate single-step polymerization method [269, 270]. Both methods offer an effective means of combining high-performance materials with varying characteristics. Overall, the integration of MXenes and conductive polymers can markedly enhance their electrochemical performance, making MXene/conductive polymer composites a promising avenue for applications. Table 5.2 provides an overview of the applications of MXene/conductive polymer composite electrodes in SCs.

5.4 Supercapacitors

We present a compilation of common formulas used to perform calculations related to SCs, particularly in the context of sandwich-type face-to-face SCs [46, 271].

Gravimetric capacitance is calculated using CV curves:

$$C_g = \frac{S}{m \times \nu \times \Delta u}, \tag{5.4}$$

where C_g (F g^{-1}) stands for gravimetric capacitance, S (cm^2) is the area of the CV curve, m (g) denotes the mass of the active substance loaded by the electrode, ν (mV s^{-1}) represents the scan rate, and Δu (V) refers to the voltage window of the test.

Table 5.2. The applications of MXene/conductive polymer composite electrodes in SCs mentioned in this review.

Method	Electrode	Electrolyte	Capacity	Scan rate and current density	Cycles	References
Ultrasonic stripping	Ti_3C_2 paper	1 M KOH	450 F cm^{-3} 280 F cm^{-3}	2 mV s^{-1} 100 mV s^{-1}	—	[69]
Mechanical rolling	Ti_3C_2	1 M H_2SO_4	900 F cm^{-3} 730 F cm^{-3}	2 mV s^{-1} 100 mV s^{-1}	—	[70]
Vacuum-assisted filtration	$Ti_3C_2T_x$/P-100-H	1 M H_2SO_4	1065 F cm^{-3}	2 mV s^{-1}	10 000, 100%	[71]
One-step coelectrodeposition	MXene/PPy	1 M H_2SO_4	416 F g^{-1}	0.5 A g^{-1}	—	[212]
Electrophoretic deposition, electrochemical polymerization	Ti_3C_2/PPy	2 M H_2SO_4 PVA/H_2SO_4	109 mF cm^{-2} 87 mF cm^{-2}	1 mA cm^{-2} 1 mA cm^{-2}	10 000, 96%	[215]
Electrochemical polymerization, codoping	PEDOT:PSS:MXene	10 mM PBS	692 F cm^{-3}	100 mV s^{-1}	500, 78%	[221]
Mechanical compression	Wavy Ti_3C_2	3 M H_2SO_4	1277 F cm^{-3} 790 F cm^{-3}	10 mV s^{-1} 5000 mV s^{-1}	10 000, 98.8%	[255]
Two-step electrochemical polymerization	N-Ti_3C_2/PANI	0.5 M H_2SO_4	228 mF g^{-1}	5 mV s^{-1}	1000, 85%	[258]
Hydrothermal reaction	MXene/PANI	6 M KOH	563 F g^{-1} 477 F g^{-1}	0.5 A g^{-1} 20 A g^{-1}	10 000, 95.15%	[259]
Chemical oxidation polymerization	TDP	1 M H_2SO_4	452 F g^{-1}	1 A g^{-1}	2000, 61%	[261]
Low-temperature chemical oxidation	$Ti_3C_2T_x$/PPy	1 M Na_2SO_4	184 F g^{-1}	2 mV s^{-1}	4000, 83.33%	[262]
Electrochemical deposition, vacuum-assisted filtration	$Ti_3C_2T_x$/PPy	PVA/H_2SO_4	420 F g^{-1}	1 A g^{-1}	10 000, 86%	[263]
In situ synthesis	3D $Ti_3C_2T_x$@PPy	3 M KOH	610 F g^{-1}	0.5 A g^{-1}	14 000, 100%	[264]

(*Continued*)

Table 5.2. *(Continued)*

Method	Electrode	Electrolyte	Capacity	Scan rate and current density	Cycles	References
Hydrothermal reaction, vacuum-assisted filtration	$Mo_{1.33}C$ MXene/ PEDOT:PSS	1 M H_2SO_4	1310 F cm^{-3}	2 mV s^{-1}	—	[266]
—	Ti_2CT_x	1 M KOH	517 F cm^{-3} 307 F cm^{-3}	2 mV s^{-1} 100 mV s^{-1}	—	[278]
—	$Ti_3C_2T_x$/PVA	1 M KOH	528 F cm^{-3} 306 F cm^{-3}	2 mV s^{-1} 100 mV s^{-1}	—	[279]
Vacuum-assisted filtration	Reassembled MXene (RAMX)	3 M H_2SO_4	978 F cm^{-3} 736 F cm^{-3}	10 mV s^{-1} 2000 mV s^{-1}	—	[280]
One-pot *in situ* polymerization	$Ti_3C_2T_x$/PANI-NTs	1 M H_2SO_4	597 F g^{-1}	0.1 A g^{-1}	5000, 94.7%	[290]
Doctor blade coating technology	PANI/MXene	1 M H_2SO_4	1167 F cm^{-3}	5 mV s^{-1}	—	[292]
Electrostatic self-assembly	Ti_3C_2/Fe-15%	1 M Li_2SO_4	485 mF cm^{-2}	2 mV s^{-1}	5000, 94.8%	[297]
In situ polymerization	Graphene-encapsulated Ti2CTx MXene@PANI (GMP)	1 M H_2SO_4	635 F g^{-1}	1 A g^{-1}	10 000, 97.54%	[302]
Chemical oxidation polymerization, vacuum-assisted filtration	PANI@MXene	3 M H_2SO_4	1632 F cm^{-3} 827 F cm^{-3}	10 mV s^{-1} 5000 mV s^{-1}	20 000, 85.7%	[304]

Gravimetric capacitance is calculated using GCD curves:

$$C_g = \frac{I \times \Delta t}{m \times \Delta u}, \tag{5.5}$$

where C_g (F g^{-1}) stands for gravimetric capacitance, I (A) is the current of the GCD test, Δt (s) represents the discharge time of a charge–discharge test, m (g) represents the mass of the active substance loaded by the electrodes, and Δu (V) refers to the voltage window of the test.

$$C_A = \frac{m}{A} \times C_g \tag{5.6}$$

$$C_V = \frac{m}{V} \times C_g \tag{5.7}$$

Here, C_A (F cm^{-2}) and C_V (F cm^{-3}) represent the areal specific capacitance and the volume specific capacitance, respectively; m (g) represents the mass of the active substance loaded by the electrode; A (cm^2) represents the electrode area; V (cm^3) stands for the electrode volume; and C_g (F g^{-1}) is the mass specific capacitance.

The energy density and power density of SCs are calculated based on the specific capacitance:

$$E = \frac{1}{2} \times C_g \times (\Delta u)^2 \tag{5.8}$$

$$P = \frac{E}{\Delta t} \times 3600, \tag{5.9}$$

where E (Wh kg^{-1}) stands for energy density, P (W kg^{-1}) represents power density, C_g (F g^{-1}) is the abovementioned gravimetric capacitance, and Δu (V) and Δt (s) separately denote the voltage window and the discharge time of the charge–discharge test.

It is important to emphasize that when constructing ASCs, energy matching should be taken into account to optimize device performance. This consideration can be expressed by the following equation:

$$C_+ \times \Delta u_+ \times m_+ = C_- \times \Delta u_- \times m_-, \tag{5.10}$$

where C_+ (F g^{-1}) and C_- (F g^{-1}) represent the gravimetric capacitances of the positive and the negative electrodes, respectively; Δu_+ (V) and Δu_- (V) stand for the voltage windows of the positive and the negative electrodes, respectively; and m_+ (g) and m_- (g) represent the masses of the active materials of the positive and the negative electrodes, respectively. Here, the areas of the positive and negative electrodes are the same.

Furthermore, the formula for calculating the pseudocapacitance percentage from the CV curve is as follows:

$$i_{cap} = k_1 \times \nu \tag{5.11}$$

$$i_{\mathrm{diff}} = k_2 \times \nu^{0.5} \tag{5.12}$$

$$i = i_{\mathrm{cap}} + i_{\mathrm{diff}} \tag{5.13}$$

$$\frac{i}{\nu^{0.5}} = k_1 \times \nu^{0.5} + k_2, \tag{5.14}$$

where i_{cap} (A), i_{diff} (A), and i (A) denote the current in the capacitive part corresponding to the surface control (pseudocapacitance), the current in the capacitive part corresponding to the diffusion control, and the total current obtained from the calculation, respectively; k_1 and k_2 are the constant coefficients in the calculation process, respectively; and ν (mV s^{-1}) represents the range of the voltage window.

5.4.1 Symmetric supercapacitors

5.4.1.1 The use of MXenes in symmetric supercapacitors

MXenes have garnered significant attention since their inception. Due to their high specific surface area and excellent conductivity, resembling those of metallic materials, MXenes have found widespread applications in battery and SC research [85, 86, 272–274]. As shown in figures 5.13(a) and (b) [275], Xia *et al* [275] used the very typical $Ti_3C_2T_x$ MXene material to prepare $Ti_3C_2T_x//Ti_3C_2T_x$ SSCs with an unconventional electrolyte (a simulated seawater electrolyte). At 0.25 A g^{-1}, the SSC device exhibited a volumetric capacitance of 27.4 F cm^{-3}. At 3 A g^{-1}, the device still exhibited a high volume capacitance of 12.6 F cm^{-3} (figures 5.13(c) and (d) [275]). The device demonstrated volume energy densities of 1.74×10^{-3} and 0.68×10^{-3} Wh cm^{-3} when operating at power densities of 0.15 and 1.53 W cm^{-3}, respectively. Furthermore, a $Ti_3C_2T_x$ electrode exhibited excellent cycle stability, maintaining a capacitance retention rate of 96.6% after 5000 charge–discharge cycles. This performance surpassed that of similar previously reported systems.

Yang *et al* [276] fabricated a flexible, independent, and ultradense delaminated $Ti_3C_2T_x$ film by mechanically pressing a conventional $Ti_3C_2T_x$ film directly into a binderless electrode that was used to assemble SCs. The mechanical high pressure increased the density (figure 5.13(e) [276]), electrical conductivity (figure 5.13(f) [276]), and wettability of the $Ti_3C_2T_x$ film. Simultaneously, the interconnected and enriched mesoporous channels facilitated the insertion and deintercalation of cations during the charging and discharging processes. The experimental results revealed that the volume performance of SCs assembled from $Ti_3C_2T_x$ films increased with the application of mechanical pressure. However, this improvement was not infinite, and the thin-film electrode exhibited its optimal electrochemical performance when prepared under 40 MPa conditions. This is illustrated in figures 5.13(g) and (h) [276]. We observe that a typical SSC derived from the $Ti_3C_2T_x$ film at 40 MPa, employing a 1 M Li_2SO_4 electrolyte, delivered a volumetric capacitance of 633 F cm^{-3} at 2 mV s^{-1} and a volumetric energy density of 22 Wh l^{-1} and retained a capacitance of 95.3% after 10 000 cycles. It is worth highlighting that when we replaced the 1 M Li_2SO_4 with 1 M 1-ethyl-3-methylimidazolium tetrafluoroborate/acetonitrile

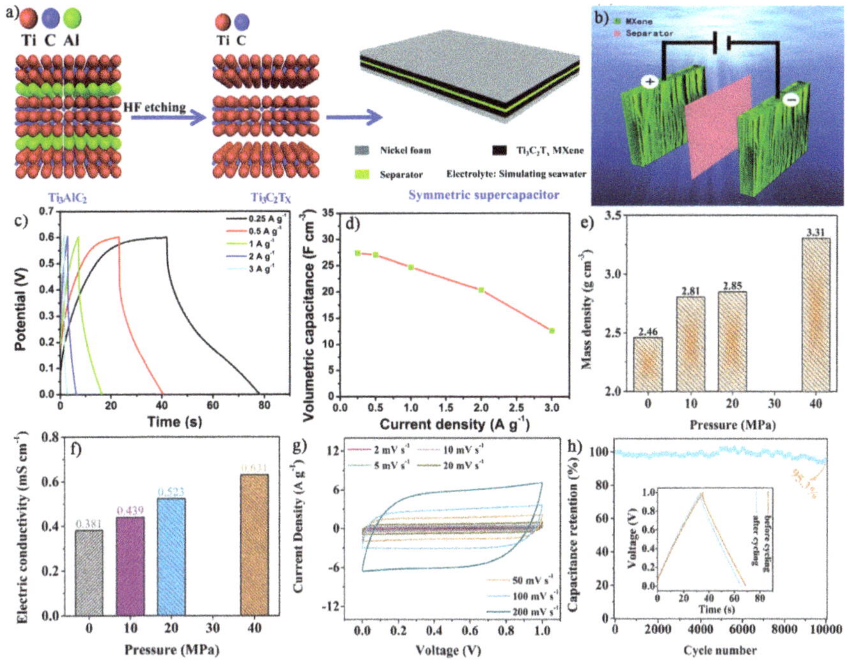

Figure 5.13. (a) A schematic diagram of the $Ti_3C_2T_x//Ti_3C_2T_x$ SSC showing the process used. (b) A schematic diagram of $Ti_3C_2T_x//Ti_3C_2T_x$ SSC simulation. GCD curves (c) and specific capacitance (d) at different current densities. ((a)–(d) Reproduced from [275] with permission from the Royal Society of Chemistry.) The mass density (e) and electrical conductivity (f) of $Ti_3C_2T_x$ films under different pressures. (g) CV curves of $Ti_3C_2T_x$ films prepared under 40 MPa at different scan rates. (h) The cyclic performance of $Ti_3C_2T_x$ film prepared under 40 MPa at 2 A g^{-1} (the inset shows the GCD curves at the beginning and end at 1 A g^{-1}). ((e)–(h) [276] John Wiley & Sons. © 2018 WILEY-VCH Verlag GmbH & Co. KGaA, Weinheim.)

(EMIMBF4/AN), the SC device extended its voltage window to 2.2 V. Furthermore, it demonstrated an exceptionally high volumetric energy density of 41 Wh l^{-1}. This achievement represents the highest value attained to date using organic electrolytes for SCs based on MXene materials.

Zhang *et al* [277] proposed an *in situ* ice template method for the preparation of independent and flexible 3D porous $Ti_3C_2T_x$/CNTs SC film (3D-PMCF) electrodes, as shown in figure 5.14(a) [277]. In contrast to the conventional vacuum drying method used for $Ti_3C_2T_x$ MXene preparation, the *in situ* ice template method avoided the issue of MXene material stacking. This technique involved freeze-drying a water film containing $Ti_3C_2T_x$/CNTs, wherein the water molecules between the MXene layers transformed into small ice particles. Subsequently, these small ice particles acted as sacrificial templates, preserving the 3D porous structure. This distinction is evident when comparing figures 5.14(b) and (c) [277], as it can be seen that the MXene films were stacked in layers, while 3D-PMCF had many interactive channels. The 3D-PMCF had a large number of active sites, which realized rapid ion transmission, thus leading to excellent electrochemical performance. As shown in figure 5.14(d) [277], in 3 M H_2SO_4, the flexible 3D-PMCF achieved a specific

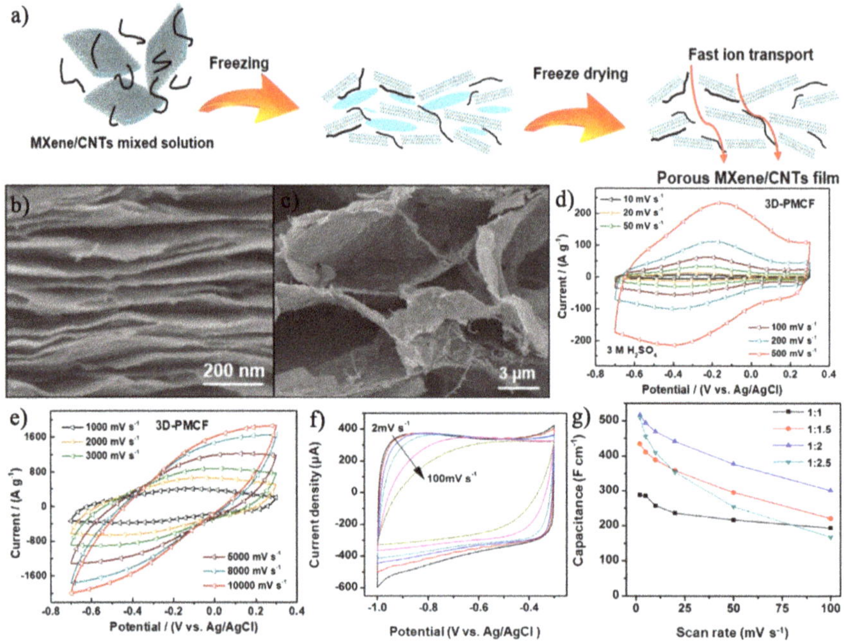

Figure 5.14. (a) A schematic illustration showing the preparation of the 3D-PMCF film. SEM images of MXene films (b) and 3D-PMCF (c). CV curves of 3D-PMCF at various scan rates: (d) 10–500 mV s^{-1}; (e) 1000–10 000 mV s^{-1}. ((a)–(e) [277] John Wiley & Sons. © 2020 WILEY-VCH Verlag GmbH & Co. KGaA, Weinheim.) CV curves (f) and specific capacitance (g) at different scan rates. ((f) and (g) Reprinted from [278], Copyright (2019), with permission from Elsevier.)

capacitance of 375 F g^{-1} at 5 mV s^{-1}. Even at 10 000 mV s^{-1}, 3D-PMCF still displayed a specific capacitance of 92.0 F g^{-1} (figure 5.14(e)) [277]. Furthermore, at a rate of 10 A g^{-1}, even after 10 000 cycles, the 3D-PMCF electrode retained a remarkable capacitance of 95.9%. Upon integrating the 3D-PMCF into an SSC, the device exhibited an energy density of 9.2 Wh kg^{-1}.

Zhu *et al* [278] successfully prepared a Ti_2CT_x//Ti_2CT_x SSC characterized by outstanding energy and power density. Their experiments demonstrated that when the mass ratio of Ti_2AlC to HF was 1:2, the etched Ti_2CT_x exhibited the best performance. As shown in figures 5.14(f) and (g) [278], in a 1 M KOH electrolyte, the 1:2 sample provided specific volume capacitances of 517 F cm^{-3} at 2 mV s^{-1} and 307 F cm^{-3} at 100 mV s^{-1}. These results are comparable to the performance of the Ti3C2Tx/PVA electrode reported by Ling *et al* [279] in 1 M KOH (2 mV s^{-1}, 528 F cm^{-3}; 100 mV s^{-1}, 306 F cm^{-3}). Notably, the Ti_2CT_x//Ti_2CT_x SSC exhibited a capacitance retention rate of 100% even after 3000 charge–discharge cycles at 20 A g^{-1}, demonstrating its exceptional cycle stability.

Wu *et al* [280] reconstructed $Ti_3C_2T_x$ microgels and $Ti_3C_2T_x$ monolayer nanosheets to construct independent flexible reassembled MXene (RAMX) film electrodes with a tunable porous structure, as shown in figure 5.15(a) [280]. This description pertains to the film's macroscopic morphology. However, at the microscopic level,

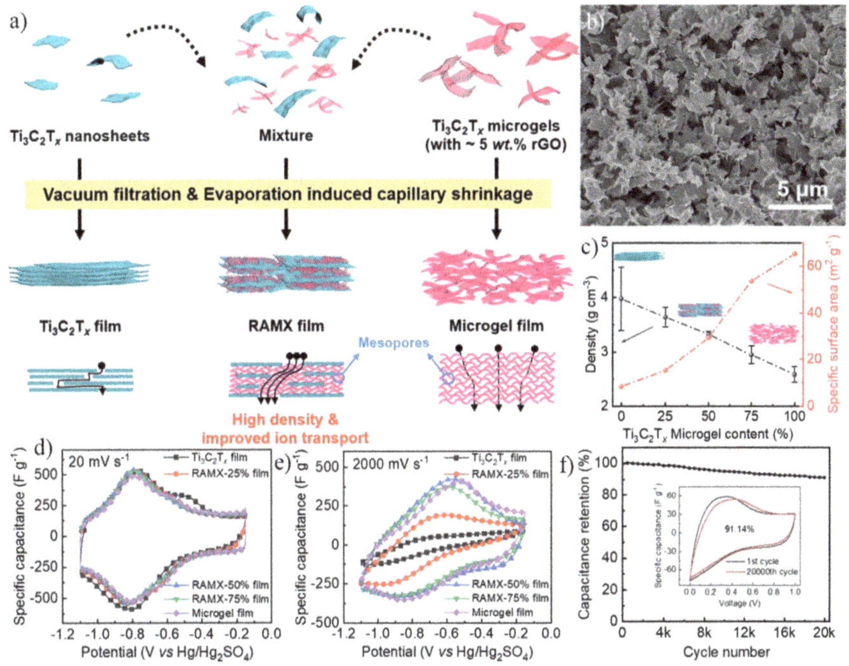

Figure 5.15. (a) A schematic diagram of the preparation of the RAMX film. (b) An SEM image of the RAMX film. (c) The RAMX film density and specific surface area versus the $Ti_3C_2T_x$ microgel content. CV curves of electrodes with different RAMX contents at different scanning rates: (d) 20 mV s^{-1} and (e) 2000 mV s^{-1}. (f) Cycling performance after 20 000 cycles at 1000 mV s^{-1}. (The inset shows the CV curves of the 1st and 20 000th cycles). ((a)–(f) [280] John Wiley & Sons. © 2021 Wiley-VCH GmbH.)

the $Ti_3C_2T_x$ MXene sheets, which were initially stacked, transformed into a 3D network structure with closely-packed mesopores (figure 5.15(b) [280]). Upon observing the $Ti_3C_2T_x$ microgels and the $Ti_3C_2T_x$ monolayer nanosheets reassembled with different mass ratios in figure 5.15(c) [280], it is apparent that this approach optimizes space utilization. Specifically, when the mass ratio of microgels reached 50%, it maximized the rate capability of the MXene film. This MXene film exhibited a specific capacitance of 978 F cm^{-3} at 10 mV s^{-1} in 3 M H_2SO_4, and even at 2000 mV s^{-1}, it still maintained a specific capacitance of 736 F cm^{-3} (figure 5.15(d) [280] and figure 5.15(e) [280]). An SSC that used the thin-film electrode assembly $Ti_3C_2T_x$//$Ti_3C_2T_x$ delivered a high energy density of 40 Wh l^{-1} at a power density of 0.83 kW l^{-1}. Even at 41.5 kW l^{-1}, it still produced 21 Wh l^{-1}. In addition, at 1000 mV s^{-1}, the SSC still retained 91.14% of its capacitance after 20 000 charge–discharge cycles (figure 5.15(f) [280]), which is the highest value achieved by an SSCs using an aqueous electrolyte.

5.4.1.2 The use of MXenes and conductive polymer composites in symmetric supercapacitors

MXenes possess a 2D sheet structure similar to that of graphene and exhibit high conductivity similar to those of metals. Consequently, they hold significant promise for applications in electrochemical energy storage [281–283]. Conductive polymers have garnered significant attention in SC research due to their exceptional stability, high conductivity in the doped state, and rapid redox reaction kinetics [248, 284]. To enhance the performance of electrode materials, researchers are actively investigating composite materials. Combining the exceptional properties of MXenes and conductive polymers and their distinctive characteristics has allowed MXene/conductive polymer composites to find widespread applications as electrode materials in SCs [285–288].

Boota *et al* [208] prepared MXene/PPy composites by the *in situ* polymerization of pyrrole between $Ti_3C_2T_x$ MXene nanosheets. Their MXene/PPy composite electrode had a capacitance of nearly 1000 F cm^{-3} and 92% of the capacitance remained after 25 000 cycles. Jian *et al* [212] used a one-step coelectrodeposition method to prepare MXene/PPy composite films. As shown in figure 5.16(a) [212], in 1 M H_2SO_4, one of their MXene/PPy composite films provided a mass capacitance of 416 F g^{-1} at 0.5 A g^{-1}. Furthermore, an SSC assembled with ITO glass electrodes covered with MXene/PPy film showed a high specific capacitance of 184 F g^{-1} at 10 mV s^{-1} as well as a good capacitance retention rate of about 86.4% after 5000 cycles at 5 A g^{-1} (figure 5.16(b)) [212]. Tong *et al* [263] assembled a 400PPy175/$Ti_3C_2T_x$ composite electrode into an all-solid-state flexible SSC and obtained a mass specific capacitance of 258.3 F g^{-1} in 1 M H_2SO_4 at 0.5 A g^{-1}. (In '400PPy175,' 400 and 175 denote 400 μl of pyrrole monomer and an electrochemical polymerization time of 175 s, respectively.) Furthermore, devices that had 400PPy175/$Ti_3C_2T_x$// 400PPy175/$Ti_3C_2T_x$ configurations exhibited energy densities of 10.82 μWh mg^{-1} and 3.99 μWh mg^{-1} at power densities of 0.11 and 1.89 mW mg^{-1}, respectively.

Wu *et al* [289] introduced a way of fabricating composite films characterized by excellent cycle stability. These films involved the combination of dispersed conjugated polymer (PDT) with layered $Ti_3C_2T_x$ MXene, as depicted in figure 5.16(c) [289]. A stable chemical bond formed between the dispersed PDT chains and the $Ti_3C_2T_x$ nanosheets (figure 5.16(d) [289]), effectively addressing performance issues arising from volume expansion and contraction changes within the polymer chain.

Simultaneously, the material's mechanical performance was enhanced, offering a rapid pathway for ion/charge transfer. In 0.5 M H_2SO_4, a PDT/$Ti_3C_2T_x$ composite electrode demonstrated a surface capacitance of 284 mF cm^{-2} at 0.5 mA cm^{-2} (refer to figure 5.16(e)) [289] and retained nearly 100% of its capacitance after 10 000 charge–discharge cycles (see figure 5.16(f)) [289]. Furthermore, an all-solid SSC constructed using the PDT/$Ti_3C_2T_x$ film exhibited exceptional flexibility, enduring 10 000 cycles of 0–90° static bending (as illustrated in figure 5.16(g)) [289], and attained a surface capacitance of up to 52.4 mF cm^{-2} at 0.1 mA cm^{-2}.

Wu and colleagues [290] fabricated Ti_3C_2/PANI nanotube (Ti_3C_2/PANI-NT) composites that exhibited a well-defined hierarchical structure (as depicted in figure 5.17(a)) [290] via the one-pot *in situ* polymerization process illustrated in figure 5.17(b) [290]. In this composite, Ti_3C_2 MXene nanosheets contributed to mechanical stability and high conductivity, while PANI-NTs served as one-dimensional high-speed transmission channels, creating additional active sites. The incorporation of the PANI-NTs effectively mitigated the self-stacking tendency of

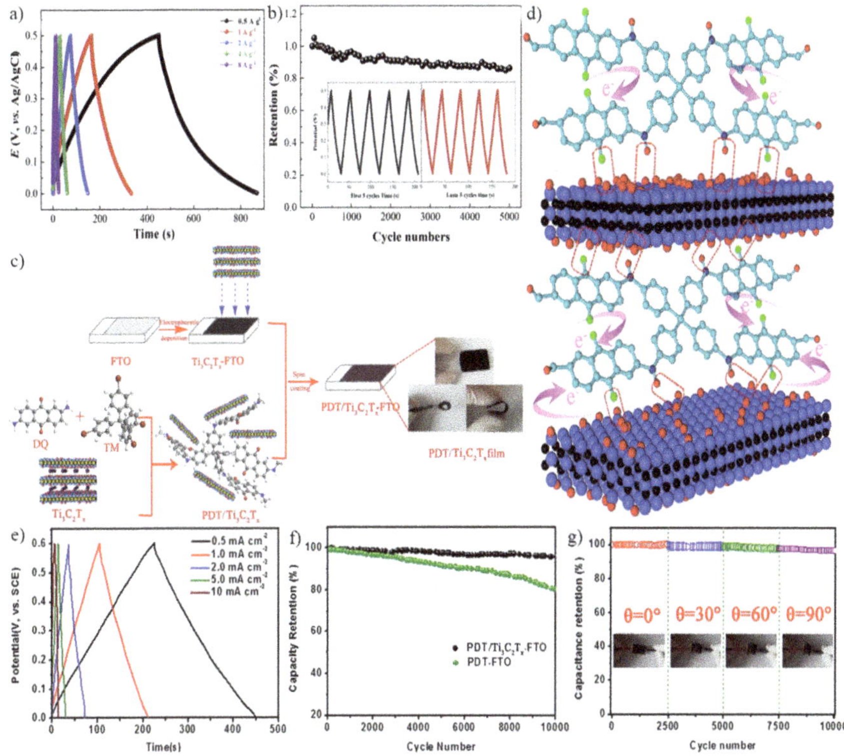

Figure 5.16. (a) GCD curves of MXene/PPy electrodes at different current densities [212]. (b) The cycle stability curve of the MXene/PPy//MXene/PPy SSC (the inset shows the GCD curves of the first five cycles and the last five cycles). ((a) and (b) Reprinted from [212], Copyright (2019), with permission from Elsevier.) (c) A schematic of the fabrication of the PDT/$Ti_3C_2T_x$ composite film. (d) Atomic-scale schematic diagram of $Ti_3C_2T_x$ MXene and PDT chains binding. (e) GCD curves of the PDT/$Ti_3C_2T_x$ electrodes at various current densities. (f) Cyclic stability of PDT-FTO and PDT/$Ti_3C_2T_x$-FTO electrodes. (g) Cycling performance of the all-solid SSC assembled based on PDT/$Ti_3C_2T_x$ film for 10 000 cycles under different static bending angles. ((c)–(g) Reprinted from [289], Copyright (2019), with permission from Elsevier.)

Ti_3C_2 nanosheets, leading to an increased interlayer spacing in the MXene layers and a greater accessible surface area for ions.

Electrochemical testing in a three-electrode system using a 1 M H_2SO_4 electrolyte revealed that the Ti_3C_2/PANI-NT composite electrode achieved a specific capacitance of 596.6 F g^{-1} at 0.1 A g^{-1} (as depicted in figure 5.17(c)) [290] and a specific capacitance retention rate of 94.7% after 5000 cycles (figure 5.17(d)) [290]. Notably, when compared to previously reported Ti_3C_2 MXene-based SSCs, the SSC assembled using Ti_3C_2/PANI-NTs demonstrated an energy density of 25.6 Wh kg^{-1} at a power density of 153.2 W kg^{-1} in 1 M H_2SO_4. Even at a power density of 1610.8 W kg^{-1}, it retained an energy density of 13.2 Wh kg^{-1}. Moreover, as shown in figure 5.17(e) [290], the Ti_3C_2/PANI-NTs//Ti_3C_2/PANI-NT SSC device exhibited a capacitance retention of 81.1% after 4000 cycles at 1 A g^{-1}.

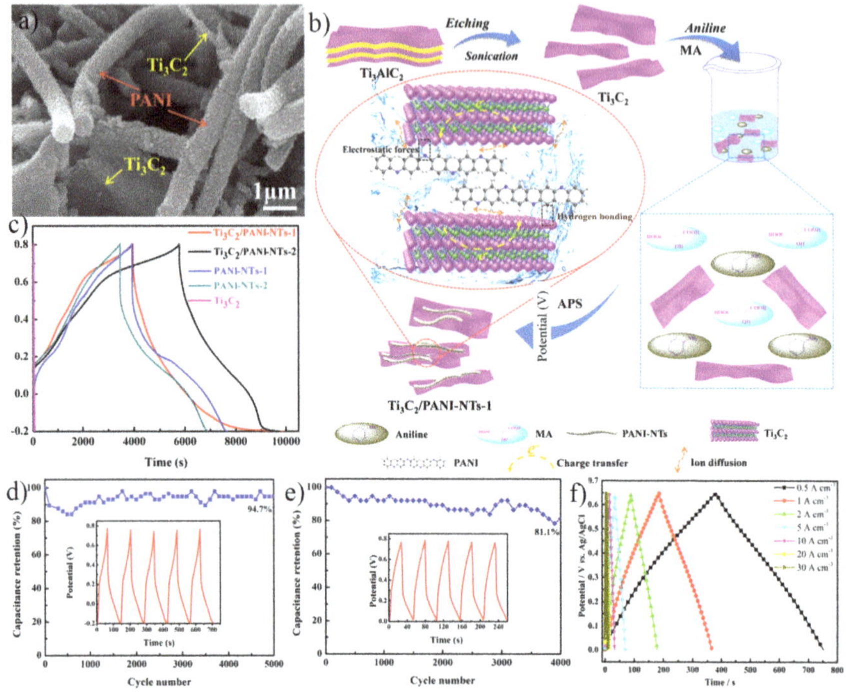

Figure 5.17. (a) An SEM image of Ti_3C_2/PANI-NTs. (b) A schematic illustration of the preparation of the Ti_3C_2/PANI-NTs composites. (c) GCD curves of the electrodes at 0.1 A g^{-1}. Cycle performance of the Ti_3C_2/PANI-NT electrode (d) and the Ti_3C_2/PANI-NT//Ti_3C_2/PANI-NT SSC device (e) (the inset shows the GCD curves of the last five cycles). ((a)–(e) Reprinted from [290], Copyright (2020), with permission from Elsevier.) (f) GCD curves of the $Mo_{1.33}C$ MXene/PEDOT:PSS flexible device [266] John Wiley & Sons. © 2017 WILEY-VCH Verlag GmbH & Co. KGaA, Weinheim.

Qin *et al* [266] employed a self-assembly method to fabricate $Mo_{1.33}C$ MXene/PEDOT:PSS composite electrodes. These composites exhibited exceptional conductivity due to the synergistic effect of the acidified PEDOT:PSS and the MXene. With a 1 M H_2SO_4 electrolyte, a flexible device assembled using $Mo_{1.33}C$ MXene/PEDOT:PSS achieved a remarkable volume capacitance of 568 F cm^{-3} at 0.5 A g^{-1} (as depicted in figure 5.17(f)) [266]. In addition, it achieved an impressive energy density of 33.2 mWh cm^{-3} at an ultrahigh power density of 19 470 mW cm^{-3}.

Zhou *et al* [291] pioneered the fabrication of i-PANI@$Ti_3C_2T_x$ composites through an *in situ* non-oxidative polymerization process. Subsequently, they produced Lig (lignosulfate)@$Ti_3C_2T_x$ and Lig@$Ti_3C_2T_x$/i-PANI@$Ti_3C_2T_x$ films using vacuum-assisted filtration, as illustrated in figure 5.18(a) [291]. The SEM image in figure 5.18(b) [291] and the transmission electron microscopy (TEM) image in figure 5.18(c) [291] reveal a compact film with a well-structured hierarchical arrangement.

The incorporation of electroactive polymers, i-PANI (prepared using p-phenylenediamine as the initiator for PANI) and lig (lignosulfonate), serves multiple purposes. It not only increases the spacing between $Ti_3C_2T_x$ layers, preventing self-stacking issues, but also contributes substantial pseudocapacitance. Furthermore, it

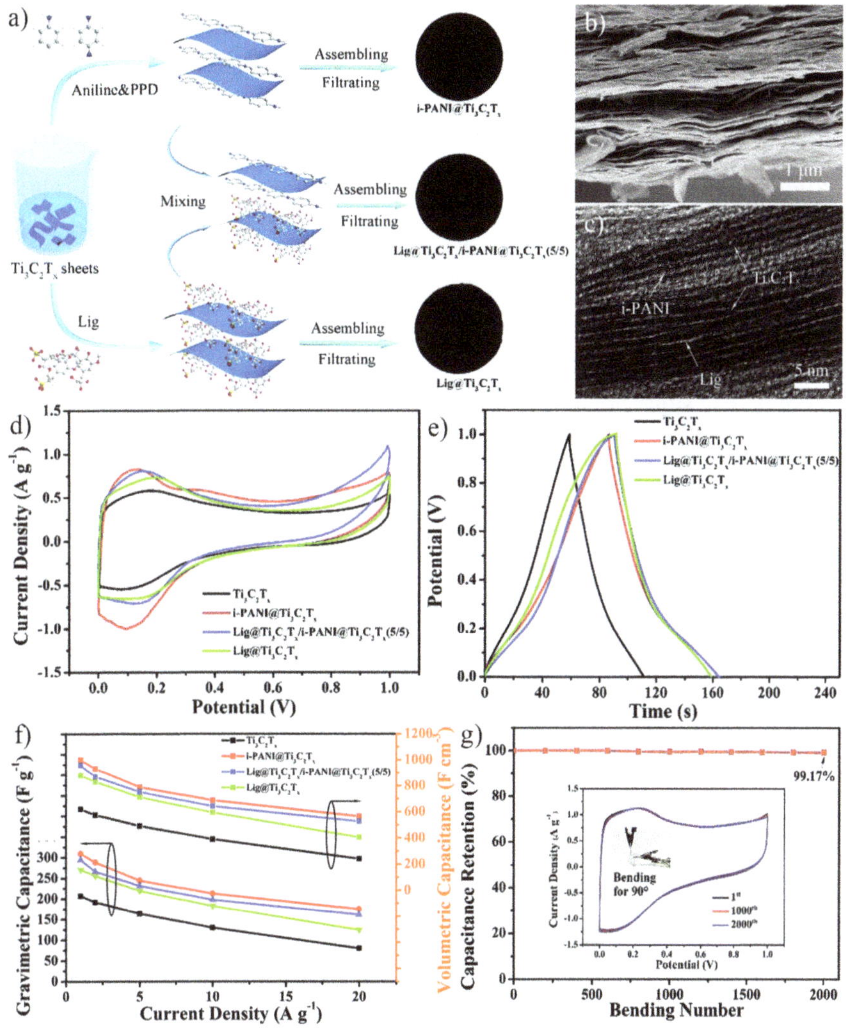

Figure 5.18. (a) A schematic diagram of the preparation of i-PANI@$Ti_3C_2T_x$, Lig@$Ti_3C_2T_x$, and Lig@$Ti_3C_2T_x$/i-PANI@$Ti_3C_2T_x$ composite films. An SEM image (b) and a TEM image (c) of Lig@$Ti_3C_2T_x$/i-PANI@$Ti_3C_2T_x$ film. (d) CV curves of different electrodes at 5 mV s^{-1}. (e) GCD curves of different electrodes at 1 A g^{-1}. (f) Specific gravimetric and volumetric capacitance at various current densities. (g) Cycle performance after 2000 cycles of 90° bends measured at 10 mV s^{-1} (the inset shows CV curves after various numbers of 90° bending cycles). ((a)–(g) Reproduced from [291] with permission from the Royal Society of Chemistry.)

facilitates the rapid and efficient transfer of ions and charges. Consequently, as depicted in figure 5.18(d–f) [291], in a PVA/H_2SO_4 electrolyte, flexible SSCs assembled with i-PANI@$Ti_3C_2T_x$, Lig@$Ti_3C_2T_x$, and Lig@$Ti_3C_2T_x$/i-PANI@$Ti_3C_2T_x$ individually achieved specific capacitances of 310 F g^{-1} (~1001 F cm^{-3}), 271 F g^{-1} (~881 F cm^{-3}), and 295 F g^{-1} (~959 F cm^{-3}) at 1 A g^{-1}.

Moreover, i-PANI@$Ti_3C_2T_x$//i-PANI@$Ti_3C_2T_x$, Lig@$Ti_3C_2T_x$//Lig@$Ti_3C_2T_x$, and Lig@$Ti_3C_2T_x$/i-PANI@Ti3C2Tx//Lig@$Ti_3C_2T_x$/i-PANI@$Ti_3C_2T_x$ SSC devices separately exhibited impressive energy densities of 34.8, 30.6, and 33.3 Wh l^{-1} at a power density of 1625 W l^{-1}. Notably, the SSC device assembled using Lig@$Ti_3C_2T_x$/i-PANI@ $Ti_3C_2T_x$ displayed exceptional mechanical durability. Even after undergoing 2000 0°–90° static bending cycles, it still retained a remarkable capacitance of 99.17%, as showcased in figure 5.18(g) [291]. This underscores the practical utility of the remarkably flexible Lig@$Ti_3C_2T_x$/i-PANI@$Ti_3C_2T_x$ composite film as an electrode material for flexible and wearable devices.

5.4.2 Asymmetric supercapacitors

5.4.2.1 The use of MXenes in asymmetric supercapacitors

In contrast to MXene-based SSCs, in which an MXene serves as the negative electrode, MXenes have been used in combination with positive electrode materials to design ASCs. This approach widens the voltage window of the device and further boosts the energy density of the capacitor [292–295]. However, finding ideal flexible cathodes that perfectly complement the original MXene membrane electrodes has been a challenge. Consequently, researchers have conducted a series of investigations to identify an ideal flexible positive electrode that can be paired with an MXene. Xia *et al* [296] employed a hydrothermal method to create $Ti_3C_2T_x$ MXene nanocomposites decorated with NiO (nickel-oxide-decorated MXene nanocomposites, Ni-dMXNCs), as depicted in figure 5.19(a) [296]. From figure 5.19(b) [296], it is evident that the NiO nanosheets successfully adhered to the MXene sheets. Compared to conventionally prepared $Ti_3C_2T_x$ MXene, TiO_2/C-$Ti_3C_2T_x$–MXene, which includes carbon-supported TiO_2 obtained through high-temperature

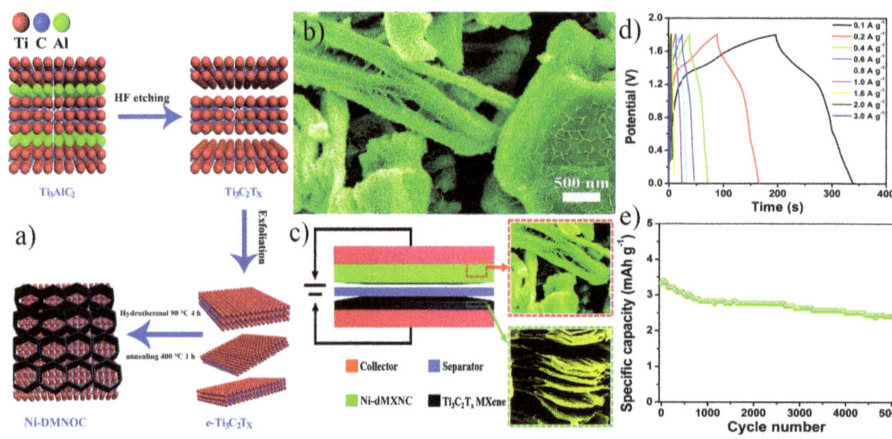

Figure 5.19. (a) An illustration of the preparation of the Ni-dMXNC by a hydrothermal method. (b) An SEM image of the Ni-dMXNC. (c) An illustration of the Ni-dMXNC//$Ti_3C_2T_x$ ASC. (d) GCD curves of the Ni-dMXNC//$Ti_3C_2T_x$ ASC at various densities. (e) The cycle performance of the Ni-dMXNC//$Ti_3C_2T_x$ ASC at 1 A g^{-1}. ((a)–(e) Reproduced from [296] with permission from the Royal Society of Chemistry.)

annealing, boasted higher conductivity and a larger specific surface area. The unique structure of Ni-dMXNC, with its high specific surface area, the enhanced surface activity of the NiO layer, and the synergistic effects of traditional $Ti_3C_2T_x$ MXene as the negative electrode, collectively enhanced the capacitive performance of the device. Consequently, ASCs were assembled using Ni-dMXNC as the positive electrode and $Ti_3C_2T_x$ MXene as the negative electrode (figure 5.19(c)) [296]. When compared to previously reported pure $Ti_3C_2T_x$ MXene SSCs, these ASCs exhibited a remarkable energy density of 1.04×10^{-2} at 0.22 W cm^{-3}. In addition, as illustrated in figures 5.19(d) and (e) [296], the device's cycling performance within the range of 0–1.8 V was assessed at 1 A g^{-1}, and it achieved a capacitance retention rate of 72.1% after 5000 charge–discharge cycles.

Zhao *et al* [297] devised and fabricated independent Ti_3C_2/FeOOH quantum dot (QD) hybrid films using electrostatic self-assembly technology (figure 5.20(a)) [297]. The amorphous FeOOH QDs served a dual purpose: they acted as interlayer pillars, preventing the self-stacking of Ti_3C_2 nanosheets, and they served as active materials with high theoretical capacitance values [297]. The TEM image in figure 5.20(b) [297] clearly shows that FeOOH QDs are uniformly distributed around the edges of the MXene nanosheets [298]. By combining figure 5.20(c) [297] and figure 5.20(d)

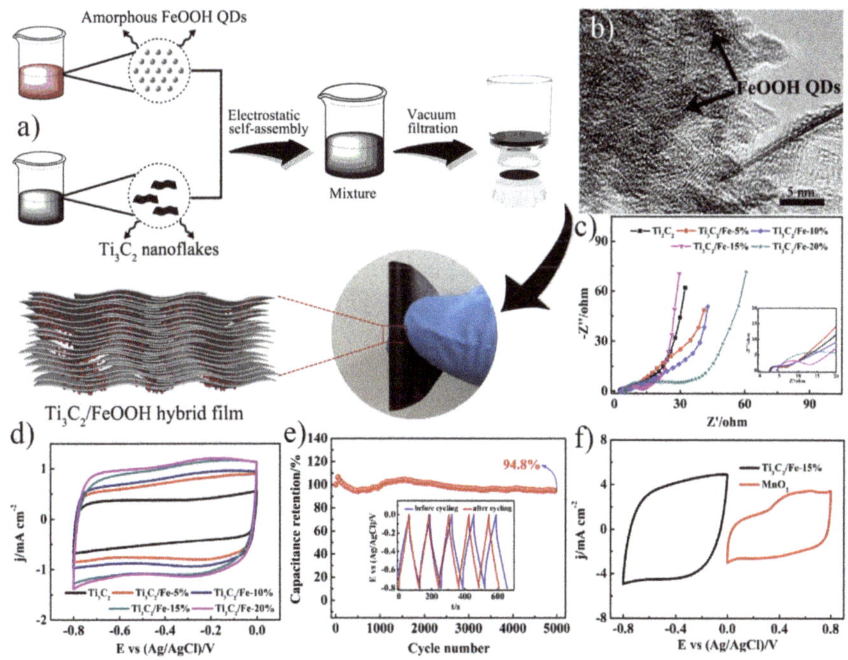

Figure 5.20. (a) A schematic diagram of the preparation of the Ti_3C_2/FeOOH hybrid films. (b) A TEM image of the Ti_3C_2/FeOOH hybrid films. (c) Nyquist plots of the different electrodes. (d) CV curves of different Ti_3C_2 and Ti_3C_2/FeOOH hybrid electrodes at 2 mV s^{-1} in a 1 M Li_2SO_4 electrolyte. (e) The cycling performance of the Ti_3C_2/Fe-15% electrode at 4 mA cm^{-2} (the inset shows the GCD curves before and after 5000 cycles). (f) CV curves of the Ti_3C_2/Fe-15% and MnO_2/CC electrodes at 10 mV s^{-1}.((a)–(f) Reprinted from [297], Copyright (2019), with permission from Elsevier.)

[297], it becomes apparent that the Ti_3C_2/Fe-15% hybrid film created at a 15% doping level of FeOOH QDs exhibited exceptional electrochemical performance. Notably, in a 1 M Li_2SO_4 electrolyte, the Ti_3C_2/Fe-15% electrode achieved a surface capacitance of 485 mF cm^{-2} at a scan rate of 2 mV s^{-1}, which was 2.3 times higher than that of a pure Ti_3C_2 thin film electrode (213 mF cm^{-2}), as depicted in figure 5.20(d) [297]. Furthermore, the capacitance of the Ti_3C_2/Fe-15% electrode remained at 94.8% even after 5000 cycles (figure 5.20(e)) [297]. Moreover, a flexible Ti_3C_2/Fe-15%//MnO_2/carbon cloth (CC) ASC was assembled, employing the Ti_3C_2/Fe-15% hybrid film as the negative electrode and MnO_2 grown on carbon cloth as the positive electrode. This ASC device boasted a wide voltage window of 1.6V (figure 5.20(f)) [297], providing an energy density of 40 mWh cm^{-2} and a power density of 8.2 mW cm^{-2}. Impressively, it retained 82% of its capacitance after 3000 cycles.

5.4.2.2 The use of MXene/conductive polymer composites in asymmetric supercapacitors

Researchers have not been entirely satisfied with the performance offered by MXene/conductive polymer SSCs and have turned their attention to MXene/conductive polymer ASCs, which offer greater operability [292, 299–301]. For instance, Fu *et al* [302] fabricated a graphene-encapsulated Ti_2CT_x MXene@PANI (GMP) structure, as depicted in figure 5.21(a) [302]. Upon examining the SEM images in figures 5.21(b)–(d) [302], it becomes evident that the chemically inert graphene bolstered the MXene framework, while the incorporation of conductive PANI provided additional active sites. The GMP electrode exhibited remarkable electrochemical properties, owing to its hierarchical nanostructure and complementary synergistic effects. In the presence of 1 M H_2SO_4, the GMP electrode achieved a mass specific capacitance of 635 F g^{-1} (~1143 F cm^{-3}) at 1 A g^{-1}, and even after 10 000 cycles, the retained capacitance was 97.54%, as illustrated in figure 5.21(e–g) [302]. Furthermore, an ASC was assembled with GMP and graphene serving as the positive and negative electrodes, respectively. This ASC device delivered an energy density of 42.3 Wh kg^{-1} at a power density of 950 W kg^{-1}, while maintaining a cycle stability of 94.25% after 10 000 cycles at 10 A g^{-1}.

Boota *et al* [303] fabricated composite electrodes, CP@rGO, by depositing PANI, PPy, and PEDOT onto reduced graphene oxide (rGO). These CP@rGO materials were utilized as both positive and negative electrodes in the assembly of a complete pseudocapacitive ASC, as depicted in figure 5.22(a) [303]. Among them, $Ti_3C_2T_x$//PANI@rGO exhibited the best performance. Operating in a 3 M H_2SO_4 electrolyte, the voltage window of the $Ti_3C_2T_x$//PANI@rGO device extended to 1.45 V, and it retained 88.42% of its capacitance after enduring 20 000 charge–discharge cycles, as illustrated in figure 5.22(b) [303]. Furthermore, the voltage windows of $Ti_3C_2T_x$//PPy@rGO and $Ti_3C_2T_x$//PEDOT@rGO also reached 1.4 V.

Li *et al* [304] synthesized 3D macroporous composites of PANI on $Ti_3C_2T_x$ MXene (designated as PANI@3D M-$Ti_3C_2T_x$) with an exceptional rate capability. This was achieved by depositing PANI onto the 3D structure of the $Ti_3C_2T_x$ MXene. Operating in a 3 M H_2SO_4 electrolyte, this electrode exhibited a remarkable volume capacitance of 1632 F cm^{-3} at a scan rate of 10 mV s^{-1}. Impressively, even

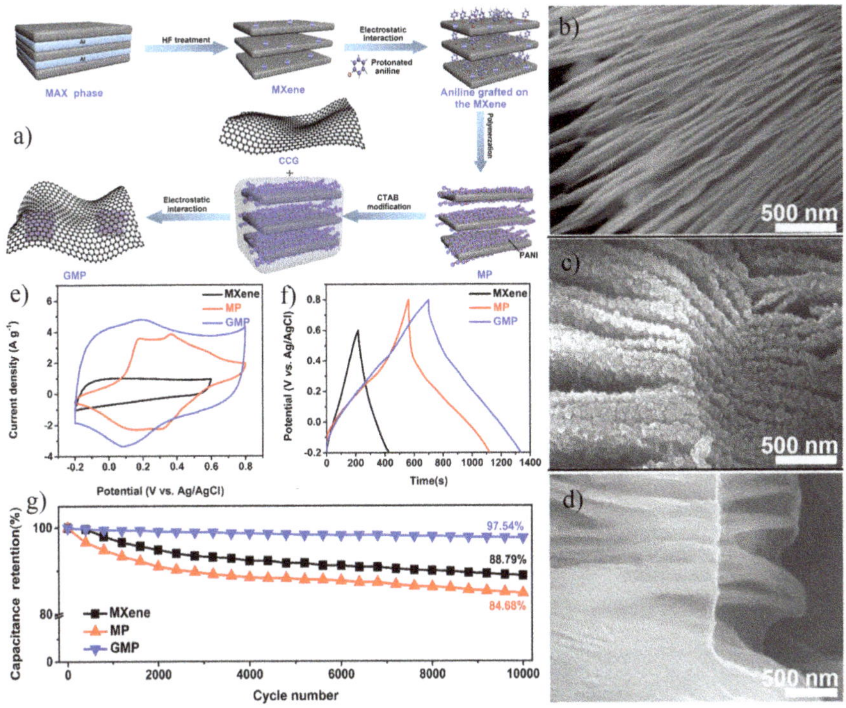

Figure 5.21. (a) A schematic diagram of the preparation of the GMP composite electrode. SEM images of the MXene (b), MXenePANI MP (c), and GMP (d). (e) CV curves of the GMP electrode at 5 mV s^{-1} in 1 M H_2SO_4. (f) GCD curves of the GMP electrode at 1 A g^{-1} in 1 M H_2SO_4. (g) Cycle performance of the GMP electrode after 10 000 charge–discharge cycles at 10 A g^{-1}. ((a)–(g) Reprinted with permission from [302], Copyright (2018), American Chemical Society.)

at a high scan rate of 5000 mV s^{-1}, it retained a substantial capacitance of 827 F cm^{-3}, as depicted in figure 5.22(c) [304]. Furthermore, to achieve energy matching, an MXene//PANI@M-$Ti_3C_2T_x$ ASC was constructed using PANI@M-$Ti_3C_2T_x$ as the positive electrode and MXene as the negative electrode. This ASC device delivered volumetric energy densities of 50.6 and 24.4 Wh l^{-1} at energy densities of 1.7 and 127 kW l^{-1}, respectively.

Li *et al* [255] employed a composite of rGO, carbon nanotubes (CNTs), and PANI alongside wavy $Ti_3C_2T_x$ MXene as the positive and negative electrodes, respectively. They devised an all-pseudocapacitive ASC, $Ti_3C_2T_x$//rGO/CNT/PANI, as depicted in figure 5.22(d) [255]. From figure 5.22(e) [255], it is evident that the CNTs were uniformly dispersed within the rGO/PANI layers, serving as spacers. In addition, figure 5.22(f) [255] illustrates the dense and folded structure of the wavy $Ti_3C_2T_x$ film, which provided numerous rapid ion transmission pathways. The high-density structures of both positive and negative electrodes, combined with a pseudocapacitive energy storage mechanism, contributed to a substantial volume capacitance for asymmetric devices. Operated in a 3 M H_2SO_4 electrolyte, the $Ti_3C_2T_x$//rGO/CNT/PANI ASC devices yielded energy densities of 70 Wh l^{-1} and

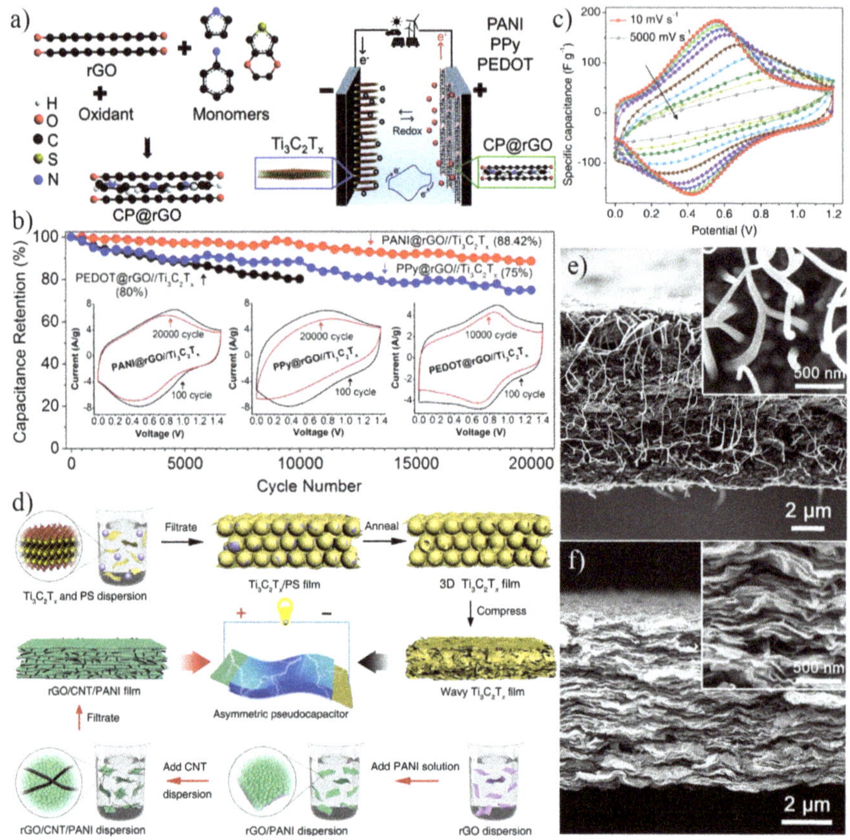

Figure 5.22. (a) An illustration of the preparation of CP@rGO composite electrodes by electrodeposition and the assembly of an ASC. (b) The cycle performance of the $Ti_3C_2T_x$//CP@rGO ASC (the inset shows the CV curves of different electrodes at different numbers of charge–discharge cycles). ((a) and (b) [303] John Wiley & Sons. © 2018 WILEY-VCH Verlag GmbH & Co. KGaA, Weinheim.) (c) CV curves of the $Ti_3C_2T_x$//CP@rGO ASC at different scan rates [304] John Wiley & Sons © 2019 WILEY-VCH Verlag GmbH & Co. KGaA, Weinheim. (d) An illustration of the preparation of all the pseudocapacitive $Ti_3C_2T_x$//rGO/CNT/PANI ASC. SEM images of the wavy $Ti_3C_2T_x$ film (e) and the rGO/CNT/PANI film (f). ((d)–(f) Reprinted from [255], Copyright (2020), with permission from Elsevier.)

34.8 Wh l^{-1} at power densities of 1.4 and 111 kW l^{-1}, respectively. In a separate study, Li *et al* [259] constructed ASC devices using MXene/PANI as the positive electrode and activated carbon (AC) as the negative electrode. This configuration achieved an energy density of 22.67 Wh kg^{-1} at a power density of 217 W kg^{-1}.

Wang *et al* [292] successfully prepared PANI/MXene inks and utilized scraper coating technology to create independent self-supporting PANI/MXene films, as illustrated in figure 5.23(a). The inclusion of nanoscale PANI particles enhanced the contact with the MXene substrate, thereby exposing a greater number of ion-accessible active sites, leading to the achievement of ideal pseudocapacitance (figure 5.23(b)) [292]. Figure 5.23(c) [292] demonstrates that the PANI/MXene composite film exhibited optimal performance in 1 M H_2SO_4, with a volumetric

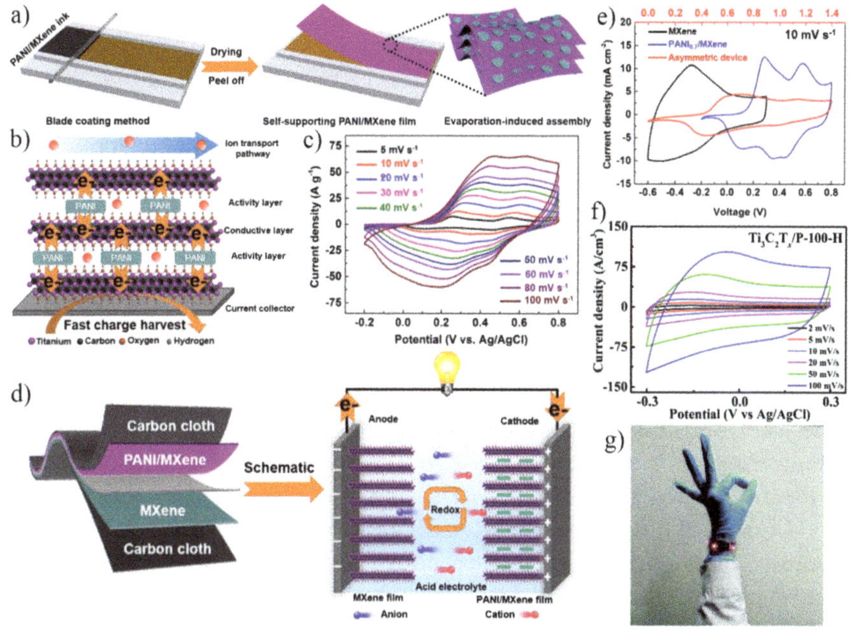

Figure 5.23. (a) A schematic of the independent self-supporting PANI/MXene film prepared using a scraper coating technology. (b) A schematic diagram of the transmission paths of ions and electrons in the PANI/MXene membrane electrode. (c) CV curves of the PANI/MXene at various scan rates. (d) A schematic diagram of the MXene//PANI/MXene ASC. (e) CV curves of the MXene electrode, the PANI/MXene electrode, and the MXene//PANI/MXene ASC at 10 mV s^{-1}. ((a)–(e) Reprinted from [292], Copyright (2021), with permission from Elsevier.) (f) CV curves of the $Ti_3C_2T_x$/P-100-H electrode at various scan rates [71]. (g) An optical image of a simple bracelet prepared using several $Ti_3C_2T_x$/P-100-H//rGO ASCs in series. ((f) and (g) Reproduced from [71] with permission from the Royal Society of Chemistry.)

capacitance of 1167 F cm^{-3} at 5 mV s^{-1}. Importantly, the presence of numerous PANI particles extended the operating voltage window of the composite electrode to 0.8 V (figure 5.23(d)) [292]. Furthermore, an all-pseudocapacitive ASC device was fabricated with the MXene and PANI/MXene serving as the negative and positive electrodes, respectively, as shown in figure 5.23(e) [292]. This ASC device delivered an energy density of 65.6 Wh l^{-1} at a power density of 1687.3 W l^{-1}.

Li *et al* [71] synthesized a high-performance $Ti_3C_2T_x$/PEDOT:PSS ($Ti_3C_2T_x$/P-100-H) composite film through a straightforward concentrated H_2SO_4 treatment. As depicted in figure 5.23(f), the results of three-electrode tests in a 1 M H_2SO_4 electrolyte revealed that the composite exhibited an impressive volumetric capacitance of 1065 F cm^{-3} at 2 mV s^{-1}. Furthermore, an exceptionally robust $Ti_3C_2T_x$/P-100-H//rGO ASC was assembled utilizing an rGO film as the positive electrode and the prepared composite film as the negative electrode. This ASC device delivered an energy density of 5.23 mWh cm^{-3} at a high power density of 7659 mW cm^{-3}. Given that a single ASC produced a voltage of 1.2 V, multiple ASCs connected in series were employed to create simple light-emitting bracelets and other applications, as illustrated in figure 5.23(g) [71]. The collective findings presented in this article underscore the substantial potential of MXene/conductive polymer composite films as electrodes for SCs.

Overall, SCs constructed using MXene/conductive polymer composite electrodes demonstrated excellent electrochemical performance and stable cycling characteristics, highlighting the promising prospects of MXene/conductive polymer-based SCs. Tables 5.2 and 5.3 provide a summary of the performance of MXene/conductive polymers in SC devices.

5.5 Methods used to modify hybrid LBHs

The modification of LBHs plays a pivotal role in addressing their inherent limitations, such as low specific capacitance and poor cycling stability. This can be achieved through a range of synthetic methods that leverage the fundamental properties of LBH materials and by making compositional and structural adjustments that enhance their performance in electrochemical applications [184–187]. Due to the 2D layered structure of LBHs, they are amenable to hybridization with various 1D and 2D nanomaterials such as CNTs, graphene, and MoS_2. The incorporation of polymers can further boost their conductivity [305–307]. When LBHs are integrated into hybrid nanostructures, the introduction of ionic vacancy defects and porous configurations can reshape their crystalline structure, providing additional active sites that lead to superior electrochemical performance. In addition, LBHs can be combined with materials such as metal sapphires and phosphides, which exhibit superior metallic properties compared to those of LBHs, resulting in composite materials [308, 309]. Generally, the enhanced performance of hybrid nanostructures is attributed to the synergistic effects arising from the heterogeneous interfaces formed between LBHs and other components. This chapter categorizes the modification methods for hybridized LBH nanostructures into four groups: the addition of components, the construction of defects, the generation of heterogeneous structures, and direct growth on substrates [310, 311]. Each of these techniques contributes to enlarging the surface area and exposing a greater number of active sites. It is worth noting, however, that each method induces distinct structural effects, which, in turn, exert varying impacts on the electrochemical properties of the hybridized LBHs.

5.5.1 The addition of components

The introduction of additional components into a pristine LBH can lead to alterations in the material's composition and structure [312]. This behavior, observed during the synthetic processes, enhances the synergy between different substances and can also modify the material's structure. In some cases, intercalation substances may improve the electrochemical properties of the material [95, 313–316]. However, it is important to consider compatibility between the raw materials when applying this modification method, as a well-thought-out theoretical study can yield unexpected benefits [317]. The preparation of composites with various components is a meaningful approach for obtaining novel and efficient electrode materials [318]. As an example, Ma *et al* [319] successfully produced Ni-embedded

Table 5.3. A summary of the research into MXene/conductive polymers in SC devices mentioned in this review.

Device	Electrolyte	Capacitance	Energy density at power density	Cycles, retained capacitance	References
$Ti_3C_2T_x$/P-100-H//rGO	1 M H_2SO_4	117 F cm^{-3}, 1.5 mA cm^{-2}	23 mWh cm^{-3} at 7659 mW cm^{-3}	—	[71]
MXene/PPy//MXene/PPy	1 M H_2SO_4	184 F g^{-1}, 10 mV s^{-1}	—	5000, 86.4%	[212]
$Ti_3C_2T_x$//rGO/CNT/PANI	3 M H_2SO_4	117 F g^{-1}, 10 mV s^{-1}	70 Wh l^{-1} at 1.4 kW l^{-1}	10 000, 80%	[255]
		85 F g^{-1}, 1000 mV s^{-1}	34.8 Wh l^{-1} at 1.1 kW l^{-1}		
AC//MXene/PANI	7 M KOH	262 F g^{-1}, 0.5 A g^{-1}	22.67 Wh kg^{-1} at 217 W kg^{-1}	10 000, 90.82%	[259]
400PPy175/$Ti_3C_2T_x$// 400PPy175/$Ti_3C_2T_x$	1 M H_2SO_4	258 F g^{-1}, 0.5 A g^{-1}	10.82 μWh mg^{-1} at 0.11 mW mg^{-1} 3.99 μWh mg^{-1} at 1.89 mW mg^{-1}	5000, 85.6%	[263]
$Mo_{1.33}C$ MXene/PEDOT:PSS// $Mo_{1.33}C$ MXene/PEDOT: PSS	1 M H_2SO_4	568 F g^{-1}, 0.5 A g^{-1}	33.3 mWh cm^{-3} at 19 470 mW cm^{-3}	10 000, 90%	[266]
$Ti_3C_2T_x$//$Ti_3C_2T_x$	Seawater	27 F cm^{-3}, 0.25 A g^{-1}	1.74×10^{-3} Wh cm^{-3} at 0.15 W cm^{-3}	5000, 96.6%	[275]
		13 F cm^{-3}, 3 A g^{-1}	0.68×10^{-3} Wh cm^{-3} at 1.53 W cm^{-3}		
$Ti_3C_2T_x$//$Ti_3C_2T_x$	1 M Li_2SO_4	633 F cm^{-3}, 2 mV s^{-1}	22 Wh l^{-1}	10 000, 95.3%	[276]
3D-PMCF//3D-PMCF	3 M H_2SO_4	375 F g^{-1}, 5 mV s^{-1} 92 F g^{-1}, 10 000 mV s^{-1}	9.2 Wh kg^{-1}	10 000, 95.9%	[277]

(*Continued*)

Table 5.3. (*Continued*)

Device	Electrolyte	Capacitance	Energy density at power density	Cycles, retained capacitance	References
Ti_2CT_x//Ti_2CT_x	1 M KOH	452 F cm^{-3}, 2 mV s^{-1} 209 F cm^{-3}, 100 mV s^{-1}	35 mWh cm^{-3} at 0.49 W cm^{-3} 17 mWh cm^{-3} at 191 W cm^{-3}	3000, 100%	[278]
RAMX//RAMX	3 M H_2SO_4	—	40 Wh l^{-1} at 0.83 kW l^{-1}21 Wh l^{-1} at 41.5 kW l^{-1}	20 000, 91.14%	[280]
PDT/$Ti_3C_2T_x$//PDT/$Ti_3C_2T_x$	0.5 M H_2SO_4	284 mF cm^{-2}, 0.5mA cm^{-2}	24 mWh cm^{-3} at 502 W cm^{-3}	10 000, 100%	[289]
Ti_3C_2/PANI-NTs//Ti_3C_2/PANI-NTs	1 M H_2SO_4	586 F g^{-1}, 0.1 A g^{-1}	25.6 Wh kg^{-1} at 153.2 W kg^{-1} 13.2 Wh kg^{-1} at 1610.8 W kg^{-1}	4000, 81.1%	[290]
MXene//PANI/MXene	1 M H_2SO_4	231 F cm^{-3}, 10 mV s^{-1}	65.6 Wh l^{-1} at 1687.3 W l^{-1} 40.3 Wh l^{-1} at 10 354 W l^{-1}	5000, 87.5%	[292]
Ni-dMXNC//$Ti_3C_3T_x$	1 M KOH	—	1.04×10^{-3} Wh cm^{-3} at 0.22 W cm^{-3}	5000, 72.1%	[296]
$Ti_3C_2T_x$/Fe-15%//MnO_2/CC	1 M Li_2SO_4	115 mF cm^{-2}, 2 mA cm^{-2}	40 mWh cm^{-2} at 8.2 mW cm^{-2}	3000, 82%	[297]
GMP//graphene	1 M H_2SO_4	68 F g^{-1}, 10 A g^{-1}	42.3 Wh kg^{-1} at 950 W kg^{-1} 25 Wh kg^{-1} at 18 000 W kg^{-1}	10 000, 94.25%	[302]
MXene//PANI@MXene	3 M H_2SO_4	87 F g^{-1}, 10 mV s^{-1}	50.6 Wh l^{-1} at 1.7 kW l^{-1} 24.4 Wh l^{-1} at 127 kW l^{-1}	—	[304]

carbon nanofiber/NiAl-LBH hybrids. Initially, they fabricated Ni-containing carbon nanofibers using electrostatic spinning and thermal treatment, which generated numerous active sites and increased ion transport pathways. Subsequently, the Ni-embedded carbon nanofibers were combined with nanostructured NiAl-LBH through a hydrothermal process (figure 5.24(a)). Their experimental results indicated that the former optimized the microstructure of NiAl-LBH, reduced its aggregation, and improved its electrochemical performance. A specific capacitance of 1228.2 C g^{-1} at 1 A g^{-1} was achieved by 3% Ni-embedded carbon nanofiber/NiAl-LBH (3%-Ni-C/NiAl-LBH) (figure 5.24(b)). This material exhibited an initial capacity retention of 88.6% and, remarkably, an ASC produced using this material achieved an energy density of 74.9 Wh kg^{-1} at 800 W kg^{-1}. Furthermore, the device retained a capacity of 91.4% after 10 000 cycles at 6 A g^{-1} (figure 5.24(c)).

NiCo-LBHs exhibit large layer spacing and a high ion exchange capacity, but their poor electrical conductivity, pronounced agglomeration, and structural defects limit their energy storage capacity. To address these issues, Wu *et al* [320] innovatively prepared zeolite imidazole framework-67 (ZIF-67) sulfur-doped

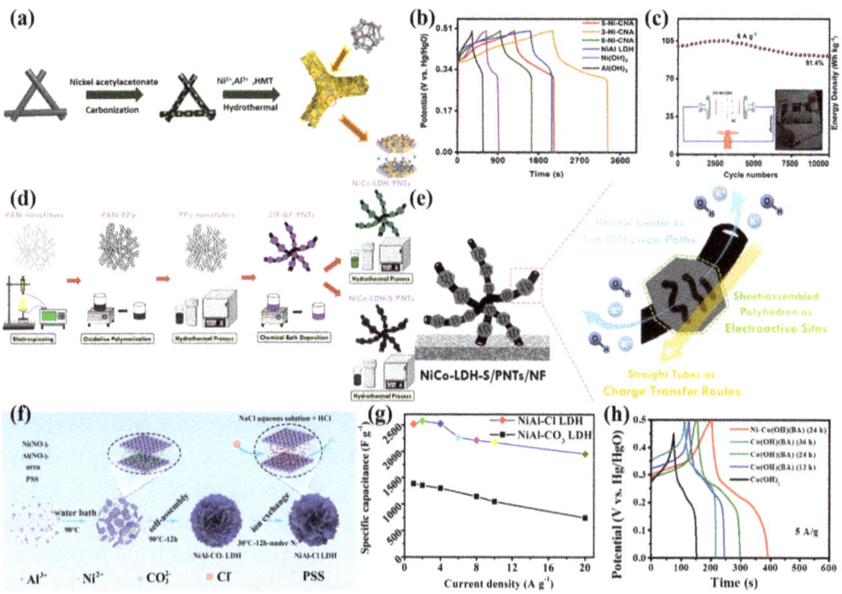

Figure 5.24. (a) Details of the preparation strategy used to obtain hierarchical structured Ni-CAN, (b) the GCD curves of the obtained samples at 1 A g^{-1} with a potential window of 0–0.5 V, (c) the cycling performance of 3%-Ni-C/NiAl-LBH//AC at 10 A g^{-1}. ((a)–(c) Reprinted from [319], Copyright (2022), with permission from Elsevier.) (d) An illustration of the synthetic processes used to obtain NiCo–LBH/polypyrrole nanotubes (PNTs) and NiCo-LBH-S/PNTs, (e) an illustration of the morphological features of NiCo-LBH-S/PNT/nanofiber (NF). ((d) and (e) Reprinted from [320], Copyright (2022), with permission from Elsevier.) (f) The specific capacitances of the NiAl-CO_3 LBH and NiAl-Cl LBH electrodes at a variety of current densities. (g) The specific capacitances of two electrodes at different current densities. ((f) and (g) reprinted from [321], Copyright (2022), with permission from Elsevier.) (h) GCD curves of those electrodes at a current density of 5 A g^{-1} (reprinted from [322], Copyright (2022), with permission from Elsevier).

NiCo-LBH and PPy nanotube composites (NiCo-LBH-S/PNTs) using electrospinning and hydrothermal methods (figure 5.24(d)). The 1D hollow PPy, characterized by a high aspect ratio, provided direct charge transfer pathways and extensive contact with the electrolyte. When the sulfur content was 7%, the NiCo-LBH-S/PNTs demonstrated a specific capacitance of 1936.3 F g^{-1}. A device assembled with a graphene anode and a NiCo-LBH-S/PNT cathode attained an energy density of 16.28 Wh kg^{-1} at 650 W kg^{-1}. Remarkably, it achieved a capacity retention rate of 74% after 8000 cycles.

Alterations in the anions positioned between LBH layers significantly impact the resulting performance. Various anion intercalations in LBHs can lead to distinct layer spacings; a larger layer spacing facilitates contact between the active atoms on the LBH plate layer and OH^- ions, ultimately enhancing electrochemical properties. Lv *et al* [321] prepared spherical NiAl-Cl LBH using chloride ions as the interlayer anions [321]. NiAl-Cl LBH, with its wider interlayer spacing, enhanced contact between the active atoms in the hydromagnesite-like layers and OH^- ions, resulting in improved electron transport kinetics, increased utilization of the active material, and better multiplicative performance. The specific capacitance of NiAl-Cl LBH at 20 A g^{-1} was 77.5% of that at 1 A g^{-1}. In addition, the gaps within the interlaced hierarchical structure facilitated the accommodation of volume changes during the reaction, ensuring the structural stability of NiAl-Cl LBH during redox reactions. Compared to materials with carbonate ions as the interlayer anions, NiAl-Cl LBH demonstrated superior performance (figure 5.24(g)). The energy density of a NiAl-Cl LBH//AC ASC reached 53.9 Wh kg^{-1} at 1540 W kg^{-1} and the SC maintained a specific capacitance of 94.1% even after 1000 cycles.

Deng *et al* [322] employed a hydrothermal method to synthesize ultrathin nickel-doped inorganic–organic cobalt hydroxide nanoribbons with benzoate anion intercalation (NiCo(OH)(BA)). Importantly, this synthetic approach did not require the use of binders or surfactants, leading to enhanced electrochemical performance attributed to the enlarged interlayer spacing and improved ion flow efficiency. The results revealed a specific capacitance of 1664 F g^{-1} at 5 A g^{-1} (figure 5.24(h)) and an 83% capacity retention after 8000 cycles. Furthermore, an ASC device (NiCo(OH)(BA)//AC) that incorporated NiCo(OH)(BA) achieved an energy density of 47.5 Wh kg^{-1} at 850 W kg^{-1} and maintained a capacity of 91% even after 8000 cycles. Saber *et al* [323] incorporated silicon into Co-LBH nanospheres to create Si/Co-LBH nanofibers, utilizing cyanate anions as the structural framework to construct these nanospheres. The structure of these nanofibers could be adjusted by controlling the synthetic conditions and silicon content to further enhance their properties. The results revealed that the morphological transformation from nanoparticles or flat plates to nanofibers significantly improved the specific capacitance of Si/Co-LBH to 621.5 F g^{-1}, along with an impressive cycling stability of 84.5%. These outcomes were attributed to the unique nanofiber morphology and the cooperative effects arising from the capacitive properties of silicon and the pseudocapacitive properties of carbon.

The size of the layer spacing of LBHs is one of the key factors affecting their electrochemical properties, and increasing the layer spacing by suitable methods can greatly improve their performance. Composites based on NiCr-LBH and polyoxotungstate nanoclusters (NiCr-LBH-POW) were fabricated by Padalkar *et al* [324] using exfoliative recombination. The interlayer intercalation hybridization of POW nanoclusters in NiCr-LBH created a cumulate frame (figure 5.25(a)), significantly expanding the spacing between layers. An Ni–Cr–LBH–POW (NCW)-2//rGO aqueous hybrid SC (AHSC) device (figure 5.25(b)) was constructed using the NiCr-LBH-POW nanohybrid material as the positive electrode. It achieved an energy density of 34 Wh kg^{-1} at 1.32 kW kg^{-1} and retained 86% of its capacitance after 10 000 charge–discharge cycles. Mahmood *et al* [325] developed a unique synthetic strategy based on a PANI-doped 2D cobalt–iron LBH (CoFe-LBH/P) nanomaterial. Their results showed that among all the polymers tested, the optimal concentration of PANI created nanopores on the CoFe-LBH nanoflakes. These ordered pores increased redox sites and promoted efficient ion movement. The optimized CoFe-LBH/P2 displayed a specific capacitance of 1686 F g^{-1} at 1 A g^{-1} (figure 5.25(d)) and exhibited excellent cycling performance (98% retention over 10 000 cycles).

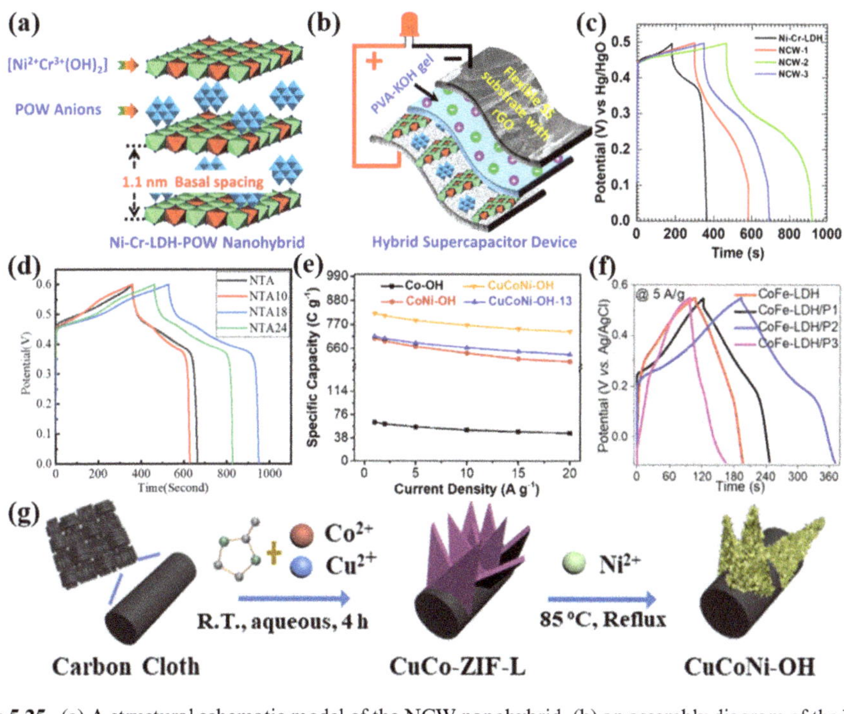

Figure 5.25. (a) A structural schematic model of the NCW nanohybrid, (b) an assembly diagram of the NCW-2//rGO AHSC, (c) GCD curves of pristine NiCr-LBH and NCW nanohybrids at 1 A g^{-1}. ((a)–(c) Reprinted from [324], Copyright (2022), with permission from Elsevier.) (d) GCD curves of different materials at 5 mA cm^{-2} (reprinted from [327], Copyright (2021), with permission from Elsevier). (e) The corresponding specific capacitances of various hydroxide electrodes at 1–20 A g^{-1}. (f) GCD curves of the CoFe-LBH/P2 electrode (reproduced from [326], Copyright (2022), with permission from Elsevier). (g) A schematic of the synthetic procedures of CuCo-ZIF-L and CuCoNi–OH. ((e) and (g) [325] John Wiley & Sons. © 2022 Wiley-VCH GmbH.)

Furthermore, an asymmetric aqueous device (CoFe-LBH/P2//AC) was prepared that achieved an energy density of 75.9 Wh kg^{-1}at 1124 W kg^{-1} and 97.5% stability after 10 000 cycles.

Deng *et al* [326] synthesized a hierarchical array of scaled trimetallic hydroxides (CuCoNi–OH) using a moderate alkaline hydrolysis strategy and rational nanostructure design. They employed a bimetallic 2D zeolite imidazole framework (CuCo-ZIF-L) as a template for this process (figure 5.25(g)). The resulting hierarchical porous structure provided a large surface area for active site exposure and facilitated rapid ion diffusion. In addition, the synergistic effect of the multiple metals in CuCoNi–OH enhanced electrical conductivity and supported redox reactions, leading to improved electrochemical kinetics when used in a SC. When used as a battery-type electrode, the CuCoNi–OH electrode exhibited a specific capacitance of 821.6 C g^{-1} at 1 A g^{-1} and retained 89.8% of its capacity at 20 A g^{-1} (figure 5.25(e)). The assembled device demonstrated remarkable energy density and power density. This synthetic strategy can be used to prepare various bimetallic zeolite imidazole frameworks and their corresponding metal hydroxides, offering an effective approach for the rational design of materials for electrochemical energy storage and conversion. Wang *et al* [327] developed a synthetic method for NiTiAl-LBH that consisted of incorporating a small amount of aluminum into the NiTi-LBH substrate layer and subsequently etching some of the aluminum using a sodium hydroxide solution. This process resulted in materials with increased specific surface area, specific capacitance, and rate performance. After 18 h of etching, the samples achieved a specific surface area of 203 m^2 g^{-1}, and the specific capacitance at 5 mA cm^{-2} reached 3483 mF cm^{-2} (figure 5.25(f)). These materials also exhibited improved structural stability compared to that of NiTi-LBH. Hybrid devices assembled using the etched samples demonstrated an energy density of 45.1 Wh kg^{-1} at 16 000 W kg^{-1}. This alkaline etching method effectively enhanced the porosity of aluminum-containing layered dihydroxy talc, leading to improved specific capacitance and rate performance; it holds promise for various applications.

Creating highly porous structures with large specific surface areas is a well-established strategy for enhancing ion and electron transport and insertion/de-insertion processes in materials intended for electrochemical applications. However, challenges still exist in this approach, including limitations in energy storage capacity due to the simplicity of the porous structure and the potential for severe performance degradation during long-term electrochemical cycling. These challenges need to be addressed to develop advanced materials with improved energy storage capabilities and long-term stability. Inspired by the natural geographical structure of forests, Liu *et al* [328] designed a Ni/Co-LBH on a metal–organic framework of ZnO nanotubes grown on transparent conductive substrates with different porous structures to simulate a 'rock-soil-tree-leaf' system ((a)). The enhanced specific surface area of 3D ZnO@Ni/Co-LBH, combined with the improved OH^- trapping ability of ZnO, the increased electrochemical activity due to Ni/Co doping, and the hybrid charge storage behavior, collectively contributed to the excellent specific capacitance and durability of these materials. Among the five ZnO@Ni/Co-LBH films studied, LBH-3 stood out for its exceptional conductivity and energy storage performance, boasting a charge capacity of 507.2 C g^{-1} at 0.1 mA cm^{-2} and a capacity retention of 72.1% after 10 000 cycles. Devices based on LBH-3 also demonstrated outstanding durability, achieving an energy density of 7.7 uWh cm^{-2} at 375.0 pW cm^{-2}. Furthermore, the ZnO@Ni/Co-LBH device had the capability to automatically switch optical functions through solar energy harvesting and charge storage/release. These findings open up exciting possibilities for the development of next-generation smart technologies, contributing to a sustainable and habitable future.

Zhou *et al* [329] synthesized ultrathin cobalt-nickel-magnesium LBH (CoNiMg-LBH) nanosheets with abundant oxygen vacancies at room temperature through a sacrificial magnesium-based replacement reaction, which represented a significant achievement (figure 5.26(b)). This process introduced self-doping and the mild reduction of magnesium, leading to an increased concentration of oxygen vacancies in the material. These oxygen vacancies, in turn, enhanced the electrochemical charge transfer efficiency and improved the adsorption capacity of the electrolyte. Density functional theory (DFT) calculations further supported this by showing that Mg^{2+} doping reduced the energy required to generate oxygen vacancies, thereby increasing their concentration. In terms of practical performance, a CoNiMg-LBH//AC device exhibited a specific capacitance of 333 C g^{-1} at 1 A g^{-1} and an energy density of 73.9 Wh kg^{-1} at 0.8 kW kg^{-1}. Even after 5000 cycles, the device experienced only a 13% capacity loss. This discovery underscores the beneficial role of magnesium in regulating oxygen vacancies to enhance the performance of SCs. It opens up opportunities to expand the range of high-quality materials used in SCs, potentially leading to improved energy storage devices. Wang *et al* [330] achieved

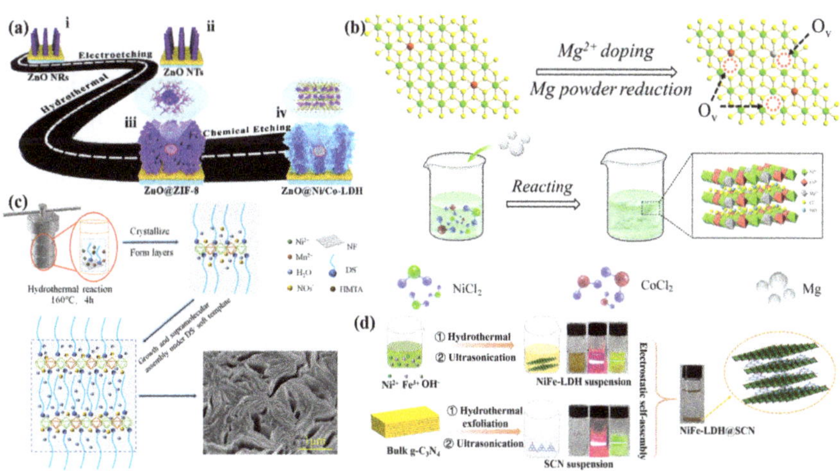

Figure 5.26. (a) The synthetic scheme of the ZnO@Ni/Co-LBH film (reproduced from [328] with permission from the Royal Society of Chemistry). (b) A schematic illustration of the fabrication strategy used to produce oxygen-vacancy-abundant CoNiMg-LBH (reprinted from [329], Copyright (2022), with permission from Elsevier). (c) The growth mechanism of LBH using SDS as a soft template (reprinted from [330], Copyright (2022), with permission from Elsevier). (d) The synthetic process of the NiFe-LBH@SCN (reprinted from [331], Copyright (2022), with permission from Elsevier).

significant improvements in loading and capacitive performance through a one-step hydrothermal process, loading NiMn–LBH onto a conductive nanofiber (NF) substrate. They utilized sodium dodecyl sulfate (SDS) as both an intercalator and a soft template. An electrode treated with 4 mM SDS exhibited an impressive areal capacitance of 6311 mF cm^{-2} at an operating current density of 5 mA cm^{-2}. When this electrode was integrated into a hybrid SC, it demonstrated an outstanding energy density of 34.61 Wh kg^{-1} at a power density of 831 W kg^{-1}. Moreover, it exhibited remarkable capacitance retention, reaching 129% after 5000 charge–discharge cycles at 4 A g^{-1} and 85% after 10 000 cycles at 10 A g^{-1}. Li *et al* [331] successfully fabricated soluble graphite nitride (SCN) nanosheet-supported NiFe-LBH through electrostatic self-assembly, as depicted in figure 5.14(d). This resulted in a one-layer, high-performance electrode (denoted by NiFe-LBH@SCN) designed for SCs. The optimized structure exhibited a remarkable specific capacitance of 1060.4 F g^{-1} at an operating current density of 1 A g^{-1}. When integrated into a hybrid SC, this electrode yielded an impressive energy density of 68.7 Wh kg^{-1} at a power density of 827.5 W kg^{-1}. In addition, it displayed excellent durability, with a capacitance retention rate of 83.3% after 8000 charge–discharge cycles.

5.5.2 Creating defects in materials

Creating defects is another promising approach for enhancing the electrochemical properties of LBHs. This method involves modifying the electronic structure and increasing the number of active sites within the LBH structure [332]. This method introduces additional pore structures into the material, making it easier for the electrolyte to penetrate the electrode material. However, controlling the conditions for this method can be challenging, and it may lead to the formation of excessive defects. Therefore, it is important to carefully select appropriate raw materials and control the conditions when creating material defects [333]. Chu *et al* [334] utilized copper (Cu) as a dopant in the preparation of CuCo-LBH. The resulting structure was grown on an NF substrate through an *in situ* hydrolytic process. Electron images of the CuCo-LBH are depicted in figures 5.27(a) and (b). The addition of Cu significantly increased the local electron density, thereby enhancing electronic conductivity and facilitating charge transfer. CuCo-LBH electrodes exhibited excellent capacitive performance, as shown in figures 5.27(c) and (d). A CuCo-LBH//AC device demonstrated an energy density of 22 Wh kg^{-1} and maintained 91.3% stability after 10 000 cycles. This study demonstrates the structural tuning of LBH materials to introduce lattice defects for the enhancement of their performance, which holds great promise for the development of superior SC electrode materials in the future.

Lei *et al* [335] synthesized NiCo-LBH on a nanofiber substrate (D-NiCo-LBHs/NF) utilizing the memory effect, as depicted in figure 5.27(e). DFT calculations revealed that the presence of Co vacancies led more electrons to approach the Fermi energy level more closely, thereby enhancing conductivity and promoting charge transfer. The SEM image in figure 5.27(f) illustrates the microscopic morphology of D-NiCo-LBH/NF, in which vertically aligned NiCo-LBH nanosheets uniformly

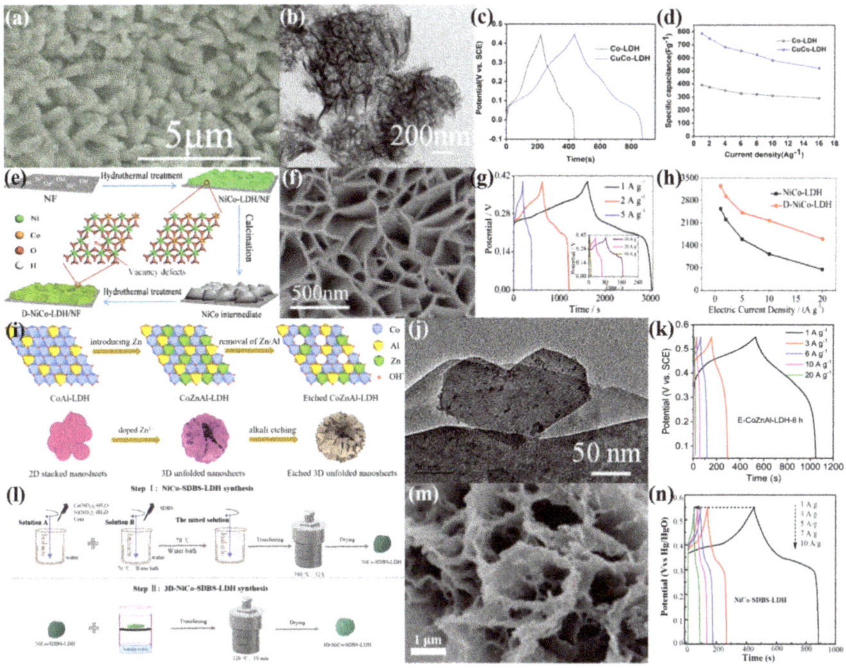

Figure 5.27. (a) An SEM image of CuCo-LBH, (b) a TEM image of CuCo-LBH, (c) GCD curves, and (d) rate capability of Co-LBH and CuCo-LBH. ((a)–(d) Reprinted with permission from [334], Copyright (2022), American Chemical Society.) (e) A schematic illustration of the introduction of vacancy defects into NiCo-LBH through the memory effect, (f) an SEM image of D-NiCo-LBH/NF nanosheet arrays, (g) GCD curves of the D-NiCo-LBH/NF electrode at different current densities, (h) capacitance retention of D-NiCo-LBH/NF and NiCo-LBH/NF electrodes at different current densities. ((e)–(h) Reprinted from [335], Copyright (2022), with permission from Elsevier.) (i) Zn^{2+}-dopant-induced morphological change in CoAl-LBH and Al/Zn dual ion etching of CoZnAl-LBH in an alkaline solution, (j) a TEM image of E-CoZnAl-LBH-8 h, (k) GCD curves of E-CoZnAl-LBH-8 h at different scan rates and current densities. ((i)–(k) Reprinted from [336], Copyright (2021), with permission from Elsevier.) (l) A diagram of the synthesis of 3D-NiCo-SDBS-LBH, (m) SEM images of 3D-NiCo-SDBS-LBH at different magnifications, (n) GCD curves of NiCo-SDBS-LBH at different scan rates and current densities. ((l)–(n) Reproduced from [337], CC BY 4.0.)

cover the surface of the nanofiber. The synthesized D-NiCo-LBH/NF exhibited a specific capacitance of 3200 F g^{-1} at 1 A g^{-1}, as shown in figures 5.27(g) and (h). An ASC constructed using D-NiCo-LBH/NF achieved an energy density of 53 Wh kg^{-1} at 752 W kg^{-1} and retained 94.7% of its capacity after 5000 cycles. Yang *et al* [336] initiated their research by transforming 2D dense CoAl-LBH into 3D loosely stacked CoZnAl-LBH through Zn^{2+} doping, resulting in morphological changes in the LBH, as illustrated in figures 5.27(i) and (j). In addition, the partial dissolution of Zn/Al double ions between the LBH lamellae in an alkaline solution led to a significant alteration in the electronic environment of the Co surface and the generation of a certain concentration of oxygen defects in CoZnAl-LBH. This transformation improved the multiplicative performance and cycling stability of CoZnAl-LBH nanosheets. In comparison to unetched CoZnAl-LBH, E-CoZnAl-LBH-8 hours exhibited a specific capacitance of 946 F g^{-1} at 1 A g^{-1}, as shown in figure 5.27(k), and a 92.3% cycle life after undergoing 4000 cycles. An E-CoZnAl-LBH-8 h//AC ASC was assembled, which achieved an energy density of 36.75 Wh kg^{-1} at 400 W kg^{-1} and a 72.7% cycle life after 8000 cycles. The doping and double ion etching strategies proposed in this study provide theoretical guidance and an experimental basis for the development of SCs with excellent properties.

Porous structures and surface defects are important factors in improving the performance of SCs. Zhong *et al* [337] employed a one-step hydrothermal method to prepare NiCo-SDBS-LBH, utilizing sodium dodecylbenzene sulfonate (SDBS) as the anionic surfactant. Subsequently, they designed and synthesized 3D connected porous flower-like 3D-NiCo-SDBS-LBH microspheres using a gas-phase hydrazine hydrate reduction method, as depicted in figure 5.27(l). The outcome demonstrated that hydrazine hydrate reduction not only introduced numerous pores, leading to the formation of oxygen vacancies, but also roughened the surface of the microspheres, as illustrated in figure 5.27(m). These alterations collectively contributed to the electrochemical activity of 3D-NiCo-SDBS-LBH, resulting in a specific capacitance of 1148 F g^{-1} at 1 A g^{-1}, as shown in figure 5.27(n) (approximately 1.46 times higher than that of NiCo-SDBS-LBH). Furthermore, the retention rate after 4000 cycles was 94%. In addition, an assembled 3D-NiCo-SDBS-LBH//AC device exhibited an energy density of 73.14 Wh kg^{-1} at 800 W kg^{-1} and a cycle life of 95.5% after undergoing 10 000 cycles.

For electrochemical materials, domain boundaries are considered to work as active sites because of their defect enrichment. Nevertheless, LBHs can easily form single-crystal nanosheets because of their 2D lattice [338]. Many studies have designed layered structures to provide abundant active sites and speed up mass transfer [339]. Ban *et al* [339] proposed a method to introduce low-angle grain boundaries (LAGB) into NiCo-LBH flakes. The defect-rich nanoflakes ultimately formed cages with a hollow structure, as depicted in figure 5.28(a). The hierarchical structure and the generation of grain boundaries was attributed to the Ni^{2+}/Co^{2+} ratio during the ‘etch growth’ process. Domain boundary defects also contributed to the preferential formation of oxygen vacancies. Furthermore, DFT calculations revealed that Co substitution played a pivotal role in creating lattice defects and establishing domain boundaries. Consequently, the NiCo-LBH-2 electrode material exhibited a substantial increase in specific capacitance, reaching 899 C g^{-1} at 1 A g^{-1}. Notably, a NiCo-LBH-2//AC asymmetric capacitor achieved a maximum energy density of 101.1 Wh kg^{-1} at 1.5 kW kg^{-1}.

Designing metal cation defects with the desired structure to improve electrochemical performance is a major challenge. Wu *et al* [332] devised a method to design ultrathin ZnNi-LBH nanosheets with Zn-rich vacancies uniformly anchored on a CuO nanowire backbone, creating high-performance capacitive electrodes. This was achieved through ZIF-8 derivatization, as depicted in figure 5.28(b). The optimized V_{Zn}-deficient electrode exhibited an areal capacitance of 3967 mF cm^{-2} at 2.0 mA cm^{-2}. Furthermore, a device composed of V_{Zn}-deficient samples and AC achieved a maximum energy density of 1.03 mWh cm^{-3} and a power density of 9.3 mW cm^{-3}. A mechanistic study revealed that V_{Zn} played a crucial role in modulating the electronic structure of the ZnNi-LBH nanosheets, thereby enhancing

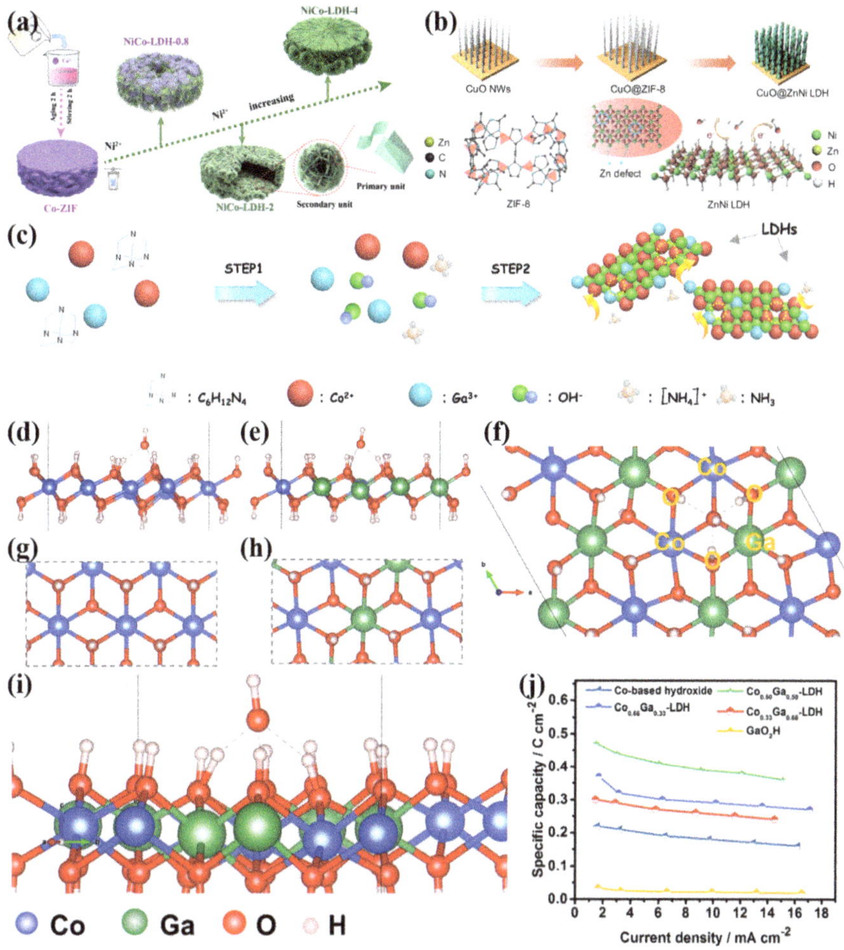

Figure 5.28. (a) The synthetic strategy used to produce NiCo-LBH (reprinted from [339], Copyright (2022), with permission from Springer). (b) A schematic illustration of the fabrication of the hierarchical ZnNi-LBH architectures with rich V_{Zn} through a ZIF-derived method (reprinted from [332], Copyright (2021), with permission from Elsevier). (c) The synthetic mechanism of defect-rich porous ultrathin LBH. The crystal structure and adsorption sites visible on the side and top views of (d, g) bulk Co-LBH, (e, h) bulk CoGa-LBH, and (f, i) oxygen-defect-rich $Co_{0.50}Ga_{0.50}$-LBH. (j) The rate performance of GaOOH, Co-based hydroxide, and Co_xGa_y-LBH electrodes. ((c)–(j) Reprinted from [341], Copyright (2022), with permission from Elsevier.)

electronic conductivity and surface faradaic reactions. This research sheds light on the influence of metal cation defects on electrochemical activity at the atomic level. Coincidentally, Kim *et al* [340] devised a lattice engineering method to simultaneously control defects and porosity by adjusting the elastic deformation and chemical interactions of nanosheets during restacking. This approach increased intercalation size and reduced charge density, effectively raising the oxygen vacancy content and enhancing porosity. The resulting defect-rich CoAl-LBH-NO_3 nanohybrids demonstrated excellent performance as electrodes, achieving a specific capacitance of 2230 F g^{-1} at 1 A g^{-1}. In addition, DFT calculations revealed a strong correlation between overpotential (capacitance) and defect content (stacking) number, underscoring the significance of defect (stacking) structure optimization in enhancing energy function. Zhang *et al* [341] developed a straightforward one-step method to create oxygen-rich 3D $Co_{0.50}Ga_{0.50}$-LBH assembled into porous ultrathin nanosheets (figure 5.28(c)). This synthetic approach introduced numerous holes into the ultrathin LBH nanosheets, leading to a high concentration of oxygen vacancies in $Co_{0.50}Ga_{0.50}$-LBH. The synergistic effect of the oxygen vacancies and the incorporation of Ga ions enhanced the OH^- adsorption capacity of the LBH nanosheets, resulting in exceptional SC properties. These LBH nanosheets achieved a specific capacitance of 0.62 C cm^{-2} at 10 mV s^{-1}. Furthermore, $Co_{0.50}Ga_{0.50}$-LBH//AC ASCs demonstrated outstanding energy density and long-term performance. This discovery also paves the way for broader applications of porous ultrathin LBH nanosheets in energy storage and related fields.

5.5.3 The generation of heterogeneous structures

The construction of heterogeneous structures is another important strategy for enhancing the electrochemical properties of LBHs [342]. The interfacial regions of heterogeneous materials can provide opportunities to enrich the number of active sites [343] and promote electron transfer [344, 345]. In addition, strong electronic interactions in the heterogeneous interfaces facilitate the enhancement of electronic and ionic conductivity [346] and redox reaction kinetics [347]. However, the tightness of the heterogeneous interface is difficult to control, and the existence of voids between the interfaces affects the electrochemical properties. Adding certain ingredients such as binders in the synthetic process may solve the tightness problem. Luo *et al* [348] successfully synthesized layered $NiCo_2O_4$@NiFe-LBH heterostructures using a sequential series of hydrothermal methods, heat treatment, and electrodeposition. In this unique structure, $NiCo_2O_4$ served as a scaffold, enhancing conductivity and accelerating electron transfer. Simultaneously, the NiFe-LBH nanosheets offered a high surface area, providing numerous active sites for electrochemical reactions. The 3D layered structure also facilitated the diffusion of electrolyte ions, as depicted in the SEM images shown in figure 5.29(a). Consequently, the synergistic collaboration between $NiCo_2O_4$ and NiFe-LBH resulted in exceptional electrochemical performance for $NiCo_2O_4$@NiFe-LBH-150/CC, including a notable area specific capacitance (1.09 F cm^{-2} at 1 mA cm^{-2}) (figures 5.29(b) and (c)), low charge transfer resistance (0.35 Ω), and excellent cycling stability.

Huang *et al* [349] obtained a unique array of intercalated pseudocapacitive properties and battery-type electrode materials in the form of LBH nanosheets using a simple and environmentally friendly two-step electrodeposition technique. The electrode materials comprised MoO_3 and NiCo-LBH, which were directly grown on a 3D conductive NF substrate, resulting in a binder-free 2D ultrathin cross-layered heterogeneous structure known as NiCo-LBH@MoO_3/NF. This heterojunction exhibited a specific capacitance of 952.2 C g^{-1} at 1 A g^{-1} and maintained an

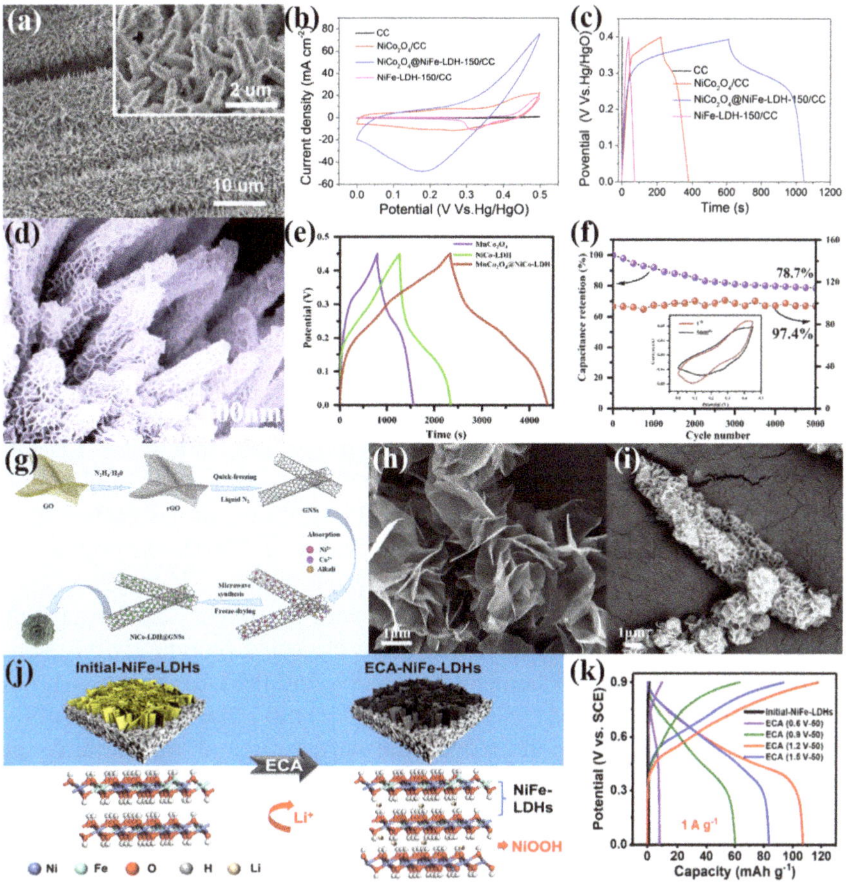

Figure 5.29. (a) An SEM image of $NiCo_2O_4$@NiFe-LBH-150/CC, (b) CV curves of the CC, $NiCo_2O_4$/CC, $NiCo_2O_4$@NiFe-LBH-150/CC, and NiFe-LBH-150/CC at 100 mV s^{-1}, (c) GCD curves of the CC, $NiCo_2O_4$/CC, $NiCo_2O_4$@NiFe-LBH-150/CC, and NiFe-LBH-150/CC at 1 mA cm^{-2}. ((a)–(c) [348] John Wiley & Sons. © 2022 Wiley-VCH GmbH.) (d) An SEM image of $MnCo_2O_4$@NiCo-LBH/NF, (e) GCD curves of $MnCo_2O_4$/NF, NiCo-LBH/NF, and $MnCo_2O_4$@NiCo-LBH/NF at 1 A g^{-1}, (f) the cycling performance at 10 A g^{-1} (the inset shows CV curves recorded at the 1st and 5000th cycles measured at 10 mV s^{-1}). ((d)–(f) Reprinted from [350], Copyright (2022), with permission from Elsevier.) (g) A schematic illustration of the synthetic procedures of NiCo-LBH@graphene nanosheets (GNSs), SEM images of (h) pristine NiCo-LBH and (i) NiCo-LBH@GNSs. ((g)–(i) Reprinted from [351], Copyright (2022), with permission from Elsevier.) (j) A schematic illustration of the electrochemical cycle activation (ECA) fabrication process (1.2 V-50), (k) GCD curves of ECA (1.2 V-50) at 1 A g^{-1}. ((j) and (k) Reprinted from [353], Copyright (2022), with permission from Elsevier.)

impressive 86.42% capacity retention after 10 000 cycles at 20 A g^{-1}. Wang *et al* [350] fabricated a core–sheath heterostructure ($MnCo_2O_4$@NiCo-LBH/NF) consisting of NiCo-LBH encapsulating $MnCo_2O_4$ nanowires on an NF substrate. The core–sheath structure, as shown in figure 5.29(d), has a diameter of approximately 65 nm and was anchored on the NF backbone, with NiCo-LBH nanosheets serving as the sheath material. This heterogeneous structure combined the advantages of interconnection between NiCo-LBH nanosheets, high electrical conductivity, and the mechanical strength of the $MnCo_2O_4$. Moreover, the $MnCo_2O_4$@NiCo-LBH/NF composite exhibited a specific capacitance of 4555.0 F g^{-1} at 1 A g^{-1} (figure 5.29(e)) and retained 78.7% of its capacitance after 5000 cycles (figure 5.29(f)).

Exploiting the synergistic effect of bilayer capacitance and pseudocapacitance and modifying nanostructures are also common strategies. Kuang *et al* [351] devised and synthesized a core–shell heterostructured graphene nanoscroll array composite. In this structure, petal-like NiCo-LBH nanoflakes were vertically anchored to a 3D interconnected skeleton of graphene nanosheets (GNSs), which was achieved using a highly convenient microwave-assisted method (figure 5.29(g)). This design offered several advantages, such as providing more active sites, facilitating electron and ion collection/transport, and mitigating volume variations during cycling. Due to its superior nanostructure (figures 5.29(h) and (i)), the resulting NiCo-LBH@GNS electrode exhibited a specific capacitance of 1470 F g^{-1} at 1 A g^{-1} and maintained an 81.6% retention rate after 1000 cycles. A multilayer multiwalled carbon nanotube and graphene nanoribbon/CoNi–LBH (MWGR/CoNi–LBH) composite was prepared by Ma *et al* [352] using a rapid microwave method. The synergistic effect between the multiwalled CNTs decorated with graphene oxide nanoribbons (MWCNTs-GONRs), which exhibited high electrical conductivity, structural stability, and favorable electrochemical properties, and the LBH with a p–n junction structure enhanced the redox reaction. Consequently, the MWGR/CoNi–LBH composite featured a unique heterogeneous structure and outstanding electrochemical properties, including a specific capacitance of 1030.2 C g^{-1} at 1 A g^{-1} (figures 5.29(b) and (c)). When used in MWGR/CoNi–LBH//AC devices, it achieved an energy density of 47.2 Wh kg^{-1} at 0.85 kW kg^{-1} and retained 88.8% of its capacity after 10 000 cycles at 10 A g^{-1}.

An intriguing heterostructured NiFe-LBH electrode material was fabricated by Zhang *et al* [353] using a high-voltage electrochemical cycle activation (ECA) technique (figure 5.29(j)). During the high-voltage ECA process, the surface of NiFe-LBH underwent *in situ* reconstitution, transforming into a low-crystalline NiOOH phase, ultimately forming a unique NiFe-LBH/NiOOH heterostructure. This surface reconstruction process generated numerous inhomogeneous interfaces, significantly increasing active sites for reversible cation adsorption and intercalation, thereby substantially enhancing electrochemical performance in neutral electrolytes. In a neutral electrolyte (2 M $LiNO_3$ solution), the ECA (1.2 V-50) electrode exhibited a specific capacitance of 107 mAh g^{-1} at 1 A g^{-1} (figure 5.29(k)), a 50-fold improvement over the pristine NiFe-LBH (2.1 mAh g^{-1} at 1 A g^{-1}). When coupled with MoS_2/rGO electrodes, the resulting ECA (1.2 V-50)//MoS_2/rGO hybrid SC device demonstrated an energy density of 48.1 Wh kg^{-1} at 432.9 W kg^{-1}.

Zeolite imidazolium frameworks (ZIFs) are excellent templates for the synthesis of functional materials and have extensive applications [354]. However, when directly used as electrode materials, ZIFs have limited exposed electroactive sites, poor chemical stability, slow charging kinetics, and unsatisfactory electrochemical performances [355]. Finding ways to combine their advantages and give full play to their performance becomes a key issue. Liao *et al* [356] demonstrated the *in situ* transformation of ZnCo-ZIFs which were modified on ZnCo nanorod arrays (ZnCo-NA) into 3D spatially distributed ZnCo-LBH/ZnCo-NA heterostructures (figure 5.30(a)). This structure boasted a substantial specific surface area, as depicted in the SEM image in figure 5.30(b). Due to its abundant electroactive sites and the presence of ion migration pathways in all directions, its electrochemical performance was significantly enhanced; it achieved a specific capacitance of 1576 F g^{-1} at 2 A g^{-1}. As illustrated in figure 5.30(c), the ZnCo-LBH/ZnCo-NA electrode exhibited the longest discharge time in the GCD curve, indicative of the highest energy storage capacity among the three samples. An asymmetric coin cell SC was assembled, which demonstrated an 88.1% capacitance retention after 5000 cycles. Importantly, this straightforward *in situ* mimetic transformation of a ZIF template into an intriguing LBH showcases the potential of this new generation of bimetallic heterostructures in energy-related applications.

Wan *et al* [357] conducted *in situ* modifications of arrays of highly porous $FeCoSe_2$@NiCo-LBH core–shell nanosheets on the surface of CC using an electrodeposition method followed by a salinization treatment. This hierarchical

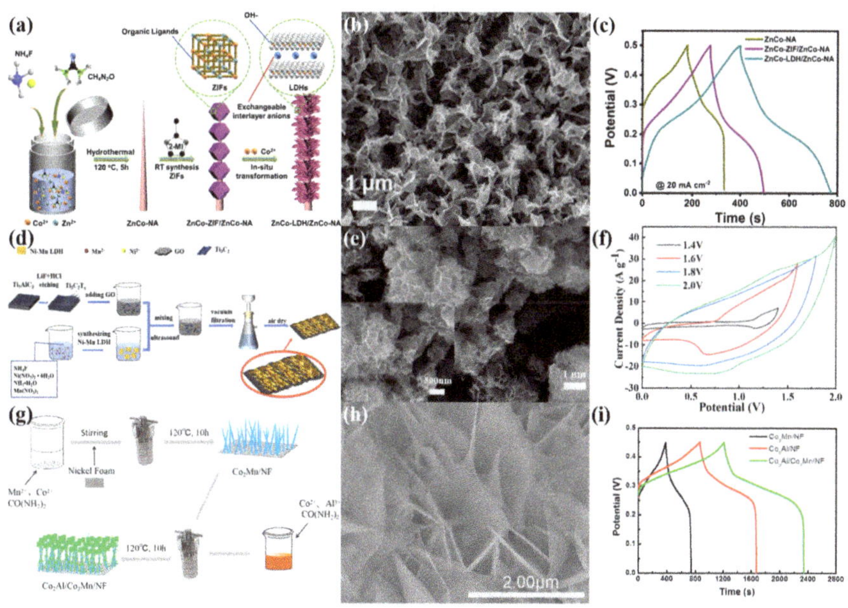

Figure 5.30. (a) A schematic illustration of the synthetic processes of the ZnCo-LBH/ZnCo-NA hybrid, (b) an SEM image of ZnCo-LBH/ZnCo-NA, (c) GCD curves of the ZnCo-NA, ZnCo-ZIF/ZnCo-NA, and ZnCo-LBH/ZnCo-NA electrodes. ((a)–(c) Reprinted from [356], Copyright (2022), with permission from Elsevier.) (d) A schematic illustration of the fabrication of the MXene/graphene oxide/NiMn–LBH (MGL) composite, (e) an SEM image of the MGL composite, (f) CV curves of MGL//AC in different scan potential windows at a scan rate of 100 mV s^{-1}. ((d)–(f) Reprinted from [274], Copyright (2021), with permission from Elsevier.) (g) A schematic diagram of the synthesis of $Co_2Al/Co_2Mn/NF$, (h) an SEM image of $Co_2Al/Co_2Mn/NF$, (i) GCD curves of the Co_2Mn/NF, Co_2Al/NF, and $Co_2Al/Co_2Mn/NF$ at a current density of 1 A g^{-1}. ((g)–(i) Reprinted from [236], Copyright (2021), with permission from Elsevier.)

heterogeneous structure, composed of two vertically aligned interconnected 2D nanosheets, offered a high surface area and an effective pathway for rapid electron/ion transport. It generated a heterogeneous interface rich in electron structure alteration. The results demonstrated that the well-designed $FeCoSe_2$@NiCo-LBH electrode exhibited a significantly higher specific capacitance of 220.9 mA h g^{-1} at 1 A g^{-1} and retained 83.5% of its capacity at 20 A g^{-1}. Moreover, it displayed better cycling stability compared to a single-component electrode. Furthermore, an ASC assembled using $FeCoSe_2$@NiCo-LBH electrodes and layered porous carbon electrodes achieved an impressive energy density of 65.9 Wh kg^{-1} at 1.248 kW kg^{-1} and maintained 87.6% of its capacity after more than 10 000 cycles. These outstanding performances highlight the promising prospects of integrated electrodes; and the concept of synthesizing heterostructures on a substrate's surface to enhance performance is highly feasible.

The MXene/graphene oxide/NiMn–LBH (MGL) material prepared by Chen *et al* [274] also made reasonable use of the advantages of heterostructures (figure 5.30(d)). The use of a heterostructure effectively mitigated the MXene stacking issue. An SEM image of the MGL revealed that nano-LBH aggregated on the surface of the MXene flakes, forming blocky porous arrays (figure 5.30(e)). This configuration ensured the structural stability of the matrix, preventing the LBH from morphologically collapsing, and significantly increasing its specific capacitance. The presence of graphene oxide accelerated charge transfer and boosted electron density. The coexistence of various components in this heterogeneous structure greatly enhanced its number of active sites and electrochemical capacity. When employed as an anode material, MGL achieved a specific capacitance of 241.9 mAh g^{-1} and maintained a cycling stability of 90.9% at 1 A g^{-1}, even in the presence of multivalent (Mn, Ni) hydroxides and stabilized carbon materials. The combination of the surface graphene oxide conductivity and the MXene substrate increased the availability of electrons on the hydroxide root. The assembled asymmetric device operated within a 2.0 V voltage window (figure 5.30(f)). This once again underscores the significance of constructing heterogeneous structures to enhance electrochemical properties and demonstrates the wide range of applications for the synergy between these three materials. Moreover, Zhu *et al* [236] directly prepared Co_2Mn bimetallic hydroxide nanofins on NF using a hydrothermal method. Subsequently, they prepared Co_2Al-LBH nanosheets on Co_2Mn nanofins through a hydrothermal method to obtain heterostructured nanocomposites (Co_2Al/Co_2Mn/NF). Figure 5.30(h) illustrates the structure of the Co_2Al/Co_2Mn/NF surface nanosheets. These nanocomposites exhibited a remarkable specific capacitance of 2502.0 F g^{-1} at 1 A g^{-1}. After 7000 cycles, 92.21% of the specific capacitance was retained, and the energy density reached 64.58 Wh kg^{-1} at 412.73 W kg^{-1}.

Stable MnO_2 nanowire@NiCo-LBH heterostructures were fabricated via a liquid-phase approach by Ma *et al* [238]. The NiCo-LBH nanosheets were uniformly grown within stable channels on the surface of ultralong MnO_2 nanowires, as depicted in the synthetic schematic shown in figure 5.31(a). In figure 5.31(b), SEM images of the MnO_2@LBH-2 sample illustrate NiCo-LBH nanosheets evenly

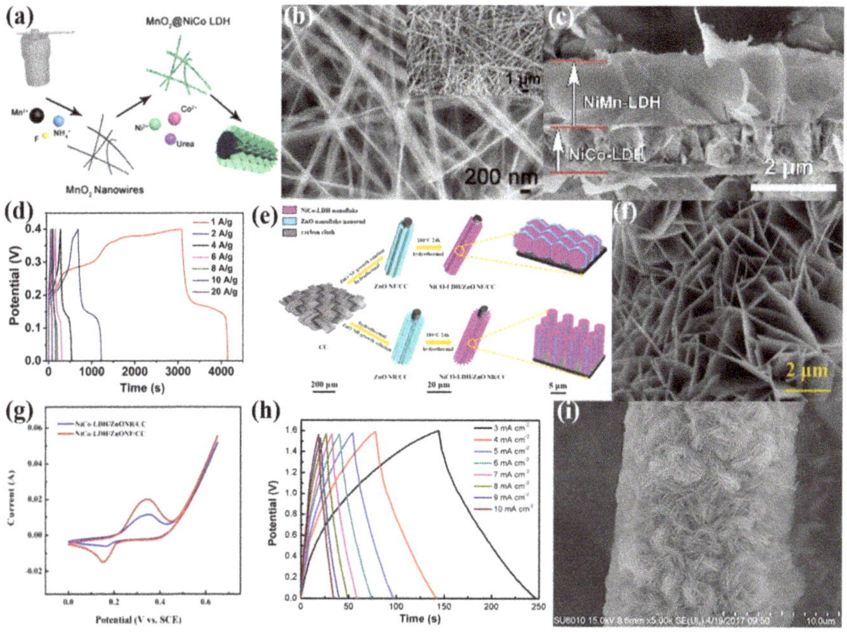

Figure 5.31. (a) A schematic illustration of the formation processes used to produce the stable MnO_2 nanowire@NiCo-LBH nanosheet core–shell heterostructure, (b) an SEM image of the MnO_2@LBH-2 samples. ((a) and (b) Reprinted with permission from [238], Copyright (2021), American Chemical Society.) (c) A cross-sectional diagram of NiCo-LBH and NiCo/NiMn–LBH, (d) GCD curves of the NiCo/NiMn–LBH electrode at various current densities. ((c) and (d) Reprinted from [358], Copyright (2021), with permission from Elsevier.) (e) The synthetic procedures of NiCo-LBH NF on ZnO NF/CC and ZnO NR/CC substrates, (f) an SEM image of NiCo-LBH/ZnO NF/CC, (g) CV profiles of NiCo-LBH/ZnO NR/CC and NiCo-LBH/ZnO NF/CC electrodes at a scanning speed of 5 mV s^{-1}. ((e)–(g) Reprinted from [359], Copyright (2021), with permission from Elsevier.) (h) GCD curves at different current densities, (i) an SEM image of the $NiCo_2S_4$/NiCo-LBH sample. ((h) and (i) Reprinted from [270], Copyright (2021), with permission from Springer.)

distributed on the periphery of the MnO_2, facilitating electron transfer and ion diffusion during electrochemical reactions. Electrochemical testing revealed that the core–shell heterostructure exhibited specific capacitances of 708 and 630 C g^{-1} at 1 and 10 A g^{-1}, respectively, along with an impressive capacitance retention rate of 82.3% after 2000 cycles. Further characterization using Raman spectroscopy showed that the prepared electrode underwent a transition from the α-phase to the β-phase during cycling, which is distinct from the behavior of NiCo-LBH. This transition was attributed to the heterogeneous structure's ability to buffer material collapse. Moreover, an ASC assembled with this electrode exhibited a capacitance retention rate of 72.4% after 10 000 cycles. Chen *et al* [358] fabricated a bilayer LBH nanosheet array using a hydrothermal method, as depicted in figure 5.31(c). The resulting bilayer electrode material, containing Ni, Co, and Mn elements, possessed a high surface area, enhancing the contact between the electrolyte and the prepared material. This bilayer electrode exhibited excellent capacitive performance, achieving a specific capacitance of 2950 F g^{-1} at 1 A g^{-1}, as shown in figure 5.31(d). Furthermore, it displayed good stability, retaining 79% of its capacity after 10 000 cycles at 10 A g^{-1}. In addition, asymmetric NiCo/NiMn–LBH//AC devices were prepared, which demonstrated good capacity and 82.2% cycling stability after 10 000 cycles. The development of this double-LBH array offers a novel approach for increasing the active sites of electrode materials.

Two nanostructures, namely ZnO nanorods (NRs) and nanosheets were prepared by a hydrothermal method on conductive flexible CC by Xiong *et al* [359]. Subsequently, NiCo-LBH nanosheets were formed on these nanostructures to prepare NiCo-LBH/ZnO NR/CC and NiCo-LBH/ZnO NF/CC heterostructures, as illustrated in figure 5.31(e). The influence of the ZnO morphology on electrochemical properties was examined in detail. It was observed that the latter heterostructure was denser and more homogeneous than the former, as depicted in the SEM image shown in figure 5.31(f). The latter heterostructure exhibited superior electrochemical properties compared to the former heterostructure, as shown in figure 5.31(g). It achieved a 2.6 times higher specific capacitance (1577.6 F g^{-1} at 1 A g^{-1}), a 2.2 times higher multiplicative capacity, and a 1.5 times higher cycling stability. Furthermore, an NiCo-LBH/ZnO NF//AC asymmetric solid-state flexible device demonstrated a maximum energy density of 51.39 Wh kg^{-1} at 800 W kg^{-1} and 87.3% capacitance retention after 1000 cycles. Finally, both packaged devices were successfully connected in series with a red 2.2 V LED, highlighting their potential for practical applications. Zhou *et al* [270] prepared a porous heterostructure of $NiCo_2S_4$/NiCo-LBH on carbon fiber paper using a straightforward solvothermal method. The active material was vertically deposited on the carbon fiber paper, as depicted in the SEM image presented in figure 5.31(i). This composite assembled into an interconnected 3D structure composed of abundant microporous dimensions, resulting in excellent electrochemical properties. The unique structure of this composite facilitated the acceleration of electron transfer and electrolyte transport during electrochemical processes. It achieved a specific capacitance of 1403 F g^{-1} at 10 mA cm^{-2}. Moreover, at 30 mA cm^{-2}, the capacitance retention after 5000 cycles reached an impressive 111.1%, indicating excellent cycling stability. SSCs assembled using this material exhibited an areal capacitance of 0.19 F cm^{-2} at 3 mA cm^{-2}, as shown in figure 5.31(h).

5.5.4 The preparation of binder-free materials

Growing metal precursors, such as NF, copper foam, stainless steel mesh, and carbon-based materials, directly on the surface of the current collector offers significant advantages [360–362]. This modification strategy reduces the mass of inactive materials such as conductive polymer binders, increasing the total energy density and allowing lighter devices to be assembled [363]. This process effectively reduces charge transfer resistance and internal resistance. Furthermore, eliminating the need for a polymer binder exposes more electroactive sites, enhancing electrical conductivity and expediting electron transfer rates. Crucially, this strategy enables the integration of various types of active materials, opening the door to synergistic effects between components that can enhance overall electrochemical performance [364]. However, a suitable base material needs to be selected. Pretreatment of the base material is beneficial for the growth of the active material.

Li *et al* [43] addressed the challenges faced by transition-metal LBHs in SC applications, such as aggregation and low conductivity, by synthesizing nickel iron sulfide nanosheets ($NiFeS_x$) and CNTs on diatomite using chemical vapor deposition and a two-step hydrothermal method. This composite was designed to harness the synergistic effects of multiple materials to enhance its electrochemical properties (figure 5.32(a)). An SEM image of NiFeSx@CNT@MnS is presented in figure 5.32(b). The simultaneous sulfidation process maintained the ortho-hexagonal nanosheet morphology of $NiFeS_x$ and assured the presence of CNTs on its surface. Diatomite served as a suitable matrix, promoting the uniform dispersion of nanomaterials on its surface, expanding the active sites in contact with the electrolyte, and significantly improving electrochemical performance. Due to its high conductivity and simultaneous sulfurization, the NiFeSx@CNT@MnS@diatomite structure

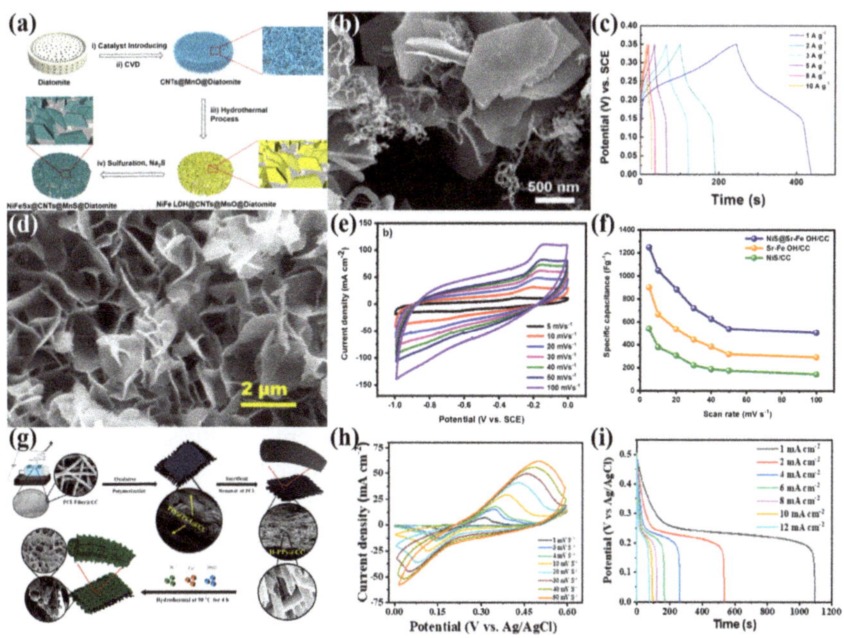

Figure 5.32. (a) A schematic illustration of the preparation processes used to produce NiFeSx@CNT@MnS@diatomite, (b) an SEM image of NiFeSx@CNTs@MnS@diatomite, and (c) GCD curves at different current densities. ((a)–(c) Reprinted from [43], Copyright (2021), with permission from Elsevier.) (d) SEM images of the NiS@SrFe OH/CC nanostructure. (e) CV curves at different scan rates, (f) the calculated specific capacitances of SrFe OH/CC, NiS/CC, and NiS@SrFe OH/CC at 5–100 mV s^{-1}. ((d)–(f) Reprinted from [365], Copyright (2022), with permission from Elsevier.) (g) A schematic representation of the step-by-step synthesis of NiCo-LBH@H-PPy@CC, (h) CV curves at different scan rates, (i) GCD profile of the NiCo-LBH@H-PPy@CC electrode at different current densities. ((g)–(i) Reprinted with permission from [366], Copyright (2022), American Chemical Society.)

exhibited a specific capacitance of 552 F g^{-1} at 1 A g^{-1} (figure 5.32(c)), with a retention rate of 68.4% at 10 A g^{-1} and a remarkable 89.8% cycling stability after 5000 cycles at 5 A g^{-1}. Furthermore, an ASC assembled using this composite and graphene achieved an energy density of 28.9 Wh kg^{-1} at 9375 W kg^{-1}. Rajapriya *et al* [365] employed a hydrothermal method to synthesize laminated NiS nanoflowers on flexible CC substrates. Subsequently, they vertically immobilized SrFe-LBH nanosheets on these highly conductive and flexible NiS/CC electrodes using electrodeposition without disrupting the original structure. This resulted in a 3D hybridized NiS@SrFe OH/CC nanostructure with an abundance of active nucleation sites. The superior structural features, as shown in figure 5.32(d), and the morphological advantages of the reticular flakes significantly enhanced the performance of NiS/CC, SrFe OH/CC, and NiS@SrFe OH/CC flexible electrodes, yielding specific capacitances of 556, 1151, and 1553 F g^{-1} (figures 5.32(e) and (f)) at 1 A g^{-1}, respectively. The NiS@SrFe OH/CC//AC/CC device demonstrated an energy density of 53.07 Wh kg^{-1} at 4.4 kW kg^{-1}.

Lohani *et al* [366] developed an electrode material consisting of thin LBH nanosheets arranged within the lumen and on the luminal portions of a PPy tunnel. SEM images revealed that the NiCo-LBH@H-PPy@CC electrode was constructed by integrating NiCo-LBHs nanosheets both inside and outside the lumen on long PPy tunnels on CC (figure 5.32(g)). The capacitance of the sample at 1.0 mA cm^{-2} reached 149.16 mAh g^{-1} (figures 5.32(h) and (i)). Furthermore, a device composed of NiCo-LBH@H-PPy@CC and vanadium phosphate carbon nanofiber (VPO@CNF900) exhibited a specific energy density of 32.42 Wh kg^{-1} at 3 mA cm^{-2}. Using facile and feasible *in situ* oxidation combined with a potentiation electrodeposition method, Wang *et al* [367] constructed densely distributed, core–shell structured Cu $(OH)_2$@NiFe-LBH nanoarrays (COH@NF-LBH/CF) on copper foam. The unique core–shell structure and the synergy between $Cu(OH)_2$ and NiFe-LBH offered significant advantages, including ample chemically active sites and efficient electron and ion transfer pathways that enhanced electrochemical performance. Specifically, at 5 mA cm^{-2}, the capacitance of the synthesized COH@NF-LBH/CF reached 4.139 F cm^{-2}, which was notably superior to those of discrete $Cu(OH)_2$ (198 mF cm^{-2} at the same current density) and NiFe-LBH/CF (71 mF cm^{-2}). Moreover, COH@NF-LBH/CF exhibited outstanding stability, achieving a retention rate of 86.47% over 5000 cycles. An ASC assembled using this material achieved a high energy density of 65.56 Wh kg^{-1} at 750 W kg^{-1}.

NF is also frequently employed as a substrate material in SCs due to its porous structure, low density, and excellent conductivity. Growing LBHs on its surface for use as SC electrode materials offers various advantages: (i) the 3D mesh structure of the NF substrate facilitates the efficient deposition of active materials and enhances charge transfer; (ii) the direct deposition of LBH materials on NF eliminates the necessity for pressing during electrochemical performance testing and the use of adhesives, rendering the test results more representative. Cao *et al* [368] conducted the chemical etching of NF using nitric acid solutions containing varying ratios of transition metals (NiCo-based). After rinsing and drying, the etched NF was employed as the anode, while a platinum mesh served as the cathode in an alkaline solution. Under constant voltage conditions, NiCo-LBH self-assembled on the NF substrate (figure 5.33(a)). Among the different compositions tested, the Ni1Co_2/NF monolithic electrode exhibited the best electrochemical performance, with a specific capacitance of 3.01 C cm^{-2} at 1 mA cm^{-2} (figure 5.33(b)). A hybrid device that used this material demonstrated an energy density of 97.4 pWh cm^{-2} at 800.5 μW cm^{-2}, and it retained 85.0% of its initial capacity after 5000 cycles (figure 5.33(c)).

A CoMn-LBH nanostructured high-performance self-contained SC electrode was prepared on a NF surface by Emin *et al* [369] via electrochemical deposition, as shown in figure 5.33(e). The electrode exhibited an open interconnected thin-layered structure (figure 5.33(d)) with a high capacitance of 2673.6 F g^{-1} at 1 A g^{-1} (figure 5.33(f)) and displayed good cycling stability (86.7% retention after 5000 cycles at 12 A g^{-1}).

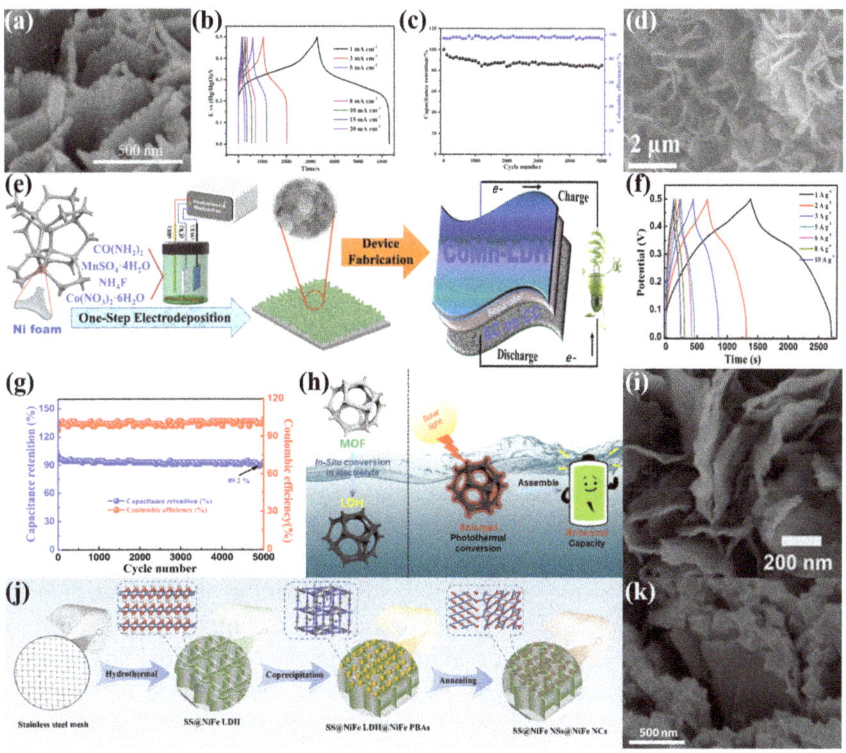

Figure 5.33. (a) An SEM image of NiCo-LBH hierarchical nanosheets on NF, (b) GCD curves of the Ni_1Co_2/NF electrode at current densities from 1 to 20 mA cm^{-2}, (c) the cycling performance of the hybrid SC (HSC) device. ((a)–(c) Reprinted from [368], Copyright (2022), with permission from Elsevier.) (d) An SEM image of CoMn-LBH, (e) the main fabrication procedures of the CoMn-LBH cathodes and aqueous asymmetric supercapacitors (AASCs), (f) GCD curves at various current densities from 1–10 A g^{-1}, (g) the cycling performance and coulombic efficiency during 5000 cycles at 5 A g^{-1}. ((d)–(g) Reprinted from [369], Copyright (2022), with permission from Elsevier.) (h) A synthetic and working process diagram, (i) an SEM image of NC37. ((h) and (i) Reprinted from [370], Copyright (2022), with permission from Elsevier.) (j) A schematic illustration of the formation of SS@NiFe-LBH@NiFe nanocubes (NCs), (k) an SEM image of SS@NiFe nanosheet (NS) @NiFe NCs. ((j) and (k) Reprinted from [371], Copyright (2021), with permission from Elsevier.)

An asymmetric device produced using this material demonstrated an energy density of 97.5 Wh kg^{-1} at 800.0 W kg^{-1}, maintained a capacitance retention of 89.2% after 5000 cycles at 5 A g^{-1}, and achieved a coulombic efficiency of approximately 100% (figure 5.33(g)). Lu *et al* [370] developed a binder-free NiCo-LBH energy storage device on NF through *in situ* electrochemically triggered MOF hydrolysis. The device exhibited a remarkable energy storage capacity under solar irradiation (figure 5.33(h)). Through electrochemically controlled hydrolysis, the ligands within the MOF were replaced by OH^-, and the resulting NiCo-LBH retained the original layered porous structure of the MOF. The NiCo-LBH electrode possessed abundant oxygen vacancies and a large surface area (figure 5.33(i)), achieving a capacity of 5.4 C cm^{-2} at 1.25 mA cm^{-2}, which was 64.3 times higher than that of the MOF template. Importantly, this electrode material also demonstrated excellent photothermal conversion capabilities, with a temperature increase of 52.9 °C in just 30 s. Furthermore, the energy density of an ASC prepared using NiCo-LBH increased by 329.2% after 15 min of sunlight irradiation at low temperatures (−4 °C).

Stainless steel (SS) mesh is also a suitable substrate material for the growth of LBHs. Wang *et al* [371] fabricated evenly distributed 3D NiFe Prussian blue analog (NiFe PBA) nanocubes on SS and converted them into 3D oxide arrays (SS@NiFe nanosheets (NSs)@NiFe NCs) by infiltrating 2D NiFe-LBH and thermally annealing it in air (figure 5.33(j)). This 3D array exhibited a nanocubic structure (figure 5.33(k)) with a high specific surface area and excellent electrochemical properties. Furthermore, Wang *et al* assembled hybrid SCs (HSCs) with SS@NiFe NSs@NiFe NCs as the positive electrode and SS@Fe_2O_3 as the negative electrode, which demonstrated impressive electrochemical performance.

5.5.5 Applications of LBH-based supercapacitors

LHD-based SCs are widely used in flexible wearables or integrated with other intelligent devices. Li *et al* [372] fabricated NiCo-LBH on flexible CC and $Ti_3C_2T_x$-functionalized CC using a high-magnetic-field electrodeposition method (figure 5.34(a)). The resulting flexible hybrid SC demonstrated excellent energy density and cyclic stability. The device exhibited minimal polarization during bending, and its total capacitance remained nearly constant, highlighting its impressive flexibility (figures 5.34(b) and (c)). Moreover, three of these devices connected in series were able to power red and yellow LED lights in parallel, demonstrating their potential for practical applications (figure 5.34(d)). Nagaraju *et al* [373] employed a hot-air oven-based method to grow aligned nickel–cobalt (NiCo) LBH nanoflake arrays (NFAs) on nickel fabric, followed by a simple electrochemical deposition process to decorate the NiCo–LBH NFAs with fluffy NC–LBH nanosheet branches (figure 5.34(e)). The CV curves obtained under various bending conditions at 50 mV s^{-1} exhibited a similar shape to the normal curves and did not display any distortion (figures 5.34(f) and (g)), indicating that the device maintained good capacitance and flexibility. Figure 5.34(h) suggests its potential suitability for wearable electronic applications. By adjusting the ratio of Ni and Co, Liu *et al* [374] obtained an optimized porous nanoflower-like

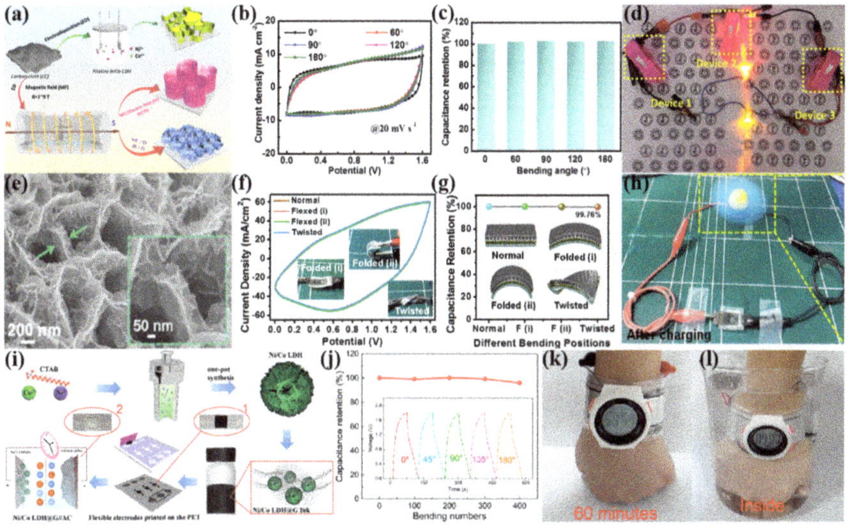

Figure 5.34. (a) A schematic illustration of the preparation of NiCo-LBH on CC, (b) CV curves and (c) the capacitance retention of the device subjected to bending at different angles, (d) a digital picture of three devices connected in series lighting up red and yellow LED lights connected in parallel. ((a)–(d) [372] John Wiley & Sons. © 2022 Wiley-VCH GmbH.) (e) Core–shell-like NC–LBH NFAs@NSs/Ni fabric, (f) CV curves and (g) the capacitance retention of the flexible hybrid SC under various flexed conditions (the corresponding insets show photographic and schematic diagrams of the device in flexed states), (h) the potential suitability of the device for wearable electronic applications. ((e)–(h) Reprinted with permission from [373], Copyright (2017), American Chemical Society.) (i) A schematic diagram of the preparation of screen-printed flexible NiCo–LBH-based electrodes, (j) the capacitance retention of the flexible Ni_3Co_1 LBH@G//AC ASC after different numbers of bending cycles (the inset shows GCD curves tested under different bending angles), (k) two devices connected in series to power an electronic watch, and (l) two devices connected in series in water to power an electronic watch. ((i)–(k) Reprinted with permission from [374], Copyright (2022), American Chemical Society.)

NiCo–LBH (figure 5.34(i)). A flexible ASC composed of Ni_3Co_1 LBH@graphene// activated carbon (AC) was prepared using screen printing, and the device demonstrated excellent flexibility, maintaining 95.8% of its capacitance after undergoing bending at different angles for 400 cycles (figure 5.34(j)). When two of these devices were connected in series, they powered a watch for over 60 min after only 50 seconds of charging, and the watch continued to function normally even when worn on the wrist and fully submerged in water (figures 5.34(k) and (l)). To provide a comprehensive overview of the performance data presented in this chapter, table 5.4 summarizes the properties of devices based on LBHs and their composite electrodes.

5.6 Conclusions and outlook

Conductive polymers have found widespread utility in SCs due to their notable characteristics, including high theoretical capacity, the ready availability of the raw materials, cost-effective preparation, and straightforward synthetic procedures. In just a decade, MXenes, exhibiting a substantial ionic surface area

Table 5.4. The properties of devices that include the LBHs and their composites mentioned in this chapter.

Devices	Electrolyte	Capacitance	Energy density at power density	Cycles	References
3%-Ni-C/NiAl-LBH//AC	6 M KOH	210.8 C g^{-1}, 1 A g^{-1}	74.9 Wh kg^{-1} at 800 W kg^{-1}	10 000, 91.4%	[319]
NiCo-LBH-S/PNT//GF-LBH@NF	3 M KOH	98 F g^{-1}, 1 A g^{-1}	16.28 Wh kg^{-1} at 650 W kg^{-1}	8000, 74%	[320]
CoFe-LBH/P2//AC	6 M KOH	1686 F g^{-1}, 1 A g^{-1}	75.9 Wh kg^{-1} at 1124 W kg^{-1}	10 000, 97.5%	[325]
CoNiMg-LBH//AC	—	333 C g^{-1}, 1 A g^{-1}	73.9 Wh kg^{-1} at 0.8 kW kg^{-1}	5000, 87%	[329]
NCW-2//rGO	PVA-KOH	98 F g^{-1}, 2 A g^{-1}	34 Wh kg^{-1} at 1.32 kW kg^{-1}	10 000, 86%	[324]
NiCo(OH)(BA)//AC	—	118 F g^{-1}, 1 A g^{-1}	47.5 Wh kg^{-1} at 850 W kg^{-1}	8000, 91%	[322]
ECA(1.2 V-50)//MoS_2/rGO	—	134 F g^{-1}, 0.5 A g^{-1}	48.1 Wh kg^{-1} at 432.9 W kg^{-1}	—	[353]
NiAl-Cl LBH//AC	6 M KOH	81.82 F g^{-1}, 1 A g^{-1}	53.9 Wh kg^{-1} at 1540 W kg^{-1}	1000, 94.1%	[321]
NMHS-4//AC	1 M KOH	97.3 F g^{-1}, 1 A g^{-1}	34.61 Wh kg^{-1} at 831 W kg^{-1}	10 000, 85%	[330]
ZnO@Ni/Co-LBH//AC	PVA-KOH	24.6 mF cm^{-2}, 0.5 mA cm^{-2}	7.7 ∝W h cm^{-2} at 375.0 pW cm^{-2}	—	[328]
CuCoNi–OH//HPC	3 M KOH	180 C g^{-1}, 0.5 A g^{-1}	39.67 Wh kg^{-1} at 400 W kg^{-1}	—	[326]
NiFe-LBH@SCN-32//AC	6 M KOH	386 F g^{-1}, 1 A g^{-1}	68.7 Wh kg^{-1} at 827.5 W kg^{-1}	8000, 83.3%	[331]
NTA18//AC	1 M KOH	126 F g^{-1}, 1 A g^{-1}	45.1 Wh kg^{-1} at 16 000 W kg^{-1}	5000, 59%	[327]
CuCo-LBH//AC	1 M KOH	76 F g^{-1}, 1 A g^{-1}	22 Wh kg^{-1} at 23 200 W kg^{-1}	10 000, 91.3%	[334]
D-NiCo-LBH/NF//AC	3 M KOH	267 F g^{-1}, 1 A g^{-1}	53 Wh kg^{-1} at 752 W kg^{-1}	5000, 94.7%	[335]
3D-NiCo-SDBS-LBH//AC	1 M KOH	205.7 F g^{-1}, 1 A g^{-1}	73.14 Wh kg^{-1} at 800 W kg^{-1}	10 000, 95.5%	[337]
NiCo-LBH-2//AC	6 M KOH	559 C g^{-1}, 1 A g^{-1}	101.1 Wh kg^{-1} at 1500 W kg^{-1}	5000, 87.8%	[339]
V_{Zn}-defect sample//AC	1.2 M LiOH	528.5 mF cm^{-2}, 2 mA cm^{-2}	1.03 mW h cm^{-3} at 9.35 mW cm^{-3}	2000, 85.6%	[332]
$Co_{0.50}Ga_{0.50}$-LBH//AC	6 M KOH	187 mF cm^{-2}, 3.15 mA cm^{-2}	33.38 Wh kg^{-1} at 920 W kg^{-1}	10 000, 95.8%	[341]

Table 5.4. (*Continued*)

Devices	Electrolyte	Capacitance	Energy density at power density	Cycles	References
E-CoZnAl-LBH-8 h//AC	1 M KOH	114 F g^{-1}, 1 A g^{-1}	36.75 Wh kg^{-1} at 400 W kg^{-1}	8000, 72.7%	[336]
NiCo-LBH@MoO_3/NF//AC	2 M KOH	952.2 C g^{-1}, 1 A g^{-1}	58.06 Wh kg^{-1} at 800 W kg^{-1}	10 000, 86.42%	[349]
NiCo-LBH@GNSs//AC	3 M KOH	102.6 F g^{-1}, 1 A g^{-1}	32.1 Wh kg^{-1} at 750.4 W kg^{-1}	—	[351]
MWGR/CoNi–LBH//AC	6 M KOH	132.9 F g^{-1}, 1 A g^{-1}	47.2 Wh kg^{-1} at 850 W kg^{-1}	10 000, 88.8%	[352]
$MnCo_2O_4$@NiCo-LBH/NF//AC	6 M KOH	60 F g^{-1}, 1 A g^{-1}	21.3 Wh kg^{-1} at 160 W kg^{-1}	5000, 86.6%	[350]
ZnCo-LBH/ZnCo-NA//AC	6 M KOH	68.4 F g^{-1}, 0.2 A g^{-1}	21.3 Wh kg^{-1} at 900 W kg^{-1}	5000, 88.1%	[356]
$FeCoSe_2$@NiCo-LBH//PPC-2	2 M KOH	95.2 mAh g^{-1}, 1 A g^{-1}	1.248 kW kg^{-1} at 65.9 Wh kg^{-1}	10 000, 87.6%	[357]
MXene/graphene oxide (GO)/NiMn–LBH//AC	PVA-KOH	69.1 mAh g^{-1}, 1 A g^{-1}	55.3 Wh kg^{-1} at 800 W kg^{-1}	4000, 94.7%	[274]
Co_2Al/Co_2Mn/NFs//AC	6 M KOH	281.7 C g^{-1}, 0.5 A g^{-1}	64.58 Wh kg^{-1} at 412.7 W kg^{-1}	7000, 92.21%	[236]
MnO_2@LBH-2//AC	6 M KOH	95.5 F g^{-1}, 0.5 A g^{-1}	31.9 Wh kg^{-1} at 502.7 W kg^{-1}	10 000, 72.4%	[238]
NiCo/NiMn–LBH//AC	6 M KOH	185.1 F g^{-1}, 1 A g^{-1}	45.16 Wh kg^{-1} at 1400 W kg^{-1}	10 000, 82.2%	[358]
NiCo-LBH/ZnO NFs//AC	1 M KOH	144.5 F g^{-1}, 1 A g^{-1}	51.39 Wh kg^{-1} at 800 W kg^{-1}	1000, 87.3%	[359]
NiFeSx@CNTs@MnS@diatomite//graphene	6 M KOH	92.3 F g^{-1}, 0.5 A g^{-1}	28.9 Wh kg^{-1} at 9375 W kg^{-1}	5000, 80.8%	[43]
NiS@SrFe OH/CC//AC	1 M KOH	146.21 F g^{-1}, 1 A g^{-1}	53.07 Wh kg^{-1} at 4.4 kW kg^{-1}	—	[365]
NiCo-LBH@H-PPy@CC//VPO@CNFs900	2 M KOH	40.53 mAh g^{-1}, 3 mA cm^{-2}	32.42 Wh kg^{-1} at 359.16 W kg^{-1}	10 000, 94.09%	[366]
COH@NF-LBH/CF//AC	6 M KOH	195.7 F g^{-1}, 1 A g^{-1}	65.56 Wh kg^{-1} at 750 W kg^{-1}	5000, 88.93%	[367]

(*Continued*)

Table 5.4. (*Continued*)

Devices	Electrolyte	Capacitance	Energy density at power density	Cycles	References
Ni_1Co_2/NF//AC	1 M KOH	273.8 mF cm^{-2}, 1 mA cm^{-2}	97.4 pWh cm^{-2} at 800.5 μW cm^{-2}	5000, 85%	[368]
CoMn-LBH//AC	2 M KOH	274.26 F g^{-1}, 1 A g^{-1}	97.5 Wh kg^{-1} at 800 W kg^{-1}	5000, 89.2%	[369]
NiCo-LBH//GO/AC	1 M KOH	1.4 C cm^{-2}, 1.25 mA cm^{-2}	1.06 mWh cm^{-2} at 1.03 mW cm^{-2}	—	[370]
NSs@NiFe NCs//SS@Fe_2O_3	1 M Na_2SO_4	102 F g^{-1}, 1 A g^{-1}	45.9 Wh kg^{-1} at 902.7 W kg^{-1}	2000, 89.7%	[371]
$Ti_3C_2T_x$/NiCo-LBH-3 T (B//E)//AC	PVA-KOH	3.12 C cm^{-2}, 1 mA cm^{-2}	0.134 mWh cm^{-2} at 1.61 mW cm^{-2}	6000, 82.3%	[372]
NC–LBH NFAs@NSs/Ni fabric//AC@CF	1 M KOH	1147.23 mF cm^{-2}, 3 mA cm^{-2}	46.15 Wh kg^{-2} at 2604.42 W kg^{-1}	2000, 86.49%	[373]
Ni_3Co_1 LBH@G//AC	PVA-KOH	599 mF cm^{-2}, 1 mA cm^{-2}	0.27 mWh cm^{-2} at 0.9 mW cm^{-2}	10 000, 123%	[374]

PPC—hierarchical porous carbon obtained by applying pine pollen as a carbon source and $CuCl_2$ as a chemical activator.

akin to that of graphene and exceptional metal-like conductivity, have swiftly gained favor among researchers.

This chapter presented a comprehensive review of recent advancements in hybrid materials comprising LBHs, MXenes, and LBHs/MXenes coupled with conductive polymers in the realm of SCs. It covered various aspects, including composite fabrication, electrode materials, SSCs, and ASCs. Despite the extensive research conducted on MXenes and conductive polymers, certain challenges persist, impeding their practical applications. The limited conductivity of LBHs inevitably hinders efficient electron transfer during redox processes, consequently affecting capacitive performance. Notably, the CV curves deviate from the ideal rectangular shape, resulting in suboptimal current–voltage profiles during constant GCD processes. Furthermore, prevailing methods for LBH synthesis and modification often entail intricate reaction conditions and generate environmental pollutants. Consequently, it is imperative to judiciously select techniques for synthesis and modification that align with practical considerations in subsequent research endeavors.

From a materials preparation perspective, conductive polymers exhibit various morphologies and tend to agglomerate. Therefore, it is imperative to meticulously devise the preparation scheme and precisely control the polymerization duration. The etching process of MXene materials involves the use of highly corrosive acids such as HF and concentrated hydrochloric acid. As evident from earlier reports in this chapter, different etching methods yield MXenes with varying surface functional groups, sizes, and morphologies. When considering volume production of MXenes, it becomes essential to explore gentler and more convenient etching methods. In addition, there is room for improvement in enhancing the utilization efficiency of the MAX phase in MXene precursors.

From the perspective of materials storage, although low temperatures can mitigate the oxidation of MXenes, they cannot prevent oxygen and water molecules from oxidizing titanium-based MXene into TiO_2. To address the issue of more effectively utilizing MXene materials, researchers have employed end-group protection methods to reduce the contact between MXene surface functional groups and oxygen and/or water molecules. For example, Wu *et al* [254] used sodium citrate for the end-group protection of an MXene, and the results showed a significant improvement in the storage time of the MXene. Nevertheless, the storage methods for MXenes still remain a challenging issue that hinders their practical applications.

From the perspective of electrode material applications, MXenes are prone to self-aggregation due to the lack of support material after the removal of the transition metals. Conductive polymers undergo expansion during charge and discharge processes, which can impact electrochemical performance. The effective integration of MXenes with conductive polymers not only addresses the self-aggregation issue of MXenes and the expansion problem of conductive polymers during charge and discharge but also imparts enhanced stability and broader practical characteristics to electrode materials, such as temperature and pH tolerance. Furthermore, since their inception, MXenes have gradually found applications in various fields, including SCs, batteries, sensors, electromagnetic shielding, and adsorption. However, MXenes come in numerous variations, each with different elemental compositions and surface functional groups. Conductive polymers also exhibit varying structures, compositions, and morphologies. The full potential of MXenes has yet to be fully harnessed, and many of their aspects remain unexplored.

In the future, in addition to fully leveraging the structural advantages of LBHs, it is essential to further enhance their conductivity and structural stability. The true potential of 2D materials and heterogeneous structures is enormous. This provides a substantial foundation for understanding the interaction kinetics and energy storage of other 2D materials, such as hexagonal boron nitride, MOFs, covalent organic frameworks, and more. MXenes are being utilized to create heterogeneous structures, driving the development of optimal electrode materials. Consequently, MXene/LBH heterogeneous structures are poised to become the electrode materials of choice for next-generation advanced energy storage applications.

Conflicts of interest

The authors declare that they have no conflicts of interest.

Acknowledgments

Certain sections of text in this chapter have been reproduced with permission from Elsevier [375] and the Royal Society of Chemistry [376].

References

[1] Bilgen S 2014 Structure and environmental impact of global energy consumption *Renew. Sustain. Energy Rev.* **38** 890–902

[2] Kannan N and Vakeesan D 2016 Solar energy for future world—a review *Renew. Sustain. Energy Rev.* **62** 1092–105

[3] Liu W J, Jiang H and Yu H Q 2019 Emerging applications of biochar-based materials for energy storage and conversion *Energy Environ. Sci.* **12** 1751–79

[4] Liu C C, Zheng K H, Zhou Y *et al* 2021 Experimental thermal hazard investigation of pressure and EC/PC/EMC mass ratio on electrolyte *Energies* **14** 2511

[5] Tan L, Wei C, Zhang Y *et al* 2022 Long-life and dendrite-free zinc metal anode enabled by a flexible, green and self-assembled zincophilic biomass engineered MXene based interface *Chem. Eng. J.* **431** 134277

[6] Sun J, Zhang X, Du Q *et al* 2021 The contribution of conductive network conversion in thermal conductivity enhancement of polymer composite: a theoretical and experimental study *ES Mater. Manuf.* **13** 53–65

[7] Ma Y, Zhuang Z, Ma M L *et al* 2019 Solid polyaniline dendrites consisting of high aspect ratio branches self-assembled using sodium lauryl sulfonate as soft templates: synthesis and electrochemical performance *Polymer* **182** 121808

[8] Zheng M, Wei Y, Ren J *et al* 2021 2-Aminopyridine functionalized magnetic core–shell Fe_3O_4@polypyrrole composite for removal of Mn (VII) from aqueous solution by double-layer adsorption *Sep. Purif. Technol.* **277** 119455

[9] Xu D J, Huang G, Guo L *et al* 2021 Enhancement of catalytic combustion and thermolysis for treating polyethylene plastic waste *Adv. Compos. Hybrid Mater.* **5** 113–29

[10] Dong X, Zhao X, Chen Y *et al* 2021 Investigations about the influence of different carbon matrixes on the electrochemical performance of $Na_3V_2(PO_4)_3$ cathode material for sodium ion batteries *Adv. Compos. Hybrid Mater.* **4** 1070–81

[11] Liu C, Xu D, Weng J *et al* 2020 Phase change materials application in battery thermal management system: a review *Materials* **13** 4622

[12] Ruan J, Sun H, Song Y *et al* 2021 Constructing 1D/2D interwoven carbonous matrix to enable high-efficiency sulfur immobilization in Li-S battery *Energy Mater.* **1** 100018

[13] Hu H, Ding F, Ding H *et al* 2020 Sulfonated poly(fluorenyl ether ketone)/sulfonated α-zirconium phosphate nanocomposite membranes for proton exchange membrane fuel cells *Adv. Compos. Hybrid Mater.* **3** 498–507

[14] Mahadik S A, Patil A, Pathan H M *et al* 2020 Thionaphthoquinones as photosensitizers for TiO_2 nanorods and ZnO nanograin based dye-sensitized solar cells: effect of nanostructures on charge transport and photovoltaic performance *Eng. Sci.* **14** 46–58

[15] Zhao J, Wei D, Zhang C *et al* 2021 An overview of oxygen reduction electrocatalysts for rechargeable zinc-air batteries enabled by carbon and carbon composites *Eng. Sci.* **15** 1–19

[16] Shi Y, Wang J Z, Chou S L *et al* 2013 Hollow structured Li_3VO_4 wrapped with graphene nanosheets *in situ* prepared by a one-pot template-free method as an anode for lithium-ion batteries *Nano Lett.* **13** 4715–20

[17] Peng X, Wang C, Liu Y *et al* 2021 Critical advances in re-engineering the cathode-electrolyte interface in alkali metal-oxygen batteries *Energy Mater.* **1** 100011

[18] Zhan C, Zeng X, Ren X *et al* 2020 Dual-ion hybrid supercapacitor: integration of Li-ion hybrid supercapacitor and dual-ion battery realized by porous graphitic carbon *J. Energy Chem.* **42** 180–4

[19] Al-Ghussain L, Samu R, Taylan O *et al* 2020 Sizing renewable energy systems with energy storage systems in microgrids for maximum cost-efficient utilization of renewable energy resources *Sustain. Cities Soc.* **55** 102059

[20] Dong Z, Kennedy S J and Wu Y 2011 Electrospinning materials for energy-related applications and devices *J. Power Sources* **196** 4886–904

[21] Sun Z, Qu K, Li J *et al* 2021 Self-template biomass-derived nitrogen and oxygen co-doped porous carbon for symmetrical supercapacitor and dye adsorption *Adv. Compos. Hybrid Mater.* **4** 1413–24

[22] Patil S S, Bhat T S, Teli A M *et al* 2020 Hybrid solid state supercapacitors (HSSC's) for high energy and power density: an overview *Eng. Sci.* **12** 38–51

[23] Dong H, Li Y, Chai H *et al* 2019 Hydrothermal synthesis of $CuCo_2S_4$ nano-structure and N-doped graphene for high-performance aqueous asymmetric supercapacitors *ES Energy Environ.* **4** 19–26

[24] Pu L Y, Zhang J X, Jiresse N K L *et al* 2021 N-doped MXene derived from chitosan for the highly effective electrochemical properties as supercapacitor *Adv. Compos. Hybrid Mater.* **5** 356–69

[25] Qu K Q, Wang W C, Shi C *et al* 2021 Fungus bran-derived nanoporous carbon with layered structure and rime-like support for enhanced symmetric supercapacitors *J. Nanostruct. Chem.* **11** 769–84

[26] Balogun M S, Qiu W, Wang W *et al* 2014 Recent advances in metal nitrides as high-performance electrode materials for energy storage devices *J. Mater. Chem.* A **3** 1364–87

[27] Lu Y, Zhang Q, Li L *et al* 2018 Design strategies toward enhancing the performance of organic electrode materials in metal-ion batteries *Chem.* **4** 2786–813

[28] Wei Y, Luo W, Li X *et al* 2022 PANI-MnO_2 and $Ti_3C_2T_x$ (MXene) as electrodes for high-performance flexible asymmetric supercapacitors *Electrochim. Acta* **406** 139874
[29] Shayeh J S, Sadeghinia M, Siadat S O R *et al* 2017 A novel route for electrosynthesis of $CuCr_2O_4$ nanocomposite with p-type conductive polymer as a high performance material for electrochemical supercapacitors *J. Colloid Interface Sci.* **496** 401–6
[30] Wang T, Li K, Le Q *et al* 2021 Tuning parallel manganese dioxide to hollow parallel hydroxyl oxidize iron replicas for high-performance asymmetric supercapacitors *J. Colloid Interface Sci.* **594** 812–23
[31] Xiao L, Qi H, Qu K *et al* 2021 Layer-by-layer assembled free-standing and flexible nanocellulose/porous Co_3O_4 polyhedron hybrid film as supercapacitor electrodes *Adv. Compos. Hybrid Mater.* **4** 306–16
[32] Li K L, Teng H, Dai X J *et al* 2022 Atomic scale modulation strategies and crystal phase transition of flower-like CoAl layered double hydroxides for supercapacitors *Crystengcomm* **24** 2081–8
[33] Wu D, Yao Z, Sun X *et al* 2022 Mussel-tailored carbon fiber/carbon nanotubes interface for elevated interfacial properties of carbon fiber/epoxy composites *Chem. Eng. J.* **429** 132449
[34] Gu W and Yushin G 2014 Review of nanostructured carbon materials for electrochemical capacitor applications: advantages and limitations of activated carbon, carbide-derived carbon, zeolite-templated carbon, carbon aerogels, carbon nanotubes, onion-like carbon, and graphene *Wiley Interdiscip. Rev. Energy Environ.* **3** 424–73
[35] Yan J, Wang Q, Wei T *et al* 2014 Recent advances in design and fabrication of electrochemical supercapacitors with high energy densities *Adv. Energy Mater.* **4** 1300816
[36] Zhi M, Xiang C, Li J *et al* 2013 Nanostructured carbon-metal oxide composite electrodes for supercapacitors: a review *Nanoscale* **5** 72–88
[37] Ma M, Yang Y, Li W *et al* 2019 Synthesis of yolk-shell structure Fe_3O_4/P(MAA-MBAA)-PPy/Au/void/TiO_2 magnetic microspheres as visible light active photocatalyst for degradation of organic pollutants *J. Alloys Compd.* **810** 151807
[38] Yan J, Fan Z, T W *et al* 2010 Fast and reversible surface redox reaction of graphene–MnO_2 composites as supercapacitor electrodes *Carbon* **48** 3825–33
[39] Gujar T P, Shinde V R, Lokhande C D *et al* 2007 Spray deposited amorphous RuO_2 for an effective use in electrochemical supercapacitor *Electrochem. Commun.* **9** 504–10
[40] Fisher R A, Watt M R and Ready W J 2013 Functionalized carbon nanotube supercapacitor electrodes: a review on pseudocapacitive materials *Ecs J. Solid State Sci. Technol.* **2** M3170–7
[41] Zhang L, Song T, Shi L *et al* 2021 Recent progress for silver nanowires conducting film for flexible electronics *J. Nanostruct Chem.* **11** 323–41
[42] Wang F, Wu X, Yuan X *et al* 2017 Latest advances in supercapacitors: from new electrode materials to novel device designs *Chem. Soc. Rev.* **46** 6816–54
[43] Li K, Hu Z, Zhao R *et al* 2021 A multidimensional rational design of nickel-iron sulfide and carbon nanotubes on diatomite via synergistic modulation strategy for supercapacitors *J. Colloid Interface Sci.* **603** 799–809
[44] Liang J, Li X, Zuo J *et al* 2021 Hybrid 0D/2D heterostructures: *in situ* growth of 0D g-C_3N_4 on 2D BiOI for efficient photocatalyst *Advanced Composites and Hybrid Materials* **4** 1122–36
[45] Li Q-Y, Hao Q, Zhu T *et al* 2020 Nanostructured and heterostructured 2D materials for thermoelectrics *Eng. Sci.* **13** 24–50

[46] Wei Y D, Zheng M M, Luo W L *et al* 2022 All pseudocapacitive MXene–MnO_2 flexible asymmetric supercapacitor *J. Energy Storage* **45** 103715
[47] Xu M, Liang T, Shi M *et al* 2013 Graphene-like two-dimensional materials *Chem. Rev.* **113** 3766–98
[48] Ehsani A, Heidari A A and Shiri H M 2019 Electrochemical pseudocapacitors based on ternary nanocomposite of conductive polymer/graphene/metal oxide: an introduction and review to it in recent studies *Chem. Rec.* **19** 908–26
[49] Li K, Feng S, Jing C *et al* 2019 Assembling a double shell on a diatomite skeleton ternary complex with conductive polypyrrole for the enhancement of supercapacitors *Chem, Commun, (Camb)* **55** 13773–6
[50] Gao Q, Pan Y, Zheng G *et al* 2021 Flexible multilayered MXene/thermoplastic polyurethane films with excellent electromagnetic interference shielding, thermal conductivity, and management performances *Adv. Compos. Hybrid Mater.* **4** 274–85
[51] Wang Y, Liu Y, Wang C *et al* 2020 Significantly enhanced ultrathin NiCo-based MOF nanosheet electrodes hybrided with $Ti_3C_2T_x$ MXene for high performance asymmetric supercapacitors *Eng. Sci.* **9** 50–9
[52] Kovalska E and Kocabas C 2016 Organic electrolytes for graphene-based supercapacitor: Liquid, gel or solid *Mater. Today Commun.* **7** 155–60
[53] Young C, Kim J, Kaneti Y V *et al* 2018 One-step synthetic strategy of hybrid materials from bimetallic metal–organic frameworks (MOFs) for supercapacitor applications *ACS Appl. Energy Mater.* **1** 2007–15
[54] Ehsani A, Bigdeloo M, Assefi F *et al* 2020 Ternary nanocomposite of conductive polymer/chitosan biopolymer/metal organic framework: synthesis, characterization and electrochemical performance as effective electrode materials in pseudocapacitors *Inorg. Chem. Commun.* **115** 107885
[55] Du D F, Wu X Z, Li S *et al* 2017 Remarkable supercapacitor performance of petal-like LBHs vertically grown on graphene/polypyrrole nanoflakes *J. Mater. Chem.* A **5** 8964–71
[56] Luo W, Wei Y, Zhuang Z *et al* 2022 Fabrication of $Ti_3C_2T_x$ MXene/polyaniline composite films with adjustable thickness for high-performance flexible all-solid-state symmetric supercapacitors *Electrochim. Acta* **406** 139871
[57] Yz A, Jie D, Ch B *et al* 2020 Co-doped Mo–Mo_2C cocatalyst for enhanced g-C_3N_4 photocatalytic H_2 evolution *Appl. Catalysis* B **260** 118220–0
[58] Zhang S J, Zhuo H, Li S Q *et al* 2021 Effects of surface functionalization of mxene-based nanocatalysts on hydrogen evolution reaction performance *Catal. Today* **368** 187–95
[59] Ahmed B, Anjum D H, Hedhili M N *et al* 2016 H_2O_2 assisted room temperature oxidation of Ti_2C MXene for Li-ion battery anodes *Nanoscale* **8** 7580–7
[60] Urbankowski P, Anasori B, Hantanasirisakul K *et al* 2017 2D molybdenum and vanadium nitrides synthesized by ammoniation of 2D transition metal carbides (MXenes) *Nanoscale* **9** 17722–30
[61] Pan Z, Cao F, Hu X *et al* 2019 A facile method for synthesizing CuS decorated Ti_3C_2 MXene with enhanced performance for asymmetric supercapacitors *J. Mater. Chem.* A **7** 8984–92
[62] Vahidmohammadi A, Mojtabavi M, Caffrey N M *et al* 2019 Assembling 2D MXenes into highly stable pseudocapacitive electrodes with high power and energy densities *Adv. Mater.* **31** 1806931

[63] Wang J, Kang H, Ma H *et al* 2021 Super-fast fabrication of MXene film through a combination of ion induced gelation and vacuum-assisted filtration *Eng. Sci.* **15** 57–66
[64] Xie Y and Kent P R C 2013 Hybrid density functional study of structural and electronic properties of functionalized $Ti_{n+1}X_n$ (X=C, N) monolayers *Phys. Rev.* B **87** 235441
[65] Zang X, Wang J, Qin Y *et al* 2020 Enhancing capacitance performance of $Ti_3C_2T_x$ MXene as electrode materials of supercapacitor: from controlled preparation to composite structure construction *Nanomicro Lett.* **12** 77
[66] Chen J, Huang Q, Huang H *et al* 2020 Recent progress and advances in the environmental applications of MXene related materials *Nanoscale* **12** 3574–92
[67] Hope M A, Forse A C, Griffith K J *et al* 2016 NMR reveals the surface functionalisation of Ti_3C_2 MXene *Phys. Chem. Chem. Phys.* **18** 5099–102
[68] Naguib M, Kurtoglu M, Presser V *et al* 2011 Two-dimensional nanocrystals: two-dimensional nanocrystals produced by exfoliation of Ti_3AlC_2 (*Adv. Mater.* 37/2011) *Adv. Mater.* **23** 4207 4207
[69] Lukatskaya M R, Mashtalir O, Ren C E *et al* 2013 Cation intercalation and high volumetric capacitance of two-dimensional titanium carbide *Science* **341** 1502–5
[70] Ghidiu M, Lukatskaya M R, Zhao M Q *et al* 2014 Conductive two-dimensional titanium carbide 'clay' with high volumetric capacitance *Nature* **516** 78–81
[71] Li L, Zhang N, Zhang M *et al* 2019 Flexible $Ti_3C_2T_x$/PEDOT:PSS films with outstanding volumetric capacitance for asymmetric supercapacitors *Dalton Trans.* **48** 1747–56
[72] Liu C C, Huang Q, Zheng K H *et al* 2020 Impact of lithium salts on the combustion characteristics of electrolyte under diverse pressures *Energies* **13** 5373
[73] Shown I, Ganguly A, Chen L C *et al* 2014 Conducting polymer-based flexible supercapacitor *Energy Sci. Eng.* **3** 2–26
[74] Wei Y D, Luo W L, Zhuang Z *et al* 2021 Fabrication of ternary MXene/MnO_2/polyaniline nanostructure with good electrochemical performances *Adv. Compos. Hybrid Mater.* **4** 1082–91
[75] Cheng H, Pan Y, Chen Q *et al* 2021 Ultrathin flexible poly(vinylidene fluoride)/MXene/silver nanowire film with outstanding specific EMI shielding and high heat dissipation *Adv. Compos. Hybrid Mater.* **4** 505–13
[76] Rehman S U, Ahmed R, Ma K *et al* 2020 Composite of strip-shaped ZIF-67 with polypyrrole: a conductive polymer–MOF electrode system for stable and high specific capacitance *Eng. Sci.* **13** 71–8
[77] Hassan M, Reddy K R, Haque E *et al* 2014 Hierarchical assembly of graphene/polyaniline nanostructures to synthesize free-standing supercapacitor electrode *Compos. Sci. Technol.* **98** 1–8
[78] Ramya R, Sivasubramanian R and Sangaranarayanan M V 2013 Conducting polymers-based electrochemical supercapacitors—progress and prospects *Electrochim. Acta* **101** 109–29
[79] Snook G A, Kao P and Best A S 2011 Conducting-polymer-based supercapacitor devices and electrodes *J. Power Sources* **196** 1–12
[80] Sivakkumar S R and Saraswathi R 2004 Performance evaluation of poly(*N*-methylaniline) and polyisothianaphthene in charge-storage devices *J. Power Sources* **137** 322–8
[81] Luo X, Yang G and Schubert D W 2021 Electrically conductive polymer composite containing hybrid graphene nanoplatelets and carbon nanotubes: synergistic effect and tunable conductivity anisotropy *Adv. Compos. Hybrid Mater.* **5** 250–62

[82] Lota K, Khomenko V and Frackowiak E 2004 Capacitance properties of poly(3,4-ethylenedioxythiophene)/carbon nanotubes composites *J. Phys. Chem. Solids* **65** 295–301
[83] Wang X, Zeng X and Cao D 2018 Biomass-derived nitrogen-doped porous carbons (NPC) and NPC/polyaniline composites as high performance supercapacitor materials *Eng. Sci.* **1** 55–63
[84] Cao S, Ge W, Yang Y *et al* 2021 High strength, flexible, and conductive graphene/polypropylene fiber paper fabricated via papermaking process *Adv. Compos. Hybrid Mater.*
[85] Nan J, Guo X, Xiao J *et al* 2021 Nanoengineering of 2D MXene-based materials for energy storage applications *Small* **17** 1902085
[86] Wu Y J, Sun Y J, Zheng J F *et al* 2021 MXenes: advanced materials in potassium ion batteries *Chem. Eng. J.* **404** 126565
[87] Aslam M K, Algarni T S, Javed M S *et al* 2021 2D MXene materials for sodium ion batteries: a review on energy storage *J. Energy Storage* **37** 102478
[88] Kshetri T, Tran D T, Le H T *et al* 2021 Recent advances in MXene-based nanocomposites for electrochemical energy storage applications *Prog. Mater Sci.* **117** 100733
[89] Chen Z, Asif M, Wang R *et al* 2022 Recent trends in synthesis and applications of porous MXene assemblies: a topical review *Chem. Rec.* **22** e202100261
[90] Zhang C F and Nicolosi V 2019 Graphene and MXene-based transparent conductive electrodes and supercapacitors *Energy Storage Mater.* **16** 102–25
[91] Ma R, Chen Z T, Zhao D N *et al* 2021 $Ti_3C_2T_x$ MXene for electrode materials of supercapacitors *J. Mater. Chem.* A **9** 11501–29
[92] Nasrin K, Sudharshan V, Subramani K *et al* 2022 Insights into 2D/2D MXene heterostructures for improved synergy in structure toward next-generation supercapacitors: a review *Adv. Funct. Mater.* **32** 2110267
[93] Qu K, Sun Z, Shi C *et al* 2021 Dual-acting cellulose nanocomposites filled with carbon nanotubes and zeolitic imidazolate framework-67 (ZIF-67)-derived polyhedral porous Co_3O_4 for symmetric supercapacitors *Adv. Compos. Hybrid Mater.* **4** 670–83
[94] Sarwar S, Lin M-C, Ahasan M R *et al* 2022 Direct growth of cobalt-doped molybdenum disulfide on graphene nanohybrids through microwave irradiation with enhanced electrocatalytic properties for hydrogen evolution reaction *Adv. Compos. Hybrid Mater.* **5** 2339–52
[95] Xue R, Guo H, Yang W *et al* 2022 Cooperation between covalent organic frameworks (COFs) and metal organic frameworks (MOFs): application of COFs–MOFs hybrids *Adv. Compos. Hybrid Mater.* **5** 1595–611
[96] Mao S D, Zhang M, Lin F H *et al* 2022 Attapulgite structure reset to accelerate the crystal transformation of isotactic polybutene *Polymers* **14** 3820
[97] Mao S D, Lin F H, Zhao Y Y *et al* 2022 Preparation of the polyvinyl alcohol thermal energy storage film containing the waste fly ash based on the phase change material *Polymer Eng. Sci.* **62** 3433–40
[98] Wang B, Lin F H, Li X Y *et al* 2019 Transcrystallization of isotactic polypropylene/bacterial cellulose hamburger composite *Polymers* **11** 508
[99] Zhang Y P, Xu H F and Lu S 2021 Preparation and application of layered double hydroxide nanosheets *RSC Adv.* **11** 24254–81
[100] Cao X, Jia Z, Hu D *et al* 2022 Synergistic construction of three-dimensional conductive network and double heterointerface polarization via magnetic FeNi for broadband microwave absorption *Adv. Compos. Hybrid Mater.* **5** 1030–43

[101] Ge R and Chen L 2022 Ultra-small RuO_2 nanoparticles supported on carbon cloth as a high-performance pseudocapacitive electrode *Adv. Compos. Hybrid Mater.* **5** 696–703

[102] Li X, Lin Z, Wei Y *et al* 2022 MXene–MnO_2–CoNi layered double hydroxides//activated carbon flexible asymmetric supercapacitor *J. Energy Storage* **22** 105668

[103] Shan Z, Archana P S, Shen G *et al* 2015 NanoCOT: low-cost nanostructured electrode containing carbon, oxygen, and titanium for efficient oxygen evolution reaction *J. Am. Chem. Soc.* **137** 11996–2005

[104] Gu Z, Atherton J J and Xu Z P 2015 Hierarchical layered double hydroxide nanocomposites: structure, synthesis and applications *Chem. Commun.* **51** 3024–36

[105] Hu P, Dong S, Yuan F *et al* 2022 Hollow carbon microspheres modified with $NiCo_2S_4$ nanosheets as a high-performance microwave absorber *Adv. Compos. Hybrid Mater.* **5** 469–80

[106] Wang B, Nie K, Xue X R *et al* 2018 Preparation of maleic anhydride grafted polybutene and its application in isotactic polybutene−1/microcrystalline cellulose composites *Polymers* **10** 393

[107] Sun H, Wang X, Li H *et al* 2020 Selenium modulates cadmium-induced ultrastructural and metabolic changes in cucumber seedlings *RSC Adv.* **10** 17892–905

[108] Tomboc G M, Kim J, Wang Y *et al* 2021 Hybrid layered double hydroxides as multifunctional nanomaterials for overall water splitting and supercapacitor applications *J. Mater. Chem.* A **9** 4528–57

[109] Xie P, Liu Y, Feng M *et al* 2021 Hierarchically porous Co/C nanocomposites for ultralight high-performance microwave absorption *Adv. Compos. Hybrid Mater.* **4** 173–85

[110] Song L, Dai C, Jin X *et al* 2022 Pure aqueous planar microsupercapacitors with ultrahigh energy density under wide temperature ranges *Adv. Funct. Mater.* **32** 2203270

[111] Wang Y, Zhang Y, Lu H *et al* 2018 Novel N-doped ZrO_2 with enhanced visible-light photocatalytic activity for hydrogen production and degradation of organic dyes *RSC Adv.* **8** 6752–8

[112] Lu H, Zhao B, Pan R *et al* 2014 Safe and facile hydrogenation of commercial Degussa P25 at room temperature with enhanced photocatalytic activity *RSC Adv.* **4** 1128–32

[113] Lu H, Wang Y, Wang Y *et al* 2015 Adjusting phase transition of titania-based nanotubes via hydrothermal and post treatment *RSC Adv.* **5** 89777–82

[114] Song S, Hao Z, Dong L *et al* 2016 A bubble-based EMMS model for pressurized fluidization and its validation with data from a jetting fluidized bed *RSC Adv.* **6** 111041–51

[115] Guzman-Vargas A, Vazquez-Samperio J, Oliver-Tolentino M A *et al* 2017 Influence on the electrocatalytic water oxidation of M^{2+}/M^{3+} cation arrangement in NiFe LBH: experimental and theoretical DFT evidences *Electrocatalysis* **8** 383–91

[116] Guo Y, Wang D, Bai T *et al* 2021 Electrostatic self-assembled $NiFe_2O_4/Ti_3C_2T_x$ MXene nanocomposites for efficient electromagnetic wave absorption at ultralow loading level *Adv. Compos. Hybrid Mater.* **4** 602–13

[117] Feng S, Zhai F, Su H *et al* 2023 Progress of metal organic frameworks-based composites in electromagnetic wave absorption *Mater. Today Phys.* **30** 100950

[118] Cheng W, Wang Y, Ge S *et al* 2021 One-step microwave hydrothermal preparation of Cd/Zr-bimetallic metal–organic frameworks for enhanced photochemical properties *Adv. Compos. Hybrid Mater.* **4** 150–61

[119] Guo L, Zhang Y, Zheng J *et al* 2021 Synthesis and characterization of ZnNiCr-layered double hydroxides with high adsorption activities for Cr(VI) *Adv. Compos. Hybrid Mater.* **4** 819–29

[120] Jia Z, Zhang X, Gu Z *et al* 2023 MOF-derived Ni–Co bimetal/porous carbon composites as electromagnetic wave absorber *Adv. Compos. Hybrid Mater.* **6** 28

[121] Wang B, Lin F H, Li X Y *et al* 2018 Isothermal crystallization and rheology properties of isotactic polypropylene/bacterial cellulose composite *Polymers* **10** 1284

[122] Wang B, Mao S, Lin F *et al* 2021 Interfacial compatibility on the crystal transformation of isotactic poly (1-butene)/herb residue composite *Polymers* **13** 1654

[123] Li X, He J, Liu M *et al* 2022 Interaction between coal and biomass during co-gasification: a perspective based on the separation of blended char *Processes* **10** 286

[124] Pathak M and Rout C S 2022 Hierarchical $NiCo_2S_4$ nanostructures anchored on nanocarbons and $Ti_3C_2T_x$ MXene for high-performance flexible solid-state asymmetric supercapacitors *Adv. Compos. Hybrid Mater.* **5** 1404–22

[125] Yan Z, Sun Z, Li A *et al* 2021 Three-dimensional porous flower-like S-doped Fe_2O_3 for superior lithium storage *Adv. Compos. Hybrid Mater.* **4** 716–24

[126] Zheng M, Ren J, Wang C *et al* 2022 Magnetite@poly(p-phenylenediamine) core–shell composite modified with salicylaldehyde for adsorption and separation of Mn (VII) from polluted water *J. Nanostruct. Chem.* **139** e52515

[127] Wang D S, Mukhtar A, Wu K M *et al* 2019 Multi-segmented nanowires: a high tech bright future *Materials* **12** 3908

[128] Wang B, Zhang H-R, Huang C *et al* 2017 Study on non-isothermal crystallization behavior of isotactic polypropylene/bacterial cellulose composites *RSC Adv.* **7** 42113–22

[129] Wang Y, Lu H, Wang Y *et al* 2016 Facile synthesis of TaO_xN_y photocatalysts with enhanced visible photocatalytic activity *RSC Adv.* **6** 1860–4

[130] Dai B, Dong F, Wang H *et al* 2023 Fabrication of CuS/Fe_3O_4@polypyrrole flower-like composites for excellent electromagnetic wave absorption *J. Colloid Interface Sci.* **634** 481–94

[131] Sarfraz M and Shakir I 2017 Recent advances in layered double hydroxides as electrode materials for high-performance electrochemical energy storage devices *J. Energy Storage* **13** 103–22

[132] Zhao M, Zhao Q, Li B *et al* 2017 Recent progress in layered double hydroxide based materials for electrochemical capacitors: design, synthesis and performance *Nanoscale* **9** 15206–25

[133] Yan A L, Wang X C and Cheng J P 2018 Research progress of NiMn layered double hydroxides for supercapacitors: a review *Nanomaterials* **8** 747

[134] Li G, Ji Y, Zuo D *et al* 2019 Carbon electrodes with double conductive networks for high-performance electrical double-layer capacitors *Adv. Compos. Hybrid Mater.* **2** 456–61

[135] Wang P, Song T, Abo-Dief H M *et al* 2022 Effect of carbon nanotubes on the interface evolution and dielectric properties of polylactic acid/ethylene-vinyl acetate copolymer nanocomposites *Adv. Compos. Hybrid Mater.* **5** 1100–10

[136] Wu Q X, Feng Z P, Cai Z M *et al* 2022 Poly(methyl methacrylate)-based ferroelectric/dielectric laminated films with enhanced energy storage performances *Adv. Compos. Hybrid Mater.* **5** 1137–44

[137] Liu H, Xu T, Liang Q *et al* 2022 Compressible cellulose nanofibrils/reduced graphene oxide composite carbon aerogel for solid-state supercapacitor *Adv. Compos. Hybrid Mater.* **5** 1168–79

[138] Xu M and Wei M 2018 Layered double hydroxide-based catalysts: recent advances in preparation, structure, and applications *Adv. Funct. Mater.* **28** 1802943

[139] Zhang H, Liang X B, Hu Y L *et al* 2021 Correlation of C/C preform density and microstructure and mechanical properties of C/C–ZrC-based ultra-high-temperature ceramic matrix composites *Adv. Compos. Hybrid Mater.* **4** 743–50

[140] Du H Y, An Y L, Wei Y H *et al* 2019 Experimental and numerical studies on strength and ductility of gradient-structured iron plate obtained by surface mechanical-attrition treatment *Mater. Sci. Eng. A* **744** 471–80

[141] Wang Y Y, Yan D F, El Hankari S *et al* 2018 Recent progress on layered double hydroxides and their derivatives for electrocatalytic water splitting *Adv. Sci.* **5** 1800064

[142] Fan G L, Li F, Evans D G *et al* 2014 Catalytic applications of layered double hydroxides: recent advances and perspectives *Chem. Soc. Rev.* **43** 7040–66

[143] Guo Y T, Meng N, Xu J *et al* 2019 Microstructure and dielectric properties of $Ba_{0.6}Sr_{0.4}TiO_3$/(acrylonitrile-butadiene-styrene)-poly(vinylidene fluoride) composites *Adv. Compos. Hybrid Mater.* **2** 681–9

[144] Li G, Wang L, Lei X *et al* 2022 Flexible, yet robust polyaniline coated foamed polylactic acid composite electrodes for high-performance supercapacitors *Adv. Compos. Hybrid Mater.* **5** 853–63

[145] Srivastava P, Mishra A, Mizuseki H *et al* 2016 Mechanistic insight into the chemical exfoliation and functionalization of Ti_3C_2 MXene *ACS Appl. Mater. Interfaces* **8** 24256–64

[146] Yu L Y, Hu L F, Anasori B *et al* 2018 MXene-bonded activated carbon as a flexible electrode for high-performance supercapacitors *ACS Energy Lett.* **3** 1597–603

[147] Naguib M, Mashtalir O, Carle J *et al* 2012 Two-dimensional transition metal carbides *ACS Nano* **6** 1322–31

[148] Naguib M, Kurtoglu M, Presser V *et al* 2011 Two-dimensional nanocrystals produced by exfoliation of Ti_3AlC_2 *Adv. Mater.* **23** 4248–53

[149] Sang X, Xie Y, Lin M W *et al* 2016 Atomic defects in monolayer titanium carbide ($Ti_3C_2T_x$) MXene *ACS Nano* **10** 9193–200

[150] Hope M A, Forse A C, Griffith K J *et al* 2016 NMR reveals the surface functionalisation of Ti 3 C 2 MXene *Phys. Chem. Chem. Phys.* **18**

[151] Wang F, Yang C, Duan C *et al* 2015 An organ-like titanium carbide material (MXene) with multilayer structure encapsulating hemoglobin for a mediator-free biosensor *J. Electrochem. Soc.* **162** B16–21

[152] Zhang T, Pan L M, Tang H *et al* 2017 Synthesis of two-dimensional $Ti_3C_2T_x$=MXene using HCl+LiF etchant: enhanced exfoliation and delamination *J. Alloys Compd.* **695** 818–26

[153] Hou C, Hou J, Zhang H *et al* 2020 Facile synthesis of $LiMn_{0.75}Fe_{0.25}PO_4$/C nanocomposite cathode materials of lithium-ion batteries through microwave sintering *Eng. Sci.* **11** 36–43

[154] Hou C, Wang B, Murugadoss V *et al* 2020 Recent advances in Co_3O_4 as anode materials for high-performance lithium-ion batteries *Eng. Sci.* **11** 19–30

[155] Cao M-S, Cai Y-Z, He P *et al* 2019 2D MXenes: electromagnetic property for microwave absorption and electromagnetic interference shielding *Chem. Eng. J.* **359** 1265–302

[156] Li H, Hou Y, Wang F *et al* 2017 Flexible all-solid-state supercapacitors with high volumetric capacitances boosted by solution processable MXene and electrochemically exfoliated graphene *Adv. Energy Mater.* **7** 1601847

[157] Bubnova O, Khan Z U, Malti A *et al* 2011 Optimization of the thermoelectric figure of merit in the conducting polymer poly(3,4-ethylenedioxythiophene) *Nat. Mater.* **10** 429

[158] Wan M 2008 *Conducting Polymers with Micro or Nanometer Structure* (Berlin: Springer)

[159] Wu D, Xu F, Sun B *et al* 2012 Design and preparation of porous polymers *Chem. Rev.* **112** 3959
[160] Xu F, Bao D, Cui Y *et al* 2021 Copper nanoparticle-deposited graphite sheets for highly thermally conductive polymer composites with reduced interfacial thermal resistance *Adv. Compos. Hybrid Mater.* **5** 2235–46
[161] Zhang D, Hu S, Sun Y *et al* 2020 XTe (X = Ge, Sn, Pb) monolayers: promising thermoelectric materials with ultralow lattice thermal conductivity and high-power factor *ES Energy Environ.* **10** 59–65
[162] Li S D, Fan J C, Li S Y *et al* 2021 *In situ*-grown Co_3O_4 nanorods on carbon cloth for efficient electrocatalytic oxidation of urea *J. Nanostruct. Chem.* **11** 735–49
[163] Ehsani A, Mahjani M G, Bordbar M *et al* 2013 Poly ortho aminophenol/TiO_2 nanocomposite: Electrosynthesis and characterization *Synth. Met.* **165** 51–5
[164] Ambade R B, Ambade S B, Shrestha N K *et al* 2017 Controlled growth of polythiophene nanofibers in TiO_2 nanotube arrays for supercapacitor applications *J. Mater. Chem.* A **5** 172–80
[165] Ma Y, Ma M L, Yin X Q *et al* 2018 Tuning polyaniline nanostructures via end group substitutions and their morphology dependent electrochemical performances *Polymer* **156** 128–35
[166] Huang K, Liu J, Lin S *et al* 2021 Flexible silver nanowire dry electrodes for long-term electrocardiographic monitoring *Adv. Compos. Hybrid Mater.* **5** 220–8
[167] Xue Y, Chen S, Yu J *et al* 2020 Nanostructured conducting polymers and their composites: synthesis methodologies, morphologies and applications *J. Mater. Chem.* C **8** 10136–59
[168] Qu K, Sun Z, Shi C *et al* 2021 Dual-acting cellulose nanocomposites filled with carbon nanotubes and zeolitic imidazolate framework-67 (ZIF-67)–derived polyhedral porous Co_3O_4 for symmetric supercapacitors *Adv. Compos. Hybrid Mater.* **4** 670–83
[169] Pan L, Qiu H, Dou C *et al* 2010 Conducting polymer nanostructures: template synthesis and applications in energy storage *Int. J. Mol. Sci.* **11** 2636–57
[170] Xue M, Li F, D C *et al* 2016 Gas sensors: high-oriented polypyrrole nanotubes for next-generation gas sensor *Adv. Mater.* **28** 8067 8067
[171] Choi I Y, Lee J, Ahn H *et al* 2015 High-conductivity two-dimensional polyaniline nanosheets developed on ice surfaces *Angew. Chem. Int. Ed. Engl.* **54** 10497–501
[172] Pan D, Dong J, Yang G *et al* 2021 Ice template method assists in obtaining carbonized cellulose/boron nitride aerogel with 3D spatial network structure to enhance the thermal conductivity and flame retardancy of epoxy-based composites *Adv. Compos. Hybrid Mater.* **5** 58–70
[173] Ma Y, Zhang C Y, Hou C P *et al* 2017 Cetyl trimethyl ammonium bromide (CTAB) micellar templates directed synthesis of water-dispersible polyaniline rhombic plates with excellent processability and flow-induced color variation *Polymer* **117** 30–6
[174] Wang W, Ma Y, Zhuang Z *et al* 2021 Synthesis of walnut-like polyaniline by using polyvinyl alcohol micellar template with excellent film transmission *J. Appl. Polym. Sci.* **138** 50701
[175] Zhuang J, Sun J, Wu D *et al* 2021 Multi-factor analysis on thermal conductive property of metal-polymer composite microstructure heat exchanger *Adv. Compos. Hybrid Mater.* **4** 27–35
[176] Chen J, Zhu Y, Guo Z *et al* 2020 Recent progress on thermo-electrical properties of conductive polymer composites and their application in temperature sensors *Eng. Sci.* **12** 13–22

[177] Zhao C Q, Jing T, Tian J Z *et al* 2021 Visible light-driven photoelectrochemical enzyme biosensor based on reduced graphene oxide/titania for detection of glucose *J. Nanostruct. Chem.* **12** 193–205
[178] Yong M, Hou C, Hao Z *et al* 2017 Morphology-dependent electrochemical supercapacitors in multi-dimensional polyanilinenanostructures *J. Mater. Chem.* A **5** 14041–52
[179] Ehsani A, Parsimehr H, Nourmohammadi H *et al* 2019 Environment-friendly electrodes using biopolymer chitosan/poly ortho aminophenol with enhanced electrochemical behavior for use in energy storage devices *Polym. Compos.* **40** 4629–37
[180] Li K L, Liu X Y, Zheng T X *et al* 2019 Tuning MnO_2 to FeOOH replicas with bio-template 3D morphology as electrodes for high performance asymmetric supercapacitors *Chem. Eng. J.* **370** 136–47
[181] Deokate R J 2020 Chemically deposited $NiCo_2O_4$ thin films for electrochemical study *ES Mater. Manuf.* **11** 16–9
[182] Sayyed S G, Mahadik M A, Shaikh A V *et al* 2019 Nano-metal oxide based supercapacitor via electrochemical deposition *ES Energy Environ.* **3** 25–44
[183] Gite A B, Palve B M, Gaikwad V B *et al* 2020 Physicochemical properties and thermoelectric studies of electrochemically deposited lead telluride films *ES Mater. Manuf.* **11** 40–9
[184] Borate H, Bhorde A, Waghmare A *et al* 2020 Single-step electrochemical deposition of czts thin films with enhanced photoactivity *ES Mater. Manuf.* **11** 30–9
[185] Shiri H M, Ehsani A, Behjatmanesh-Ardakani R *et al* 2019 Electrosynthesis of Y_2O_3 nanoparticles and its nanocomposite with POAP as high efficient electrode materials in energy storage device: Surface, density of state and electrochemical investigation *Solid State Ionics* **338** 87–95
[186] Debiemme-Chouvy C, Fakhry A and Pillier F 2018 Electrosynthesis of polypyrrole nano/micro structures using an electrogenerated oriented polypyrrole nanowire array as framework *Electrochim. Acta* **268** 66–72
[187] Berggren M, Crispin X, Fabiano S *et al* 2019 Organic electrochemical devices: ion electron–coupled functionality in materials and devices based on conjugated polymers *Adv. Mater.* **31** 1970160
[188] Mana P M, Bhujbal P K and Pathan H M 2020 Fabrication and characterization of ZnS based Photoelectrochemical solar cell *ES Energy Environ.* **12** 77–85
[189] Hajian A, Rafati A A, Afraz A *et al* 2014 Electrosynthesis of high-density polythiophene nanotube arrays and their application for sensing of riboflavin *J. Mol. Liq.* **199** 150–5
[190] Raaijmakers M and Benes N E 2016 Current trends in interfacial polymerization chemistry *Prog. Polym. Sci.* **63** 86–142
[191] Zhang L, Du W, Nautiyal A *et al* 2018 Recent progress on nanostructured conducting polymers and composites: synthesis, application and future aspects *Sci. China Mater.* **61** 303–52
[192] Ingle R V, Shaikh S F, Bhujbal P K *et al* 2020 Polyaniline doped with protonic acids: optical and morphological studies *ES Mater. Manuf.* **8** 54–9
[193] Chen L, Lan C, Xu B *et al* 2021 Progress on material characterization methods under big data environment *Adv. Compos. Hybrid Mater.* **4** 235–47
[194] Tang X, Guo X, Wu W *et al* 2018 2D metal carbides and nitrides (MXenes) as high-performance electrode materials for lithium-based batteries *Adv. Energy Mater.* **8** 1801897
[195] Yu L, Fan Z, Shao Y *et al* 2019 Versatile N-doped MXene ink for printed electrochemical energy storage application *Adv. Energy Mater.* **9** 1901839

[196] Huang J H, Cheng X Q, Wu Y D *et al* 2021 Critical operation factors and proposed testing protocol of nanofiltration membranes for developing advanced membrane materials *Adv. Compos. Hybrid Mater.* **4** 1092–101

[197] Buczek S, Barsoum M L, Uzun S *et al* 2020 Rational design of titanium carbide MXene electrode architectures for hybrid capacitive deionization *Energy Environ. Mater.* **3** 398–404

[198] Zhang Y Z, El-Demellawi J K, Jiang Q *et al* 2020 MXene hydrogels: fundamentals and applications *Chem. Soc. Rev.* **49** 7229–51

[199] Zhao M Q, Ren C E, Ling Z *et al* 2015 Flexible MXene/carbon nanotube composite paper with high volumetric capacitance *Adv. Mater.* **27** 339–45

[200] Yan J, Ren C E, Maleski K *et al* 2017 Flexible MXene/graphene films for ultrafast supercapacitors with outstanding volumetric capacitance *Adv. Funct. Mater.* **27** 1701264

[201] Lukatskaya M R, Kota S, Lin Z *et al* 2017 Ultra-high-rate pseudocapacitive energy storage in two-dimensional transition metal carbides *Nat. Energy* **2** 17105

[202] Ehsani A 2015 Influence of counter ions in electrochemical properties and kinetic parameters of poly tyramine electroactive film *Prog. Org. Coat.* **78** 133–9

[203] Vahidmohammadi A, Moncada J, Chen H *et al* 2018 Thick and freestanding MXene/PANI pseudocapacitive electrodes with ultrahigh specific capacitance *J. Mater. Chem.* A **6** 22123–33

[204] Zhu Q, Huang Y, Li Y *et al* 2021 Aluminum dihydric tripolyphosphate/polypyrrole-functionalized graphene oxide waterborne epoxy composite coatings for impermeability and corrosion protection performance of metals *Adv. Compos. Hybrid Mater.* **4** 780–92

[205] Li X, Zhao W, Yin R *et al* 2018 A highly porous polyaniline-graphene composite used for electrochemical supercapacitors *Eng. Sci.* **3** 89–95

[206] Ma Y, Hou C P, Zhang H P *et al* 2019 Three-dimensional core–shell Fe_3O_4/polyaniline coaxial heterogeneous nanonets: preparation and high performance supercapacitor electrodes *Electrochim. Acta* **315** 114–23

[207] Xu H, Zheng D, Liu F *et al* 2020 Synthesis of an MXene/polyaniline composite with excellent electrochemical properties *J. Mater. Chem.* A **8** 5853–8

[208] Boota M, Anasori B, Voigt C *et al* 2016 Pseudocapacitive electrodes produced by oxidant-free polymerization of pyrrole between the layers of 2D titanium carbide (MXene) *Adv. Mater.* **28** 1517–22

[209] Zhao L, Wang K, Wei W *et al* 2019 High-performance flexible sensing devices based on polyaniline/MXene nanocomposites *InfoMat* **1** 407–16

[210] Gu H, Zhang H, Gao C *et al* 2018 New functions of polyaniline *ES Mater. Manuf.* **1** 3–12

[211] Guo J, Chen Z, Abdul W *et al* 2021 Tunable positive magnetoresistance of magnetic polyaniline nanocomposites *Adv. Compos. Hybrid Mater.* **4** 534–42

[212] Jian X, He M, Chen L *et al* 2019 Three-dimensional carambola-like MXene/polypyrrole composite produced by one-step co-electrodeposition method for electrochemical energy storage *Electrochim. Acta* **318** 820–7

[213] Huang Y, Li H F, Wang Z F *et al* 2016 Nanostructured polypyrrole as a flexible electrode material of supercapacitor *Nano Energy* **22** 422–38

[214] Elayappan V, Murugadoss V, Fei Z *et al* 2020 Influence of polypyrrole incorporated electrospun poly(vinylidene fluoride-co-hexafluoropropylene) nanofibrous composite membrane electrolyte on the photovoltaic performance of dye sensitized solar cell *Eng. Sci.* **10** 78–84

[215] Zhang C, Xu S K, Cai D *et al* 2020 Planar supercapacitor with high areal capacitance based on Ti_3C_2/polypyrrole composite film *Electrochim. Acta* **330** 135277

[216] Wei D, Wu W, Zhu J *et al* 2020 A facile strategy of polypyrrole nanospheres grown on Ti_3C_2–MXene nanosheets as advanced supercapacitor electrodes *J. Electroanal. Chem.* **877** 114538
[217] Wei H, Li A, Kong D *et al* 2021 Polypyrrole/reduced graphene aerogel film for wearable piezoresisitic sensors with high sensing performances *Adv. Compos. Hybrid Mater.* **4** 86–95
[218] Guo J, Li X, Liu H *et al* 2021 Tunable magnetoresistance of core–shell structured polyaniline nanocomposites with 0-, 1-, and 2-dimensional nanocarbons *Adv. Compos. Hybrid Mater.* **4** 51–64
[219] Gao L F, Li C, Huang W C *et al* 2020 MXene/polymer membranes: synthesis, properties, and emerging applications *Chem. Mater.* **32** 1703–47
[220] Wan Y J, Li X M, Zhu P L *et al* 2020 Lightweight, flexible MXene/polymer film with simultaneously excellent mechanical property and high-performance electromagnetic interference shielding *Compos.Appl. Sci. Manuf.* **130** 105764
[221] Wustoni S, Saleh A, El-Demellawi J K *et al* 2020 MXene improves the stability and electrochemical performance of electropolymerized PEDOT films *APL Mater.* **8** 121105
[222] Hou C L and Yu H Z 2020 Modifying the nanostructures of PEDOT:PSS/$Ti_3C_2T_x$ composite hole transport layers for highly efficient polymer solar cells *J. Mater. Chem.* C **8** 4169–80
[223] Guan X, Feng W, Wang X *et al* 2020 Significant enhancement in the seebeck coefficient and power factor of p-type poly(3,4-ethylenedioxythiophene):poly(styrenesulfonate) through the incorporation of n-type MXene *ACS Appl. Mater. Interfaces* **12** 13013–20
[224] Chen Z, Han Y, Li T *et al* 2018 Preparation and electrochemical performances of doped MXene/poly(3,4-ethylenedioxythiophene) composites *Mater. Lett.* **220** 305–8
[225] Li X, Yan J and Zhu K 2021 Fabrication and characterization of Pt doped Ti/Sb–SnO_2 electrode and its efficient electro-catalytic activity toward phenol *Eng. Sci.* **15** 38–46
[226] Chen C, Boota M, Xie X *et al* 2017 Charge transfer induced polymerization of EDOT confined between 2D titanium carbide layers *J. Mater. Chem.* A **5** 5260–5
[227] Chiu W W, Travas-Sejdic J, Cooney R P *et al* 2010 Studies of dopant effects in poly(3,4-ethylenedi-oxythiophene) using Raman spectroscopy *J. Raman Spectrosc.* **37** 1354–61
[228] Umar A, Kumar R, Algadi H *et al* 2021 Highly sensitive and selective 2-nitroaniline chemical sensor based on Ce-doped SnO_2 nanosheets/Nafion-modified glassy carbon electrode *Adv. Compos. Hybrid Mater.* **4** 1015–26
[229] Zheng X, Shen J, Hu Q *et al* 2021 Vapor phase polymerized conducting polymer/MXene textiles for wearable electronics *Nanoscale* **13** 1832–41
[230] Hwang S, Park N I, Choi Y J *et al* 2019 PEDOT:PSS nanocomposite via partial intercalation of monomer into colloidal graphite prepared by *in situ* polymerization *J. Ind. Eng. Chem.* **76** 116–21
[231] Zhang J, Seyedin S, Gu Z *et al* 2017 MXene: a potential candidate for yarn supercapacitors *Nanoscale* **9** 18604–8
[232] Hussain S, Liu H L, Vikraman D *et al* 2021 Characteristics of Mo_2C-CNTs hybrid blended hole transport layer in the perovskite solar cells and X-ray detectors *J. Alloys Compd.* **885** 161039
[233] Zuo W, Yang C, Zang L *et al* 2020 One-step hydrothermal synthesis of flower-like MnO_2–$Ni_xCo_{1-x}(OH)_2$@carbon felt positive electrodes with 3D open porous structure for high-performance asymmetric supercapacitors *J. Mater. Sci.* **56** 4797–809

[234] Zhang K, Ma Z, Deng H *et al* 2021 Improving high-temperature energy storage performance of PI dielectric capacitor films through boron nitride interlayer *Adv. Compos. Hybrid Mater.*
[235] Najib S and Erdem E 2019 Current progress achieved in novel materials for supercapacitor electrodes: mini review *Nanoscale Adv.* **1** 2817–27
[236] Zhu X, Li X, Tao H *et al* 2021 Preparation of Co_2Al layered double hydroxide nanosheet/Co_2Mn bimetallic hydroxide nanoneedle nanocomposites on nickel foam for supercapacitors *J. Alloys Compd.* **851** 156868
[237] Kasap S, Kaya I I, Repp S *et al* 2019 Superbat: battery-like supercapacitor utilized by graphene foam and zinc oxide (ZnO) electrodes induced by structural defects *Nanoscale Adv.* **1** 2586–97
[238] Ma Z, Fan L, Jing F *et al* 2021 MnO_2 nanowires@NiCo-LBH nanosheet core–shell heterostructure: a slow irreversible transition of hydrotalcite phase for high-performance pseudocapacitance electrode *ACS Appl. Energy Mater.* **4** 3983–92
[239] Sahoo R, Pham D T, Lee T H *et al* 2018 Redox-driven route for widening voltage window in asymmetric supercapacitor *ACS Nano* **12** 8494–505
[240] Tu L L and Jia C Y 2010 Conducting polymers as electrode materials for supercapacitors *Progress Chem.* **22** 1610–8
[241] Xia C, Xie Y B, Du H X *et al* 2015 Ternary nanocomposite of polyaniline/manganese dioxide/titanium nitride nanowire array for supercapacitor electrode *J. Nanopart. Res.* **17** 1–12
[242] Wang P, Yang L, Gao S *et al* 2021 Enhanced dielectric properties of high glass transition temperature PDCPD/CNT composites by frontal ring-opening metathesis polymerization *Adv. Compos. Hybrid Mater.* **4** 639–46
[243] Kazemi S H, Hosseinzadeh B, Kazemi H *et al* 2018 Facile synthesis of mixed metal-organic frameworks: electrode materials for supercapacitors with excellent areal capacitance and operational stability *ACS Appl. Mater. Interfaces* **10** 23063–73
[244] Sun S Y, Huang M J, Wang P C *et al* 2019 Controllable hydrothermal synthesis of Ni/Co MOF as hybrid advanced electrode materials for supercapacitor *J. Electrochem. Soc.* **166** A1799–805
[245] Wang Y, Hu Y-J, Hao X *et al* 2020 Hydrothermal synthesis and applications of advanced carbonaceous materials from biomass: a review *Adv. Compos. Hybrid Mater.* **3** 267–84
[246] Saxena P and Shukla P 2021 A comprehensive review on fundamental properties and applications of poly(vinylidene fluoride) (PVDF) *Adv. Compos. Hybrid Mater.* **4** 8–26
[247] He X, Ou D, Wu S *et al* 2021 A mini review on factors affecting network in thermally enhanced polymer composites: filler content, shape, size, and tailoring methods *Adv. Compos. Hybrid Mater.* **5** 21–38
[248] Yang J, Liu Y, Liu S L *et al* 2017 Conducting polymer composites: material synthesis and applications in electrochemical capacitive energy storage *Mater. Chem. Front.* **1** 251–68
[249] Bag P P, Singh G P, Singha S *et al* 2020 Synthesis of metal-organic frameworks (MOFs) and their applications to biology, catalysis and electrochemical charge storage: a mini review *Eng. Sci.* **13** 1–10
[250] Ma Y, Xie X, Yang W *et al* 2021 Recent advances in transition metal oxides with different dimensions as electrodes for high-performance supercapacitors *Adv. Compos. Hybrid Mater.* **4** 906–24

[251] Zhang C J, Anasori B, Seral-Ascaso A *et al* 2017 Transparent, flexible, and conductive 2D titanium carbide (MXene) films with high volumetric capacitance *Adv. Mater.* **29** 1702678
[252] Fafarman , Ghidiu A T *et al* 2016 Highly conductive optical quality solution-processed films of 2D titanium carbide *Adv. Funct. Mater.* **26** 4162–8
[253] Jiang H M, Wang Z G, Yang Q *et al* 2018 A novel $MnO_2/Ti_3C_2T_x$ MXene nanocomposite as high performance electrode materials for flexible supercapacitors *Electrochim. Acta* **290** 695–703
[254] Lu C X, Li A R, Zhai T F *et al* 2020 Interface design based on Ti_3C_2 MXene atomic layers of advanced battery-type material for supercapacitors *Energy Storage Mater.* **26** 472–82
[255] Li K, Wang X H, Wang X F *et al* 2020 All-pseudocapacitive asymmetric MXene-carbon-conducting polymer supercapacitors *Nano Energy* **75** 104971
[256] Fan Z, Wang J, Kang H *et al* 2020 A compact MXene film with folded structure for advanced supercapacitor electrode material *ACS Appl. Energy Mater.* **3** 1811–20
[257] Nie R, Wang Q, Sun P *et al* 2018 Pulsed laser deposition of $NiSe_2$ film on carbon nanotubes for high-performance supercapacitor *Eng. Sci.* **6** 22–9
[258] Wu W L, Niu D J, Zhu J F *et al* 2019 Organ-like Ti_3C_2 MXenes/polyaniline composites by chemical grafting as high-performance supercapacitors *J. Electroanal. Chem.* **847** 113203
[259] Li Y, Kamdem P and Jin X J 2021 Hierarchical architecture of MXene/PANI hybrid electrode for advanced asymmetric supercapacitors *J. Alloys Compd.* **850** 156608
[260] Ma Y, Hou C, Zhang H *et al* 2017 Morphology-dependent electrochemical supercapacitors in multi-dimensional polyaniline nanostructures *J. Mater. Chem.* A **5** 14041–52
[261] Chen Z, Wang Y, Han J *et al* 2020 Preparation of polyaniline onto dl-tartaric acid assembled MXene surface as an electrode material for supercapacitors *ACS Appl. Energy Mater.* **3** 9326–36
[262] Wu W L, Wei D, Zhu J F *et al* 2019 Enhanced electrochemical performances of organ-like Ti_3C_2 MXenes/polypyrrole composites as supercapacitors electrode materials *Ceram. Int.* **45** 7328–37
[263] Tong L, Jiang C, Cai K F *et al* 2020 High-performance and freestanding $PPy/Ti_3C_2T_x$ composite film for flexible all-solid-state supercapacitors *J. Power Sources* **465** 228267
[264] Le T A, Tran N Q, Hong Y *et al* 2019 Intertwined titanium carbide MXene within a 3D tangled polypyrrole nanowires matrix for enhanced supercapacitor performances *Chemistry* **25** 1037–43
[265] Tao Q, Dahlqvist M, Lu J *et al* 2017 Two-dimensional $Mo_{1.33}C$ MXene with divacancy ordering prepared from parent 3D laminate with in-plane chemical ordering *Nat. Commun.* **8** 14949
[266] Qin L, Tao Q, El Ghazaly A *et al* 2018 High-Performance ultrathin flexible solid-state supercapacitors based on solution processable $Mo_{1.33}C$ MXene and PEDOT:PSS *Adv. Funct. Mater.* **28** 1703808
[267] Wang X, Kajiyama S, Iinuma H *et al* 2015 Pseudocapacitance of MXene nanosheets for high-power sodium-ion hybrid capacitors *Nat. Commun.* **6** 6544
[268] Cai J, Xu W, Liu Y *et al* 2018 Robust construction of flexible bacterial cellulose@Ni(OH) paper: toward high 2 capacitance and sensitive H_2O_2 detection *Eng. Sci.* **5** 21–9
[269] Huang J, Chen Q, Chen S *et al* 2021 Al^{3+}-doped $FeNb_{11}O_{29}$ anode materials with enhanced lithium-storage performance *Adv. Compos. Hybrid Mater.* **4** 733–42
[270] Zhou Z, Tie J, Yang H *et al* 2021 3D hierarchical $NiCo_2S_4$/Ni–Co LBH architecture for high-performance supercapacitor *J. Mater. Sci. Mater. Electron.* **32** 3843–53

[271] Jiang Q, Kurra N, Alhabeb M *et al* 2018 All pseudocapacitive MXene–RuO_2 asymmetric supercapacitors *Adv. Energy Mater.* **8** 1703043
[272] De S, Maity C K, Sahoo S *et al* 2021 Polyindole booster for $Ti_3C_2T_x$ MXene based symmetric and asymmetric supercapacitor devices *ACS Appl. Energy Mater.* **4** 3712–23
[273] Ding F, Hu H, Ding H *et al* 2020 Sulfonated poly(fluorene ether ketone) (SPFEK)/α-zirconium phosphate (ZrP) nanocomposite membranes for fuel cell applications *Adv. Compos. Hybrid Mater.* **3** 546–50
[274] Chen M, Chen J, Tan X *et al* 2021 Facile self-assembly of sandwich-like MXene/graphene oxide/nickel-manganese layered double hydroxide nanocomposite for high performance supercapacitor *J. Energy Storage* **44** 103456
[275] Xia Q X, Shinde N M, Zhang T *et al* 2018 Seawater electrolyte-mediated high volumetric MXene-based electrochemical symmetric supercapacitors *Dalton Trans.* **47** 8676–82
[276] Yang C H, Tang Y, Tian Y P *et al* 2018 Achieving of flexible, free-standing, ultracompact delaminated titanium carbide films for high volumetric performance and heat-resistant symmetric supercapacitors *Adv. Funct. Mater.* **28** 1705487
[277] Zhang P, Zhu Q, Soomro R A *et al* 2020 *In situ* ice template approach to fabricate 3D flexible MXene film-based electrode for high performance supercapacitors *Adv. Funct. Mater.* **30** 2000922
[278] Zhu K, Jin Y, Du F *et al* 2019 Synthesis of Ti_2CT_x MXene as electrode materials for symmetric supercapacitor with capable volumetric capacitance *J. Energy Chem.* **31** 11–8
[279] Ling Z, Ren C E, Zhao M Q *et al* 2014 Flexible and conductive MXene films and nanocomposites with high capacitance *Proc. Natl Acad. Sci. USA* **111** 16676–81
[280] Wu Z T, Liu X C, Shang T X *et al* 2021 Reassembly of MXene hydrogels into flexible films towards compact and ultrafast supercapacitors *Adv. Funct. Mater.* **31** 2102874
[281] Zhao M Q, Xie X, Ren C E *et al* 2017 Hollow MXene spheres and 3D macroporous MXene frameworks for na-ion storage *Adv. Mater.* **29** 1702410
[282] Das P and Wu Z S 2020 MXene for energy storage: present status and future perspectives *J. Phys.Energy* **2** 032004
[283] Rahane G K, Jathar S B, Rondiya S R *et al* 2020 Photoelectrochemical investigation on the cadmium sulfide (CdS) thin films prepared using spin coating technique *ES Mater. Manuf.* **11** 57–64
[284] Wang J, Wang J, Kong Z *et al* 2017 Conducting-polymer-based materials for electrochemical energy conversion and storage *Adv. Mater.* **29** 1703044
[285] Gund G S, Park J H, Harpalsinh R *et al* 2019 MXene/polymer hybrid materials for flexible AC-filtering electrochemical capacitors *Joule* **3** 164–76
[286] Li J M, Levitt A, Kurra N *et al* 2019 MXene-conducting polymer electrochromic microsupercapacitors *Energy Storage Mater.* **20** 455–61
[287] Faruk M O, Ahmed A, Adak B *et al* 2021 High performance 2D MXene based conducting polymer hybrids: synthesis to emerging applications *J. Mater. Chem.* C **9** 10193–215
[288] Wang B, Sun X, Xie X *et al* 2021 Experimental investigation of a novel cathode matrix flow field in proton exchange membrane fuel cell *ES Energy Environ.* **12** 95–107
[289] Wu X M, Huang B, Lv R Y *et al* 2019 Highly flexible and low capacitance loss supercapacitor electrode based on hybridizing decentralized conjugated polymer chains with MXene *Chem. Eng. J.* **378** 122246

[290] Wu W, Wang C, Zhao C *et al* 2020 Facile strategy of hollow polyaniline nanotubes supported on Ti_3C_2–MXene nanosheets for High-performance symmetric supercapacitors *J. Colloid Interface Sci.* **580** 601–13
[291] Zhou Y, Zou Y, Peng Z *et al* 2020 Arbitrary deformable and high-strength electroactive polymer/MXene anti-exfoliative composite films assembled into high performance, flexible all-solid-state supercapacitors *Nanoscale* **12** 20797–810
[292] Wang Y, Wang X, Li X *et al* 2021 Scalable fabrication of polyaniline nanodots decorated MXene film electrodes enabled by viscous functional inks for high-energy-density asymmetric supercapacitors *Chem. Eng. J.* **405** 126664
[293] Couly C, Alhabeb M, Van Aken K L *et al* 2017 Asymmetric flexible MXene-reduced graphene oxide micro-supercapacitor *Adv. Electron. Mater.* **4** 1700339
[294] Boota M, Bécuwe M and Gogotsi Y 2020 Phenothiazine–MXene aqueous asymmetric Pseudocapacitors *ACS Appl. Energy Mater.* **3** 3144–9
[295] Su Y, Yin H, Wang X *et al* 2021 Preparation and properties of ethylene-acrylate salt ionomer/polypropylene antistatic alloy *Adv. Compos. Hybrid Mater.* **4** 104–13
[296] Xia Q X, Fu J, Yun J M *et al* 2017 High volumetric energy density annealed-MXene-nickel oxide/MXene asymmetric supercapacitor *RSC Adv.* **7** 11000–11
[297] Zhao K, Wang H, Zhu C *et al* 2019 Free-standing MXene film modified by amorphous FeOOH quantum dots for high-performance asymmetric supercapacitor *Electrochim. Acta* **308** 1–8
[298] Guo Y, Jiang Q, Peng B *et al* 2020 Investigating dynamic processes of nanomaterials using *in situ* liquid phase TEM technologies: 2014–2019 *Eng. Sci.* **9** 17–24
[299] Qin L, Tao Q, Liu X *et al* 2019 Polymer-MXene composite films formed by MXene-facilitated electrochemical polymerization for flexible solid-state microsupercapacitors *Nano Energy* **60** 734–42
[300] Boota M, Rajesh M and Becuwe M 2020 Multi-electron redox asymmetric supercapacitors based on quinone-coupled viologen derivatives and $Ti_3C_2T_x$ MXene *Mater. Today Energy* **18** 100532
[301] Liang W and Zhitomirsky I 2021 MXene–carbon nanotube composite electrodes for high active mass asymmetric supercapacitors *J. Mater. Chem.* A **9** 10335–44
[302] Fu J, Yun J, Wu S *et al* 2018 Architecturally robust graphene-encapsulated MXene Ti_2CT_x@polyaniline composite for high-performance pouch-type asymmetric supercapacitor *ACS Appl. Mater. Interfaces* **10** 34212–21
[303] Boota M and Gogotsi Y 2019 MXene-conducting polymer asymmetric pseudocapacitors *Adv. Energy Mater.* **9** 1802917
[304] Li K, Wang X, Li S *et al* 2020 An ultrafast conducting polymer@MXene positive electrode with high volumetric capacitance for advanced asymmetric supercapacitors *Small* **16** 1906851
[305] Zhao Y H, Liu K X, Hou H *et al* 2022 Role of interfacial energy anisotropy in dendrite orientation in Al-Zn alloys: a phase field study *Mater. Des.* **216** 110555
[306] Xin T Z, Tang S, Ji F *et al* 2022 Phase transformations in an ultralight BCC Mg alloy during anisothermal ageing *Acta Mater.* **239** 118248
[307] Cai J C, Murugadoss V, Jiang J Y *et al* 2022 Waterborne polyurethane and its nanocomposites: a mini-review for anti-corrosion coating, flame retardancy, and biomedical applications *Adv. Compos. Hybrid Mater.* **5** 641–50

[308] Mamaril G S S, De Luna M D G, Bindumadhavan K *et al* 2021 Nitrogen and fluorine co-doped 3-dimensional reduced graphene oxide architectures as high-performance electrode material for capacitive deionization of copper ions *Sep. Purif. Technol.* **272** 117559

[309] Zhang Y F, Zheng J J, Nan J J *et al* 2023 Influence of mass ratio and calcination temperature on physical and photoelectrochemical properties of ZnFe-layered double oxide/cobalt oxide heterojunction semiconductor for dye degradation applications *Particuology* **74** 141–55

[310] Pan D, Yang G, Abo-Dief H M *et al* 2022 Vertically aligned silicon carbide nanowires/boron nitride cellulose aerogel networks enhanced thermal conductivity and electromagnetic absorbing of epoxy composites *Nano-Micro Lett.* **14** 118

[311] Feng Y, Li Y C, Ye X M *et al* 2022 Synthesis and characterization of 2,5-furandicarboxylic acid poly(butanediol sebacate-butanediol) terephthalate (PBSeT) segment copolyesters with excellent water vapor barrier and good mechanical properties *J. Mater. Sci.* **57** 10997–1012

[312] Suryanti , Suryani L, Hartatiek S E I *et al* 2019 The effect of Mn_2O_3 nanoparticles on its specific capacitance of symmetric supercapacitors FC–ZnO–x(Mn_2O_3) *7th Int. Conf. of Adv. Mater. Sci. Technol. (ICAMST)* vol 44 pp 3355–60

[313] Taer E, Putri A, Farma R *et al* 2019 The effect of potassium iodide (KI) addition to aqueous-based electrolyte (sulfuric acid/H_2SO_4) for increase the performance of supercapacitor cells *7th Int. Conf. of Adv. Mater. Sci. Technol. (ICAMST)* vol 44 pp 3241–4

[314] Zhao Y, Liu F, Zhao Z *et al* 2022 Direct ink printing reduced graphene oxide/KCu_7S_4 electrodes for high-performance supercapacitors *Adv. Compos. Hybrid Mater.* **5** 1516–26

[315] Zhao Y, Liu F, Zhu K *et al* 2022 Three-dimensional printing of the copper sulfate hybrid composites for supercapacitor electrodes with ultra-high areal and volumetric capacitances *Adv. Compos.Hybrid Mater.* **5** 1537–47

[316] Lian M, Huang Y, Liu Y *et al* 2022 An overview of regenerable wood-based composites: preparation and applications for flame retardancy, enhanced mechanical properties, biomimicry, and transparency energy saving *Adv. Compos. Hybrid Mater.* **5** 1612–57

[317] Wang R, Lei W, Wang L *et al* 2021 N-doped carbon nanofibrous film with unique wettability, enhanced supercapacitive property, and facile capacity to demulsify surfactant free oil-in-water emulsions *Chem. Res. Chin. Univ.* **37** 436–42

[318] Wang J, Tang H, Ren H *et al* 2014 pH-regulated synthesis of multi-shelled manganese oxide hollow microspheres as supercapacitor electrodes using carbonaceous microspheres as templates *Adv. Sci.* **1** 1400011

[319] Ma Q, Han X, Cui J *et al* 2022 Ni embedded carbon nanofibers/Ni–Al LBHs with multicomponent synergy for hybrid supercapacitor electrodes *Colloids Surf.*, A **649** 129270

[320] Wu Y F, Hsiao Y C, Liao C H *et al* 2022 Novel design of sulfur-doped nickel cobalt layered double hydroxide and polypyrrole nanotube composites from zeolitic imidazolate Framework-67 as efficient active material of battery supercapacitor hybrids *J. Colloid Interface Sci.* **628** 540–52

[321] Lv H, Rao H, Liu Z *et al* 2022 NiAl layered double hydroxides with enhanced interlayer spacing via ion-exchange as ultra-high performance supercapacitors electrode materials *J. Energy Storage* **52** 104940

[322] Deng H, Liu T, Liao W *et al* 2022 Double metal ions synergistic effect in the Ni-doped Co (OH)(BA) nanobelts for enhanced supercapacitor performance *J. Phys. Chem. Solids* **164** 110641

[323] Saber O, Ansari S A, Osama A *et al* 2022 One-dimensional nanoscale Si/Co based on layered double hydroxides towards electrochemical supercapacitor electrodes *Nanomaterials* **12** 1404
[324] Padalkar N S, Sadavar S V, Shinde R B *et al* 2022 Layer-by-layer nanohybrids of Ni–Cr-LBH intercalated with 0D polyoxotungstate for highly efficient hybrid supercapacitor *J. Colloid Interface Sci.* **616** 548–59
[325] Mahmood A, Zhao B, Javed M S *et al* 2022 Unprecedented dual role of polyaniline for enhanced pseudocapacitance of cobalt-iron layered double hydroxide *Macromol. Rapid Commun.* **43** e2100905
[326] Deng X Y, Qin H Y, Liu X Y *et al* 2022 Hierarchically porous trimetallic hydroxide arrays for aqueous energy storage and oxygen evolution with enhanced redox kinetics *J. Alloys Compd.* **918** 165650
[327] Wang X, Wu F, Fan J *et al* 2021 High specific surface area NiTiAl layered double hydroxide derived via alkali etching for high performance supercapacitor electrode *J. Alloys Compd.* **888** 161502
[328] Liu X-A, Wang J, Tang D *et al* 2022 A forest geotexture-inspired ZnO@Ni/Co layered double hydroxide-based device with superior electrochromic and energy storage performance *J. Mater. Chem.* A **10** 12643–55
[329] Zhou G, Gao X, Wen S *et al* 2022 Magnesium-regulated oxygen vacancies of cobalt-nickel layered double hydroxide nanosheets for ultrahigh performance asymmetric supercapacitors *J. Colloid Interface Sci.* **612** 772–81
[330] Wang X, Cheng Y, Qiao X *et al* 2022 High-loading and high-performance NiMn layered double hydroxide nanosheets supported on nickel foam for supercapacitor via sodium dodecyl sulfonate intercalation *J. Energy Storage* **52** 104834
[331] Li Z, Yao M, Hu Z *et al* 2022 g-C_3N_4 promoted NiFe-LBH self-assemble high performance supercapacitor composites *J. Alloys Compd.* **919** 165805
[332] Wu H, Zhang X, Xue J *et al* 2021 Engineering active sites on hierarchical ZnNi layered double hydroxide architectures with rich Zn vacancies boosting battery-type supercapacitor performances *Electrochim. Acta* **374** 137932
[333] Bi R, Xu N, Ren H *et al* 2020 A hollow multi-shelled structure for charge transport and active sites in lithium-ion capacitors *Angew. Chem. Int. Ed. Engl.* **59** 4865–8
[334] Chu X, Meng F, Yang H *et al* 2022 Cu-doped layered double hydroxide constructs the performance-enhanced supercapacitor via band gap reduction and defect triggering *ACS Appl. Energy Mater.* **5** 2192–201
[335] Lei G, Chen D, Li Q *et al* 2022 NiCo-layered double hydroxide with cation vacancy defects for high-performance supercapacitors *Electrochim. Acta* **413** 140143
[336] Caihong Y, Zhang B, Xie X *et al* 2021 Three-dimensional independent CoZnAl-LBH nanosheets via asymmetric etching of Zn/Al dual ions for high-performance supercapacitors *J. Alloys Compd.* **861** 157933
[337] Zhong L, Yan Z, Wang H *et al* 2022 Hydrazine hydrate induced three-dimensional interconnected porous flower-like 3D-NiCo-SDBS-LBH microspheres for high-performance supercapacitor *Materials* **15** 1405
[338] Kim J B, Koo S H, Kim I H *et al* 2022 Characteristic dual-domain composite structure of reduced graphene oxide and its application to higher specific capacitance *Chem. Eng. J.* **446** 137390

[339] Ban J J, Wen X H, Lei H H *et al* 2022 In-plane grain boundary induced defect state in hierarchical NiCo-LBH and effect on battery-type charge storage *Nano Res.* **16** 4908–16
[340] Kim N, Gu T H, Shin D *et al* 2021 Lattice engineering to simultaneously control the defect/stacking structures of layered double hydroxide nanosheets to optimize their energy functionalities *ACS Nano* **15** 8306–18
[341] Zhang H, Bai Y, Chen H *et al* 2022 Oxygen-defect-rich 3D porous cobalt–gallium layered double hydroxide for high-performance supercapacitor application *J. Colloid Interface Sci.* **608** 1837–45
[342] Gao W, Li Y, Zhao J *et al* 2022 Design and preparation of graphene/Fe_2O_3 nanocomposite as negative material for supercapacitor *Chem. Res. Chin. Univ.* **38** 1097–104
[343] Sadavar S V, Padalkar N S, Shinde R B *et al* 2022 Lattice engineering exfoliation-restacking route for 2D layered double hydroxide hybridized with 0D polyoxotungstate anions: cathode for hybrid asymmetric supercapacitors *Energy Storage Mater.* **48** 101–13
[344] Raut S D D, Shinde N M M, Ghule B G G *et al* 2022 Room-temperature solution-processed sharp-edged nanoshapes of molybdenum oxide for supercapacitor and electro-catalysis applications *Chem. Eng. J.* **433** 133627
[345] Ding C, Zhu N, Wang X *et al* 2022 Experimental study on the burning behaviors of 21700 lithium-ion batteries with high specific energy after different immersion duration *Adv. Compos. Hybrid Mater.* **5** 2575–88
[346] Feng J Z, Zhang X Q, Xu Y T *et al* 2022 Regulating the electrolyte ion types and exposed crystal facets for pseudocapacitive energy storage of transition metal nitrides *Energy Storage Mater.* **46** 278–88
[347] Helal A, Shah S S, Usman M *et al* 2022 Potential applications of nickel-based metal-organic frameworks and their derivatives *Chem. Rec.* **22** e202200055
[348] Luo D, Yong Y, Hou J *et al* 2022 Construction of hierarchical $NiCo_2O_4$@NiFe-LBH core–shell heterostructure for high-performance positive electrode for supercapacitor *ChemNanoMat* **8** e202200086
[349] Huang X, Sun R, Li Y *et al* 2022 Two-step electrodeposition synthesis of heterogeneous NiCo-layered double hydroxides@MoO_3 nanocomposites on nickel foam with high performance for hybrid supercapacitors *Electrochim. Acta* **403** 139680
[350] Wang Y, Wang Z, Zheng X *et al* 2022 Core-sheath heterostructure of $MnCo_2O_4$ nanowires wrapped by NiCo-layered double hydroxide as cathode material for high-performance quasi-solid-state asymmetric supercapacitors *J. Alloys Compd.* **904** 164047
[351] Kuang H, Zhang H, Liu X *et al* 2022 Microwave-assisted synthesis of NiCo-LBH/graphene nanoscrolls composite for supercapacitor *Carbon* **190** 57–67
[352] Ma Q, Wang S, Han X *et al* 2022 Construction of three-dimensional (3D) vertical nanosheets electrode with electrochemical capacity applied to microsupercapattery *Vacuum* **198** 110914
[353] Zhang P, Deng X, Li W *et al* 2022 Electrochemical-induced surface reconstruction to NiFe-LBHs-based heterostructure as novel positive electrode for supercapacitors with enhanced performance in neutral electrolyte *Chem. Eng. J.* **449** 137886
[354] Luo L, Zhou Y L, Yan W *et al* 2022 Construction of advanced zeolitic imidazolate framework derived cobalt sulfide/MXene composites as high-performance electrodes for supercapacitors *J. Colloid Interface Sci.* **615** 282–92

[355] El-Deen A G, Abdel-Sattar M K and Allam N K 2022 High-performance solid-state supercapacitor based on Ni–Co layered double hydroxide@Co_3O_4 nanocubes and spongy graphene electrodes *Appl. Surf. Sci.* **587** 152548
[356] Liao L, Zheng K, Zhang Y *et al* 2022 Self-templated pseudomorphic transformation of ZIF into layered double hydroxides for improved supercapacitive performance *J. Colloid Interface Sci.* **622** 309–18
[357] Wan L, Chen L, Xie M *et al* 2022 Hierarchical $FeCoSe_2$@NiCo-layered double hydroxide nanosheet arrays with boosted performance for hybrid supercapacitors *J. Alloys Compd.* **901** 163567
[358] Chen X, He M, Zhou Y *et al* 2021 Design of hierarchical double-layer NiCo/NiMn-layered double hydroxide nanosheet arrays on Ni foam as electrodes for supercapacitors *Mater. Today Chem.* **21** 100507
[359] Xiong H, Liu L, Fang L *et al* 2021 3D self-supporting heterostructure NiCo-LBH/ZnO/CC electrode for flexible high-performance supercapacitor *J. Alloys Compd.* **857** 158275
[360] Wang H, Shu T, Lin C X *et al* 2022 Hierarchical construction of Co_3S_4 nanosheet coated by 2D multi-layer MoS_2 as an electrode for high performance supercapacitor *Appl. Surf. Sci.* **578** 151897
[361] Guo Y Q, Qiu H, Ruan K P *et al* 2022 Flexible and insulating silicone rubber composites with sandwich structure for thermal management and electromagnetic interference shielding *Compos. Sci. Technol.* **219** 109253
[362] Zhu Q S, Zhao Y, Miao B J *et al* 2022 Hydrothermally synthesized ZnO-RGO-PPy for water-borne epoxy nanocomposite coating with anticorrosive reinforcement *Prog. Org. Coat.* **172** 107153
[363] Kumar D, Joshi A, Singh G *et al* 2022 Polyoxometalate/ZIF-67 composite with exposed active sites as aqueous supercapacitor electrode *Chem. Eng. J.* **431** 134085
[364] Chen Y S, Yin Z, Huang D L *et al* 2022 Uniform polypyrrole electrodeposition triggered by phytic acid-guided interface engineering for high energy density flexible supercapacitor *J. Colloid Interface Sci.* **611** 356–65
[365] Rajapriya A, Keerthana S, Viswanathan C *et al* 2022 Three dimensional integrated architecture of Sr Fe LBH on hierarchical NiS framework as a flexible electrode for efficient energy storage and conversion applications *J. Energy Storage* **53** 105091
[366] Lohani P C, Tiwari A P, Chhetri K *et al* 2022 Polypyrrole nanotunnels with luminal and abluminal layered double hydroxide nanosheets grown on a carbon cloth for energy storage applications *ACS Appl. Mater. Interfaces* **14** 23285–96
[367] Wang Z H, Liu Z Q, Wang L *et al* 2022 Construction of core–shell heterostructured nanoarrays of Cu(OH)(2)@NiFe-layered double hydroxide through facile potentiostatic electrodeposition for highly efficient supercapacitors *Chemelectrochem* **9** e202101711
[368] Cao W, Xiong C, Liu Y *et al* 2022 Novel fabrication strategy of nanostructured NiCo-LBHs monolithic supercapacitor electrodes via inducing electrochemical *in situ* growth on etched nickel foams *J. Alloys Compd.* **902** 163679
[369] Emin A, Song X, Du Y *et al* 2022 One-step electrodeposited Co and Mn layered double hydroxides on Ni foam for high-performance aqueous asymmetric supercapacitors *J. Energy Storage* **50** 104667
[370] Lu Y, Guo J, He Z *et al* 2022 Direct access to NiCo-LBH nanosheets by electrochemical-scanning-mediated hydrolysis for photothermally enhanced energy storage capacity *Energy Storage Mater.* **48** 487–96

[371] Wang J, Li M, Zhai Y *et al* 2021 Construction of three-dimensional nanocube-on-sheet arrays electrode derived from Prussian blue analogue with high electrochemical performance *Appl. Surf. Sci.* **556** 149789
[372] Li H, Lin S, Li H *et al* 2022 Magneto-electrodeposition of 3D cross-linked NiCo-LBH for flexible high-performance supercapacitors *Small Methods* **6** e2101320
[373] Nagaraju G, Chandra Sekhar S, Krishna Bharat L *et al* 2017 Wearable fabrics with self-branched bimetallic layered double hydroxide coaxial nanostructures for hybrid supercapacitors *ACS Nano* **11** 10860–74
[374] Liu Z, Zhou H, Zeng F *et al* 2022 All-printed high-performance flexible supercapacitors using hierarchical porous nickel-cobalt hydroxide inks *ACS Appl. Energy Mater.* **5** 9418–28
[375] Luo W, Ma Y, Li T, Thabet H K, Hou C, Ibrahim M M *et al* 2022 Overview of MXene/conducting polymer composites for supercapacitors *J. Energy Storage* **52** 105008
[376] Li X, Ren J, Sridhar D, Xu B B, Algadi H, El-Bahy Z M *et al* 2023 Progress of layered double hydroxide-based materials for supercapacitors *Mater. Chem. Front.* **7** 1520–1561

IOP Publishing

Recent Advances in Materials for Energy Harvesting and Storage

Suresh C Pillai, Daniel M Mulvihill and Aswathy Babu

Chapter 6

A novel solar-driven energy-storage- based hybrid desalination system

Muhammad Ahmad Jamil, Ben Bin Xu, Kim Choon Ng, Nida Imtiaz, Qian Chen, Muhammad Imran, Yinzhu Jiang and Muhammad Wakil Shahzad

Global energy consumption is surging and is estimated to rise by 400 quadrillion British thermal units (qBtu) from 2018 to 2050. A sharp increase in primary energy use has been observed as a result of growing populations, fast industrialization, and other socioeconomic variables such as rising water demands and improving living standards. By 2050, the Gulf Cooperation Council (GCC) countries may have a water supply–demand gap of more than 78% that cannot be satisfied by renewable water sources; the only workable way to address this is by seawater desalination. All traditional desalination methods now in use only operate between 10% and 13% of their thermodynamic limit, which results in significant CO_2 emissions and brine rejection. Innovative methods and alternative energy sources are desperately needed from a sustainability standpoint. As a component of a hybrid multieffect desalination and adsorption desalination (MEDAD) water treatment system, an innovative MgO-based solar heat storage scheme is proposed in this chapter. Using a solar hot water storage system, the operation of the hybrid MEDAD cycle was proven in the first phase. The MgO energy storage system, which is now being installed, will replace the heating system. In addition to being useful for a thermal desalination system, the proposed arrangement can readily store thermal energy at 120 °C–140 °C during hydration and dehydration operations. Because the suggested MgO + MEDAD cycle operates entirely on solar power, it can meet sustainability objectives.

6.1 Introduction

The global primary energy demand is surging at a very high pace and will surpass 900 qBtu in the next 25 years (from 650 qBtu in 2020). This demand is mainly governed by the Gulf and Asia because of rapid industrial and economic growth and

doi:10.1088/978-0-7503-5749-4ch6

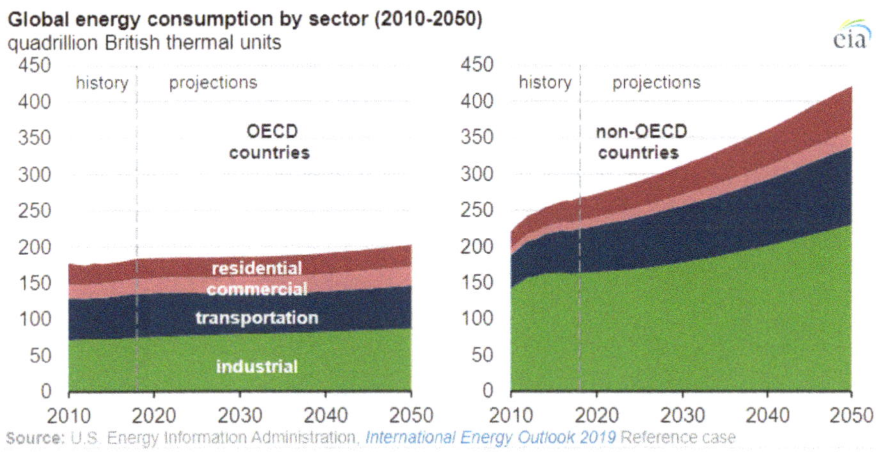

Figure 6.1. Worldwide energy consumption by sector [1]. (Source: U.S. Energy Information Administration (2019).)

is mostly generated by non-OECD countries, as shown in figure 6.1 [1]. The major energy consumption sectors include production, farming, processing, mineral extraction, and building sectors, which account for 50% of the total. In addition, it is anticipated that improving living standards and rising household demands will also create an additional 30% of energy consumption in the industrial sector in the next three decades. Consequently, corporate energy consumption is expected to exceed 315 quadrillion Btu [1].

It is also worth mentioning that this high energy consumption is particularly driven by GCC countries, which have seen massive development in the last few decades. Due to harsh climatic conditions and freshwater scarcity, cooling, and water production are the major energy-consuming sectors in GCC countries, accounting for over 50% of the total [2].

Meanwhile, it is also important to mention that despite low annual precipitation rates, GCC countries are reported to be among the top water demand areas globally and have a massively growing water demand due to ongoing growth in population, industrialization, and infrastructure [3]. An overview of the water consumption scenario for different GCC countries is presented in figure 6.2 [4]. It infers that the water availability and demand gap in GCC countries is expected to exceed 70% and cannot be filled with conventional strategies such as waste minimization, leakage elimination, conservation, or available renewable water sources [5]. The only feasible source to meet future water demands is seawater desalination, which is possible because of abundant seawater availability near populated areas [6]. Therefore, GCC countries already have the highest rate of water desalination in the world, owning more than 81% of the desalination plants, which constitute around 40% of the world's current capacity for water desalination. For this purpose, different systems including multieffect desalination (MED), multistage flash (MSF), and seawater reverse osmosis (SWRO) are typically used. Recently, hybrid desalination technologies were proposed

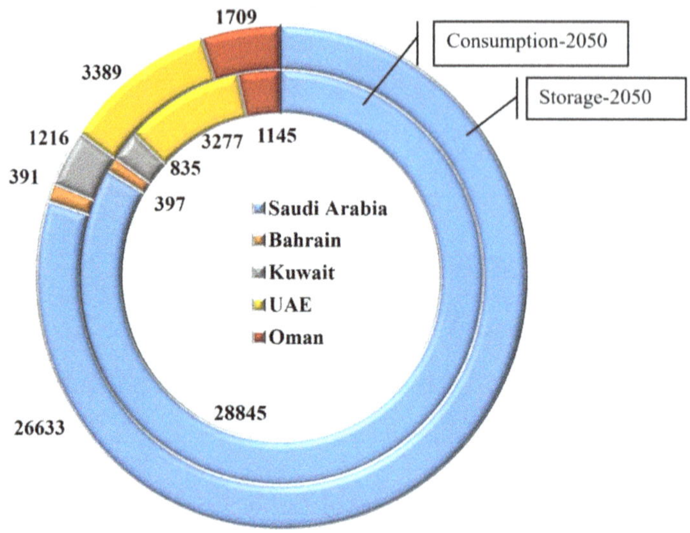

Figure 6.2. GCC water consumption and potential shortage [4].

to address limitations in conventional systems by combining two conventional systems to maximize productivity [7]. Despite all these efforts, as per 2020 estimates, the Gulf countries are expected to face a water shortage of almost 2585 million m^3/year by 2050, which is 77% higher than the current available capacity.

Though the water treatment quality and supply capacity of the existing desalination systems are satisfactory for the current demand, their massive expansion is required to meet future water demands. Moreover, their energy efficiency and environmental impacts need significant improvements. This is because the traditional desalination methods not only use a lot of energy but also harm the ecosystem and aquatic life. Currently, all traditional desalination systems only operate between 10% and 13% of their thermodynamic limit, which has a significant negative impact on brine rejection and CO_2 emissions [8]. Therefore, the adoption of alternate energy sources and novel methods is critically needed for sustainable development. In this regard, the most practical alternative energy solutions are to shift to renewable energy-based sources. In the GCC region, solar and wind energy are abundantly available, thus making these options viable for large-scale use. Meanwhile, the GCC governments have announced massive projects to investigate and use renewable energy for water treatment and electricity generation, as shown in figure 6.3 [9].

Currently, renewable energy sources only account for <1% of the freshwater production. Renewable energy's intermittent availability is the main obstacle to its widespread use. This is because desalination systems need 24 h operation throughout the year. Renewable energy can only partially handle the large-scale fluctuations that are expected in renewable energy-based systems. For instance, solar-based systems only operate during solar hours, and wind-based systems needs large amounts of infrastructure as well as continuous wind to operate the desalination plants.

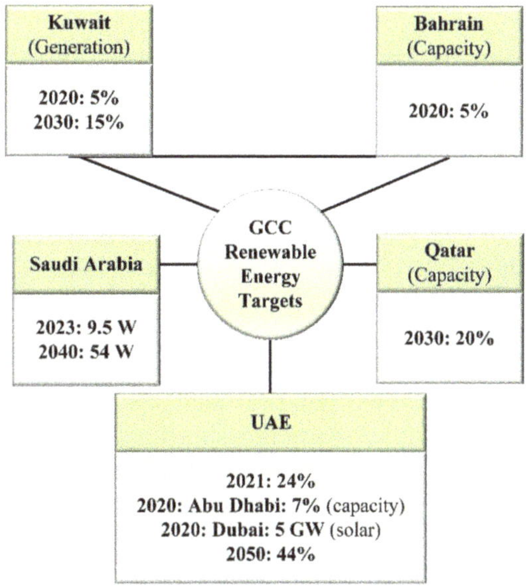

Figure 6.3. Renewable energy targets for GCC countries [9].

In this chapter, we propose a hybrid desalination system integrated with a solar heat storage arrangement that can run continuously throughout the day. In the recommended hybrid system, solar heat storage is employed to enable the 24 h operation of a MEDAD cycle. The proposed arrangement is expected to increase desalination system performance by more than 20% by addressing the limits associated with typical MED operational temperatures. As a proof of concept, the experimental results are presented.

6.2 An overview of alternative energy storage options

The above discussion suggests that energy storage may be a viable solution that would allow desalination systems to be operated using renewable energy. This is because energy storage offers 24 h operational flexibility for desalination systems. In this regard, solar energy is a lucrative source due to its abundant supply and free availability. However, its daytime availability and large-scale fluctuations in solar irradiance are the major obstacles to its widespread use, particularly for desalination [10]. Therefore, any solar thermal energy storage (TES) system development would be a breakthrough for continuous solar usage for desalination applications. Meanwhile, it is also important to emphasize that such a system can not only collect thermal energy for use during the nighttime but can also address the supply–demand difference by redirecting energy into storage. This improves overall system availability and efficiency [11].

For this purpose, different schemes have been developed and tested over the years by the research community. For instance, one of the systems is the sensible heat storage system (a) which stores heat purely based on a temperature differential between its initial and final stages. It does not involve any chemical kinetics, reactions, or any phase transformations. The temperature differential is maintained across the sink and source to keep heat flowing and the thermal energy is supplied for useful purposes. The heat accumulation process at the source is known as system charging, while the heat rejection at the sink is known as discharging. The most important characteristic of these systems is that the chemical composition and phase of the storage system remain the same [12]. The performance of these systems is governed by their thermodynamic property, i.e. heat capacity (kWh kg^{-1} K^{-1}). So, in these systems, a higher heat capacity is desirable because of a direct link to heat storage. Some of the common examples in solids include rocks, earth, and bulky metal objects. Similarly, in liquid-phase heating or cooling, oils, water, and molten salts are commonly used. The major advantages of these systems include faster charging and discharging speeds and the unchanged chemical composition and phase of the system. Meanwhile, the major shortcomings of these systems include higher energy losses, poor energy-to-weight ratios (energy density), and inferior efficiency, which rule them out for many applications [13, 14].

In contrast, the second type of system includes phase-change-based arrangements (b). In these systems, energy is stored during the phase-change process and retrieved during the restoration of the original phase (the energies of storage and retrieval may differ). The most critical thermodynamic property governing the energy storage performance of these systems is the latent heat of fusion (kJ kg^{-1} K^{-1}). Since the values of latent heat are significantly larger than the system heat capacity, the energy storage capability per unit volume of these systems is 3–4 times that of the sensible heat storage systems. It is important to mention that these systems operate at constant temperatures because of the phase-change process; however, large-scale pressure variations hinder the large-scale expansion of these systems. Some common examples of phase-change-based energy storage materials include paraffin and nonparaffin compounds.

The third type of system used for energy storage is thermochemical energy storage systems (c). These are particularly preferred because they have very high energy storage capacity per unit volume. These systems operate by receiving heat that is consumed (stored) by powering chemical reactions that disintegrate a material into smaller subcomponents. To regenerate heat, these subcomponents are recombined, and the heat is used for beneficial applications. Some of the major benefits of these systems include higher storage capacity (5–10 times), efficient storage, and safe transportation for a longer time because of the use of separate storage vessels. Based on these advantages, thermochemical energy storage is the most viable option among all the options discussed here. However, this technology is in its initial stages of development and need extensive research into materials reliability, system safety, recyclability, possible toxicity, and durability for long-term and large-scale operation.

It is also worth mentioning that a comparison of the different characteristics related to the energy storage of these systems shows the superiority of thermochemical systems. For instance, the volumetric energy density is the highest for thermochemical systems (C), which achieve 500 kWh m^{-3}, followed by latent heat systems (B) with 100 kWh m^{-3}, and then sensible heat systems (A) at 50 kWh m^{-3}. Similarly, the gravimetric energy density follows the order C>B>A and has values of 1, 0.075, and 0.02 kWh kg^{-1}, respectively. Moreover, thermochemical systems store energy at ambient temperatures, while the other two require a charging-step temperature. Similarly, the sensible and latent energy storage systems have high thermal losses, which reduces the energy storage period compared to thermochemical systems. Also, the transportation of thermochemical systems is safe and easy for longer distances because of the use of sealed chemical storage in different tanks. However, sensible heat storage systems are already established on an industrial scale, latent heat systems are established on a pilot scale, while the thermochemical systems have only been developed at the lab scale. Finally, the application areas for sensible systems include low and medium-temperature applications; for phase-change materials, the application area is high-temperature applications; while thermochemical systems can support low-, medium-, and high-temperature applications.

Overall, it is evident that thermochemical systems have a lot of benefits over their counterparts. These units are currently in the initial phases of innovation, and extensive work will be needed to test and implement them. The proper material and reversible reaction must be found to create a system that is efficient and economical [15]. The selection of suitable kinetics and analysis of the proposed material's thermophysical properties constitute the initial steps in the construction of these systems, which are categorized as shown in figure 6.4 [16]. Each of these systems has its benefits and limitations, which are governed by the chemical reactions involved. The conversion of components during a chemical reaction also affects the enthalpy of the process. Table 6.1 presents a comprehensive comparison of different thermal energy storage systems [17].

The above table shows that the volumetric energy density of thermochemical systems has been determined to be around 8–10 times greater than that of sensible systems and 3–5 times more than that of latent heat systems. Among the different available options, MgO is the best choice among thermochemical systems because of its minimal heat loss, high storage density, smaller environmental footprint, non-toxicity, and negligible corrosion effects. Moreover, its operational characteristics can be improved by improving the base material. Due to these characteristics, a small-scale MgO-based system can be developed for remote desalination systems. However, significant research is needed to ensure system reliability and performance for long-term operation.

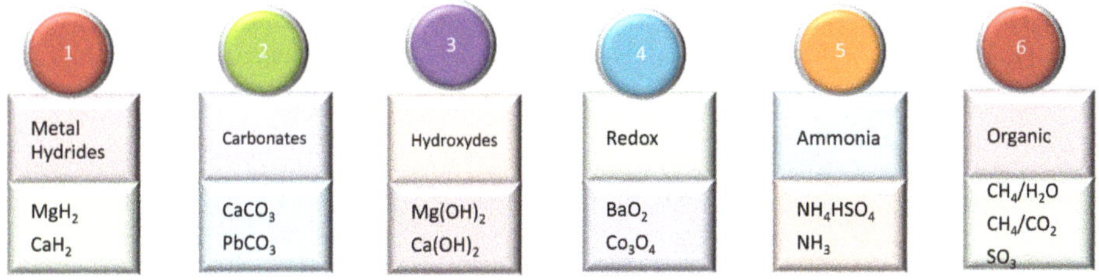

Figure 6.4. Different types of chemical reactions for thermal heating systems (THSs) [15–17].

Table 6.1. A detailed summary of thermal heating system (THS) reactions [15–17].

	Reaction	Advantages	Drawbacks
1	**$MgH_2 + \Delta H_r \leftrightarrow Mg + H_2$** *Hydrogen storage is performed by reversible metallic-hydride reactions.	• $\Delta H_r = 75.1$ kJ mol^{-1} • Insignificant by-products	• The rate of reaction is slow • Dopants required • Frittage required • Excessive operating pressure (49–101 bar) • Significant heat leakage
2	**$CaCO_3 + \Delta H_r \leftrightarrow CaO + CO_2$** *Limestone can be generated by the reversible carbonate reaction or CO_2 sequestration.	• $\Delta H_r = 179$ kJ mol^{-1} • By-product not used • No stimulant usage	• Frittage and clustering • Transformation of volume >100% • Ti used as the dopant • Inadequate and sluggish rate of reaction
3	**$Mg(OH)_2 + \Delta H_r \leftrightarrow MgO + H_2O$** * It is recommended to use the above hydroxide reaction with chemical heat pump technology for preserving and transforming heat at temperatures between 91 °C and 150 °C.	• $\Delta H_r = 80$ kJ mol^{-1} • No by-product formation • Stimulant not used • High energy density materials used • Mechanisms for pressure are ambient • Outstandingly reversible reaction • Nontoxic • Accessibility and low cost of materials	• No sufficient experimental data • Thermally less conductive
4	**$BaO_2 + \Delta H_r \leftrightarrow BaO + O_2$** * The above reaction may be used in THS systems, but the reaction is not fully accomplished.	• $\Delta H_r = 76$ kJ mol^{-1} • No by-products • O_2 reactant	• Reaction is inadequate • 50% reversibility • Reaction is delayed • Substantial operating pressure (up to 10 bar)
5	**$NH_4HSO_4 + \Delta H_r \leftrightarrow NH_3 + H_2O + SO_3$** *Can be applied in a capable THS system but produces poisonous gas, i.e. sulfur	• $\Delta H_r = 337$ kJ mol^{-1} • No catalyst required	• Causes corrosion • Detrimental • Not enough data from experiments

(Continued)

Table 6.1. (*Continued*)

	Reaction	Advantages	Drawbacks
6	**$CH_4 + H_2O + \Delta H_r \leftrightarrow CO + H_2O$** Side reaction: **$CO + H_2O + \Delta H_r \leftrightarrow CO_2 + H_2 + \Delta H_r$** It has been proposed that methane steam reforming, an organic and more irreversible reaction, be utilized for transporting thermal energy from nuclear plants. Due to the intricate side reaction, it is not appropriate for commercial use.	• $\Delta H_r = 249$ kJ mol_{CH4}^{-1} • $\Delta H_r = -40.3$ kJ mol^{-1} • Gaseous appearance	• Side reaction • Price and accessibility of CH_4 • Stimulant desirable • Irreversible nature of the process • Involves H_2

6.3 Conventional desalination systems and their limitations

The basic operation of a desalination system begins with the entry of feed water, which is subsequently separated into freshwater and brine [18]. The additional prerequisites for this, including pretreatment, posttreatment, temperature, pressure, etc. are, however, system specific [19]. For instance, to extend membrane life, membrane-based systems require extensive pretreatment of the feed. Thermal systems do not need any special pretreatment; however, it is necessary to preheat the feed for the system to operate more efficiently in terms of energy [20]. Brine management, energy recovery, cogeneration facilities, and other extra components are also used in advanced systems to increase system performance while minimizing environmental effects [21, 22]. The two main categories of desalination process—membrane-based and evaporation-based—involve different technologies. SWRO, membrane distillation, ultrafiltration, nanofiltration, forward osmosis, etc. are examples of membrane-based processes. MED, MSF, vapor-compression-based systems, and adsorption desalination (AD) systems are categorized as evaporation-based systems [23, 24], In the GCC countries, the thermal process is preferred because the harsh feeds restrict the functionality of membrane-based devices [25, 26].

The biggest significance for seawater desalination systems is shown by the fact that >60% of desalination processes are implemented for seawater, while the remainder is applied to brackish water, river water, or wastewater. [27]. It is also crucial to note that several hybrid techniques that have been developed over time have significantly outperformed traditional technologies [28]. The list below provides a brief overview of the desalination systems now in operation around the globe.

Reverse osmosis (RO): the most widely used membrane method, RO, uses a partially permeable membrane to separate freshwater from highly saline source water [29]. As shown in figure 6.5 [30], the key parts of a typical RO system are the pumps and RO trains. However, an energy recovery component is also incorporated into contemporary RO systems for greater energy efficiency. A Pelton turbine (PT) and a pressure exchanger (PX), which drives a booster pump to increase the feed pressure before entering the HPP, are typical components of the energy recovery section. Energy recovery lowers the size and energy requirements for high-pressure pumps (HPPs), increasing the effectiveness of the entire system. However, the membrane type and water salinity determine the usual working pressure for RO performed using HPPs, which is 6–8 bar [6]. Additional improvements to RO systems include hybridization with other desalination and energy production systems and solar-powered operation [31]. Nevertheless, RO systems have other drawbacks, including membrane fouling, the short operational life of the module, high maintenance requirements, and low operational availability.

Membrane distillation (MD): As shown in figure 6.6, in membrane distillation, freshwater is extracted from the feed using a microporous hydrophobic membrane

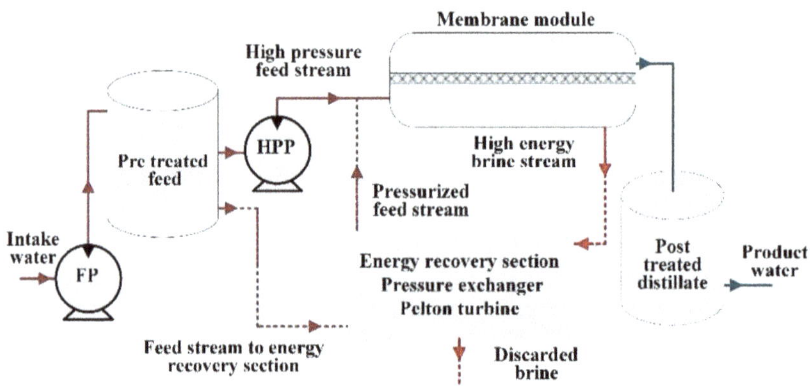

Figure 6.5. A schematic diagram of an RO system with an energy recovery section [30].

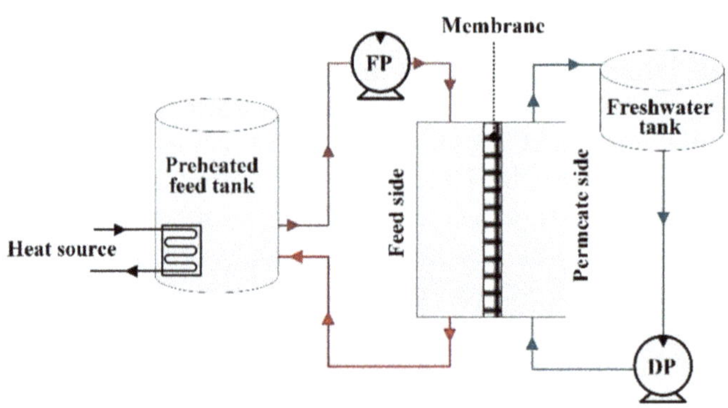

Figure 6.6. A schematic of the direct contact MD process [32].

[32]. The membrane's hydrophobicity hinders the mass transfer of feed water by forming a gas–liquid contact. The permeate side of the membrane experiences vapor formation as a result of the temperature and pressure differential, which is followed by condensation in the low-pressure compartment to create freshwater [33]. Due to the membrane's variable pore size, which ranges from 10 nm to 1 m, a pressure differential across the membrane must be maintained and is further strengthened by the diffusion effect [34, 35]. MD has several advantages over its competitors, including a modular design, straightforward operation, a high membrane surface area compared to its volumetric size, and the ability to function with a range of feeds [36].

Multistage flash is one of the earliest desalination techniques still used, having been developed by Westinghouse in 1957 and later refined by Silver [37, 38]. MED was also developed at the same time, but it was unable to gain market share due to serious heat exchanger fouling problems. Systems that rely on flashing, on the other hand, employ flash chambers without fouling problems. The temperature and pressure of the preheated feed created vapor as the feed is showered into the flash chambers of these systems. Figure 6.7 illustrates how the condensate is collected for freshwater after these vapors are used to warm up the entering seawater. The system can be used in a variety of configurations, such as a once-through configuration and brine recirculation. The recirculation arrangement reportedly performs better than the other arrangements [39]. According to reports, the system's performance parameters are a gain output ratio of 2–7 and a second-law efficiency of 1.8%–2.3% [40]. Highly saline effluent, the recycling of energy, and larger area footprints are some of the system's drawbacks [41].

Vapor compression desalination: in place of a condenser (as used in MED), the system in figure 6.8 [43] uses a thermal vapor compressor or mechanical vapor compressor. The vapors produced by the previous effect are directed toward the vapor compressor, where they are compressed to produce high temperatures and pressures [44]. To create steam for the evaporator coils, these high-temperature, high-pressure vapors are then directed to the first evaporator. The incoming feed is subsequently evaporated by this steam, producing vapors that proceed in the same manner, and the condensate from each evaporator is collected to create freshwater [45, 46]. The use of a compressor reduces the need for a separate steam production

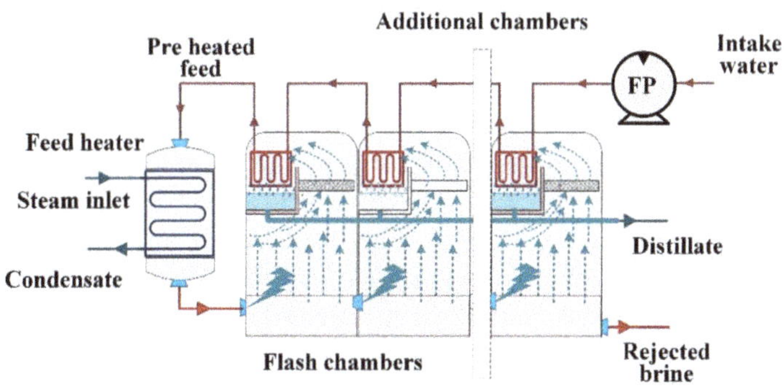

Figure 6.7. A schematic of an MSF system with brine circulation [42].

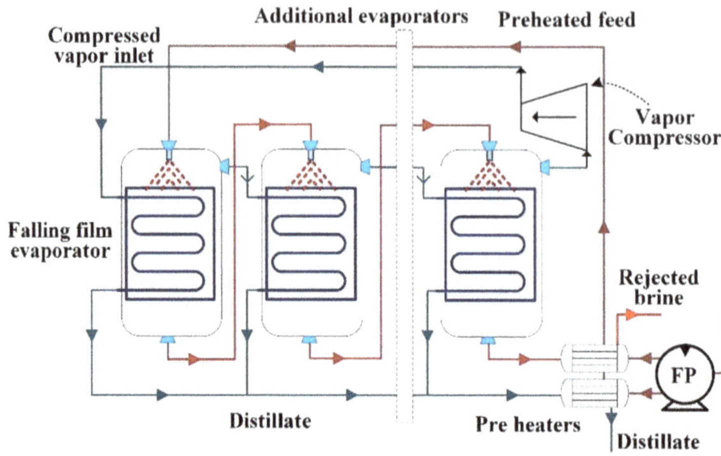

Figure 6.8. A schematic of an MED-MVC system [43].

plant to operate the system; instead, mechanical vapor compression (MVC) and thermal vapor compression (TVC) can operate using electrical energy and bleed steam. MVC systems are, however, favored for remote locations with medium water production capacities of 5000 m^3 d^{-1} or less [47–49].

Direct contact spray evaporation and condensation (DCSEC): in comparison to traditional MED and MSF systems, this system has fewer operational expenses and maintenance difficulties. Its efficiency is higher and the scaling problem is reduced when the feed water is showered directly into the chamber and vapors are used to preheat the incoming feed (as shown in figure 6.9) [50]. The procedure involves preheating saltwater before injecting it into the evaporation chamber through a nozzle, causing evaporation as a result of changes in temperature and pressure. Then, at various pressure and temperature levels, the vapors are sent into a linked condensing chamber where they condense. For large-scale production, the numbers of evaporator and condenser chambers can be increased. Afterward, the freshwater is supplied for drinking and the brine is rejected [51].

Humidification–dehumidification: figure 6.10 illustrates the components of a dehumidification system, which include a humidifier, dehumidifier, heater, pump, and fan. In the humidifier, water is used to mix and humidify the air before it is transferred to the dehumidifier. The intake water is preheated by condensing freshwater in the dehumidifier. As the process continues, the humidifier is sprayed with the preheated feed [52]. The system can be operated in several configurations, such as open air, closed air, open water, or a mix of these configurations [53, 54].

Electrodialysis: For the majority of industrial wastewater, brackish water, and drinking water quality treatment, electrodialysis is chosen [56]. As seen in figure 6.11 [57], these systems use an electrical potential to separate the cations and anions in the salts, removing the unwanted elements. Ions can only travel across the membrane to the electrodes that are countercharged [58, 59]. The electrodes are subsequently cleaned using a reverse current flow, continuing the process [60].

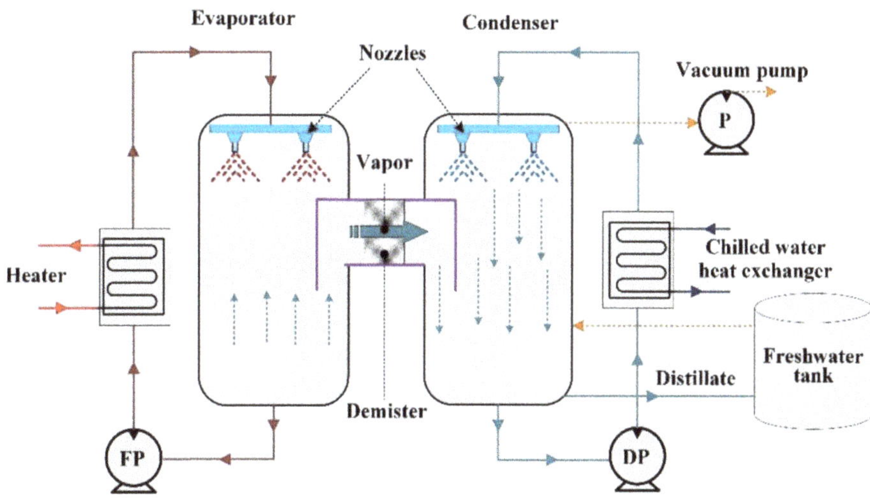

Figure 6.9. The direct contact spray evaporation and condensation process [50].

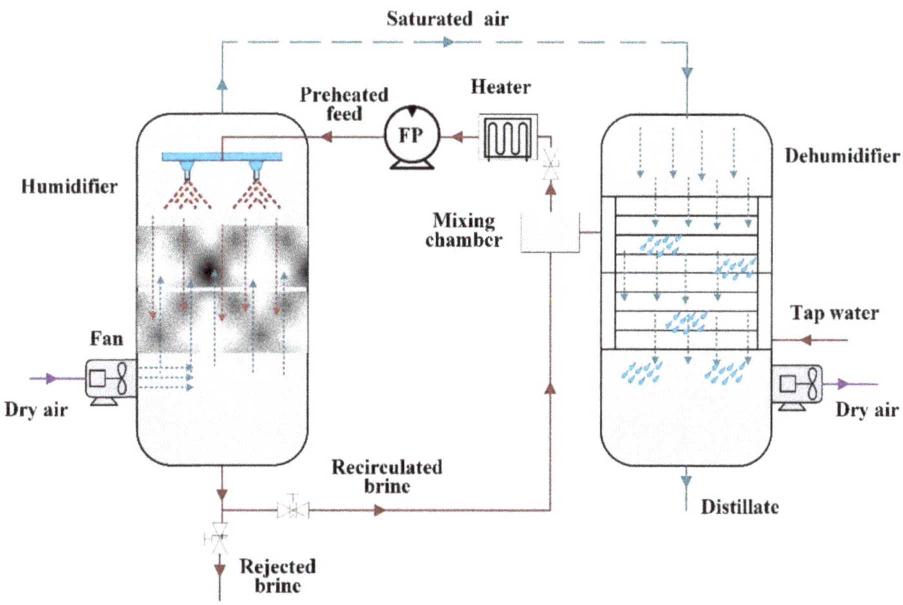

Figure 6.10. A humidification–dehumidification system [55].

Multieffect desalination: due to its use of low temperatures and cheap products, multieffect evaporation (MEE) and MED have acquired great popularity [61]. These systems work by spraying seawater over evaporator tubes filled with steam [62]. The first evaporator's coils are filled with steam that can be obtained from power plants, waste heat, or special steam-producing facilities [18]. The process continues until the vapor from the last stage is condensed in a condenser, as indicated in figure 6.12 [63], at which point the vapors from the first stage are employed collectively in the coils of

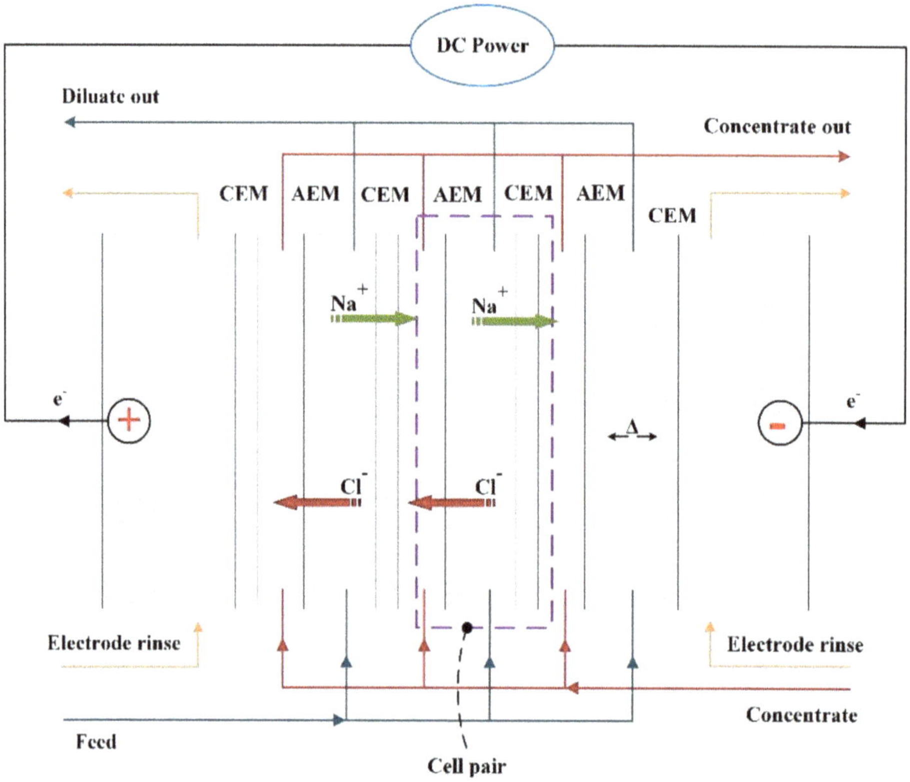

Figure 6.11. An electrodialysis system with an anion-cation exchange membrane [56].

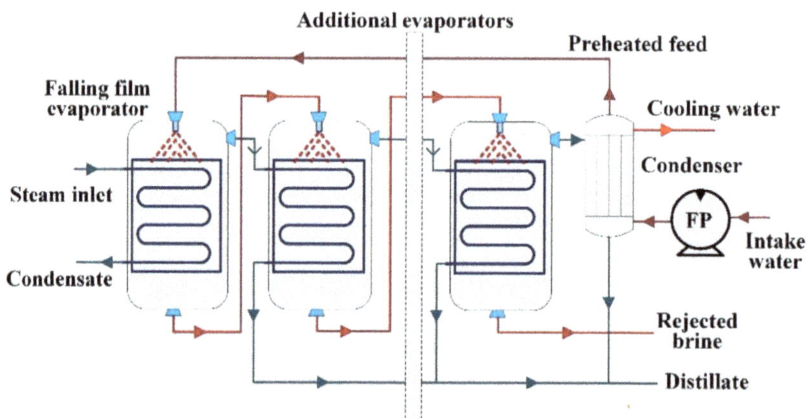

Figure 6.12. A forward-feed MED system [63, 64].

the following evaporator. In addition to this configuration, the system is also capable of operating in parallel and parallel cross-feed topologies, in which the feed is processed equally in each module [64]. The reported system performance for MED is as follows: 2 kWh m^{-3} of energy usage, a performance ratio of six, and a

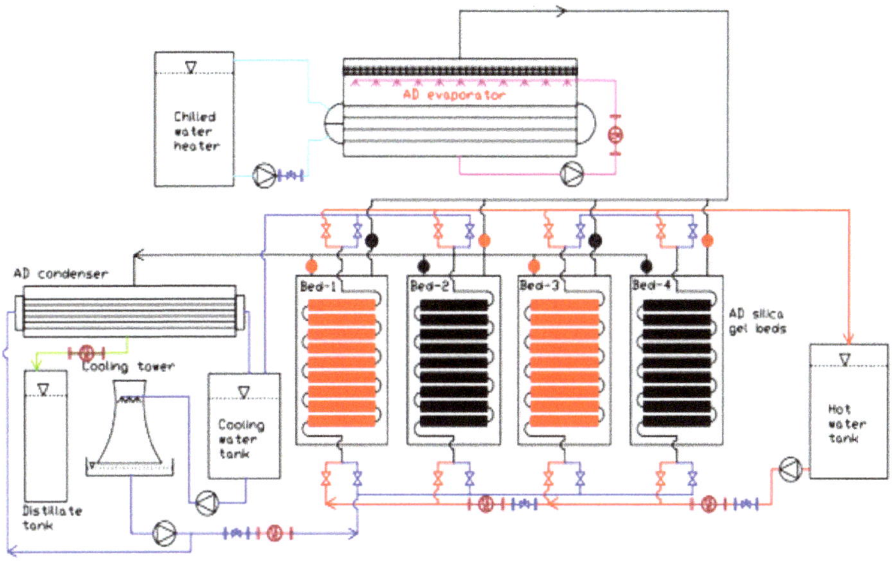

Figure 6.13. A schematic of the AD cycle [66].

second-law efficiency of 7%. According to reports, these performances are improved when the MED system is connected to other systems [65].

Adsorption desalination: heat evaporative heat exchangers, AD, condensing heat exchangers, pumps, and an initial chemical treatment section are the five key parts of an AD system. The entire process diagram is displayed in figure 6.13 [66]. Since the system operates discontinuously (in a batch mode), two alternative processes take place:

Cycle 1: adsorption and evaporation: in this step, the vapors produced by evaporation are captured by the AD section. Seawater is sprayed over the tubes in the evaporative heat exchanger and the heat source is passed through the tubes. The evaporation is started by a heat source; however, throughout the AD process, the adsorbent's high attraction for water vapor lowers the evaporator pressure and aids in evaporation. The thermal energy is typically exchanged using chilled water, and the source temperature can be used to adjust the AD evaporator's operational temperature. The AD evaporator can function at a variety of chilled water temperatures, ranging from 5 °C to 30 °C, and still create a cooling effect. The process goes on until the saturation condition is achieved [67].

Cycle 2: desorption and condensation: in this stage, the extracted vapor passes through the condensing heat exchanger, drinking water is collected, and the AD system soaked with vapor is then prepared again using thermal energy [65].

Two outputs are obtained from the AD system, including cooling via the first process, adsorption-aided evaporation, and freshwater via the second process, desorption-aided condensation, both of which transform seawater. Utilizing the multibed approach, benefits such as cooling and water generation can be generated concurrently.

The operating and flipping method is utilized in multibed AD systems. During operation, one or more AD beds go through the AD process while another one or more adsorbent reactor beds go through the desorption process. The thermal energy temperature and the amount of adsorbent in a bed determine how long an adsorbent reactor bed operates in either an adsorption or desorption mode. To improve cycle performance, adsorber beds and desorber beds are preheated and precooled during a brief period known as switching before the reactor's tasks are changed. Adsorption and desorption processes in AD cycles are operated and swapped by a smart control system that can switch on and off the warm or cool water supplies for a bed. All vapor valves are closed during the switching operation, preventing any adsorption or desorption from occurring [68].

Overall, the specific energy needs of conventional desalination systems for the production of freshwater are reportedly 2.8–8 kWh m^{-3} for membrane systems, 15–22 kWh m^{-3} for evaporation systems, and 18–27 kWh m^{-3} for flashing systems [69, 70]. Compared to the minimum amount of energy required for these processes, these energy requirements are significantly higher (by around 5–30 times) [71]. This suggests that a technological breakthrough will be necessary to significantly outperform the current systems because current operating techniques cannot avoid operational cycle limitations. To address the drawbacks encountered by each system on its own, integrating two systems has emerged as a workable approach. A successful method for achieving this goal was developed by combining an established MED system with a recently developed, inexpensive thermally driven AD system [65, 72]. An AD system's capacity to function at low temperatures and with low-grade or renewable energy is one of its important features. According to its description, the hybrid system is a MEDAD system that performs noticeably better than conventional standalone MED and AD systems. The performance of the hybrid system is thoroughly discussed in the next section.

6.4 The proposed system: materials and methods

6.4.1 Chemical reaction and working cycle

The proposed system involves an endothermic chemical reaction in a thermochemical material and results in the decomposition of the thermochemical material into reactants such as hydroxide and water. So, the overall process includes three steps, known as charging, storage, and discharging. The charging step takes place when the thermochemical material absorbs heat and decomposes into hydroxide and water. The reactants are then stored as a source of energy. The last step is discharging, in which the hydroxide and water are combined and heat is released, which is used as thermal energy. Our proposed system was developed using MgO, and in this case, these three processes are referred to as desiccation, storage, and hydration, as shown in figure 6.14.

The working cycle of the MgO/H_2O-based heat pump is shown in figure 6.15. It involves two major processes termed dehydration and hydration. During the hydration process, thermal energy (from solar) is supplied to $Mg(OH)_2$ which results in an endothermic chemical reaction that stores the supplied energy. In the

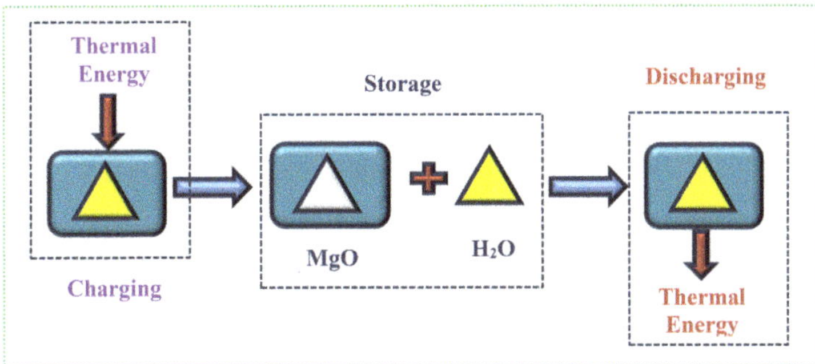

Figure 6.14. The charging–discharging cycle of the proposed system.

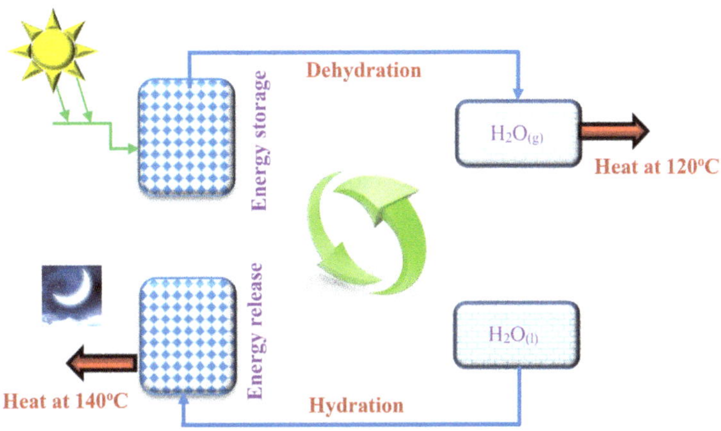

Figure 6.15. The working cycle of the MgO/H_2O-based heat pump.

second process, this energy is emitted during the soaking of MgO, which is an exothermic process. The temperature achieved is very high, ranging from 120 °C to 140 °C and an enthalpy change of 81.02 kJ mol^{-1} is achieved.

6.4.2 The proposed system

The typical operational range of a MEDAD system is 60 °C–70 °C, which is well below the temperature produced by the proposed energy storage system. The hydration/dehydration processes produce heat at 100 °C–120 °C, thus making the overall system viable for operation in a hybrid arrangement using renewable energy. Meanwhile, it is also important to note that during solar hours, the dehydration of the material produces water vapor which is used as a heat source in the first evaporative heat exchanger of the MEDAD. In addition, the vapor produced by each evaporator is then used as steam in the subsequent evaporators to produce freshwater. This is how the daytime operation of the hybrid system is achieved continuously. During nighttime, the MgO is hydrated using water, which results in an exothermic reaction and produces a large amount of heat. The heat is then reused in the MEDAD cycle for continuous operation during nighttime. In this manner, the proposed system operates continuously for 24 h on a renewable energy-based arrangement and using a thermochemical energy storage concept.

The proposed solar-operated MgO-based energy storage system hybridized with an advanced MEDAD system is shown in figure 6.16. The process starts as the solar irradiance is concentrated in the heat sink (i.e. the thermal oil tank which receives heat), increasing its temperature. The hot oil is then passed through heat exchanger tubes that are coated with MgO. This heat dehydrates the material, and the vapor produced is directed to the MEDAD system evaporator to evaporate the incoming feed water. After complete dehydration of the heat exchanger tubes, the evaporator tubes are opened to a water source. This results in exothermic hydration of the heat exchanger tubes, resulting in a large amount of heat release. This released heat is then captured by the oil flowing through the heat exchanger tube and is subsequently used to produce vapor for the nighttime operation of the MEDAD system. In this manner, the system achieves 24 h operation for water production from saline water using renewable energy and thermal energy storage arrangements.

An experimental system has been developed and tested based on the arrangement proposed above. The system's two main parts are the solar collectors and the MEDAD system. The first part is used to harness solar energy using evacuated tube collectors which are installed so as to receive maximum solar irradiance, as shown in figure 6.17. The solar thermal energy is stored in large water tanks that can be reheated by electrical heaters to maintain the required temperature.

The second part of the system includes the MEDAD hybrid desalination system as shown in figure 6.18(a) and the human–machine interface (HMI) control as presented in figure 6.18(b). The aim of hybridizing these two systems is to boost

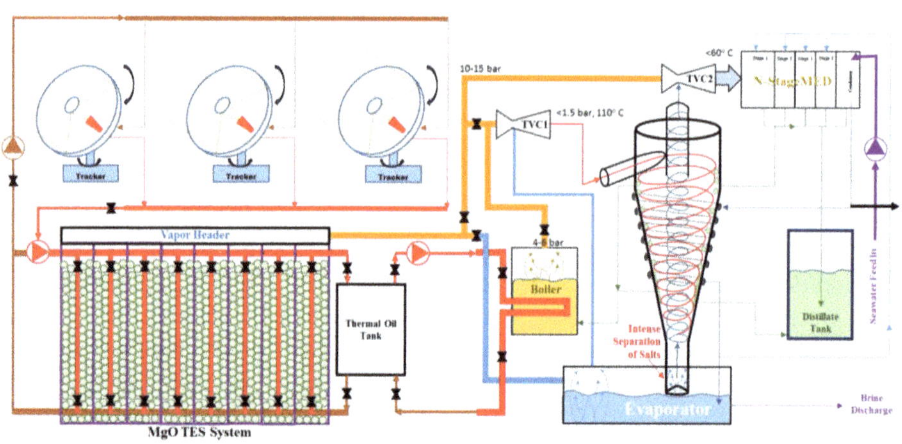

Figure 6.16. The proposed MEDAD system combined with thermal energy storage.

Figure 6.17. Solar evacuated tube collectors.

water production. This is because, in conventional MED systems, a huge fraction of the thermal energy is rejected in the condenser. In the hybrid arrangement, the AD cycles increase vapor affinity, which boosts production, and the condenser is omitted. Moreover, the system can operate a larger number of modules, which also increases productivity. The system is installed in such a way that the AD system is at ground level and the MED system is on a platform above the AD system. This arrangement makes it easy to connect these two systems with minimal vapor loss. The detailed design specifications of the system are presented in table 6.2.

6.5 Results and discussion

The most significant parameter in solar thermal-energy-based systems is the temperature achieved. It is observed that (refer to figure 6.19) the proposed system achieved the required temperature. However, to maintain the required temperature, a bypass valve is required, along with an air-cooled heat exchanger. This heat exchanger plays an important role in the safety of the solar evacuated tube collector circuit. It maintains the set temperature and avoids any steam generation during heating.

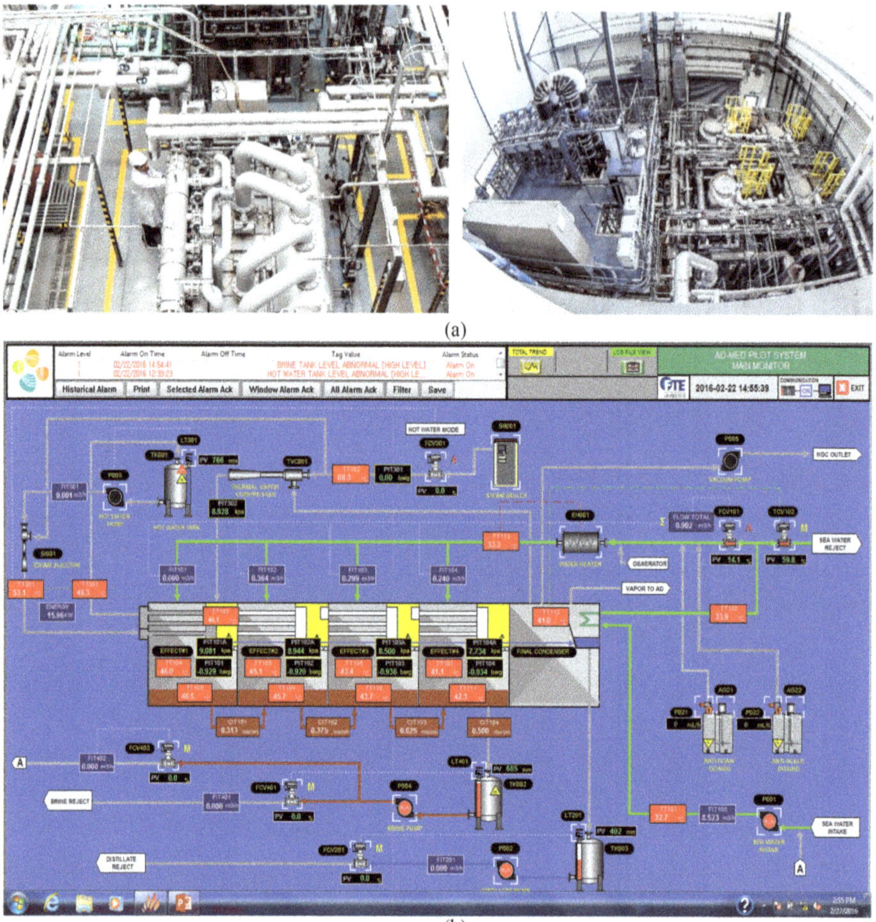

Figure 6.18. (a) The proposed MEDAD hybrid system, (b) the system HMI control window of the MEDAD hybrid system.

Table 6.2. Experimental system characteristics.

Parameter	Value
Evacuated tube collector area	352 m^2
AD cycle silica gel mass	310 kg
AD cycle silica gel area	850 $m^2\ kg^{-1}$
MED evaporator heat transfer area	4 m^2
Total number of evaporators	4
Evaporator tube dimensions (diameter, thickness, length)	19, 0.5, 63 mm

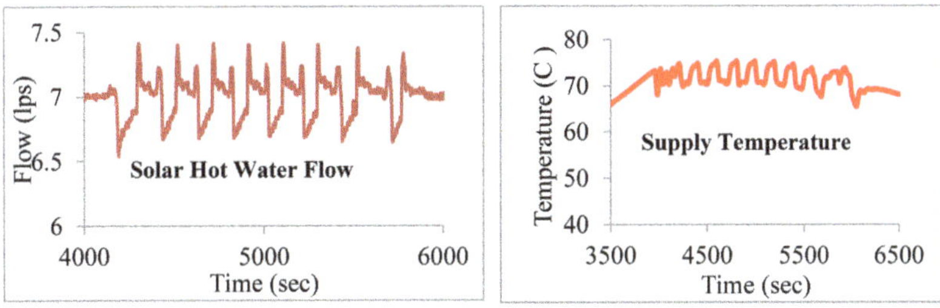

Figure 6.19. The temperature profiles for the solar evacuated tube collector circuit combined with a heat exchanger.

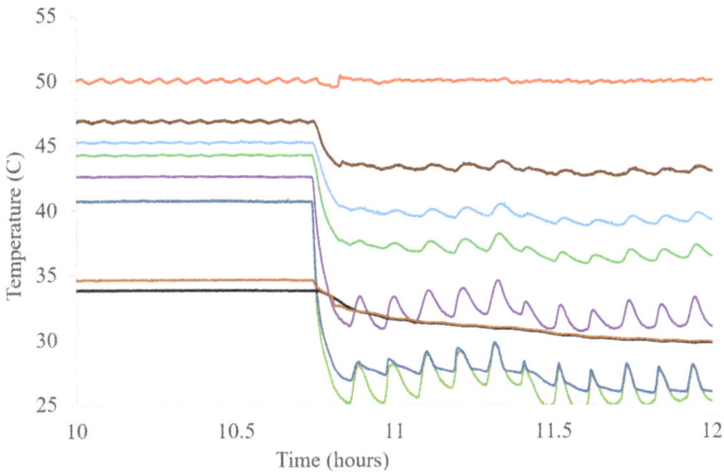

Figure 6.20. The temperature trend of the proposed system.

This is required because steam not only damages the connecting pipes but can also burst the collectors. Meanwhile, there are chances of pressure variations in the solar system due to fluctuations in the solar irradiance at different hours of the day. Therefore, to neutralize any large-scale pressure variations in the system, special tanks with nitrogen bladders are also installed in the system. Therefore, the combination of heat exchangers and nitrogen tanks makes system operation smooth and safe by neutralizing excessive pressure or heat in the system.

The temperature trends for system operation are shown in figure 6.20. To investigate the system in detail, two different operation regimes were studied. For instance, in figure 6.11, the first 10.8 h represent the standalone MED system operational temperature trend. This shows that the system operates at temperatures varying from 41 °C to 48 °C. In the second regime, the AD system is connected to the MED system through control valves. It can be observed that as soon as the AD system starts operating, the temperatures for all evaporators drop. The hybrid system MEDAD operated at 25 °C to 45 °C. Moreover, the hybridization also increased the temperature difference between the evaporators, which increased

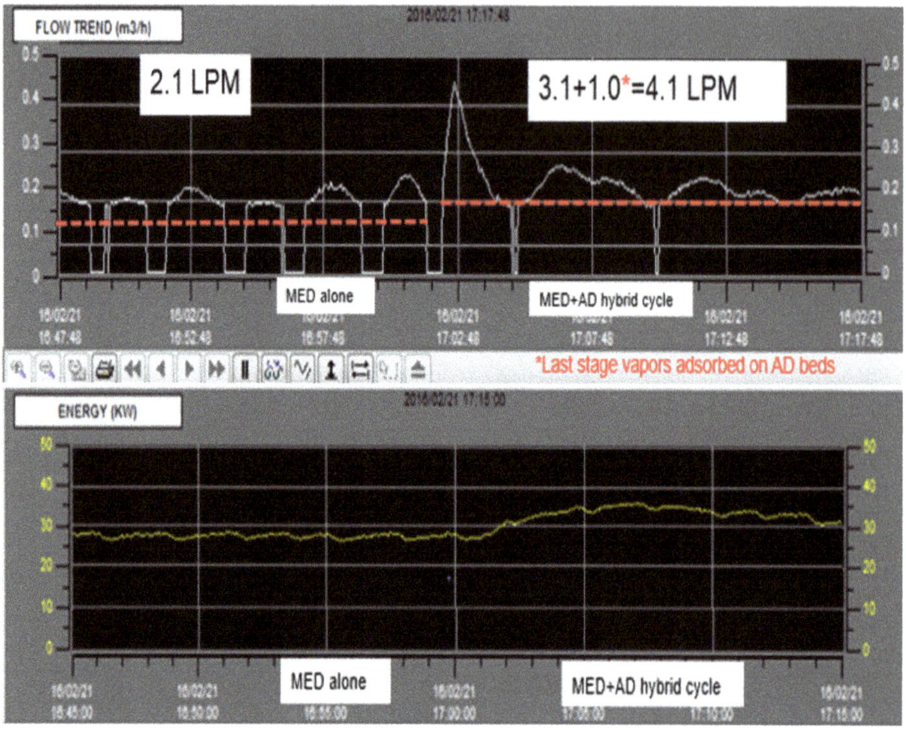

Figure 6.21. Water production and energy consumption trends of the proposed system.

flash-based water production. This hybrid arrangement offered almost double the water productivity for the same heat source, as shown in figure 6.21.

Overall, the hybrid system offers several benefits, such as low-temperature operation, a lower heat transfer area, higher productivity, and a smaller environmental footprint. Most importantly, the lower operating temperature results in a lower risk of scaling and fouling of the evaporator and condenser tubes. Moreover, an economic analysis calculated the freshwater cost to be \$0.48 m^{-3}, which is 50% lower than the cost of conventional desalination systems. The system is now under development for the replacement of hot water energy storage tanks with MgO-based thermochemical energy storage systems. The system will then be experimentally investigated in detail to confirm the reliability of the whole system under the periodic operation of the system for a longer time. The preliminary analysis results show that the proposed idea is a lucrative solution for the future sustainability of the desalination sector.

6.6 Economic analysis

We also conducted a detailed economic analysis and compared the five most feasible cases, namely: (i) a photovoltaic (PV)-driven RO desalination system, (ii) a concentrator photovoltaic (CPV)-driven RO desalination system, (iii) a concentrated solar power (CSP) production and RO desalination system, (iv) a CSP power production and RO+MED desalination system, and (v) a CSP power production,

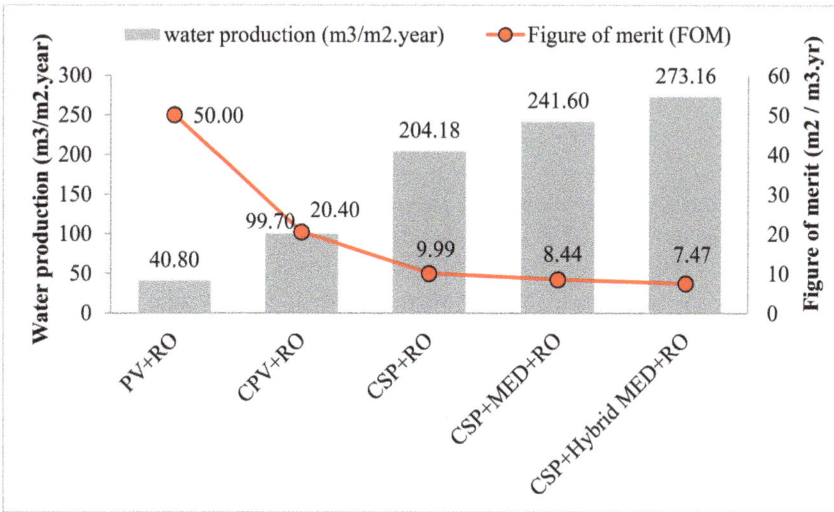

Figure 6.22. An economic analysis and comparison of different systems.

energy storage, and RO+hybrid MED desalination system. It can be seen (refer to figure 6.22) that the last case has the highest water production in terms of per-square-meter solar installation.

6.7 Conclusions

Water scarcity is among the formidable challenges the world is facing, and desalination is the only practical solution. However, the existing desalination systems are not only energy intensive but also have significant environmental impacts because of their low thermodynamic efficiency. This study proposes an innovative solar-energy-operated system. The system consists of a thermochemical energy storage system integrated with an advanced MEDAD hybrid system. The integration allows 24 h operation on solar energy by storing heat during solar hours and utilizing it during the nighttime. In the initial stage of the project, a hybrid MEDAD pilot operated by hot water from solar collectors was tested satisfactorily. The system was operated continuously for 24 h. Analysis showed that MED and AD cycle hybridization can double water productivity for the same energy input. The unit water production cost for the system is estimated to be \$0.48 m^{-3} which is almost 50% lower than the cost of traditional systems. A MgO-based system is being installed in place of the hot water storage in the next phase. The proposed system operates on solar energy, thus offering energy and environmental benefits compared to conventional desalination systems.

Acknowledgments

The authors would like to thank the Royal Academy of Engineering/Leverhulme Trust for financial support. This Fellowship was supported by the Royal Academy of Engineering under the RAEng/Leverhulme Trust Research Fellowships program awarded to Dr Muhammad Wakil Shahzad (reference #LTRF2223-19-103). The authors would also like to thank Northumbria for research support and KAUST for the use of experimental facilities and experimentation.

References

[1] Kahan A 2020 U.S. Energy Information Administration: EIA projects nearly 50% increase in world energy usage by 2050, led by growth in Asia; https://www.eia.gov/todayinenergy/detail.php?id=42342

[2] Ng K C, Burhan M, Chen Q, Ybyraiymkul D, Akhtar F H, Kumja M, Field R W and Shahzad M W 2021 A thermodynamic platform for evaluating the energy efficiency of combined power generation and desalination plants *npj Clean Water* **4** 25

[3] Ghaffour N, Missimer T M and Amy G L 2013 Technical review and evaluation of the economics of water desalination: current and future challenges for better water supply sustainability *Desalination* **309** 197–207

[4] Orient Planet Group 2023 Orient Planet Group Report: GCC doubling efforts to conserve water amid rising demand brought forth by population and economic growth https://www.orientplanet.com/PressReleasesWM.html

[5] Thavalengal M S, Jamil M A, Mehroz M, Xu B B, Yaqoob H, Sultan M, Imtiaz N and Shahzad M W 2023 Progress and prospects of air water harvesting system for remote areas: a comprehensive review *Energies* **16** 2686

[6] Qureshi B A and Zubair S M 2015 Exergetic analysis of a brackish water reverse osmosis desalination unit with various energy recovery systems *Energy* **93** 256–65

[7] Lawal D, Antar M, Khalifa A, Zubair S and Al-Sulaiman F 2018 Humidification-dehumidification desalination system operated by a heat pump *Energy Convers. Manag.* **161** 128–40

[8] Shahzad M W, Burhan M, Ang L and Choon Ng K 2017 Energy-water-environment nexus underpinning future desalination sustainability *Desalination* **413** 52–64

[9] World Economic Forum 2017 Why Gulf countries need to invest in renewable energy? https://www.weforum.org/agenda/2017/05/why-oil-rich-gulf-countries-need-to-invest-in-renewable-energy/

[10] Jamil M A, Yaqoob H, Farooq M U, Teoh Y H, Xu B B, Mahkamov K, Sultan M, Ng K C and Shahzad M W 2021 Experimental investigations of a solar water treatment system for remote desert areas of Pakistan *Water* **13** 1070

[11] M C and Yadav A 2017 Water desalination system using solar heat: a review *Renew. Sustain. Energy Rev.* **67** 1308–30

[12] Pardo P, Deydier A, Anxionnaz-Minvielle Z, Rougé S, Cabassud M and Cognet P 2014 A review on high temperature thermochemical heat energy storage *Renew. Sustain. Energy Rev.* **32** 591–610

[13] Gil A, Medrano M, Martorell I, Lázaro A, Dolado P, Zalba B and Cabeza L F 2010 State of the art on high temperature thermal energy storage for power generation. Part 1—concepts, materials and modellization *Renew. Sustain. Energy Rev.* **14** 31–55

[14] Alehosseini E and Jafari S M 2020 Nanoencapsulation of phase change materials (PCMs) and their applications in various fields for energy storage and management *Adv. Colloid Interface Sci.* **283** 102226

[15] Aydin D, Casey S P and Riffat S 2015 The latest advancements on thermochemical heat storage systems *Renew. Sustain. Energy Rev.* **41** 356–67

[16] Ding Y and Riffat S B 2013 Thermochemical energy storage technologies for building applications: a state-of-the-art review *Int. J. Low-Carbon Technol.* **8** 106–16

[17] Li J, Gao G, Kutlu C, Liu K, Pei G, Su Y, Ji J and Riffat S 2019 A novel approach to thermal storage of direct steam generation solar power systems through two-step heat discharge *Appl. Energy* **236** 81–100

[18] Jamil M A, Yaqoob H, Abid A, Farooq M U, Xu B, Bin , Dala L, Ng K C and Shahzad M W 2021 An exergoeconomic and normalized sensitivity based comprehensive investigation of a hybrid power-and-water desalination system *Sustain. Energy Technol. Assessments* **47** 101463

[19] Ng K C, Thu K, Oh S J, Ang L, Shahzad M W and Ismail A B 2015 Recent developments in thermally-driven seawater desalination: energy efficiency improvement by hybridization of the MED and AD cycles *Desalination* **356** 255–70

[20] Jamil M A, Din Z U, Goraya T S, Yaqoob H and Zubair S M 2020 Thermal-hydraulic characteristics of gasketed plate heat exchangers as a preheater for thermal desalination systems *Energy Convers. Manag.* **205** 112425

[21] Chung H W, Nayar K G, Swaminathan J, Chehayeb K M and Lienhard V J H 2017 Thermodynamic analysis of brine management methods: zero-discharge desalination and salinity-gradient power production *Desalination* **404** 291–303

[22] Chen Q, Burhan M, Shahzad M W, Ybyraiymkul D, Akhtar F H, Li Y and Ng K C 2021 A zero liquid discharge system integrating multi-effect distillation and evaporative crystallization for desalination brine treatment *Desalination* **502** 114928

[23] Nassrullah H, Anis S F, Hashaikeh R and Hilal N 2020 Energy for desalination: a state-of-the-art review *Desalination* **491** 114569

[24] Amidpour M, Jabari M, Kolahi F and Ghaebi M-R 2021 *Synergy Development in Renewables Assisted Multi-carrier Systems* (Cham: Springer)

[25] Eltawil , Mohamed A, Zhao Zhengming Z and Yuan L 2008 Renewable energy powered desalination systems: technologies and economics-state of the art *12th Int. Water Technol. Conf., IWTC12 2008 (Alexandria, Egypt)* Vol 12 pp 1–38

[26] Mabrouk A N and Fath H E S 2015 Technoeconomic study of a novel integrated thermal MSF–MED desalination technology *Desalination* **371** 115–25

[27] Shahzad M W, Burhan M and Ng K C 2017 Pushing desalination recovery to the maximum limit: membrane and thermal processes integration *Desalination* **416** 54–64

[28] Ahmed F E, Hashaikeh R and Hilal N 2020 Hybrid technologies: the future of energy efficient desalination—a review *Desalination* **495** 114659

[29] Qureshi B A and Zubair S M 2016 Energy-exergy analysis of seawater reverse osmosis plants *Desalination* **385** 138–47

[30] Jamil M A, Qureshi B A and Zubair S M 2016 Exergo-economic analysis of a seawater reverse osmosis desalination plant with various retrofit options *Desalination* **401** 88–98

[31] Abbasi H R, Pourrahmani H, Yavarinasab A, Emadi M A and Hoorfar M 2019 Exergoeconomic optimization of a solar driven system with reverse osmosis desalination unit and phase change material thermal energy storages *Energy Convers. Manag.* **199** 112042

[32] Lee J G, Jang Y, Fortunato L, Jeong S, Lee S, Leiknes T O and Ghaffour N 2018 An advanced online monitoring approach to study the scaling behavior in direct contact membrane distillation *J. Memb. Sci.* **546** 50–60

[33] Fortunato L, Jang Y, Lee J G, Jeong S, Lee S, Leiknes T O and Ghaffour N 2018 Fouling development in direct contact membrane distillation: non-invasive monitoring and destructive analysis *Water Res.* **132** 34–41

[34] Mustakeem M, Qamar A, Alpatova A and Ghaffour N 2021 Dead-end membrane distillation with localized interfacial heating for sustainable and energy-efficient desalination *Water Res.* **189** 116584
[35] Elcik H, Fortunato L, Alpatova A, Soukane S, Orfi J, Ali E, AlAnsary H, Leiknes T O and Ghaffour N 2020 Multi-effect distillation brine treatment by membrane distillation: effect of antiscalant and antifoaming agents on membrane performance and scaling control *Desalination* **493** 114653
[36] El-Bourawi M S, Ding Z, Ma R and Khayet M 2006 A framework for better understanding membrane distillation separation process *J. Memb. Sci.* **285** 4–29
[37] EI-Dessoukey H T, Shaban H I and Al-Ramadan H 1986 Steady-state analysis of multi-stage flash desalination process *Desalination* **103** 271–87
[38] Silver R S and Herd J R 1957 *Improvements in or relating to evaporators* UK GB829820A
[39] Mabrouk A A, Nafey A S and Fath H E S 2007 Thermoeconomic analysis of some existing desalination processes *Desalination* **205** 354–73
[40] Mabrouk A A, Nafey A S and Fath H E S 2007 Analysis of a new design of a multi-stage flash—mechanical vapor compression desalination process *Desalination* **204** 482–500
[41] der Bruggen B V and Vandecasteele C 2002 Distillation vs. membrane filtration: overview of process evolutions in seawater desalination *Desalination* **143** 207–18
[42] Junjie Y, Shufeng S, Jinhua W and Jiping L 2007 Improvement of a multi-stage flash seawater desalination system for cogeneration power plants *Desalination* **217** 191–202
[43] Jamil M A and Zubair S M 2017 Design and analysis of a forward feed multi-effect mechanical vapor compression desalination system: an exergo-economic approach *Energy* **140** 1107–20
[44] Farahat M A, Fath H E S, El-Sharkawy I I, Ookawara S and Ahmed M 2021 Energy/exergy analysis of solar driven mechanical vapor compression desalination system with nano-filtration pretreatment *Desalination* **509** 115078
[45] Shahzamanian B, Varga S, Soares J, Palmero-Marrero A I and Oliveira A C 2021 Performance evaluation of a variable geometry ejector applied in a multi-effect thermal vapor compression desalination system *Appl. Therm. Eng.* **195** 117177
[46] He W F, Zhu W P, Xia J R and Han D 2018 A mechanical vapor compression desalination system coupled with a transcritical carbon dioxide Rankine cycle *Desalination* **425** 1–11
[47] Ettouney H M, El-Dessouky H T, Faibish R S and Gowin P J 1986 Evaluating the economics of desalination *Chem. Eng. Process.* **98** 32–9
[48] Ettouney H 2006 Design of single-effect mechanical vapor compression *Desalination* **190** 1–15
[49] Eisavi B, Nami H, Yari M and Ranjbar F 2021 Solar-driven mechanical vapor compression desalination equipped with organic Rankine cycle to supply domestic distilled water and power—thermodynamic and exergoeconomic implications *Appl. Therm. Eng.* **193** 116997
[50] Qian C, Alrowais R, Burhan M and Ybyraiymkul D 2020 A self-sustainable solar desalination system using direct spray technology *Energy* **205** 118037
[51] Alrowais R, Qian C, Burhan M, Ybyraiymkul D, Wakil M and Choon K 2020 A greener seawater desalination method by direct-contact spray evaporation and condensation (DCSEC): experiments *Appl. Therm. Eng.* **179** 115629
[52] Jamil M A, Elmutasim S M and Zubair S M 2018 Exergo-economic analysis of a hybrid humidification dehumidification reverse osmosis (HDH-RO) system operating under different retrofits *Energy Convers. Manag.* **158** 286–97

[53] Lawal D U, Zubair S M and Antar M A 2018 Exergo-economic analysis of humidification-dehumidification (HDH) desalination systems driven by heat pump (HP) *Desalination* **443** 11–25
[54] Lawal D U, Antar M A, Khalifa A, Zubair S M and Al-Sulaiman F A 2020 Experimental investigation of heat pump driven humidification- dehumidification desalination system for water desalination and space conditioning *Desalination* **475** 114199
[55] Lawal D U and Qasem N A A 2020 Humidification-dehumidification desalination systems driven by thermal-based renewable and low-grade energy sources: a critical review *Renew. Sustain. Energy Rev.* **125** 109817
[56] Generous M M, Qasem N A A and Zubair S M 2021 Entropy generation analysis of electrodialysis desalination using multi-component groundwater *Desalination* **500** 114858
[57] Generous M M, Qasem N A A and Zubair S M 2020 Exergy-based entropy-generation analysis of electrodialysis desalination systems *Energy Convers. Manag.* **220** 113119
[58] Generous M M, Qasem N A A and Zubair S M 2021 An innovative hybridization of electrodialysis with reverse osmosis for brackish water desalination *Energy Convers. Manag.* **245** 114589
[59] Generous M M, Qasem N A A, Akbar U A and Zubair S M 2021 Techno-economic assessment of electrodialysis and reverse osmosis desalination plants *Sep. Purif. Technol.* **272** 118875
[60] Qasem N A A, Zubair S M, Qureshi B A and Generous M M 2020 The impact of thermodynamic potentials on the design of electrodialysis desalination plants *Energy Convers. Manag.* **205** 112448
[61] Rostamzadeh H, Ghiasirad H, Amidpour M and Amidpour Y 2020 Performance enhancement of a conventional multi-effect desalination (MED) system by heat pump cycles *Desalination* **477** 114261
[62] Christ A, Rahimi B, Regenauer-Lieb K and Chua H T 2017 Techno-economic analysis of geothermal desalination using hot sedimentary aquifers: a pre-feasibility study for Western Australia *Desalination* **404** 167–81
[63] Abid A, Jamil M A, Sabah N us, Farooq M U, Yaqoob H, Khan L A and Shahzad M W 2020 Exergoeconomic optimization of a forward feed multi-effect desalination system with and without energy recovery *Desalination* **499** 114808
[64] Elsayed M L, Mesalhy O, Mohammed R H and Chow L C 2018 Transient performance of MED processes with different feed configurations *Desalination* **438** 37–53
[65] Shahzad M W, Thu K, Kim Y and Ng K C 2015 An experimental investigation on MEDAD hybrid desalination cycle *Appl. Energy* **148** 273–81
[66] Shahzad M W 2013 The hybrid multi-effect desalination (MED) and the adsorption (AD) cycle for desalination *PhD Thesis* National University of Singapore https://scholarbank.nus.edu.sg/handle/10635/49640
[67] Wakil M, Choon K, Thu K and Baran B 2014 Multi effect desalination and adsorption desalination (MEDAD): a hybrid desalination method *Appl. Therm. Eng.* **72** 289–97
[68] Shahzad M W, Choon Ng K, Thu K and Saha B B 2014 Multi effect desalination and adsorption desalination (MEDAD): a hybrid desalination method *Appl. Therm. Eng.* **72** 289–97
[69] Al-Karaghouli A and Kazmerski L L 2013 Energy consumption and water production cost of conventional and renewable-energy-powered desalination processes *Renew. Sustain. Energy Rev.* **24** 343–56

[70] Shahzad M W, Burhan M, Ybyraiymkul D and Ng K C 2019 'Desalination processes' efficiency and future roadmap *Entropy* **21** 84
[71] Lienhard J H, Mistry K H, Sharqawy M H and Thiel G P 2017 Thermodynamics, exergy, and energy efficiency in desalination systems *Desalination Sustainability: A Technical, Socioeconomic, and Environmental Approach* (Amsterdam: Elsevier) ch 4
[72] Shahzad M W, Ng K C, Thu K, Saha B B and Chun W G 2014 Multi effect desalination and adsorption desalination (MEDAD): a hybrid desalination method *Appl. Therm. Eng.* **72** 289–97

IOP Publishing

Recent Advances in Materials for Energy Harvesting and Storage

Suresh C Pillai, Daniel M Mulvihill and Aswathy Babu

Chapter 7

Recent developments in electrodes and separators for high-performance lithium–sulfur batteries

Jithu Joseph, K Sreekala and J Mary Gladis

The development of high-performance rechargeable batteries with superior energy density will play a vital role in the future of electric mobility and consumer electronics. Lithium–sulfur batteries (LISBs), which have higher capacity and volumetric energy density, will enable consumers to make choices beyond Li-ion battery (LIB) technology. The high abundance of sulfur, its eco-friendly nature, and better electrochemical performance demand the design and fabrication of high-energy-density LISBs. However, the large, industrial-scale development of LISBs can only be achieved by overcoming significant challenges, such as the insulating nature of the sulfur, Li dendrite growth, low cyclability, volume expansion of the cathodes, and the polysulfide shuttle effect. The real-time solutions to the existing challenges will be the introduction of efficient sulfur host cathodes, functionalized separators, binders, and electrolytes. Introducing conductive sulfur host cathode materials, modified separators, and interlayers will enhance the conductivity of the sulfur and suppress the polysulfide shuttle effect more efficiently. Chemical and physical anchoring of the polysulfides can be achieved by porous carbon and polar-based materials as sulfur host cathodes and by modifying the other cell components. This chapter focuses on the recent development of electrodes and separators for high-performance LISB applications.

7.1 Introduction

The current energy requirements call for high-energy-density batteries to improve the performance of electric vehicles and portable consumer electronics. Developments in post-LIB technology are occurring daily to establish cost-effective, high-performance batteries. Beyond LIBs, the important new-generation high-energy-density batteries are Na-ion, Zn-ion, Al-ion, Li–S, and Na–S batteries, which have promising performances and complex challenges. Immense research efforts are happening among the global energy storage research community to address these challenges and solve them in a facile manner [1, 2]. Versatile features such as high theoretical capacity, energy density, sulfur abundance, and the environmentally benign nature of LISBs have made them potential choices for future energy storage applications. However, compared to current LIBs, the performance of LISBs is hindered by challenges such as the nonconductive nature of sulfur, cathodic volume expansion, Li dendrite formation, low cyclability, and a complex reversibility problem—the polysulfide shuttle effect [3]. Modeling and theoretical studies will improve the identification of suitable electrodes, electrolytes, and their precise electrode composition, boosting the development of high-energy-density LISBs. The severe challenges of LISBs can be resolved by the robust development of efficient electrodes, sulfur host cathode materials, functionalized separators, interlayers, and dendrite-free Li metal anodes [4].

7.2 The cell chemistry/working mechanism of Li–S batteries

The promising theoretical capacities of Li metal anodes (3861 mAh g^{-1}) and sulfur cathodes (1675 mAh g^{-1}) make LISBs an ideal rechargeable battery choice. In LISBs, the vital cell compartments consist of a Li metal anode, a sulfur cathode, an electrolyte, and a separator, and the working mechanism involves a discharging–charging process. During the discharging process, the oxidation of the Li anode yields Li ions and electrons, and elemental sulfur (S_8) cathodes undergo multistep reduction to form higher-order (Li_2S_n, $6 < n \leqslant 8$) and lower-order lithium polysulfides (Li_2S_n, $2 < n \leqslant 6$) by combining with Li ions. The overall polysulfide conversion reaction occurs within a voltage window of (1.8–2.7 V), and these reduction reactions are responsible for the higher capacity of LISBs (figure 7.1).

The overall reaction formula is: $S_8 + 16\ Li^+ + 16\ e^- \Longleftrightarrow 8\ Li_2S$

The formation of high-order polysulfides (Li_2S_8, Li_2S_6) soluble in the organic electrolyte occurs in a voltage window between 2.3 and 2 V, and these polysulfide reduction paths give an LISB a theoretical capacity of 419 mAh g^{-1}. The discharge plateaus (2.1–1.7 V) represent the generation of insoluble and insulative lower-order lithium polysulfides (Li_2S_4, Li_2S_2, Li_2S), which contribute a theoretical capacity of 1256 mAh g^{-1} toward the total capacity of the LISB. The overall mechanism involves the multistep reduction of S_8 to Li_2S during the discharging process and the reverse reduction reactions during the charging of an LISB. However, the sluggish reaction kinetics, the insulative nature of the lower-order polysulfides, and the parasitic reactions of the higher-order polysulfides with the Li anode, i.e. the polysulfide shuttle phenomena, impede the electrochemical performance of LISBs. In addition, side reactions between the higher-order polysulfides and the anodic compartment result in severe capacity decay in LISBs [5, 6].

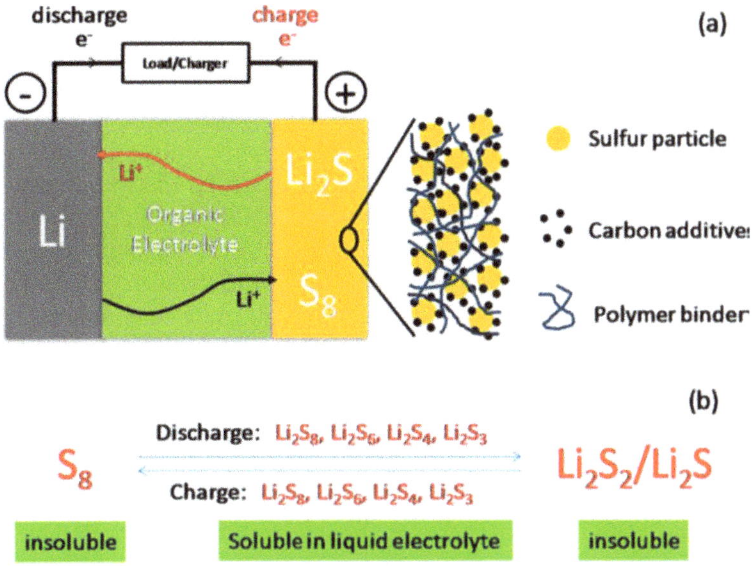

Figure 7.1. A schematic of (a) the working mechanism and (b) the electrochemical reactions of a lithium–sulfur battery (reprinted with permission from [6], Copyright (2014), American Chemical Society).

7.3 The challenges of Li–S batteries

7.3.1 Technical challenges of Li–S batteries

Three major technical challenges associated with LISBs hinder them from realizing their high theoretical capacity, as follows:

1. The insulating nature of sulfur
2. The volume expansion of sulfur upon cycling
3. Polysulfide dissolution and the shuttle effect

7.3.1.1 The insulating nature of sulfur

In practical applications, the insulating behavior of sulfur prevents the sulfur cathodes from achieving their high theoretical capacity in LISBs. Similarly, discharge products such as lower-order polysulfides also have poor conductivity. Introducing conductive additives such as carbon nanomaterials, conductive polymers, metal oxides, sulfides, etc. as sulfur host cathode materials improves the conductivity of the cathodes and the overall performance of LISBs. However, large amounts of additives lower the active material (sulfur) loading and reduce the specific capacity of LISBs.

7.3.1.2 The volume expansion of the cathodes

During LISB cycling, the sulfur cathode undergoes a volume expansion of around 80% during its conversion of Li_2S. This can lead to the structural destruction of the cathode and the loss of active material, causing capacity fade upon long-term cycling. Various materials, including carbonaceous materials, conductive polymers, metal-based oxides, and sulfides with better mechanical stability, flexibility, and other structural properties, are used as cathode hosts to overcome these structural stability issues.

7.3.1.3 Polysulfide dissolution and the shuttle effect

The sulfur (S_8) reduction reactions that occur during the discharging of LISBs to Li_2S include several intermediate polysulfide species that are highly soluble in LISB electrolytes. Polysulfides shuttle between the electrodes through the electrolyte, resulting in severe capacity decay and self-discharge problems in LISBs. Thus, sulfur confinement within the cathodic compartment is adequate to enable high performance in LISBs. This prevents the movement of the polysulfide species toward the lithium anode, impeding side reactions with the lithium anode and achieving good electrochemical results. Various physical and chemical confinement strategies have been adopted to curtail the polysulfide shuttle effect. Different functional materials are employed that have the innate capability to enhance the conversion kinetics of polysulfides, thereby restricting the polysulfide shuttle effect and improving the cycle life of LISBs [7–10].

7.4 Li–S battery components

A typical LISB consists of a Li metal anode, a sulfur cathode, separators (polypropylene), and an electrolyte (lithium bis(trifluoromethanesulfonyl)imide, LiTFSI). Robust design and development of the electrode materials are adequate to tackle challenges such as Li dendrite growth, sluggish reaction kinetics, poor sulfur conductivity, low cycling stability, and the severe polysulfide shuttle effect. Introducing modified anodes, efficient sulfur host cathodes, functionalized separators, and electrolyte designs also boosts LISB performance.

7.4.1 Lithium metal anodes

The development of efficient anodes is essential for high-performance rechargeable LIBs and post-LIBs. Li metal has unique characteristics such as a high theoretical capacity (3860 mAh g^{-1}) and a more negative electrochemical potential (−3.04 V), making it an exemplary anode choice for LISBs. However, it suffers from problems such as Li dendrite formation, capacity decay, and air sensitivity issues that prevent it from being used in a scalable manner [11]. Li dendrite growth is initiated by the formation of a passivation layer between the Li metal anode and the electrolyte, termed the solid–electrolyte interface (SEI), through electrolyte decomposition. Compared to LIB parasitic reactions, the degree of side reactions between the SEI layer and soluble polysulfides is higher in LISBs and causes severe capacity fade during cycling.

So, efficient strategies to obtain dendrite-free Li metal anodes are adequate to achieve high energy density and coulombic efficiency in LISBs. The introduction of electrolyte additives, an artificial SEI, and protective coatings can mitigate Li dendrite growth in LISBs. Electrolyte additives such as $LiNO_3$ enhance LISB efficacy by impeding dendrite formation, which results in batteries that have better cyclic stability. Potential dendrite-blocking electrolyte additives are used to improve LISB performance. An artificial SEI layer with suitable ion diffusion and transport features is one of the promising strategies for achieving dendrite-free Li metal anodes [12–16].

Deposition methods such as physical vapor deposition (PVD) and molecular layer deposition (MLD) enable precise protective layer deposition on the Li metal surface for dendrite-free anodes. Higher coulombic efficiency and better electrochemical performance can be attained by protecting the anode through the deposition of suitable layers such as lithium phosphorous oxynitride (LiPON), Cu, Sn, and MoS_2 onto the Li metal anode surface using PVD or MLD. Like thin-film deposition methods, nitridation is a feasible route for controlling the nonuniform growth of an SEI on the metal anode surfaces. A coating of protective Li_3N layers suppresses dendritic Li growth on the surface of the metal anodes. Adopting non-lithium anode materials with superior capacity and electrochemical features also boosts the performance of LISBs. High-capacity silicon and carbon-based materials are suitable anode choices for LISBs. Prelithiated Si anodes exhibit better rate capability and higher capacity in LISBs and offer promising cycling life. Hard carbon anodes also show promising cyclability and electrochemical outputs in LISBs [6, 17–23].

7.4.2 Sulfur host cathode materials

In LISBs, the inherent challenges of the sulfur cathodes are: the insulative nature of sulfur, the polysulfide shuttle effect, volume expansion, sluggish reaction kinetics, and low cyclability. These can be resolved by integrating conductive polysulfide anchoring materials into sulfur cathodes. Efficient sulfur host cathode materials, such as carbon-based materials, metal oxides, metal sulfides, single-atom catalysts, conductive polymers, and 2D layered materials, enhance the performance of LISBs. The host materials provide a conductive framework for the sulfur cathodes and mitigate the polysulfide shuttle effect through the physical and chemical confinement of the polysulfides. Carbon materials such as graphene, carbon nanotubes, and porous carbons impede the polysulfides through physisorption and increase the conductivity of the sulfur cathodes. Chemical entrapment of the polysulfides and conductivity tuning of S cathodes can also achieved by the amalgamation of polar- and Lewis-acid-based electrocatalysts. The important electrocatalysts include metal oxides, metal sulfides, conductive polymers, and single-atom catalysts with sulfur cathodes. The important sulfur host cathode materials are discussed in the following section [24–26].

7.4.2.1 Carbon-based sulfur host cathode materials

Carbon materials have versatile features such as high conductivity, large specific surface area, wide pore size distributions, enhanced surface functionalities, and electrochemical stability, making them potential sulfur host materials for LISBs. Efficient sulfur infiltration, physical confinement of polysulfides, and enhanced sulfur cathode conductivity can be achieved by integrating various allotropes of carbons such as graphene, heteroporous carbon, and CNTs into the cathode. Synthetic strategies such as melt diffusion, ball milling, and pyrolysis are employed to develop sulfur–carbon composite cathodes for LISBs. Wang *et al* noted that the insertion of porous carbon into sulfur enhanced the conductivity and electrochemical performance of LISBs compared to the performance of pristine sulfur-cathode-based batteries [27]. Another breakthrough was established by Nazar's group in 2009; they designed a mesoporous carbon–sulfur composite as an efficient sulfur host cathode for LISBs. The assembled cell showed a high specific capacity of 1005 mAh g^{-1} and outstanding cyclability [28].

7.4.2.1.1 Graphene–sulfur composites

The high conductivity, mechanical strength, flexibility, and electrochemical stability of 2D graphene makes it an efficient sulfur host cathode material for LISBs [29]. Wang *et al* synthesized graphene–sulfur cathodes using graphene and polyethylene glycol (PEG)-decorated sulfur particles. The assembled cell exhibited an excellent capacity of 600 mAh g^{-1} and improved cyclic stability [30]. Reduced graphene oxide, thermally exfoliated graphene, 3D graphene, and functionalized graphene are essential sulfur host materials that enhance conductivity, induce polysulfide entrapment, and address the volume expansion challenges of the LISBs [31–33]. The amalgamation of graphene with carbon-nanotube-functionalized porous carbon as a cathode host material enhances the electrochemical performance of LISBs by providing high conductivity and strong physisorption of polysulfides [34]. Donoro *et al* designed a multidimensional graphene/carbon nanotube/nano sulfur hybrid cathode for LISBs. The assembled cell exhibited a capacity of 1067 mAh g^{-1} at 50 mA g^{-1} and a good rate performance of 539 @ 1 A g^{-1} [35]. Modification of graphene and its composites propels LISB performance. Amino-functionalized reduced graphene oxide is an efficient sulfur host cathode material and shows strong polarity-induced confinement of polysulfides. The strong interaction between amino-functionalized rGO and the polysulfide is confirmed by its binding energy value, which ranges from 1.13 to 1.38 eV. The assembled cell delivered 80% capacity retention over 350 cycles and high rate capability [36]. Song *et al* developed template-free mesoporous, highly crumpled nitrogen-integrated graphene sheets and employed them as a sulfur host cathode and interlayer for LISBs. Their assembled cell delivered a capacity of 1000 mAh g^{-1}, resulting from the versatile pore structure and dual confinement of the polysulfides through chemical and physical adsorption. Covalently grafted graphene–polysulfur cathodes yielded a high areal capacity of 12 mAh cm^{-2} at a high very sulfur loading of 10.5 mg cm^{-2} in LISBs. This composite showed excellent conductivity and strong anchoring of polysulfides. Porphyrin-derived graphene (988 mAh g^{-1} at 2 C rate) and heteroatom-doped graphene (600 mAh g^{-1} at 0.8 A g^{-1} over 300 cycles) act as sulfur host cathodes with a strong conductive framework and enable facile confinement of polysulfides. Pristine and functionalized graphene are promising sulfur host materials that provide conductivity and impede the polysulfide shuttle through strong physisorption and chemisorption in LISBs [37–41].

7.4.2.1.2 CNT–sulfur composites

Carbon nanotubes (CNTs) are 1D materials that have the potential to be sulfur host cathodes for LISBs due to their very high aspect ratio, conductivity, and ability to form extended interconnected networks. Guo *et al* reported the superior electrochemical performance of sulfur-disordered carbon nanotubes as sulfur host cathodes in LISBs. Their conductive cathode mitigated the lithium polysulfide shuttle effect with the assistance of physisorption and yielded outstanding cyclic stability. Theoretical and experimental studies of the efficacy of CNT-S cathodes indicated that CNTs show excellent discharge capacity and durability when used as sulfur host materials. Introducing superior dopants and composite materials into CNTs enhances the conductivity and polysulfide capture of long-life LISBs. Oxygen-functionalized CNTs also showed good interaction with polysulfides but comparatively less electrochemical output owing to the lower electronic conductivity of oxidized CNTs. Thiol-functionalized CNTs (figure 7.2) were utilized as a sulfur cathode host, which delivered impressive cycling stability over 300 cycles with a low capacity loss of 0.166% per cycle at a rate of 0.1 C [42–45].

Wang *et al* developed a CNT–MXene composite as an integral sulfur immobilizer host material to mitigate the polysulfide shuttle effect; their assembled cell delivered an initial capacity of 1451 mAh g^{-1} at a rate of 0.2 C. Synergistic entrapment of polysulfides was enabled by the MXene–CNT composite, which displayed superior cyclability. The conjunction of the single- and multiwalled CNTs (SWCNTs/MWCNTs) into the sulfur cathodes boosted LISB performance. Synergic enhancement of the conductivity, enabled by the SWCNTs and the MWCNTs, allowed strong polysulfide confinement and ionic diffusion features. The assembled cell delivered a discharge capacity of 1221 mAh g^{-1} at a rate of 0.1 C and showed a retained capacity of 876 mAh g^{-1} after 100 cycles [46, 47].

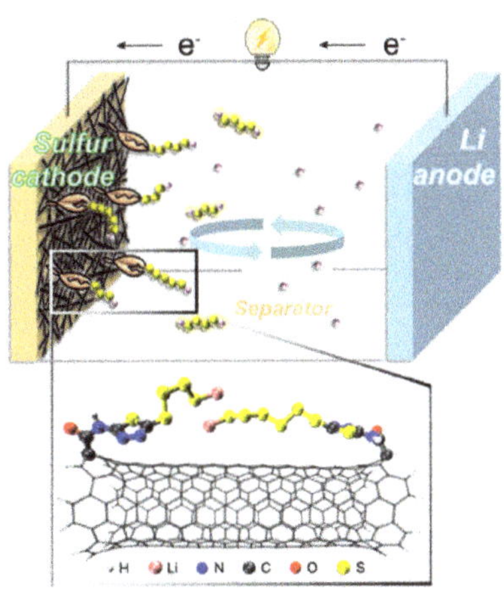

Figure 7.2. The operational scheme of the thiol-functionalized CNT-S cathode composite (reprinted with permission from [45], Copyright (2019), American Chemical Society).

7.4.2.1.3 Porous carbon–sulfur composites

Porous carbon materials are superior for sulfur encapsulation due to their large specific surface area and suitable pore size distribution. Based on the pore diameter, porous materials are broadly classified as microporous (0–2 nm), mesoporous (2–50 nm), or macroporous (>50 nm). Carbon materials with hierarchical pores (a combination of micro-, meso-, and macropores) are potential sulfur host cathode materials for LISBs. Hu *et al* showed that microporous carbon is highly efficient at physically confining polysulfide within its porous architecture. Xu *et al* studied the electrochemical performance of a microporous carbon–sulfur cathode material which delivered a capacity of 608 mAh g^{-1} over 170 cycles at a current density of 0.5 A g^{-1}. However, at higher sulfur loads, microporous carbon-based cathodes were found to show a decline in capacity. This results from the fact that the microporous structure could only load a specific amount of sulfur inside its pores [48, 49].

Mesoporous carbon with an optimum pore structure for sulfur entrapment was investigated for use as a cathode composite in LISBs. Studies showed that mesoporous carbon allowed maximum sulfur loading but was less efficient in capturing polysulfides within the pores. Therefore, bimodal hierarchical porous structures were used as cathode matrices for sulfur cathodes to perform efficient physical entrapping of polysulfides. The inner macro- and mesopores helped to load the sulfur within the material, and the outer micropores effectively confined the polysulfides physically, resulting in efficient electrochemical performance of the assembled cells [49–53]. An overview of the important carbon-based sulfur host materials is provided in table 7.1.

7.4.2.2 Metal-oxide- and metal-sulfide-based sulfur host cathodes

Metal oxides and sulfides are essential non-carbon-based electrocatalysts, which provide strong chemical polysulfide confinement and accelerate the reduction reaction kinetics of LISBs. Employing unique yolk–shell structural designs for

Table 7.1. Carbon-based sulfur host cathode materials.

Carbon–sulfur host cathode material	C rate/current density (A g^{-1} or mA g^{-1})	Specific capacity (mAh g^{-1})	Reference
Graphene–CNT–nano sulfur	50 mA g^{-1}	1067	[35]
Porphyrin-derived graphene	2 C	988	[40]
CNT–MXene	0.2 C	1451	[46]
SWCNT–MWCNT composite	0.1 C	1221	[47]
Mesoporous carbon–MXene	0.2 C	966	[53]

sulfur–TiO_2 cathodes with inner voids leads to high durability and delivers higher capacities for LISBs. The conductive composite sulfur host cathode allows strong chemisorption of the lithium polysulfides and mitigates the volume expansion of the cathodes. An assembled cell based on this design exhibited a capacity of 1030 mAh g^{-1} at a rate of 0.5 C and had good cyclability. Hollow multishelled TiO_{2-x} with engineered defects showed an exemplary initial discharge capacity of 903 mAh g^{-1} at a rate of 0.5 C when integrated into sulfur cathodes. The developed defective electrocatalyst physicochemically entrapped polysulfides and improved the conductivity of the cathode [54, 55].

Wang *et al* theoretically and experimentally investigated the electrocatalytic activity of defective TiO_2 as an efficient sulfur host cathode for LISBs. Oxygen-deficient TiO_2 boosted the sluggish polysulfide reaction kinetics and induced strong confinement of the polysulfides. Their assembled cell showed a capacity of 1472 mAh g^{-1} at a rate of 0.2 C with better cyclic stability and rate capability. Guo *et al* developed TiO_2 sulfur host cathodes with high sulfur loading, and the versatile 'room-like' structure of the TiO_2 induced superior chemisorption of the lithium polysulfides. The assembled cell delivered high discharge capacity and outstanding cyclability [56, 57].

As an alternative to TiO_2, the superior electrocatalytic activity of MnO_2-based sulfur host materials toward lithium polysulfides makes them potential cathode host choices for LISBs. Liang *et al* reported that a S–MnO_2 cathode composite exhibited a capacity of 1300 mAh g^{-1} and superior durability in LISBs. The composite cathodes rapidly converted the polysulfides and showed outstanding chemical entrapment. The mesoporous MnO_2–sulfur composite cathode induced strong chemisorption of the polysulfides, and an assembled cell based on the composite cathode showed a high initial capacity of 1349.3 mAh g^{-1} at a rate of 1 C and superior cycling stability. The functionalization of TiO_2 and MnO_2 (figure 7.3) with potential dopants and carbon moieties such as porous carbon, graphene, and CNTs improves the performance of the resulting cathodes in LISBs and enables high

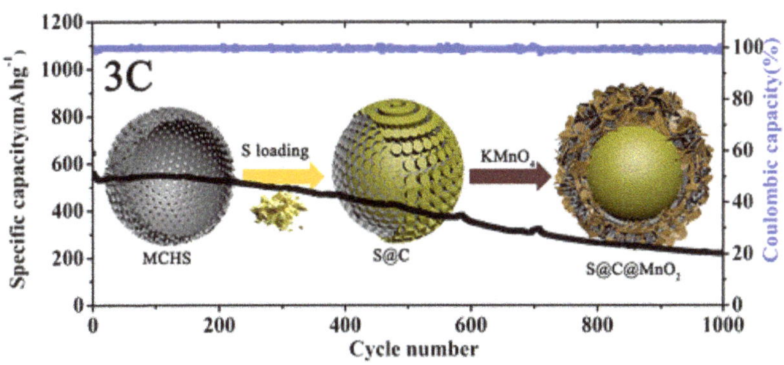

Figure 7.3. A schematic of the S@C@MnO_2 cathode composite with outstanding cycling stability (reprinted with permission from [63], Copyright (2017), American Chemical Society).

conductivity and the strong anchoring of lithium polysulfides via dual adsorption (chemical and physical adsorption) pathways [58–63].

The fascinating features of sulfiphilic metal sulfides, such as high binding strength, conductivity, and electrocatalytic activity, has made them promising cathode host materials for LISBs. Numerous theoretical studies and experimental validations have confirmed the polysulfide shuttle mitigation efficacy of the metal sulfides through chemisorption of the lithium polysulfides. Various polar sulfides of metals such as cobalt, nickel, molybdenum, titanium, iron, tungsten zinc, etc. are employed as polysulfide immobilizers in Li–S batteries. Theoretical studies of metal sulfides such as CoS_2, TiS_2, VS_2, SnS_2, and MoS_2 have revealed their superior ability to perform lithium polysulfide conversion and showed their better binding capability due to the intrinsic polarity of these metal sulfides [64].

MoS_2, a 2D dichalcogenide, possesses astonishing electrical conductivity and versatile electronic features. Metallic MoS_2 (1T MoS_2) is also a hot material as a sulfur host due to its higher electrocatalytic activity in the sulfur conversion reaction. Highly conductive 1T MoS_2-nanoflower-decorated graphene developed by Cheng *et al* displayed good electrocatalytic activity and enabled high sulfur utilization. Designing strongly polar conductive materials with better porosity improves the kinetics of the conversion reaction and the anchoring of polysulfides. Very recently, 1T MoS_2 nanosheets were grown in a regulated manner on porous carbon and employed as sulfur hosts in Li–S batteries. The precursor, porous carbon from natural plane tree fluff, had a unique microtube structure capable of forming porous carbon. The synthesized composite structure possessed highly conductive porous carbon that improved charge conduction, inhibited the polysulfide shuttle effect, and boosted the kinetics of the polysulfide conversion reaction. A cell constructed using this material delivered a capacity of 735.2 mAh g^{-1} at a rate of 1 C after 300 cycles and displayed improved stability [65, 66].

To enhance the sulfur encapsulation and sulfur loading capacity of cathodes, various other metal sulfides with hollow porous and yolk–shell structures have been chosen as sulfur hosts. These porous structures can load more sulfur and accommodate the volume change of sulfur during continuous charging and discharging. Flower-like Ni_3S_2 hollow microspheres were synthesized by the hydrothermal route and used as the sulfur host. The synthesized structures increased the efficient exposure of the active surface sulfur-carrying capacity and displayed good results. Xu *et al* reported the efficacy of Co_3S_4 polyhedra integrated into carbon nanofiber as a suitable sulfur host cathode material for LISBs. A cell assembled using this material showed a capacity of 953 mAh g^{-1} at a rate of 1 C and an excellent areal capacity of 13 mAh cm^{-2} at a rate of 0.3 C rate under very high active material loading conditions. Pristine and functionalized cobalt-based sulfides confine the polysulfides and offer better cyclability [67, 68].

Bimetal sulfides are another attractive candidate capable of giving better electrochemical results due to the synergy of two metal atoms. Defect-engineered $NiCo_2S_4$, which had hybrid orbitals, high electronic conductivity, and more exposed active sites, was used as a cathode matrix for Li–S batteries. Yolk–shell structured $NiCo_2S_4$ hollow spheres with many sulfur vacancies ($NiCo_2S_{4-x}$) were prepared by an anion

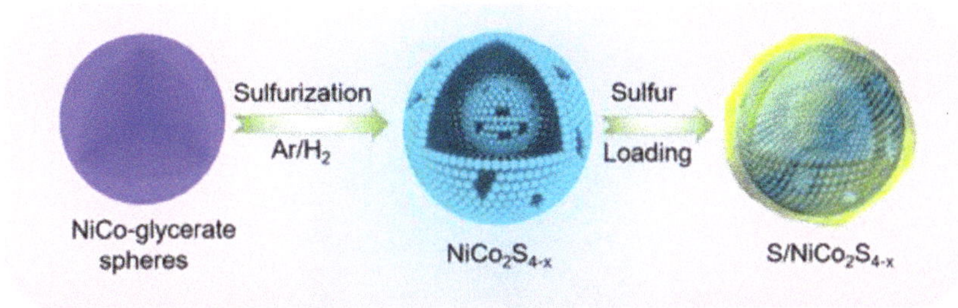

Figure 7.4. A schematic of the bimetallic S/$NiCo_2S_{4-x}$ composite cathode (reprinted with permission from [70], Copyright (2021), American Chemical Society).

exchange method. The efficiently engineered structure provided cavities for high sulfur loading and accommodated the volume changes during cycling. Furthermore, the nanoparticles within the porous shells reduced the charge transfer length and increased the electrolyte accessibility. The material was also effective in interacting with the polysulfides and acting as an electrocatalyst. A cell with a S/$NiCo_2S_{4-x}$ cathode revealed a retained capacity of 598.2 mAh g^{-1} after 500 cycles at 1 C and delivered a remarkable rate capability of 628.9 mAh g^{-1} at 5 C (figure 7.4) [69, 70].

Despite these outstanding results, there are still challenges in developing the facile large-scale synthesis of these materials, elucidating the mechanism of polysulfide interaction in detail, and commercially implementing these cathodes for advanced LISBs.

7.4.2.3 MXene-based sulfur host cathodes

Two-dimensional (2D) layered transition-metal carbides, nitrides, and carbonitrides, known as MXenes, have higher conductivity, volume-specific capacity, excellent hydrophilicity, and unique surface features that make them suitable sulfur host cathode materials for LISBs. Liang *et al* were the first to examine the use of an MXene as a sulfur host cathode material. They discussed the dominant interaction between the Ti_2C MXenes and polysulfides, which resulted in decreased overall polysulfide shuttling, better cycling performance, and a specific capacity of 1200 mAh g^{-1} at a rate of 5 C. X-ray photoelectron spectroscopy (XPS) data for the Ti_2C-integrated polysulfides confirmed the strong binding of the polysulfides with Ti atoms on the surface. The introduction of the MXene nanosheets also increased the capacity retention of the assembled LISBs by around 80% over 400 cycles at a rate of 2 C. Wang *et al* theoretically investigated the electrochemical performance of a composite of VO_2(p) (paramontroseite) and V_2C MXenes as cathode materials in Li–S batteries using first-principles calculations. Their simulations showed that the VO_2 (p)-V_2C/S cathode mitigated the polysulfide shuttle effect by strong chemisorption and displayed significant sulfur loading characteristics (figure 7.5). A cell assembled using this material showed a high capacity of 1250 mAh g^{-1} at 0.2 C with outstanding cycling stability [71, 72].

Functionalized MXenes exhibit remarkable electrochemical performance in LISBs compared to that of pristine MXenes, as a result of the synergic

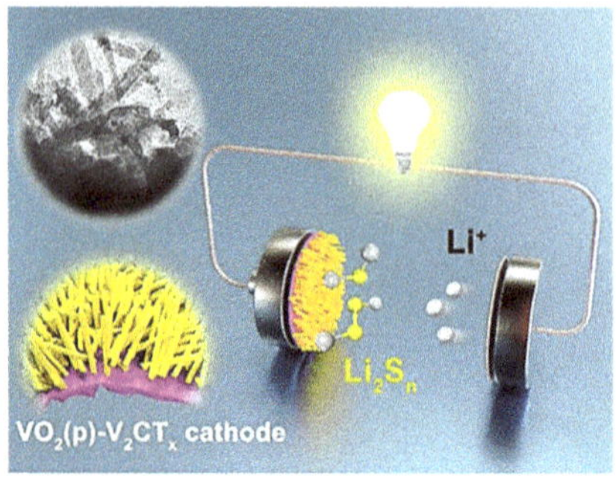

Figure 7.5. A schematic of the VO_2(p)-V_2CT_X–sulfur host cathode (reprinted with permission from [72], Copyright (2019), American Chemical Society).

electrocatalytic activity of the composite sulfur host cathode materials. Wang *et al* discussed the efficacy of MXene@hierarchical porous carbon nanotubes (MHPCs) in capturing polysulfides. This electrocatalyst exhibited a capacity of 1451 mAh g^{-1} at a rate of 0.2 C and excellent cyclability in Li–S batteries. Zhao *et al* studied the efficacy of a sulfur-loaded mesoporous carbon nanosphere (S@MCS–SiO_2) integrated chessboard-like MXene (SMMX) electrode as a sulfur host cathode for Li–S batteries. The SMMX electrode exhibited a high capacity of 1303.6 mAh g^{-1} at a rate of 0.1 C. It induced strong polysulfide confinement and displayed high electrical conductivity, and cells assembled using this material showed long cyclability characteristics over 1000 cycles at a rate of 1 C. Lieu *et al* investigated the electrochemical performance of polymorphic $CoSe_2$ nanoparticles integrated into MXene nanosheets as sulfur host cathodes (S@$CoSe_2$–MXene) in high-energy Li–S battery applications. The S@$CoSe_2$–MXene effectively induced strong chemisorption of the polysulfides through the templating of sulfur spheres, which resulted from the efficient electrocatalytic activity; a cell assembled from this material showed higher initial capacity and excellent cyclability over 1000 cycles. MXene-based electrodes and separator materials can induce strong chemisorptive anchoring of polysulfides, enhancing the conductivity of sulfur cathodes and making them suitable candidates for LISB applications. Low-cost and sustainable synthetic methods are required to develop high-quality layered MXenes, which is imperative for the fabrication of high-performance LISBs [73–75].

7.4.2.4 Others

Conductive polymers, metal–organic frameworks (MOFs), and single-atom catalysts are promising sulfur host cathode materials due to their versatile structural and electronic features. Conductive polymers (polypyrrole (Ppy), polythiophene (PTh), polyaniline (PANI), the polythiophene derivative, poly (3,4-ethylenedioxythiophene) (PEDOT)) and their composites induce potential suppression of the polysulfide shuttle effect and improve the conductivity of sulfur cathodes. Pristine (ZIF-8, Mn-MOF, HKUST-1, ZIF-67) and composite MOF-based sulfur host cathodes have tunable porosity, high specific surface area, and functionalized catalytic active sites. As a

Table 7.2. Non-carbon-based sulfur host cathode materials.

Sulfur host cathode materials	C rate	Specific capacity (mAh g^{-1})	Reference
Yolk–shell TiO_2–sulfur	0.5 C	1030	[54]
Mesoporous MnO_2–sulfur	1 C	1349.3	[59]
Co_3S_4-carbon nanofiber–sulfur	1 C	953	[68]
VO_2 (p)-V_2C MXene–sulfur	0.2 C	1250	[72]
Mesoporous carbon–SiO_2–MXene	0.1 C	1303.6	[74]

result, they offer strong anchoring of the polysulfides and enhanced LISB electrochemical performance. Single-atom catalysts are emerging as suitable sulfur host materials, which boost the performance of LISBs by inducing strong electrocatalytic activity toward the polysulfide shuttle effect [76–78]. The important non-carbon-based sulfur host cathodes are listed in table 7.2.

7.5 Li–S battery electrolytes

Electrolytes are one of the crucial components of LISBs; they allow ionic transportation and influence the batteries' electrochemical performance to a great extent. In LISBs, the most widely used liquid organic electrolyte (LiTFSI) enables the dissolution and diffusion of polysulfide and lithium ions, leading to the polysulfide shuttle [79]. Therefore, to tackle these issues, modifications to the electrolyte design are essential to achieve better LISB performance. The important electrolyte systems of LISBs are discussed below.

7.5.1 Organic liquid electrolytes

Lithium salts play a vital role in the electrolyte, mainly as a lithium-ion source. Therefore, the most widely used lithium salts in LISBs are lithium bis(fluorosulfonyl) imide (LiFSI) and LiTFSI due to their high heat tolerance, ionic conductivity, compatibility, and ability to form an SEI. A combination of 1,3-dioxalane (DOL) and 1,2-dimethoxyethane (DME) (DOL/DME) is the best solvent pair used in LISB liquid electrolytes (1.0 M LiTFSI in DOL/DME with a certain amount of $LiNO_3$ as an additive) and delivers superior electrochemical results. To improve the safety and cyclability features of LISBs, smaller volumes of additives such as metal nitrates are incorporated into the electrolyte, among which lithium nitrate is widely employed [80, 81].

7.5.2 Ionic liquid electrolytes

Ionic liquid (IL) electrolytes possess high lithium-ion conduction, wide potential windows, a stable SEI, and enhanced safety. LISBs employ IL electrolytes with bis [(trifluoromethyl)sulfonyl]imide $(TFSI)^-$ anions and a variety of cations such as 1-butyl-3-methylimidazolium (BMIM), 1-ethyl-3-methylimidazolium (EMIM), 1-butyl-1-methylpyrrolidinium (PYR14), and 1-butyl-1-methylpiperidinium

(PiP14). These electrolytes can work effectively at higher temperatures due to increased lithium-ion conduction, even if they are viscous at room temperature [82].

7.5.3 Solid-state polymer electrolytes

Solid-state polymer electrolytes exhibit good electrochemical performance in LISBs due to their low density, stability, and safety. Polymer electrolytes consisting of polyethylene oxide with a number of vinylic oxygen moieties and their derivative complex lithium salts (LiTFSI, LiFSI) are used in LISBs. Poly(ethylene oxide)–lithium trifluoride containing zirconia and a Li_2S-based solid-state electrolyte have facile Li-ion conduction and transport properties, which boost the performance of LISBs [83].

7.5.4 Gel polymer electrolytes

Gel polymer electrolytes (GPEs) stand between conventional liquid electrolytes and solid polymer electrolytes and are a good choice in LISBs. Various gel polymer electrolytes include pentaerythritol tetra acrylate (PETEA)-based GPE, polyvinylidene fluoride (PVDF)-based GPE, polydopamine (PDA)-based GPE, etc. A GPE composed of poly(ethylene oxide) (PEO), poly(acrylamido-acrylic acid) (P(AM-co-AA)), and poly(3,4-ethylenedioxythiophene):poly(styrene sulfonate) (PEDOT:PSS) with multifunctional properties has been employed, which provided good electrochemical results. The use of cross-linked polymers in the electrolyte matrix increased the mechanical properties and yielded good performance [84, 85].

7.5.5 Composite electrolytes

Composite electrolytes provide the synergistic effect of superior lithium-ion conduction and the better stability of an inorganic electrolyte combined with the better interfacial properties of an organic electrolyte. The incorporation of inorganic nanoparticles such as $Li_7La_3Zr_2O_{12}$ (LLZO), $Li_{0.33}La_{0.557}TiO_3$ (LLTO), $LiAlO_2$, and $Li_{10}SnP_2S_{12}$ (LSPS) in PEO yielded good electrochemical performance in LISBs. Ceramic-based Li-ion conductive electrolytes are also outstanding electrolyte choices, which allow the development of high-performance solid-state LISBs [86–88].

7.6 Separators for Li–S batteries

Separators are integral in batteries; they act as ionic conductors to transport ions and as electrical insulators to prevent internal short circuits. The commercially available porous polymer separators employed in batteries are Celgard membranes, usually made of polyethylene (PE), polypropylene (PP), and PE/PP. The general properties needed for a separator are good porosity, wettability, tensile strength, puncture strength, etc. In LISBs, the separator must selectively allow the movement of lithium ions and obstruct the passage of polysulfide anions [89, 90]. Carbon nanomaterials that have good electrical conductivity, low cost, tunable porosity, high surface area, good structural and thermal stability, and ease of functionalization are exemplary choices for functionalized separators in LISBs. Porous carbon materials are one of the most attractive groups of materials used as separator modifiers in LISBs. A pioneering study of microporous carbon was performed by Manthiram *et al* in 2012, which showed an initial discharge capacity of 990 mAh g^{-1} with 0.28% fading after 100 cycles at a rate of 2 C [91].

Manthiram's group demonstrated an MWCNT-modified separator for advanced high-performance LISBs. CNTs provided fast channels for the transport of electrons, which led to decreased resistance and good electrochemical activity. A cell assembled using the modified separator delivered a capacity of 1324 mAh g^{-1}, a low capacity decay of 0.14% after 300 cycles, and strong physical confinement of the lithium polysulfides [92]. Kim *et al* functionalized SWCNTs with polyacrylic acid (PAA) and employed them as separator coating for LISBs. The functionalization enhanced the polysulfide interaction and led to improved electrochemical results. Zhou *et al* demonstrated a graphene-coated separator for LISBs, which delivered an initial discharge capacity of 1006 mAh g^{-1} and a capacity decay of 0.10% after 300 cycles at a rate of 0.9 C [93, 94].

PDA is a polymer with abundant hydroxyl and amine functional groups. PDA-based separators delivered a high initial discharge capacity of 889 mAh g^{-1} and a capacity decay of 0.11% after 200 cycles at a rate of 0.5 C [95]. An MnO_2 modified separator was designed for LISBs, effectively trapping the polysulfides using strong chemical bonding. A cell with the modified separator showed a retained capacity of 665 mAh g^{-1} at 1 C after 1000 cycles [96]. MoS_2, a 2D layered metal chalcogenide, has been extensively explored in many fields. Ghazi *et al* fabricated a highly conductive MoS_2-coated Celgard separator for use in LISBs. A cell constructed using this separator showed a diffusion coefficient value of 7.6×10^{-9} cm^2 s^{-1} and a lithium-ion conductivity of 2.0×10^{-4} S cm^{-1} [97].

CoS_2, another high electronically conductive metal sulfide, was combined with rGO and applied as separator coating in LISBs. A cell constructed using this separator delivered an initial capacity of 1122.3 mAh g^{-1} at 0.2 C and 583.9 mAh g^{-1} at 2 C with superior cyclability [98]. Zinc sulfide embedded in N-doped 3D carbon nanosheets was utilized as a separator material for LISBs. A cell fabricated using this separator material delivered an impressive cyclability at a rate of 0.5 C and superior electrochemical performance [99]. A carbon–WS_2-based separator coating was developed for LISBs, which showed excellent polysulfide adsorption capability and good ionic conductivity. A cell based on the C–WS_2-modified separator exhibited an initial discharge capacity of 996 mAh g^{-1} at 1 C and retained impressive stability even after 1000 cycles. Separator modifications allow the facile suppression of the polysulfides through physicochemical confinement and enhance LISB performance. Numerous developments in functionalized separators have employed highly conductive pristine and composite MXenes, MOFs, and single-atom catalysts to enhance the electrochemical performance of LISBs [100].

7.7 Binders for Li–S batteries

The binder strongly adheres to the active materials of the current collector and connects the active material and other additives. In LISBs, electrochemically stable PVDF is a widely used binder material due to its versatile binding features. But PVDF fails to tackle the volume expansion and the interaction with the polysulfides and requires toxic solvents such as N-methyl-2-pyrrolidine (NMP) for dissolution. Therefore, multifunctional binders are employed in LISBs which can overcome the volume expansion, increase the electronic and ionic conductivities, interact with the polysulfides, and catalyze the polysulfide conversion reaction [101–104]. Various categories of binders are discussed below.

7.7.1 Binders showing superior adhesion

The weak van der Waals force, the covalent force, and the electrostatic binding force play vital roles in the binding properties [105]. Gelatin, a natural polymer with abundant COOH and NH_2 groups, can increase the binding strength. Furthermore, its hydrophilic and insoluble nature in organic-based electrolytes can increase the stability of the electrode [106]. In an early report, a cathode with a gelatin binder delivered an initial capacity of 1132 mAh g^{-1} and good cyclability for 50 cycles.

7.7.2 Binders capable of overcoming volume expansion

Binders with various structural properties, such as composite, cross-linked, and hyper-branched structures, have been developed to overcome cathodic volume expansion. A carboxymethyl cellulose (CMC)/nitrile butadiene rubber-based binder possessed superior structural features that protected the cathode during cycling. A composite binder based on this material delivered an excellent electrochemical performance and superior cyclability [107]. Another cross-linked binder for LISBs, PEDOT:PSS–Mg^{2+}, showed a capacity of 1097 mAh g^{-1} and superior capacity retention over 250 cycles [108]. A polycation (β-CD_p–N+) with a quaternary ammonium cation and a β cyclodextrin polymer with numerous branching structures were used as a binder in LISBs, which showed better cyclability and rate properties [109].

7.7.3 Electron-conductive binders

The integration of conductive binders enhances the conductivity of sulfur cathodes. A reduced graphene oxide–polyacrylic acid (GOPAA) binder was utilized in LISBs, which showed a 30% enhancement in capacity and cycling stability. These results were attributed to the electronic conductivity offered by the continuous conductive pathways of GOPAA. In another study, the conductive polymer PEDOT:PSS was applied as the binder, which showed significant electrochemical results due to improved conductivity [110, 111].

7.7.4 Ion-conductive binders

Lithium-ion conductivity is crucial in any lithium-based battery. Nafion has good ion-conductive properties and is employed as a binder in LISBs. A cell assembled using Nafion delivered good rate performance due to its superior ionic conduction. Another ion-conductive binder combined with Li-Nafion, poly(vinylpyrrolidone) (PVP), and nanosilica showed an initial discharge capacity of 1373 mAh g^{-1} at 0.2 C and high cyclability over 300 cycles at 1 C. An ion-conducting polymer (PSPEG: organo polySulfide containing polyethylene glycol) binder in LISBs consisted of ethylene oxide groups and oxygen atoms that could effectively conduct the lithium ions. The binder, which had an ionic conductivity of 2.11×10^{-4} S cm^{-1}, helped the cell to achieve stable cycling over 500 cycles [112–114].

7.7.5 Binders capable of controlling the polysulfide shuttle

Various polymers that chemically and electrostatically interact with polysulfides can be used as binders to control the polysulfide shuttle. A polybenzimidazole binder with a number of N- and O-containing functionalities was employed as a binder in LISBs. The experimental details confirmed the effective polysulfide interaction of the binder, leading to superior electrochemical results. Another polymer binder, amphiphilic Hyperbranched polymer (AHP), which has numerous amino groups, was also found to be an effective binder with excellent cyclability. A cell based on a 3D cross-linked binder that contained soybean protein, phytic acid, and poly(ethylene oxide) (SPP) and phosphate groups showed an initial discharge capacity of 629.7 mAh g^{-1} at a rate of 1 C and good cycling stability over 800 cycles. A polycation binder (β-CDp–N^{+}) consisting of a quaternary ammonium cation and a cyclodextrin polymer was proposed for LISBs. The binder electrostatically interacted with the polysulfides, thus confining them. A cell constructed using this material showed an areal capacity of 4.4 mAh cm^{-2} at 50 mA g^{-1}, even at a high sulfur loading [115–117].

7.8 Conclusions and future perspectives

Global climate change warnings have highlighted the need to develop sustainable energy storage systems such as high-performance batteries. The requirements for post-Li-ion battery technology and high-energy-density batteries with better safety features have prompted the development of high-performance LISBs. The integral design of functionalized cathodes, separators, binders, dendrite-free novel anodes, and electrode modifications will shape the future of LISBs. Potential electrocatalysts with strong polysulfide confinement, high conductivity, and rapid electrochemical conversion kinetics will resolve the current challenges of LISBs. The advancement of LISBs into pouch cells and large battery packs is vital to meet the current energy requirements in a cost-effective and sustainable manner. Many groundbreaking reports of LISB developments from both industry (Li–S Energy Ltd, Poly Plus Battery Co, Oxis Energy, etc.) and research centers (the Argonne National Laboratory) are enormously encouraging regarding the practical usage of LISBs in the coming years.

This chapter discussed the insightful developments of sulfur host cathode materials, electrolytes, separators, and binders employed in the current LISB technology.

The important future research directions and challenges are as follows:

- The development of novel sulfur host cathode materials, which enables conductivity and strong confinement of the polysulfides and boosts the polysulfide conversion kinetics.

- Pre-lithiation strategies adopted for the development of dendrite-free Li metal anodes.
- Electrolytes can be modified by designing solid-state and aqueous-based electrolytes to understand and tackle the challenges of using organic-based electrolytes.
- The advancement of high sulfur loadings and the development of the solid-state LISBs.
- The mechanistic process responsible for the reversibility features of the many polysulfide conversion electrocatalysts is unknown. Modeling and theoretical studies will offer insights into, and an in-depth understanding of the mechanistic path of polysulfide conversion kinetics.
- A demonstration of flexible and pouch cell LISB devices using existing efficient electrodes and other cell components would be adequate for the commercialization of LISBs.

List of abbreviations

LISBs	Lithium–sulfur batteries
LIB	Lithium-ion battery
LiTFSI	Lithium bis(trifluoromethanesulfonyl)imide
PEDOT	Poly(3,4-ethylenedioxythiophene)
PPy	Polypyrrole
PTh	Polythiophene
PANI	Polyaniline
PEDOT-PSS	Poly(3,4-ethylenedioxythiophene) polystyrene sulfonate

References

[1] Bashir T, Zhou S, Yang S, Ismail S A, Ali T, Wang H, Zhao J and Gao L 2023 Progress in 3D-MXene electrodes for lithium/sodium/potassium/magnesium/zinc/aluminum-ion batteries *Electrochem. Energy Rev.* **6** 5

[2] Grey C P and Hall D S 2020 Prospects for lithium-ion batteries and beyond—a 2030 vision *Nat. Commun.* **11** 6279

[3] Jiao X *et al* 2023 Multi-physical field simulation: a powerful tool for accelerating exploration of high-energy-density rechargeable lithium batteries *Adv. Energy Mater.* **13** 2301708

[4] Feng S, Fu Z H, Chen X and Zhang Q 2022 A review on theoretical models for lithium–sulfur battery cathodes *InfoMat.* **4** e12304

[5] Pan H *et al* 2023 Boosting lean electrolyte lithium–sulfur battery performance with transition metals: a comprehensive review *Nano-Micro Lett.* **15** 165

[6] Manthiram A, Fu Y, Chung S H, Zu C and Su Y S 2014 Rechargeable lithium–sulfur batteries *Chem. Rev.* **114** 11751–87

[7] Manthiram A, Fu Y and Su Y S 2013 Challenges and prospects of lithium–sulfur batteries *Acc. Chem. Res.* **46** 1125–34

[8] Shao Q, Zhu S and Chen J 2023 A review on lithium–sulfur batteries: challenge, development, and perspective *Nano Res.* **16** 8097–138

[9] Saroha R, Ahn J H and Cho J S 2021 A short review on dissolved lithium polysulfide catholytes for advanced lithium–sulfur batteries *Korean J. Chem. Eng.* **38** 461–74
[10] Wang R, Wang K, Gao S, Jiang M, Han J, Zhou M, Cheng S and Jiang K 2018 Electrocatalysis of polysulfide conversion by conductive RuO_2 nano dots for lithium–sulfur batteries *Nanoscale.* **10** 16730–7
[11] Li S Y, Wang W P, Duan H and Guo Y G 2018 Recent progress on confinement of polysulfides through physical and chemical methods *J. Energy Chem.* **27** 1555–65
[12] Li C *et al* 2022 Mapping techniques for the design of lithium–sulfur batteries *Small* **18** 2106657
[13] Tao T, Lu S, Fan Y, Lei W, Huang S and Chen Y 2017 Anode improvement in rechargeable lithium–sulfur batteries *Adv. Mater.* **29** 1700542
[14] Piao N *et al* 2021 Lithium metal batteries enabled by synergetic additives in commercial carbonate electrolytes *ACS Energy Lett.* **6** 1839–48
[15] Gu S *et al* 2021 Nitrate additives coordinated with crown ether stabilize lithium metal anodes in carbonate electrolyte *Adv. Funct. Mater.* **31** 2102128
[16] Hu Y, Chen W, Lei T, Jiao Y, Wang H, Wang X, Rao G, Wang X, Chen B and Xiong J 2020 Graphene quantum dots as the nucleation sites and interfacial regulator to suppress lithium dendrites for high-loading lithium–sulfur battery *Nano Energy* **68** 104373
[17] Sun Y *et al* 2020 Tailoring the mechanical and electrochemical properties of an artificial interphase for high-performance metallic lithium anode *Adv. Energy Mater.* **10** 2001139
[18] Castillo J, Coca-Clemente J A, Rikarte J, Sáenz de Buruaga A, Santiago A and Li C 2023 Recent progress on lithium anode protection for lithium–sulfur batteries: review and perspective *APL Mater.* **11** 010901
[19] Li J, Ma C, Chi M, Liang C and Dudney N J 2015 Solid electrolyte: the key for high-voltage lithium batteries *Adv. Energy Mater.* **5** 1401408
[20] Cha E, Patel M D, Park J, Hwang J, Prasad V, Cho K and Choi W 2018 2D MoS_2 as an efficient protective layer for lithium metal anodes in high-performance Li–S batteries *Nat. Nanotechnol.* **13** 337–44
[21] Yan K, Lu Z, Lee H W, Xiong F, Hsu P C, Li Y, Zhao J, Chu S and Cui Y 2016 Selective deposition and stable encapsulation of lithium through heterogeneous seeded growth *Nat. Energy* **1** 1–8
[22] Xia S, Zhang X, Liang C, Yu Y and Liu W 2020 Stabilized lithium metal anode by an efficient coating for high-performance Li–S batteries *Energy Storage Mater.* **24** 329–35
[23] Zhao Y *et al* 2019 Natural SEI-inspired dual-protective layers via atomic/molecular layer deposition for long-life metallic lithium anode *Matter* **1** 1215–31
[24] Baloch M, Shanmukaraj D, Bondarchuk O, Bekaert E, Rojo T and Armand M 2017 Variations on Li_3N protective coating using ex situ and *in situ* techniques for Li in sulphur batteries *Energy Storage Mater.* **9** 141–9
[25] Zeng P *et al* 2023 *In situ* reconstruction of electrocatalysts for lithium–sulfur batteries: progress and prospects *Adv. Funct. Mater.* **33** 2301743
[26] Kim J T, Hao X, Wang C and Sun X 2023 Cathode materials for single-phase solid–solid conversion Li–S batteries *Matter* **6** 316–43
[27] Dong S, Liu H, Hu Y and Chong S 2022 Cathode materials for rechargeable lithium–sulfur batteries: current progress and future prospects *ChemElectroChem.* **9** e202101564
[28] Wang J L, Yang J, Xie J Y, Xu N X and Li Y J 2002 Sulfur–carbon nano-composite as cathode for rechargeable lithium battery based on gel electrolyte *Electrochem. Commun.* **4** 499–502

[29] Ji X, Lee K T and Nazar L F 2009 A highly ordered nanostructured carbon–sulfur cathode for lithium–sulphur batteries *Nat. Mater.* **8** 500–6
[30] Yang T, Xia J, Piao Z, Yang L, Zhang S, Xing Y and Zhou G 2021 Graphene-based materials for flexible lithium–sulfur batteries *ACS Nano* **15** 13901–23
[31] Wang H, Yang Y, Liang Y, Robinson J T, Li Y, Jackson A, Cui Y and Dai H 2011 Graphene-wrapped sulfur particles as a rechargeable lithium–sulfur battery cathode material with high capacity and cycling stability *Nano Lett.* **11** 2644–7
[32] Lu L Q, Lu L J and Wang Y 2013 Sulfur film-coated reduced graphene oxide composite for lithium–sulfur batteries *J. Mater. Chem.* A **1** 9173–81
[33] Li J, Wei W and Meng L 2019 Liquid-phase exfoliated-graphene-supporting nanostructural sulfur as high-performance lithium–sulfur batteries cathode *Compos. Commun.* **15** 149–54
[34] Papandrea B *et al* 2016 Three-dimensional graphene framework with ultra-high sulfur content for a robust lithium–sulfur battery *Nano Res.* **9** 240–8
[35] Zhang Z, Kong L L, Liu S, Li G R and Gao X P 2017 A high-efficiency sulfur/carbon composite based on 3D graphene nanosheet@ carbon nanotube matrix as cathode for Lithium–sulfur battery *Adv. Energy Mater.* **7** 1602543
[36] Doñoro Á, Muñoz-Mauricio Á and Etacheri V 2021 High-performance lithium sulfur batteries based on multidimensional graphene–CNT-nanosulfur hybrid cathodes *Batteries* **7** 26
[37] Wang Z, Dong Y, Li H, Zhao Z, Bin Wu H, Hao C, Liu S, Qiu J and Lou X W 2014 Enhancing lithium–sulphur battery performance by strongly binding the discharge products on amino-functionalized reduced graphene oxide *Nat. Commun.* **5** 5002
[38] Song J, Yu Z, Gordin M L and Wang D 2016 Advanced sulfur cathode enabled by highly crumpled nitrogen-doped graphene sheets for high-energy-density lithium–sulfur batteries *Nano Lett.* **16** 864–70
[39] Chang C H and Manthiram A Covalently grafted polysul-fur-graphene nanocomposites for ultrahigh sulfur-loading lithium-polysulfur batteries *ACS Energy Lett.* **3** 72–7
[40] Kong L, Li B Q, Peng H J, Zhang R, Xie J, Huang J Q and Zhang Q 2018 Porphyrin-derived graphene-based nanosheets enabling strong polysulfide chemisorption and rapid kinetics in lithium–sulfur batteries *Adv. Energy Mater.* **8** 1800849
[41] Li J, Xue C, Xi B, Mao H, Qian Y and Xiong S 2018 Heteroatom dopings and hierarchical pores of graphene for synergistic improvement of lithium–sulfur battery performance *Inorg. Chem. Front.* **5** 1053–61
[42] Guo J, Xu Y and Wang C 2011 Sulfur-impregnated disordered carbon nanotubes cathode for lithium–sulfur batteries *Nano Lett.* **11** 4288–94
[43] Lin Y, Ticey J, Oleshko V, Zhu Y, Zhao X, Wang C, Cumings J and Qi Y 2021 Carbon-nanotube-encapsulated-sulfur cathodes for lithium–sulfur batteries: integrated computational design and experimental validation *Nano Lett.* **22** 441–7
[44] Li Z, Wu H B and Lou X W 2016 Rational designs and engineering of hollow micro-/nanostructures as sulfur hosts for advanced lithium–sulfur batteries *Energy Environ. Sci.* **9** 3061–70
[45] Xu Y W, Zhang B H, Li G R, Liu S and Gao X P 2019 Covalently bonded sulfur anchored with thiol-modified carbon nanotube as a cathode material for lithium–sulfur batteries *ACS Appl. Energy Mater.* **3** 487–94
[46] Wang X *et al* 2021 Strain engineering of a MXene/CNT hierarchical porous hollow microsphere electrocatalyst for a high-efficiency lithium polysulfide conversion process *Angew. Chem. Int. Ed.* **60** 2371–8

[47] Yahalom N, Snarski L, Maity A, Bendikov T, Leskes M, Weissman H and Rybtchinski B 2023 Durable lithium–sulfur batteries based on a composite carbon nanotube cathode *ACS Appl. Energy Mater.* **6** 4511–9

[48] Hu L, Lu Y, Li X, Liang J, Huang T, Zhu Y and Qian Y 2017 Optimization of microporous carbon structures for lithium–sulfur battery applications in carbonate-based electrolyte *Small* **13** 1603533

[49] Xu Y, Wen Y, Zhu Y, Gaskell K, Cychosz K A, Eichhorn B, Xu K and Wang C 2015 Confined sulfur in microporous carbon renders superior cycling stability in Li/S batteries *Adv. Funct. Mater.* **25** 4312–20

[50] Li X, Cao Y, Qi W, Saraf L V, Xiao J, Nie Z, Mietek J, Zhang J G, Schwenzer B and Liu J 2011 Optimization of mesoporous carbon structures for lithium–sulfur battery applications *J. Mater. Chem.* **21** 16603–10

[51] Li L, Ma Z and Li Y 2022 Accurate determination of optimal sulfur content in mesoporous carbon hosts for high-capacity stable lithium–sulfur batteries *Carbon* **197** 200–8

[52] Kim S J, Ahn M, Park J, Jeoun Y, Yu S H, Min D H and Sung Y E 2020 Enhancing the of performance of lithium-sulfur batteries through electrochemical impregnation of sulfur in hierarchical mesoporous carbon nanoparticles *ChemElectroChem.* **7** 3653–5

[53] Li X *et al* 2023 Ordered mesoporous carbon grafted MXene catalytic heterostructure as Li-ion kinetic pump toward high-efficient sulfur/sulfide conversions for Li–S battery *ACS Nano* **17** 1653–62

[54] Wei Seh Z, Li W, Cha J J, Zheng G, Yang Y, McDowell M T, Hsu P C and Cui Y 2013 Sulphur–TiO_2 yolk–shell nanoarchitecture with internal void space for long-cycle lithium–sulphur batteries *Nat. Commun.* **4** 1331

[55] Salhabi E H, Zhao J, Wang J, Yang M, Wang B and Wang D 2019 Hollow multi-shelled structural TiO_{2-x} with multiple spatial confinement for long-life Lithium–sulfur batteries *Angew. Chem.* **131** 9176–80

[56] Wang H E, Yin K, Qin N, Zhao X, Xia F J, Hu Z Y, Guo G, Cao G and Zhang W 2019 Oxygen-deficient titanium dioxide as a functional host for lithium–sulfur batteries *J. Mater. Chem.* A **7** 10346–53

[57] Guo J, Zhao S, Shen Y, Shao G and Zhang F 2020 'Room-like' TiO_2 array as a sulfur host for lithium–sulfur batteries: combining advantages of array and closed structures *ACS Sustain. Chem. Eng.* **8** 7609–16

[58] Liang X, Hart C, Pang Q, Garsuch A, Weiss T and Nazar L F 2015 A highly efficient polysulfide mediator for lithium–sulfur batteries *Nat. Commun.* **6** 5682

[59] Tu S, Zhao X, Cheng M, Sun P, He Y and Xu Y 2019 Uniform mesoporous MnO_2 nanospheres as a surface chemical adsorption and physical confinement polysulfide mediator for lithium–sulfur batteries *ACS Appl. Mater. Interfaces* **11** 10624–30

[60] Yao W, Chu C, Zheng W, Zhan L and Wang Y 2018 'Pea-pod-like' nitrogen-doped hollow porous carbon cathode hosts decorated with polar titanium dioxide nanocrystals as efficient polysulfide reservoirs for advanced lithium–sulfur batteries *J. Mater. Chem.* A **6** 18191–205

[61] Huang J Q, Wang Z, Xu Z L, Chong W G, Qin X, Wang X and Kim J K 2016 Three-dimensional porous graphene aerogel cathode with high sulfur loading and embedded TiO_2 nanoparticles for advanced lithium–sulfur batteries *ACS Appl. Mater. Interfaces* **8** 28663–70

[62] Rehman S, Tang T, Ali Z, Huang X and Hou Y 2017 Integrated design of MnO_2@ carbon hollow nanoboxes to synergistically encapsulate polysulfides for empowering lithium sulfur batteries *Small* **13** 1700087

[63] Ni L, Zhao G, Yang G, Niu G, Chen M and Diao G 2017 Dual core–shell-structured S@ C@ MnO_2 nanocomposite for highly stable lithium–sulfur batteries *ACS Appl. Mater. Interfaces* **9** 34793–803

[64] Abraham A M, Boteju T, Ponnurangam S and Thangadurai V 2022 A global design principle for polysulfide electrocatalysis in lithium–sulfur batteries—a computational perspective *Battery Energy* **1** 20220003

[65] Cheng Z, Chen Y, Yang Y, Zhang L, Pan H, Fan X, Xiang S and Zhang Z 2021 Metallic MoS_2 nanoflowers decorated graphene nanosheet catalytically boosts the volumetric capacity and cycle life of lithium–sulfur batteries *Adv. Energy Mater.* **11** 2003718

[66] Chen R, Shen J, Chen K, Tang M and Zeng T 2021 Metallic phase MoS_2 nanosheet decorated biomass carbon as sulfur hosts for advanced lithium–sulfur batteries *Appl. Surf. Sci.* **566** 150651

[67] Yan Y, Chen Y, Wang Z, Qin C, Bakenov Z and Zhao Y 2021 Flower-like Ni_3S_2 hollow microspheres as superior sulfur hosts for lithium–sulfur batteries *Microporous Mesoporous Mater.* **326** 111355

[68] Xu H and Manthiram A 2017 Hollow cobalt sulfide polyhedra-enabled long-life, high areal-capacity lithium–sulfur batteries *Nano Energy* **33** 124–9

[69] Xiong Z, Chen Q, Qin J, You J and Cheng S 2023 Tubular $NiCo_2S_4$ hierarchical architectures as sulfur hosts for advanced rechargeable lithium sulfur batteries *Int. J. Electrochem. Sci.* **18** 100159

[70] Wang W, Li J, Jin Q, Liu Y, Zhang Y, Zhao Y, Wang X, Nurpeissova A and Bakenov Z 2021 Rational construction of sulfur-deficient $NiCo_2S_{4-x}$ hollow microspheres as an effective polysulfide immobilizer toward high-performance lithium/sulfur batteries *ACS Appl. Energy Mater.* **4** 1687–95

[71] Liang X, Garsuch A and Nazar L F 2015 Sulfur cathodes based on conductive MXene nanosheets for high-performance lithium–sulfur batteries *Angew. Chem.* **127** 3979–83

[72] Wang Z, Yu K, Feng Y, Qi R, Ren J and Zhu Z 2019 VO_2 (p)-V2C (MXene) grid structure as a Lithium polysulfide catalytic host for high-performance Li–S battery *ACS Appl. Mater. Interfaces* **11** 44282–92

[73] Wang X, Luo D, Wang J, Sun Z, Cui G, Chen Y, Wang T *et al* 2021 Strain engineering of a MXene/CNT hierarchical porous hollow microsphere electrocatalyst for a high-efficiency lithium polysulfide conversion process *Angew. Chem. Int. Ed.* **60** 2371–8

[74] Zhao J, Qi Y, Yang Q, Huang T, Wang H, Wang Y, Niu Y, Liu Y, Bao S and Xu M 2022 Chessboard structured electrode design for Li–S batteries based on MXene nanosheets *Chem. Eng. J.* **429** 131997

[75] Lieu W Y, Fang D, Li Y, Li X L, Lin C, Thakur A, Wyatt B C *et al* 2022 Spherical templating of $CoSe_2$ nanoparticle-decorated MXenes for lithium–sulfur batteries *Nano Lett.* **22** 8679–87

[76] Zhang Q, Huang Q, Hao S M, Deng S, He Q, Lin Z and Yang Y 2022 Polymers in lithium–sulfur batteries *Adv. Sci.* **9** 2103798

[77] Zheng Y, Zheng S, Xue H and Pang H 2019 Metal–organic frameworks for lithium–sulfur batteries *J. Mater. Chem.* A **7** 3469–91

[78] Wang F, Li J, Zhao J, Yang Y, Su C, Zhong Y L, Yang Q H and Lu J 2020 Single-atom electrocatalysts for lithium sulfur batteries: progress, opportunities, and challenges *ACS Mater. Lett.* **2** 1450–63

[79] Liu G, Sun Q, Li Q, Zhang J and Ming J 2021 Electrolyte issues in lithium–sulfur batteries: development, prospect, and challenges *Energy Fuels* **35** 10405–27
[80] Foropoulos J and DesMarteau D D 1984 Synthesis, properties, and reactions of bis ((trifluoromethyl) sulfonyl) imide, $(CF_3SO_2)_2NH$ *Inorg. Chem.* **23** 3720–3
[81] Wang Z *et al* 2023 Highly soluble organic nitrate additives for practical lithium metal batteries *Carbon Energy* **5** e283
[82] Wang J, Chew S Y, Zhao Z W, Ashraf S, Wexler D, Chen J, Ng S H, Chou S L and Liu H K 2008 Sulfur–mesoporous carbon composites in conjunction with a novel ionic liquid electrolyte for lithium rechargeable batteries *Carbon N. Y.* **46** 229–35
[83] Marmorstein D, Yu T H, Striebel K A, McLarnon F R, Hou J and Cairns E J 2000 Electrochemical performance of lithium/sulfur cells with three different polymer electrolytes *J. Power Sources* **89** 219–26
[84] Cui Y, Li J, Yuan X, Liu J, Zhang H, Wu H and Cai Y 2022 Emerging strategies for gel polymer electrolytes with improved dual-electrode side regulation mechanisms for lithium–sulfur batteries *Chem. Asian J.* **17** e202200746
[85] Mukkabla R, Ojha M and Deepa M 2020 Poly(N-methylpyrrole) barrier coating and SiO_2 fillers based gel electrolyte for safe and reversible Li–S batteries *Electrochim. Acta* **334** 135571
[86] Liu W, Lee S W, Lin D, Shi F, Wang S, Sendek A D and Cui Y 2017 Enhancing ionic conductivity in composite polymer electrolytes with well-aligned ceramic nanowires *Nat. Energy* **2** 1–7
[87] Inaguma Y, Liquan C, Itoh M, Nakamura T, Uchida T, Ikuta H and Wakihara M 1993 High ionic conductivity in lithium lanthanum titanate *Solid State Commun.* **86** 689–93
[88] Fu K *et al* 2017 Three-dimensional bilayer garnet solid electrolyte based high energy density lithium metal–sulfur batteries *Energy Environ. Sci.* **10** 1568–75
[89] Fan W, Zhang L and Liu T 2018 Multifunctional second barrier layers for lithium–sulfur batteries *Mater. Chem. Front.* **2** 235–52
[90] Kannan S K, Hareendrakrishnakumar H, Joseph J and Joseph M G 2022 Synergistic restriction to polysulfides by a carbon nanotube/manganese sulfide-decorated separator for advanced lithium–sulfur batteries *Energy Fuels* **36** 8460–70
[91] Su Y S and Manthiram A 2012 Lithium–sulphur batteries with a microporous carbon paper as a bifunctional interlayer *Nat. Commun.* **3** 1–6
[92] Chung S H and Manthiram A 2014 High-performance Li–S batteries with an ultra-lightweight MWCNT-coated separator *J. Phys. Chem. Lett.* **5** 1978–83
[93] Kim A, Oh S H, Adhikari A, Sathe B R, Kumar S and Patel R 2023 Recent advances in modified commercial separators for lithium–sulfur batteries *J. Mater. Chem.* A*11 7833–66*
[94] Rana M, Li M, Huang X, Luo B, Gentle I and Knibbe R 2019 Recent advances in separators to mitigate technical challenges associated with rechargeable lithium sulfur batteries *J. Mater. Chem.* A **7** 6596–615
[95] Li C, Liu R, Xiao Y, Cao F and Zhang H 2021 Recent progress of separators in lithium–sulfur batteries *Energy Storage Mater.* **40** 439–60
[96] Tian Y-W, Zhang Y-J, Wu L, Dong W-D, Huang R, Dong P-Y, Yan M, Liu J, Mohamed H S H and Chen L-H 2023 Bifunctional separator with ultra-lightweight MnO_2 coating for highly stable lithium–sulfur batteries *ACS Appl. Mater. Interfaces* **15** 6877–87

[97] Ghazi Z A, He X, Khattak A M, Khan N A, Liang B, Iqbal A, Wang J, Sin H, Li L and Tang Z 2017 MoS_2/Celgard separator as efficient polysulfide barrier for long-life lithium–sulfur batteries *Adv. Mater.* **29** 1606817
[98] Feng J, Li Y, Yuan J, Zhao Y, Zhang J, Wang F, Tang J and Song J 2022 Energy-saving synthesis of functional CoS_2/RGO interlayer with enhanced conversion kinetics for high-performance lithium–sulfur batteries *Front. Chem.* **9** 830485
[99] Li Z, Zhang F, Tang L, Tao Y, Chen H, Pu X, Xu Q, Liu H, Wang Y and Xia Y 2020 High areal loading and long-life cycle stability of lithium–sulfur batteries achieved by a dual-function ZnS-modified separator *Chem. Eng. J.* **390** 124653
[100] Ali S, Waqas M, Jing X, Chen N, Chen D, Xiong J and He W 2018 Carbon–tungsten disulfide composite bilayer separator for high-performance lithium–sulfur batteries *ACS Appl. Mater. Interfaces* **10** 39417–21
[101] Kannan S K, Joseph J and Joseph M G 2023 Review and perspectives on advanced binder designs incorporating multifunctionalities for lithium–sulfur batteries *Energy Fuels* **37** 6302–22
[102] Shafique A, Rangasamy V S, Vanhulsel A, Safari M, Van Bael M K, Hardy A and Sallard S 2020 The impact of polymeric binder on the morphology and performances of sulfur electrodes in lithium–sulfur batteries *Electrochim. Acta* **360** 136993
[103] Wang H, Zheng P, Yi H, Wang Y, Yang Z, Lei Z, Chen Y, Deng Y, Wang C and Yang Y 2020 Low-cost and environmentally friendly biopolymer binders for Li–S batteries *Macromolecules* **53** 8539–47
[104] Yuan H, Huang J Q, Peng H J, Titirici M M, Xiang R, Chen R, Liu Q and Zhang Q 2018 A review of functional binders in lithium–sulfur batteries *Adv. Energy Mater.* **8** 1802107
[105] Fourche G 1995 An overview of the basic aspects of polymer adhesion. Part I: fundamentals *Polym. Eng. Sci.* **35** 957–67
[106] Akhtar N, Shao H, Ai F, Guan Y, Peng Q, Zhang H, Wang W, Wang A, Jiang B and Huang Y 2018 Gelatin-polyethylenimine composite as a functional binder for highly stable lithium–sulfur batteries *Electrochim. Acta* **282** 758–66
[107] Waluś S, Robba A, Bouchet R, Barchasz C and Alloin F 2016 Influence of the binder and preparation process on the positive electrode electrochemical response and Li/S system performances *Electrochim. Acta* **210** 492–501
[108] Yan L, Gao X, Thomas J P, Ngai J, Altounian H, Leung K T, Meng Y and Li Y 2018 Ionically cross-linked PEDOT: PSS as a multifunctional conductive binder for high-performance lithium–sulfur batteries *Sustain. Energy Fuels* **2** 1574–81
[109] Zeng F, Wang W, Wang A, Yuan K, Jin Z and Yang Y S 2015 Multidimensional polycation β-cyclodextrin polymer as an effective aqueous binder for high sulfur loading cathode in lithium–sulfur batteries *ACS Appl. Mater. Interfaces* **7** 26257–65
[110] Xu G, Yan Q B, Kushima A, Zhang X, Pan J and Li J 2017 Conductive graphene oxide-polyacrylic acid (GOPAA) binder for lithium–sulfur battery *Nano Energy* **31** 568–74
[111] Zuo C and Ding L 2017 Modified PEDOT layer makes a 1.52 V Voc for perovskite/PCBM solar cells *Adv. Energy Mater.* **7** 1601193
[112] Gao J, Sun C, Xu L, Chen J, Wang C, Guo D and Chen H 2018 Lithiated nafion as polymer electrolyte for solid-state lithium sulfur batteries using carbon–sulfur composite cathode *J. Power Sources* **382** 179–89

[113] Li G, Cai W, Liu B and Li Z 2015 A multi functional binder with Lithium ion conductive polymer and polysulfide absorbents to improve cyclability of lithium–sulfur batteries *J. Power Sources* **294** 187–92
[114] Zeng F L, Li N, Shen Y Q, Zhou X Y, Jin Z Q, Yuan N Y, Ding J N, Wang A B, Wang W K and Yang Y S 2019 Improve the electrodeposition of sulfur and lithium sulfide in lithium–sulfur batteries with a comb-like ion-conductive organo-polysulfide polymer binder *Energy Storage Mater.* **18** 190–8
[115] Li G, Wang C, Cai W, Lin Z, Li Z and Zhang S 2016 The dual actions of modified polybenzimidazole in taming the polysulfide shuttle for long-life lithium–sulfur batteries *NPG Asia Mater.* **8** e317
[116] Jiao Y, Chen W, Lei T, Dai L, Chen B, Wu C and Xiong J 2017 A novel polar copolymer design as a multifunctional binder for strong affinity of polysulfides in lithium–sulfur batteries *Nanoscale Res. Lett.* **12** 1–8
[117] Wang H, Yang Y, Zheng P, Wang Y, Ng S W, Chen Y, Deng Y, Zheng Z and Wang C 2020 Water-based phytic acid-crosslinked supramolecular binders for lithium–sulfur batteries *Chem. Eng. J.* **395** 124981

IOP Publishing

Recent Advances in Materials for Energy Harvesting and Storage

Suresh C Pillai, Daniel M Mulvihill and Aswathy Babu

Chapter 8

Recent trends in materials for sodium-ion batteries

K S Krishnendu, Tanaya Dutta and J Mary Gladis

As an alternative to LIBs, sodium-ion batteries (SIBs) are emerging as a sustainable and efficient energy storage platform due to their high energy density, abundance, cost-effectiveness, and similar electrochemical features, such as a suitable redox potential close to that of lithium. The commercialization of high-performance SIBs will need potential electrode materials and appropriate electrolyte systems. Compared with LIBs, SIBs have lower energy density (gravimetric and volumetric), lower specific capacity, and lower cyclability. However, this can be alleviated by developing high-capacity electrode materials and suitable electrolyte choices for SIBs. This chapter discusses recent developments in the electrode materials, electrolytes, and separators used in SIBs. Insertion-type, conversion-type, and alloy-type materials are the essential classes of anode materials used in SIBs. The cathode materials mainly comprise layered metal oxides, polyanionic compounds, and Prussian blue analogs. Other than electrode materials, electrolytes also play a vital role in the performance of SIBs. Liquid electrolytes mainly consist of three parts: salts, solvents, and additives. The safety issues of liquid electrolytes can be overcome by using solid electrolytes, which also provide better energy density. Like other cell components, separators are critical for the fabrication of all batteries. As they inhibit physical contact between the cathode and the anode, they ensure the safety of the battery. Although polymeric compounds such as polyolefins (polyethylene (PE) and polypropylene (PP)) are the most commonly used separators, functionalized separators are required to enhance the electrochemical performance of SIBs. Advancements in state-of-the-art electrodes, electrolytes, separators, and binders and the introduction of novel materials will boost the performance of SIBs.

doi:10.1088/978-0-7503-5749-4ch8

8.1 Introduction

Sony made the pioneering move of introducing lithium-ion batteries (LIBs) in 1991, marking their initial market entry. Since then, LIBs have become the most sophisticated and viable option for energy storage devices. Introducing LIBs to the market has decreased the dependence on fossil fuels, thereby decreasing the political instability regarding fossil fuel production. Li-ion technology outperforms the competing technologies such as nickel (Ni)–metal hydride, Ni–cadmium (Cd), and lead (Pb)–acid. Li-ion technology is considered an excellent choice for powering the next generation of electric vehicles (EVs) [1]. Even though LIBs can meet the requirements of the systems involved in energy storage, they raise concerns such as the availability of Li metal (a component element of LIBs found in the Earth's crust), high cost, and safety. Despite the global increase in total lithium (Li) consumption, the available mineable lithium sources can sustain this demand for more than 50 years [2]. However, rechargeable sodium-ion (Na^+) batteries (SIBs) can replace Li-ion battery technology based on material abundance and standard electrode potential. Sodium, the second lightest element and smallest alkali metal after lithium, has an advantage as a replacement for lithium. The high cost and environmental impacts of producing elements such as cobalt, copper, and nickel, essential in LIBs, paved the way for developing a new technology. The availability of inexpensive sodium resources has advanced the growth of SIB technology as a replacement for LIBs. The battery components and energy storage mechanisms of SIBs are similar to those of LIBs. However, LIBs have superior energy density to that of SIBs [3]. The replacement of LIBs by SIBs is still an open question, but the research developments related to SIBs may cause visible changes in the energy sector. In this chapter, we comprehensively summarize the developments in the recent and important research articles related to SIB anodes, cathodes, electrolytes, binders, and additives.

8.2 Cell chemistry: the working mechanism of Na-ion batteries

The fundamental working principle of SIBs is similar to that of LIBs. AN SIB comprises a positive electrode (cathode) and a negative electrode (anode), separated by a separator. This separator serves as an electrical insulator while allowing ionic conduction. In addition, an SIB contains an electrolyte in which a sodium salt is dissolved in a suitable solvent mixture, facilitating the flow of ions between the two electrodes. The electrolyte contacts various cell components, including current collectors, the cell casing, binders, and so on, to enhance cell performance. Figure 8.1 shows a schematic view of a Na-ion cell. The structure of the cell can be compared with that of Li-ion cells. The electrode reaction for a carbon-based anode and a layered transition-metal oxide cathode ($NaMO_2$, where M=Fe, Ni, Mn, Co, etc.) is given by:

$$Na_xC_6 + Na_{1-x}MO_2 \Leftrightarrow NaMO_2 + C_6.$$

Typically, a sodium-ion cell is assembled in the discharged state. The working principle of SIBs lies in the reversible movement of Na ions between the anode and

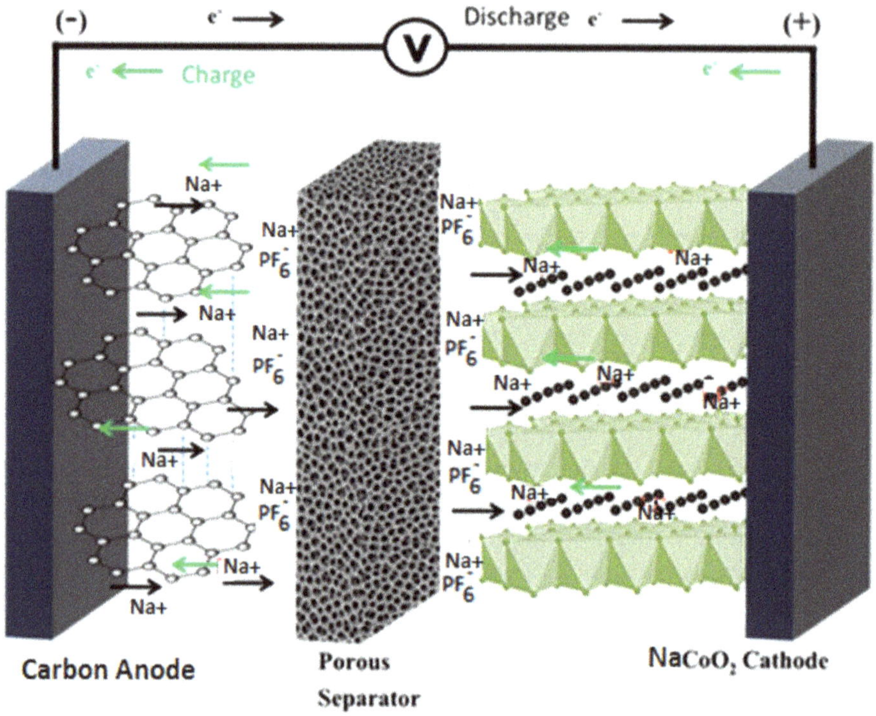

Figure 8.1. A schematic representation of a Na-ion cell (reprinted with permission from [4], Copyright (2020), American Chemical Society).

cathode while maintaining an electron flow through an external circuit. The use of hard carbon (HC) as the anode and a transition-metal oxide as the cathode increase the reversibility, stability, and capacity retention over continuous cycling processes. The charging of SIBs involves the transport of Na ions from the cathode side towards the anode side through the electrolyte medium (usually a sodium-ion conductive medium). The HC anode intercalates the Na ions between its layers. During discharging, the Na ions stored in the HC anode are released and intercalated back into the cathode structure.

8.3 Na-ion battery components

A typical SIB consists of a negative electrode (anode) that can be made from HC, metal oxides, phosphorous, etc. Positive electrode (cathode) materials include layered metal oxides, polyanion-based compounds, Prussian blue and its analogs, etc. Typically, electrolytes that include sodium-containing salts such as sodium tetrafluoroborate ($NaBF_4$), sodium trifluoromethanesulfonate (NaTF), sodium trifluoromethanesulfinate (NaTFS), sodium bis(fluorosulfonyl)imide (NaFSI), sodium hexafluorophosphate ($NaPF_6$), or sodium perchlorate ($NaClO_4$) can be dissolved in either organic or aqueous solvents. The electrode compartments are separated by a porous separator, which may be made from PE, a ceramic-coated material, or glass fiber. Binders and additives are also two essential constituents that decide the overall efficiency of SIBs. The detailed development and importance of each component are discussed in the following sections.

8.3.1 Anode materials

The anode is the negative electrode and an integral cell component of SIBs. So, the design and development of high-performance anodes can lead to advances in Na-ion technology. Graphite, the traditional anode of LIBs, cannot intercalate Na ions into its graphene layers due to their larger ionic radius, and Si-based anode materials also suffer practical challenges. The challenges mentioned above hamper the progress of SIBs to a large extent. The commercialization of SIBs is still difficult because the anode material must accommodate Na ions whose radii are 1.02 Å. A material with large interstitial spaces is required [5]. Sodium is unsuitable as an anode material because of its high risk of dendrite formation, high reactivity with electrolytes, and low melting point of 97.7 °C, which can result in safety problems. In SIBs, based on the sodiation/desodiation processes involved, reactions in anode materials are divided into three primary groups: (1) reactions in insertion-based anode materials, (2) reactions in conversion-based anode materials, and (3) alloying reactions in anode materials. Within the group of insertion-type anode materials, it is essential to consider titanium-based oxides and carbonaceous materials [6]. Recent studies have broadened the application of conversion reactions to include transition-metal sulfides (TMSs) and transition-metal oxides (TMOs). With electron-conductive carbon forms, the alloying reaction involves sodium-M (M=metal) alloying with compounds from groups 14 and 15 (figure 8.2).

8.3.1.1 Insertion materials

Based on the insertion reaction mechanism, the insertion-type materials are broadly divided into carbon-based materials and titanium oxide compounds.

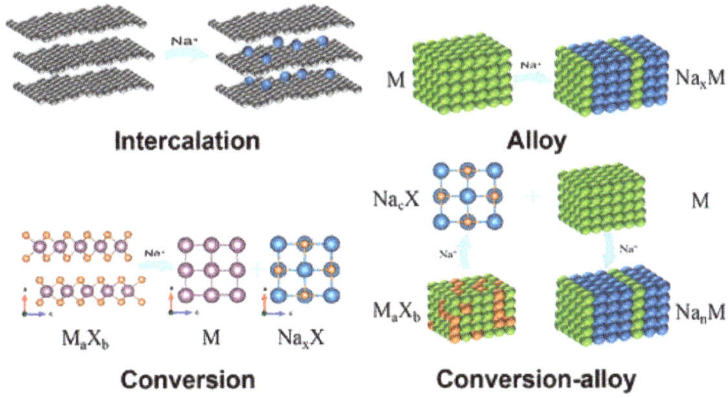

Figure 8.2. The reaction mechanisms of the anode materials of SIBs (reprinted with permission from [7], Copyright (2023), American Chemical Society).

8.3.1.1.1 Carbon-based anode materials

Graphite: the natural availability and low cost of carbon materials make them suitable for use as anode materials in SIBs. Even though graphite is chemically and thermally stable when exposed to sodium, the intercalation of Na ions is difficult due to the weak interaction between graphite and sodium. Di Vincenzo *et al* studied the formation of Na-intercalated graphite using theoretical studies, which showed that compared to the lithium-ion bond, the Na–C bond is very weak and interacts less. The small interlayer spacing of graphite anodes hinders Na ions' facile intercalation/de-intercalation into/from graphite [8]. Wen *et al* explored the possibilities of expanded graphite as an efficient anode material for SIBs. Expanded graphite has an enlarged lattice distance of 4.3 Å but retains a long-range-ordered layered structure similar to that of graphite. A cell constructed using expanded graphite delivered a reversible capacity of 284 mAh g^{-1} at a current density of 20 mAh g^{-1}. The reversible expansion and contraction of the interlayers during the insertion and removal of Na ions were studied using cyclic voltammetry and *in situ* transmission electron microscopy (TEM) [9]. Tian *et al* designed a series of high-defect graphite-type anodes for SIBs and produced them using a ball milling process. The capacity of the high-defect graphite was 12.7 mAh g^{-1} at 0.1 A g^{-1} and its capacity retention was 8000 cycles at 5 A g^{-1} (figure 8.3) [10].

Cabello *et al* synthesized expanded graphite by direct thermal expansion carried out at 1400 °C. He also compared the material with Hummers' graphene oxide (GO) and expanded graphite prepared by treating them at three different temperatures (500 °C, 1000 °C, and 1300 °C). Among these methods, thermally expanded graphite achieved the highest capacity, reaching 1205 mAh g^{-1}, surpassing the performance of other materials [11].

Hard Carbon: HC, a so-called non-graphitic carbon, consists of a random arrangement of graphene sheets, resulting in a nanoporous structure. Its disordered structure and greater interlayer spacing make HC a suitable anode choice for SIBs. Based on DFT calculations, Stevens and Dahn *et al* proposed that the Na insertion mechanism in HC occurs in two domains. Two plateaus indicate: (i) the intercalation (sodiation) of the Na ions into graphene sheets and (ii) the filling of the nanopores of the disordered carbon framework by Na ions. This mechanism of Na-

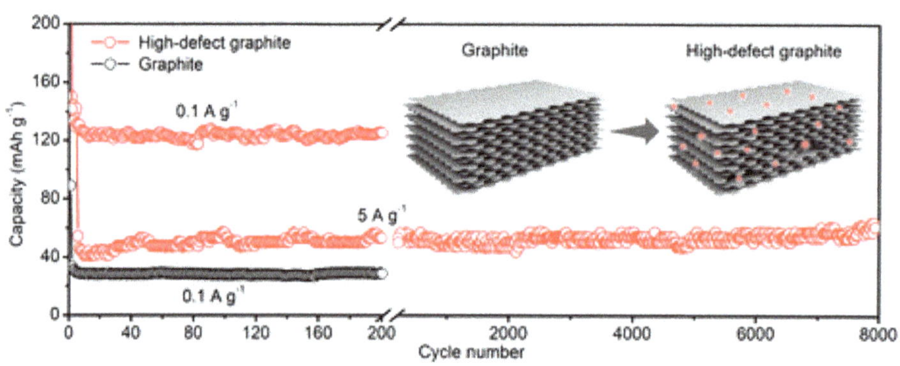

Figure 8.3. A comparison of the capacities of graphite and high-defect graphite (reprinted with permission from [10], Copyright (2023), American Chemical Society).

ion intercalation to HC was first analyzed by wide-angle *in situ* x-ray scattering. Recently, it was again confirmed by solid-state Na nuclear magnetic resonance (NMR) [12]. To understand the structural change that takes place during Na-ion insertion, Komaba *et al* demonstrated and analyzed the presence of extended interlayer spacing of the parallel graphene in HC [13]. Bommier *et al* proposed a multistep process for the intercalation of Na ions in an HC lattice, which involves: (1) defective sites within the slope-voltage region that absorbs Na ions, (2) the incorporation of Na ions into the HC lattice, and (3) a plateau region involving the absorption of Na ions at the surface's pores [14]. The reversibility of HC depends on the particle size, additives, electrolytes, and vacancy defects. Theoretical investigations (DFT calculations and *ab initio* calculations) have yielded a better understanding of the fundamental correlation between the formation of a solid electrolyte interphase (SEI) layer and the electrochemical characteristics of HC.

Graphene: graphene, a 2D carbon material, is a good choice as an anode material for SIBs and LIBs due to its extensive surface area, chemical stability, and enhanced electrical conductivity. Studies have shown that graphene and GO composites can improve Li- and Na-ion storage capacity. Wang *et al* used Hummers' method to prepare reduced GO, which exhibited a reversible capacity of 217.2 mAh g^{-1} at 40 mA g^{-1} and 95.6 mAh g^{-1} at a current density of 1000 mA g^{-1} [15]. Ding *et al* studied the mechanism of sodium-ion storage in graphene at various carbonization temperatures by preparing few-layer graphene at different temperatures. The material had a large interlayer spacing compared with that of graphite. At a carbonization temperature of 1100 °C, followed by activation at 300 °C, it exhibited a pseudo-graphitic structure with an interlayer spacing of 0.388 nm, showing faster sodium-ion diffusion and enhanced sodium-ion storage properties [16]. Datta *et al* reported that the absorption of sodium atoms in graphene sheets increased with divacancy and Stone–Wales defects. They reported that the assembled cell delivered a capacity of 1450 mAh g^{-1} for the maximum possible divacancy defect and a capacity of 1071 mAh $^{-1}$g for the Stone–Wales defect [17].

Heteroatom doping: The electrochemical characteristics of carbonaceous materials can be enhanced by doping them with heteroatoms (N, B, S, and P). The functionalization and defect sites induced by heteroatom doping of the carbon surface help to absorb more Na ions and improve the electrode–electrolyte surface interaction. Kale *et al* used the electrospinning technique to prepare self-supporting pseudo-graphitic nanofibers modified with nitrogen. Free-standing carbon nanofibers (CNFs) (in which a binder and a current collector were not required) exhibited capacities of 210 and 87 mAh g^{-1} at current densities of 20 and 1600 mA g^{-1}, respectively. This anode material showed outstanding performance for sodium storage in terms of cycling stability and rate capability (figure 8.4) [18].

Li *et al* prepared a disordered carbon modified with sulfur, resulting in a distinctive 3D structure resembling a coral-like shape. A cell assembled using this carbon material exhibited an outstanding reversible capacity of 516 mAh g^{-1} and an excellent cyclability of 1000 cycles [19]. Yang *et al* synthesized a carbon material doped with S and P through a single-step plasma-in-liquid procedure. This material, which had many macropores, exhibited a cycling lifespan of 3000 cycles at a current density of

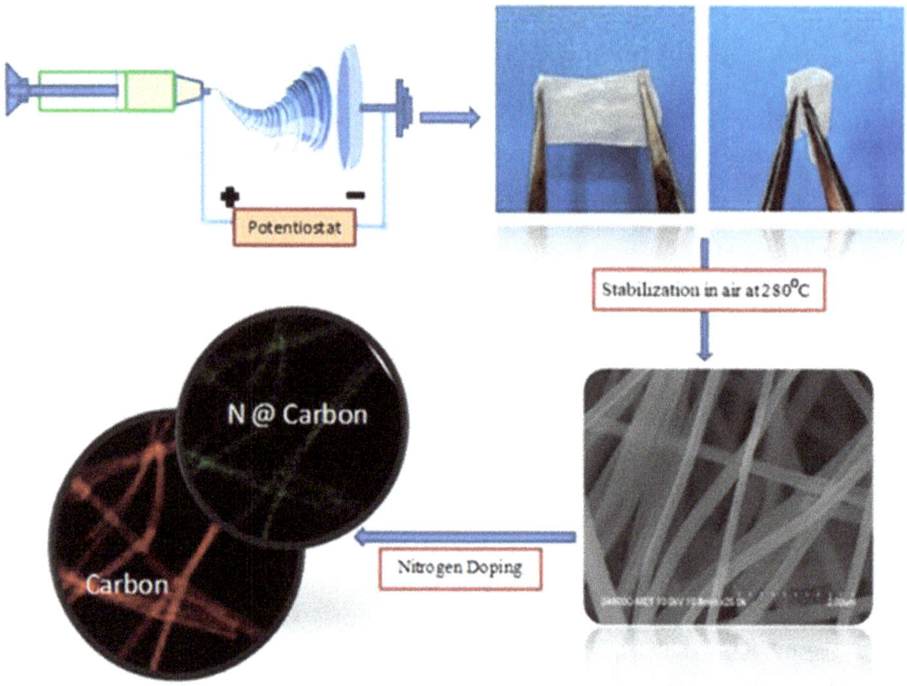

Figure 8.4. The use of the electrospinning method to synthesize self-supporting pseudo-graphitic nanofibers decorated with nitrogen (reprinted with permission from [18], Copyright (2021), American Chemical Society).

10 A g^{-1} while maintaining a high reversible capacity exceeding 125 mAh g^{-1} [20]. Li *et al* discussed a metal–organic framework (MOF)-derived method for synthesizing a porous carbon polyhedron co-doped with nitrogen and oxygen in which Na ions could adsorbed at the active sites. At 0.1 A g^{-1}, the material displayed a capacity of 313 mAh g^{-1} after 100 cycles; it exhibited a cycling stability of 228 mAh g^{-1} at 1 A g^{-1} after 2000 cycles [21].

8.3.1.2 Conversion materials

Conversion materials chemically convert some atomic species into a new compound, as opposed to the reversible shuttling of metal atoms in the host matrix. Along with conversion reactions, these materials can undergo insertion or alloying reactions, depending on the transition metal present in the compounds. However, the storage of multiple sodium atoms leads to drastic volume changes in these materials, which can damage electrodes and result in gradual capacity fading. These challenges can be surmounted by introducing carbon-based materials or performing electrolyte modification to obtain a stable SEI layer. The following section deals with recent advances in conversion-based anode materials.

8.3.1.2.1 Anode materials based on transition-metal oxides

Alcantara *et al* were the first to demonstrate the concept of conversion-type materials by preparing a mixed TMO as an anode for SIBs. The $NiCo_2O_4$ electrode material

Table 8.1. Anode materials based on TMOs.

Transition-metal oxide	Current density ($mA\ g^{-1}$ or $A\ g^{-1}$)	Specific capacity ($mAh\ g^{-1}$)	Reference
γ-Fe_2O_3 @ C nanocomposites	200 $mA\ g^{-1}$	740	[23]
(Fe_3O_4-QD/CN)	2 $A\ g^{-1}$	1030	[24]
Mesoporous cobalt oxide	90 $mA\ g^{-1}$	707	[25]
SnO microspheres	20 $mA\ g^{-1}$	580	[26]
CuO/C	200 $mA\ g^{-1}$	402	[27]

prepared by the thermal decomposition of oxalate precursors showed a capacity of 300 $mAh\ g^{-1}$ [22]. Zhang *et al* prepared a three-dimensional (3D) porous γ- Fe_2O_3@C nanocomposite by aerosol spray pyrolysis technology and studied it as a viable anode material for SIBs. The nanocomposite anode material accommodated significant volume changes and facilitated easy transfer of electrons and ions during continuous cycling. The cell exhibited high performance with excellent rate capability and cyclability, which made the nanocomposite anode material a suitable anode for SIBs [23]. Liu *et al* prepared carbon nanosheets (CNs) decorated with Fe_3O_4 quantum dots (Fe_3O_4-QD/CN), which displayed a reversible capacity of 416 $mAh\ g^{-1}$ at 0.1 $A\ g^{-1}$ for 1000 cycles and showed a cycle life retention of 70% [24].

Yang *et al* synthesized highly ordered mesoporous cobalt oxide with superior ion transport pathways, resulting in an initial capacity of 707 $mAh\ g^{-1}$ at 90 $mA\ g^{-1}$. *Ex situ* characterization techniques confirmed the occurrence of a conversion reaction during sodium uptake/extraction [25]. Lu *et al* reported their findings for the synthesis and electrochemical response of Sn-oxide-based (SnO, SnO_2, and SnO_2/C) anodes for SIBs. SnO exhibits excellent performance with low charge transfer resistance, good cycle life, and superior conductivity. The high theoretical capacity, chemical stability, and abundance of tin oxide materials makes them exemplary anode materials for SIBs [26]. Klein *et al* studied the efficacy of Cu_2O anodes for SIBs, and their assembled cell displayed a high capacity of 600 $mAh\ g^{-1}$ at 0.1 C. Lu *et al* synthesized CuO/C spheres with a micro–nanostructure, which showed a capacity of 402 $mAh\ g^{-1}$ at a current density of 200 $mA\ g^{-1}$ with better rate capabilities [27]. Notable anode materials based on TMOs are listed in table 8.1.

8.3.1.2.2 Anode materials based on transition-metal sulfides

The weaker M–S bonds of metal sufides result in higher conversion kinetics than those of metal oxides, leading to greater initial coulombic efficiency, low volume expansion, and improved reversibility. Several metal sulfides based on Fe, Co, Ni, Sn, and Mo have been studied for use as anode materials in SIBs. Hu *et al* prepared a MoS_2 nanoflower-like anode material for SIBs. At 1 and 10 $A\ g^{-1}$, their electrode exhibited discharge capacities of 300 and 195 $mAh\ g^{-1}$, respectively [28]. A study by Wang *et al* explored anode materials obtained from MOFs, specifically TMS composites, showcasing remarkable performance in their specific capacity and cycling performance [29]. Hou *et al* employed an amide linkage to bind TMSs to carbon nanotubes

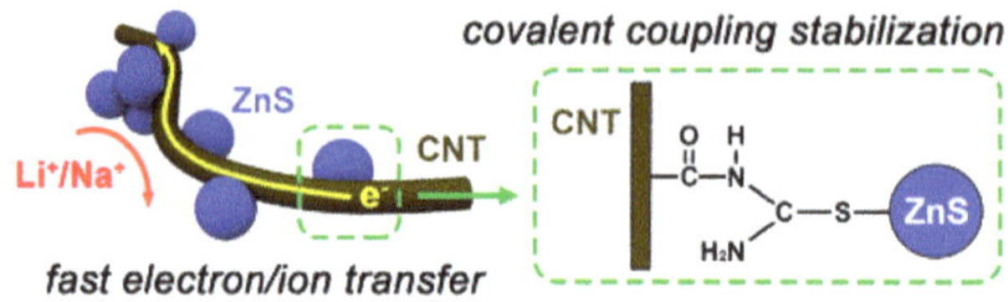

Figure 8.5. Covalent coupling stabilization for TMS-based anode materials (reprinted with permission from [30], Copyright (2021), American Chemical Society).

(CNTs) as a part of a covalent coupling strategy. The covalently coupled ZnS/CNT composite exhibited a significant reversible capacity of 314 mAh g^{-1} at 5 A g^{-1} after 500 cycles. This indicates that the covalent coupling strategy holds promise for use in the design of high-performance anode materials (figure 8.5) [30].

Zhu *et al* proposed a facile method for preparing a dual carbon-protected metal sulfide by co-precipitation and thermal pyrolysis. Their N-doped carbon-coated iron sulfides wrapped in graphene sheets displayed a high reversible capacity of 440 mAh g^{-1} at 0.05 and at 0.2 A g^{-1}; the retained capacity after cycling was 95.8% [31]. Wang *et al* studied an improvement in Na-ion storage performance by forming a nanocomposite of WS_2 and conductive carbon and then modifying the composite with nitrogen. A cell produced using their electrode displayed a reversible capacity of 360 mAh g^{-1} at 100 mA g^{-1}, which was higher than the capacities of WS_2 and a WS_2/carbon composite. Modifying the nanocomposite with nitrogen enhanced the electrode–electrolyte interfacial reaction in the WS_2 nanosheets in a low-voltage regime [32]. Deng *et al* introduced carbon as a buffer material to accommodate the volume expansion of conversion-type materials during the sodiation and desodiation processes. In this study, natural stibnite ore (Sb_2S_3) modified with carbon sheets served as the anode material and displayed significantly improved characteristics, which included a notable increase in reversible capacity. This composite exhibited outstanding electrochemical performance compared to the performance of pristine natural stibnite ore in terms of cycling stability and rate capability [33]. Yu *et al* produced a highly conductive reduced-GO-coated flower-like SnS_2 composite by a hydrothermal method to restrict volume expansion. The graphene prevented the clustering or agglomeration of SnS_2 nanoparticles, enabling a cell to achieve a specific capacity of 649 milliampere-hours per gram (mAh g^{-1}) and a discharge capacity of 509 mAh g^{-1} after 100 cycles [34].

8.3.1.2.3 Anode materials based on transition-metal phosphides

The versatile structural and electronic properties of transition-metal phosphides make them promising anode materials for use in SIBs. Nam *et al* synthesized binary zinc phosphides (ZnP_2 and Zn_3P_2) using a solid-state heat treatment process and studied their Na-ion storage characteristics. They made a carbon-modified ZnP_2-based composite based on Na reactivity for better electrochemical performance. A cell based on their composite exhibited a reversible capacity of 883 mAh g^{-1} after 130 cycles with no capacity deterioration (figure 8.6) [35].

Liang *et al* synthesized TMPs by taking metal hydroxide and glucose as precursors to convert metal hydroxides to phosphides. In a plasma environment,

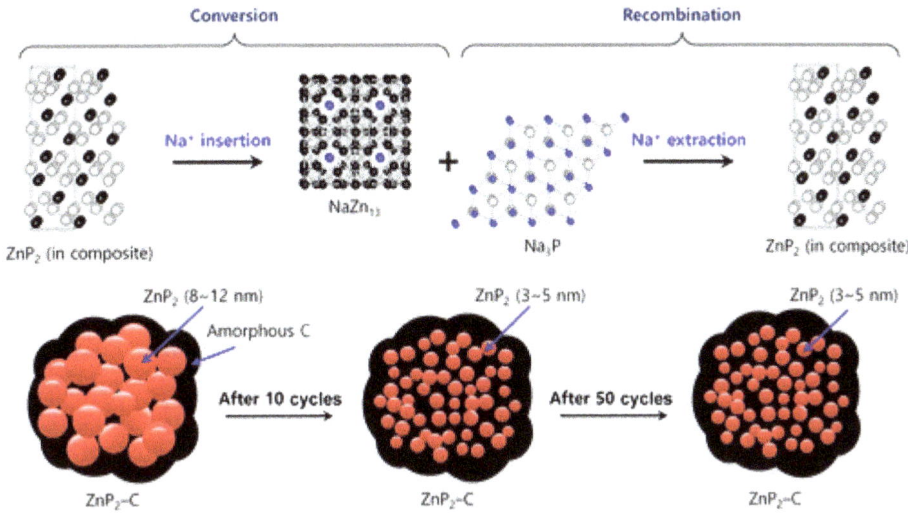

Figure 8.6. A schematic representation of the crystallite size variations of ZnP_2 in the ZnP_2–C composite during ongoing Na cycling (reprinted with permission from [35], Copyright (2020) American Chemical Society).

glucose decomposition formed carbon shells on the phosphide particles. The Ni_2P@C-based material exhibited a high capacity of 693 mAh g^{-1} after 50 cycles at 100 mA g^{-1} [36]. Lu *et al* prepared selenium phosphide (Se_4P_4) by a facile mechanical milling technique; due to the synergistic effect of Se and P, it improved the electrochemical performance of SIBs. This process entailed the initial generation of $Na_xSe_4P_4$ from Se_4P_4. The electrochemical results for this material showed a good reversible capacity of 1048 mAh g^{-1} at 50 mA g^{-1} and excellent cyclability and rate capability features [37].

Kim *et al* introduced Sn_4P_3-based anode materials that displayed a high capacity of 718 mAh g^{-1} with a minimal loss in capacity. Further studies showed that Na_3P and Sn undergo an alloying reaction in addition to the conversion reaction, forming $Na_{15}Sn_4$ [38]. Li *et al* synthesized nanosized CoP particles by ball milling, and their assembled cell delivered a high specific capacity of 770 mAh g^{-1}. Their cell showed an improved rate capability and low-voltage polarization. The reaction of P with Na in CoP was further confirmed by *ex situ* x-ray photoelectron spectroscopy (XPS) and STEM characterization techniques [39].

8.3.1.3 Alloying materials

The unique electrochemical characteristics of alloying materials, which allow them to accommodate Na ions at a very low operating voltage (below 1 V), make them efficient anode materials in SIBs. High-capacity alloy materials based on group 14 and 15 anode materials are widely employed for SIBS applications. However, the increased size of Na ions can lead to significant volume changes during the alloying and de-alloying reactions that can exert mechanical stress on the active particles in anode materials. This section discusses the alloy reaction mechanism of Na ions in group 14 and elements in group 15.

In terms of natural abundance and specific capacity, silicon (Si) is another suitable anode material for SIBs. Jung *et al* conducted an atomic-level analysis to investigate the suitability of silicon (Si) anodes in SIBs. Amorphous silicon is preferred over crystalline silicon for alloying reactions, since it can deliver a capacity of 725 mAh g^{-1}. The volume expansion exhibited by the amorphous $Na_{0.76}Si$ phase is 114%, and it has excellent Na-ion mobility. Germanium, which has a chemistry similar to that of silicon, can form a bond with one sodium atom, delivering a theoretical capacity of 369 mAh g^{-1} [40]. Using DFT and *ab initio* molecular dynamics (AIMD) calculations, Sharma *et al* studied the kinetics of the sodium insertion and desodiation processes by focusing on $Na_{64}Ge_{64}$. The experimental data for the correlation between the intercalation potential and the capacity of the intermediate equilibrium structures aligned with the empirical results [41]. Li *et al* examined phase transformations and their impact on performance across various Sn-based systems. This investigation encompassed metallic Sn and its SnO_2, SnS_2/SnS, and Sn_4P_3 alloys. A charge storage capacity of 847 mAh g^{-1} was obtained by forming an alloy of Sn with up to 3.75 Na atoms per formula unit. Sodiation and desodiation in metallic Sn progress through a sequence of both metastable amorphous and crystalline structures [42]. Sayed *et al* prepared anodes that had nanoscale multilayers of silicon and tin films, aiming to benefit from the high theoretical capacities of Si (954 mAh g^{-1} (NaSi)) and Sn (847 mAh g^{-1} ($Na_{15}Sn_4$)). A multilayer configuration of Si and Sn layers 3 nm thick demonstrated excellent coulombic efficiency. Even after 100 cycles, 97% of the initial capacity is maintained, compared to the mere 7% retention observed for Sn film. Incorporating Si as an alloying element stabilized grain growth and prevented pulverization and surface chemistry modification of the anodes. This significantly impacted SEI layer formation (figure 8.7) [43].

The elements in Group 15, such as P, As, Sb, and Bi, serve as alloying elements with sodium and are considered viable candidates for anodes in SIBs. However,

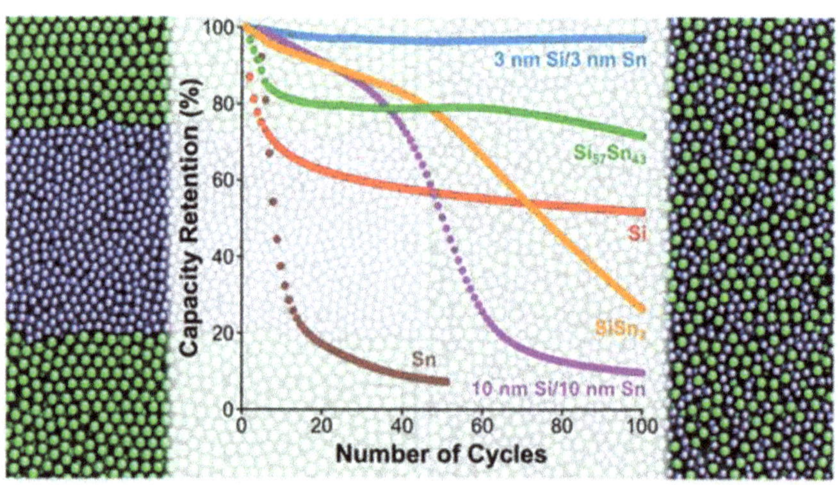

Figure 8.7. The capacities retained by different combinations of Si and Sn (reprinted with permission from [43], Copyright (2020), American Chemical Society).

volume expansion and contraction during cycling is also seen in these anode materials. The fully sodiated state of antimony (Na_3Sb) indicates it can deliver a capacity of 660 mAh g^{-1}. Hu *et al* prepared an antimony/multilayer graphene composite in which antimony was uniformly distributed on graphene by a vapor deposition technique. The synergistic effect of both antimony and graphene improved sodium-ion diffusion and electron transport, stabilizing the SEI layer during the volumetric expansion/contraction of antimony during cycling. At 100 mA g^{-1}, the composite exhibited a value of 452 mAh g^{-1} at 100 mA g^{-1} and better efficiency in cycling and rate capability and retained a capacity of 90% even after 200 cycles [44]. The lower atomic weight of P makes it an efficient anode material choice for SIBs. Here, a theoretical capacity of 2596 mAh g^{-1} is achieved. The electrochemical reaction of phosphorus with sodium forms Na_3P. The allotropes of phosphorous (white, red, and black) have versatile features such as an amorphous nature and an orthorhombic structure that make them suitable anode materials for SIBs. However, these materials suffer from a huge volume expansion of around 490% upon sodiation/desodiation. Song *et al* introduced a phosphorous–carbon nanotube (P-CNT) hybrid anode prepared through a ball milling process. A polymer binder and cell based on this material displayed a capacity of 1586.2 mAh g^{-1} with ~99% cycling efficiency [45]. Yang *et al* used the aerosol spray pyrolysis method to fabricate Bi nanoparticles incorporated within carbon spheres (Bi@C). The Bi@C microspheres delivered a high capacity of 123.5 mAh g^{-1} at 100 mA g^{-1} after 100 cycles. Their specific capacities were 299 and 90 mAh g^{-1} at 0.1 and 2 A g^{-1}, respectively [46]. Insertion-type, conversion-type, and alloying-type materials are the dominant classes of anode materials in SIBs. Strategies are urgently needed to enhance the capacity of low-volume-change insertion-type anodes and also impede the volume expansion of high-capacity alloying-type anodes, which are common in SIB technology. The functionalization and advancement of these materials with suitable dopants and composite host materials will enhance the overall electrochemical performance of SIBs.

8.3.2 Cathode materials

The high specific surface area, facile diffusion characteristics, and versatile structural features of the layered transition-metal oxides (M = transition metals) make them potential cathode materials for SIBs. Delmas *et al* classified layered transition-metal oxides into two forms based on their structures: 'Pn' and 'On' [47–49]. It is crucial to develop high-voltage, long-life cathodes for SIBs to surmount raw material and cost constraints. The rich natural abundance, high voltage and robust polyanion frameworks suitable for Na-ion storage make the alluaudite-type structured iron based sulfates an excellent cathode material for SIBs [50]. The slow kinetics and electrode–electrolyte interfacial instability of active positive-electrode materials at high voltage hinder the practical development of SIBs. Zhang *et al* proposed a way to mitigate these issues by engineering multiscale interfaces of $Na_{2.2}Fe_{1.87}(SO_4)_3$ to improve Na-ion storage performance. Physicochemical analysis and theoretical computations indicated that the structure formed by the $Na_6Fe(SO_4)_4$ phase enhanced ion transport by lowering energy barriers and densifying the migration channels of Na ions. At the cathode side, a conductive interphase of Na-ions is formed by the adsorption of ClO_4^- anions and fluoroethylene carbonate molecules in the electrolyte. The formation of this interphase and the (11–2) plane of the material are interrelated. A $Na_{2.2}Fe_{1.87}(SO_4)_3$-based cathode material exhibited an initial discharge capacity of 83.9 mAh g^{-1} at a discharge voltage of 2.35 V. At 24 mAh g^{-1}, the material showed a specific capacity retention of 97% after 40 cycles [51].

In the high-voltage region, an irreversible oxygen redox reaction in sodium-containing layered cathodes enhances the structure's instability, reducing the cycle life of cells upon cycling. To solve this problem, Guo *et al* adopted a doping technique by incorporating lightweight boron into the cathode active material to reduce the irreversible oxygen oxidation at high voltage. Covalent B–O bonds and negatively charged oxygen atoms provided a vigorous ligand framework for the $NaLi_{1/9}Ni_{2/9}Fe_{2/9}Mn_{4/9}O_2$ cathode material. This boron-doped cathode material exhibited a reversible capacity of 160.5 mAh g^{-1} at a current density of 25 mA g^{-1} and a capacity retention of 82.8% after completing 200 cycles at a current density of 250 mA g^{-1}. It displayed a reversible redox reaction of the transition metal at room temperature [52].

Layered transition-metal oxides have recently piqued the attention of researchers. However, they have been associated with limited capacity and inferior phase transition problems during Na^+ ion insertion/extraction. Wang and his co-workers observed the presence of frozen, relatively large K^+ ions occupying P2-$Na_{0.612}K_{0.056}MnO_2$ prismatic Na^+ sites, facilitating the creation of thermodynamically preferable Na^+ vacancies. The strengthening of Mn–O bonds occurred during the charging and discharging states and reduced phase transition. Their testing revealed a supreme specific capacity of 240.5 mAh g^{-1} and a capacity retention of 98.2% [53].

Xu and colleagues introduced vanadate-based compounds of alkaline earth metals as promising anode materials for SIBs. Specifically, calcium vanadate (CaV_4O_9) nanowires prepared by their group displayed remarkable electronic conductivity. The prepared nanowires showed a minimal volume expansion of <10% during battery operation. Two variations of these nanowires, namely CVO-450 and CVO-550, were prepared for a performance evaluation. When employed as a SIB anode together with excess $Na_3V_2(PO_4)_3/C$ nanoparticles, calcium vanadate displayed outstanding cycling stability and rate performance, showcasing a reversible capacity that exceeded 300 mAh g^{-1} at a voltage of approximately 1.0 V (figure 8.8(c)) [54].

For the construction of SIBs, materials characterized by an open system are preferred for sodiation/desodiation mechanisms. Ma and colleagues published their findings after observing the path of sodium-ion (Na^+) diffusion within $Na_3[Ti_2P_2O_{10}F]$ using high-temperature neutron powder diffraction experiments. $Na_3[Ti_2P_2O_{10}F]$ has a structure that comprises $TiFO_5$ octahedra and PO_4 tetrahedra (figures 8.2(a) and (b)). The octahedra and tetrahedra are alternately connected by four shared oxygen atoms. Square-net-like sheets are formed due to the shared F atoms of $TiFO_5$ octahedra. The long Ti–F–Ti distance (~4.3 Å) in $Na_3[Ti_2P_2O_{10}F]$ facilitated an open-framework structure. $Na_3[Ti_2P_2O_{10}F]$ exhibited a reversible

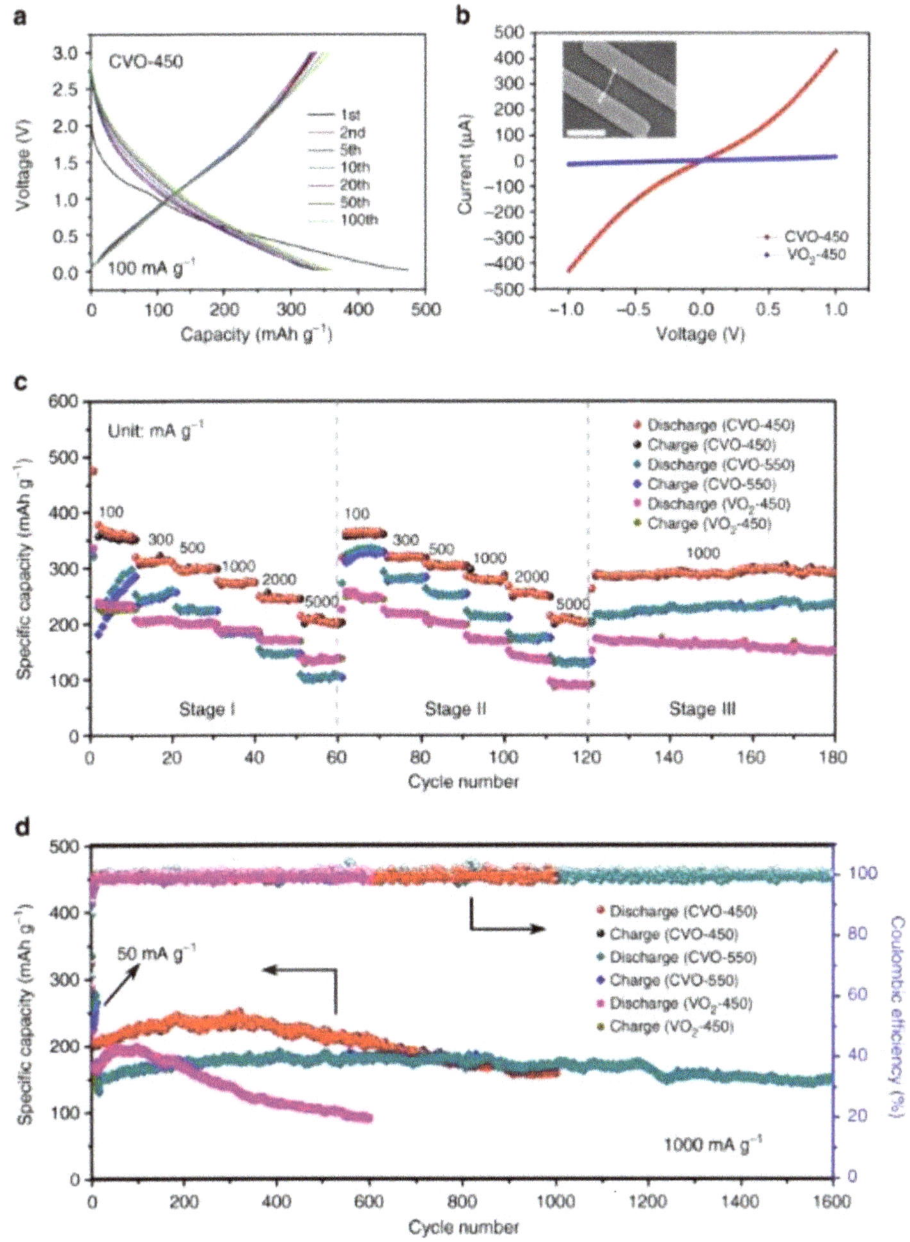

Figure 8.8. (a) The discharge and charge profiles of CVO-450 across multiple cycles at a 100 mA g^{-1} rate. (b) *I–V* curves of CVO-450 and VO_2-450 nanowires. CVO-450 single nanowire device is shown as an inset figure. (c) Rate performance evaluations of CVO-450, CVO-550, and VO_2-450 at 100, 300, 500, 1000, 2000, and 5000 mA g^{-1}. (d) Cycling performances and Coulombic efficiency of CVO-450, CVO-550, and VO_2-450 at 1000 mA g^{-1} current density. Reprinted from [54], Copyright (2017), with permission from Nature.

capacity of 100 mAh g^{-1} and a poor rate capability when employed as an anode. The reaction mechanisms associated with the cells were investigated using high-temperature neutron diffraction (HTND) from 25 °C to 600 °C. This experiment revealed an easy development of Na-ion displacement through the material. Poor stability and low rate capability are the main issues in SIBs. These issues should be addressed by a material with high mechanical strength, chemical stability, and electrically conductive networks [55]. Liu *et al* synthesized a molybdenum disulfide@CNF embedded graphene structure which was mechanically strong. An SIB constructed using this material yielded highly efficient electrochemical performance, achieving a high specific capacity of 598 mAh g^{-1} for 1000 cycles [56]. High-performance LIBs with Li_xCoO_2 have drawn the research community's attention. Therefore, in the 'post-lithium era,' researchers are casting their gaze on a structurally related material, Na_xCoO_2. Willis *et al* examined a diffusion process within $Na_{0.8}CoO_2$ using techniques that included diffuse x-ray scattering, quasi-elastic neutron scattering, and AIMD simulations [57].

Like other batteries, SIBs also have a dendrite formation issue. Lei and colleagues developed a composite material consisting of an inorganic ionic conductor and a gel polymer electrolyte, which served as a cathode. This composite was made by coating alumina nanowires with poly(vinylidene fluoride-co-hexafluoropropylene) (PVDF-HFP) electrolyte. This composition resulted in a homogeneous deposition of sodium by providing transportation channels within the nanowires. The undesired reaction between sodium metal and electrolyte was mitigated, and the formation of dendrites was also prevented. Cells constructed using this material exhibited a high discharge capacity and a long cycle life [58]. A novel titanate compound was reported as a potential anode material for SIBs. The titanate compound displayed a capacity of ~83 mAh g^{-1} after 100 cycles at a rate of 0.1 C. However, it provided a low average voltage of ~0.5 V [59]. Sahu *et al* reported three-dimensional (3D) MoS_2 nano-flowers [60]. A MoS_2/rGO composite was prepared using an ultrasonic exfoliation method. At 100 and 50 mA g^{-1}, the anode material displayed specific capacities of 575 and 218 mAh g^{-1}, respectively. An ionic liquid (IL) electrolyte composed of aluminum chloride/1-methyl-3-ethylimidazolium chloride/sodium chloride was recorded by Sun *et al.* This electrolyte was augmented by the addition of ethyl aluminum dichloride and 1-ethyl-3-methylimidazolium bis(fluorosulfonyl)imide. Batteries that used this electrolyte displayed a voltage of ~4V [61]. Sun *et al* demonstrated a spinel $Li_4Ti_5O_{12}$ with zero strain as an anode that could store sodium. They performed a first-principles calculation to confirm the three-phase separation mechanism [62].

Lee *et al* presented an open-framework crystal structure of sodium manganese hexacyanomanganate ($Na_2Mn^{II}[Mn^{II}(CN)_6]$). At a rate of C/5, their cells had a notable discharge capacity of 209 mAh g^{-1} in a propylene carbonate (PC) electrolyte [63]. The researchers Sun *et al* created a hybrid material consisting of a few phosphorene layers between graphene layers. This material exhibited a specific capacity of 2440 mAh g^{-1} at 0.05 mA g^{-1}. While operating in the potential window of 0–1.5 V, the material retained a capacity of 83% after undergoing 100 cycles (figure 8.9) [64].

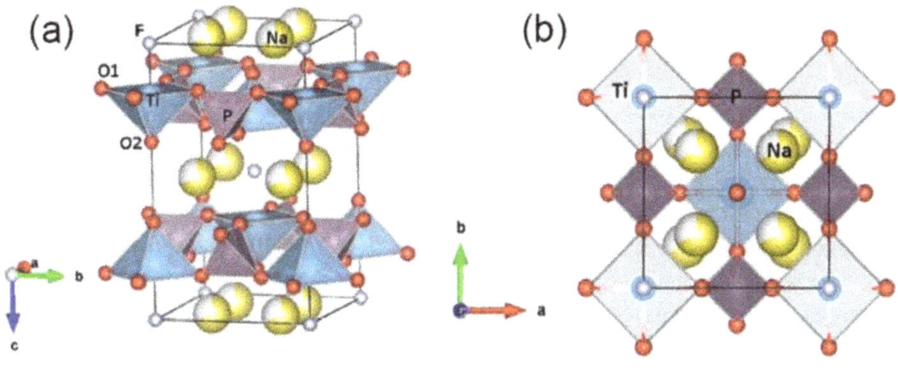

Figure 8.9. The crystal structure of $Na_3[Ti_2P_2O_{10}F]$ (a) and (b) (reprinted from [55], Copyright (2014), with permission from Nature).

8.3.3 Electrolytes, additives, and binders

The electrolyte serves as the connecting medium between the anode and cathode, enabling the ions to traverse the separator. The electrolyte must possess ionic conductivity while being an electronic insulator to prevent short-circuits and self-discharge. It should easily wet the electrode surface and be chemically inert to other battery components, such as electrodes, current collectors, etc. The electrochemical window of the electrolyte should be large to avoid the decomposition of the same during electrochemical processes. Electrolytes and binder choices are crucial for SIBs. Electrolytes and binders can form a surface layer (SL) and an SEI layer at the cathode and anode. This section mainly discusses liquid and solid electrolytes. Other electrolytes used in SIBs include ionic liquids, gel polymer electrolytes, etc. The binders and additives used in SIB technology are also briefly summarized.

8.3.3.1 Electrolytes

The electrolytes used in SIBs are broadly classified into liquid and solid electrolytes. The minimal viscosity, high ionic conductivity, and high wettability of liquid electrolytes make them a primary choice as electrolytes. However, their poor physical and chemical stability is still a problem. Another challenge faced by liquid electrolytes is dendrite formation. In liquid electrolytes, salts are dissolved in organic or aqueous solvents. The salts used in SIBs are $NaBF_4$, NaTF, NaTFS, NaFSI, $NaPF_6$, $NaClO_4$, etc. Organic solvents such as PC, ethylene carbonate (EC), dimethylene carbonate (DMC), or combinations of these solvents are used to dissolve the above sodium-based salts. The electrolyte additives fluoroethylene carbonate (FEC) and vinylene carbonate (VC) enhance electrochemical performance.

Ponrouch *et al* modified a mixture of sodium salt and EC:PC to form $EC_{0.45}$:$PC_{0.45}$:$DMC_{0.1}$ and observed enhanced ionic conductivity, minimal viscosity, and the formation of a stable SEI layer. ILs comprising organic cation Na salt are also considered to be electrolytes with better performance. The properties of ILs include a high boiling point, a large potential window, high thermal stability, etc. However, the lower ionic conductivity and expensive synthetic procedure are still points of concern [65]. Wu *et al* prepared a novel IL electrolyte by combining 1-ethyl-3-methylimidazolium-bis-tetrafluoroborate ($EMIBF_4$) and $NaBF_4$ in various concentrations. The ionic conductivity and thermal stability of the IL electrolyte and 0.1 M $NaBF_4$ combination were impressive. The wide electrochemical window of 4V was validated by quantum chemical theoretical calculation. Monti *et al* prepared an IL-based SIB by combining imidazolium–bis[(trifluoromethyl)sulfonyl]imide (TFSI) ILs (EMIm-TFSI and BMIm-TFSI) and NaTFSI for SIB applications. They found that the electrolytes exhibited high ionic conductivity between −86 °C and 150 °C [66].

Hagiwara *et al* fabricated a cell using sodium chromite/hard carbon with 1-ethyl-3-methylimidazolium and *N*-methyl-*N*-propylpyrrolidinium bis(fluorosulfonyl) amide as an electrolyte. The assembled cell exhibited a gravimetric energy density of 75 Wh kg^{-1} and a volumetric energy density of 125 Wh l^{-1} and retained 87% of its capacity after 500 cycles [67]. To comprehend the mechanism of Na-ion transport in ILs, Lourenco *et al* used a hybrid approach by combining a Canongia Lopes & Padua (CL&P) force field with DFT calculations. They concluded that the accumulation of Na^+ ions was a consequence of multiple interactions between Na^+ and anions in the IL systems, resulting in an enhancement in the transport number of Na^+ through a mechanism called hopping diffusion [68]. In addition to ILs, co-intercalating electrolytes (ether electrolytes) that can participate in redox reactions also emerge as vital electrolyte choices for SIBs. Important ether-based electrolytes such as diethylene glycol dimethyl ether–triethylene glycol dimethyl ether (TREGDME) exhibit outstanding electrochemical performance in SIBs. Lecce *et al* studied an electrolyte by dissolving sodium trifluoromethanesulfonate ($NaCF_3SO_3$) salt in TREGDME solvent for use in both Na-ion and Na-S batteries. The electrolyte showed better safety due to its lower flammability and high ionic conductivity [69].

Solid electrolytes offer better safety and high energy density and are less flammable than liquid electrolytes. Solid electrolytes also function as a separator that allows the easy movement of ions with a low chance of dendrite growth. The limited ionic conductivity, low wettability, and elevated interfacial resistance restrict their use in high-end electrochemical applications. Gel polymer electrolytes (GPEs) and ceramic electrolytes (CEs) are classed as solid electrolytes. In recent studies, a CE and a GPE were combined to improve electrochemical performance. Kim *et al* prepared a 'structurable' gel polymer electrolyte (SGPE) using a straightforward phase separation technique that did not rely on solvents. A glass-fiber (GF) separator that incorporated the SGPE exhibited improved cell performance [70]. Gao *et al* prepared an electrolyte comprising PVDF-HFP supported by a glass-fiber paper and enhanced through polydopamine modification to enhance its mechanical and surface characteristics. This composite polymer matrix displayed impressive mechanical robustness, withstood high temperatures up to 200 °C, possessed a broad electrochemical stability window, and demonstrated excellent ionic conductivity [71].

In the category of CE electrolytes, natrium superionic conductors (NASICONs) and beta alumina are mainly used. Certain NASICONs with better ionic conductivity, such as Na_3PS_4 and $Na_3Zr_2Si_2PO_{12}$, are widely employed as electrolytes in

SIBs. Wang *et al* produced a solid electrolyte, $Na_3Zr_2Si_2PO_{12}$ (NZSP), using a microwave sintering process. The performance of NZSP electrolytes was studied by constructing a cell from $Na_3V_2(PO_4)_3$/C as a cathode, NZSP as a solid-state electrolyte, and sodium as an anode. The cell exhibited a discharge-specific capacity of 75.7 mAh g^{-1} and sustained a capacity retention rate of 81.97% even after undergoing 100 cycles at a discharge/charge rate of 0.5 C [72].

8.3.3.2 Additives

A well-chosen electrolyte additive helps to form a stable SEI layer, enhances electrochemical stability, and ensures safety by preventing overcharging and reducing flammability. FEC is the most widely used additive in SIBs because it forms a stable SEI layer.

Komaba *et al* concluded that among the different additives used in SIBs, FEC improved the electrochemical performance of SIBs with hard carbon (as the anode) and $NaNi_{1/2}Mn_{1/2}O_2$ (as the cathode). They pointed out that optimizing electrolytes and other components, such as additives, binders, etc. is critical to improving the performance of SIBs [73]. Bouibes *et al* studied the concentration effect of FEC on the SEI layer in PC-based electrolytes. They confirmed that only a small amount of FEC is needed to obtain better electrochemical performance from SIBs. An increase in the FEC amount decreased the stability of the SEI layer, thereby deteriorating cell performance [74].

The performance of SIBs when VC is used as an additive has also been studied. Shi and colleagues investigated the inclusion of VC as an additive in $NaCF_3SO_3$-diglyme (DGM) to enable the use of sodium-ion full cells without the need for the preactivation of the cathode and anode. An assembled FeS@C||$Na_3V_2(PO_4)_3$@C full cell that used this electrolyte demonstrated prolonged cycling durability and sustained a high capacity over time [75]. Fang *et al* explored four distinct nonflammable additives, namely trimethyl phosphite (TMP), tri(2,2,2-trifluoroethyl) phosphite (TFEP), dimethyl methylphosphonate (DMMP), and methyl nonafluorobutyl ether (MFE), in SIB applications. Their findings revealed that when used alongside a Prussian blue cathode and a carbon nanotube anode, an electrolyte based on the nonflammable MFE demonstrated superior performance compared to the other additives [76].

Compensating additives such as NaN_3, Na_3P, Na_2O_2, Na_2NiO_2, Na_2CO_3, $Na_2C_2O_4$, and $NaCrO_2$ have also been studied for use in practical SIBs in the future. Guo *et al* developed a Na_2O_2-decorated cathode and investigated its performance in Na-ion full cells. A cathode embellished with Na_2O_2 demonstrated enhanced discharge capacity, exceptional capacity retention, and enhanced rate capability when paired with a carbon anode in a full cell, outperforming a pristine cathode [77]. Electrode materials are pivotal components of the battery domain. When used in high-energy-density SIBs, such cathode components can accommodate or expel Na ions without any structural change or phase transition. For the development of SIBs, various types of cathode materials are available. Among these, the whole discussion revolves around highly featured layered transition-metal oxides. In addition, Prussian blue analogs are commonly used as cathode materials in the fabrication of SIBs. To achieve intense market penetration of SIBs, researchers need to focus on developing highly mechanically stable, high-capacity, highly chemically stable cathode materials.

8.3.3.3 Binders

Binder materials are critical components of electrochemical cells, since they maintain electrode integrity and promote the facile transport of ions/electrons during battery cycling. However, the importance of binders in the electrochemical performance of a cell is yet to be explored. The binder acts as a 'bridge' that ensures the proper contact between the electrode material, the current collector, and the conductive agent. An appropriate choice of binder material directly influences the capacity, lifespan, and safety of a battery.

In both LIBs and SIBs, polyvinylidene fluoride (PVDF) is used as a conventional binder due to its thermal and chemical stability. Due to its limited mechanical strength and insufficient performance during continuous cycling, traditional PVDF binders cannot meet the problems encountered during the volume expansion of electrodes. Since PVDF is an electrical insulator, an additional conductive agent is needed for electrical conductivity. PVDF also requires the toxic volatile solvent N-methyl pyrrolidone (NMP), which is not environmentally friendly. Polyacrylic acid (PAA), sodium alginate (SA), and carboxymethyl cellulose (CMC) are also used as binders in SIBs. In addition, novel binders such as conductive binders, cross-linking binders, and self-healing binders are used [78].

Subramanian *et al* studied the impact of binders such as CMC, PVDF, poly(amide imide) (PAI), and electrolytes on the sol-gel synthesis of an $Na_3V_2(PO_4)_2F_3$ (NVPF) cathode. They concluded that NVPF with a CMC binder and a diethylene glycol dimethyl ether (DEGDME) electrolyte exhibited high Na^+ ion diffusion kinetics, less cell resistance, and excellent electrochemical performance. Compared with the PVDF binder, the NVPF cathode showed a coulombic efficiency of 92% and an extended life of 10 000 cycles [79]. Callegari *et al* applied a self-healing strategy to black phosphorous (BP) at the nanoscale as an anode material. They developed a green binder combining poly(ethylene oxide) and ureidopyrimidinone (UPy)-telechelic systems and their associated blends to replace previous binders such as CMC and PAA used in SIBs. As a result of better adhesion, buffering properties, and damage reparation, the self-healing electrode exhibited a higher current density of 3.5 A g^{-1}. In contrast, the performance of the BP electrode was poor at lower current densities [80].

In regard to SIB anodes, Malchik *et al* examined MXenes as effective binders in aqueous electrolyte solutions. They integrated $NaTi_2(PO_3)_4$ particles with MXene 2D titanium carbide ($Ti_3C_2T_x$) as a binding agent in a highly concentrated $NaClO_4$ electrolyte solution. An anode with 20 wt% of the $Ti_3C_2T_x$ binder displayed a high cycling efficiency of 99.1% and prolonged stability [81]. The durability of rechargeable batteries depends on the binding strength of the electrode materials used in the cell. The introduction of novel functionalized binders such as CMC to replace PVDF will enhance the adhesion and binding of the electrode materials employed in SIBs.

8.3.4 Separators

Like other battery components, the separator also plays a crucial role in battery performance and safety in SIBs. The separator helps to prohibit direct contact between the cathode and anode and allows the easy shuttling of ions inside the battery by accommodating electrolytes. The role of separators in a battery system is undeniable, and separator failure causes a dangerous threat to battery safety. The separator must include the following traits: (1) it should be resistant to heat and able to avoid shrinkage at high temperatures, (2) it should inhibit the growth of dendrites, (3) it should possess considerable mechanical strength to secure the structural integrity of the separator, (4) it should provide good wettability for electrolyte, and (5) it should be chemically compatible with the other battery components so that no further chemical reactions occur, thus safeguarding the battery system. Polymeric compounds, such as commercial polyolefins (porous PE and PP), are the most commonly used separators. However, to improve the electrochemical performance, functionalized separators would be adequate for SIBs.

Suharto *et al* reported a separator modified with a ceramic material (Z-PE separators) for use in SIBs. Easy transfer was facilitated by thoroughly wetting the Z-PE separator with an electrolyte formed by the mixture of EC/PC and 1 M $NaClO_4$, which resulted in an ionic conductivity of 7.0×10^{-4} S [82]. Arunkumar *et al* presented a ceramic-membrane (CM)-modified separator created by incorporating PVDF-HFP/ poly (butyl methacrylate) (PBMA) polymers and modifying them with barium titanate ($BaTiO_3$). The PVDF-HFP/PBMA polymers effectively adhered to the $BaTiO_3$ particles, resulting in a CM separator with consistent pore size. This separator retained more liquid electrolyte, addressing the problem of electrolyte leakage [83]. Another PP separator modified with PVDF nanofiber was reported, and it was concluded that the PVDF nanofiber enhanced the affinity of the PP separator for the EC/DEC-$NaClO_4$ electrolyte [84]. A modified cellulose acetate (MCA) separator was produced for use in $Na/Na_3V_2(PO_4)_3$ half-cell. It retained 93.78% of its capacity after completing 10 000 cycles at 10 C. Wang *et al* developed composite separators that feature a unique combination of a conductive polymer and nitrogen-deficient MXene with titanium, endowing the separators with the remarkable ability to create a uniform ion flux and enhance ion migration. This innovation effectively curtailed the growth of dendrites during electrochemical cycling [85]. In theory, modifying the separator's pore structure and enhancing its compatibility with the electrolyte makes it possible to partially inhibit the sodium dendrites' growth and expedite the sodium transport rate. Like LIBs, SIBs also suffer from serious dendrite formation which affects battery safety.

For high-energy-density SIBs, high-performance separators are needed. A strong relationship between cell performance and separator structure is required to build high-performance separators. Some high-performance separators are either expensive or not environmentally benign to manufacture. Therefore, it is difficult for them to penetrate the market. Currently, researchers are trying to modify existing separators by incorporating functional groups or varying the coating thickness. The robust development of a functionalized separator with suitable porosity and better electrolyte wettability is essential to tackle sodium dendrite formation and to ensure battery safety.

8.3.5 Conclusions and future perspectives

The applications of LIBs extend from portable devices to vehicles due to their high operating voltage and energy density. The demand for batteries for zero-emission EVs is expected to increase in the coming years. The lack of lithium resources and the high cost of cobalt or other rare metals used in the fabrication of LIB electrodes demand an alternative technology. Academic and industrial societies are speaking up about the demand for an alternative technology, as a result of which, the importance of SIBs has become clear. The widespread availability and affordability of sodium resources have facilitated developments in SIBs. Since the landscape of battery technology is dynamic, advancements can easily occur. SIBs have the potential to compete with current Li-ion technology and any other established technologies. Even though SIBs are a viable and sustainable option for future energy storage needs, energy density and cycle life obstacles should be discussed. Continued research and development efforts are crucial to unlock the potential of SIBs to achieve a more sustainable and efficient energy storage future. This chapter discussed the recent advancements and challenges of the anodes, cathodes, electrolytes, separators, and binders used in current SIB technology. Research and development of novel anode and cathode materials can improve the cycling stability, diffusion rate, and energy density of SIBs. Electrolyte optimization can also improve the overall performance of SIBs. The transition from liquid to solid electrolytes can reduce dendrite formation and enhance battery safety. But, before the widespread commercial adoption of solid-state SIBs can take place, challenges such as cost reduction and manufacturing scalability should be addressed. Electrolytes and the suitable selection of binders and additives are important for SIBs to succeed in the market.

Various battery systems are needed to fulfill the diverse applications in electric vehicles, grid storage, and portable devices. Much of the equipment used in SIBs and LIBs is the same, so it is easier for companies to switch to the new SIB technology. If lithium becomes unavailable or too expensive, the cost advantage of SIB raw materials will pave the way to a new technology that can replace current LIB technology. Na-based batteries can be a promising and viable alternative to current LIB technology in the electrochemical energy storage systems.

The future research directions and the challenges are the following:

Theoretical and machine learning studies related to Na-ion storage and the kinetics of novel electrode materials.

An in-depth understanding of the charge storage mechanism and capacity fade behavior is needed to develop advanced electrodes in SIB technology.

The design and development of high-capacity, durable electrode materials for SIBs.

The development of flexible, solid-state, and pouch cell SIBs.

References

[1] Hwang J Y, Myung S T and Sun Y K 2017 Sodium-ion batteries: present and future *Chem. Soc. Rev.* **46** 3529–614

[2] Zhao L, Zhang T, Li W, Li T, Zhang L, Zhang X and Wang Z 2022 Engineering of sodium-ion batteries: opportunities and challenges *Engineering* **24** 172–83

[3] Dunn B, Kamath H and Tarascon J M 2011 Electrical energy storage for the grid: a battery of choices *Science* **334** 928–35

[4] Abraham K M 2020 How comparable are sodium-ion batteries to their Lithium-ion counterparts? *ACS Energy Lett.* **5** 3544–7

[5] Kang H, Liu Y, Cao K, Zhao Y, Jiao L, Wang Y and Yuan H 2015 Update on anode materials for Na-ion batteries *J. Mater. Chem.* A **3** 17899–913

[6] Perveen T, Siddiq M, Shahzad N, Ihsan R, Ahmad A and Shahzad M I 2020 Prospects in anode materials for sodium ion batteries review *Renew. Sustain. Energy Rev.* **119** 109549

[7] Qiao S, Zhou Q, Ma M, Liu H K, Dou S X and Chong S 2023 Advanced anode materials for rechargeable sodium-ion batteries *ACS Nano* **17** 11220–52

[8] DiVincenzo D P and Mele E J 1985 Cohesion and structure in stage-1 graphite intercalation compounds *Phys. Rev. B: Condens. Matter* **32** 2538–53

[9] Wen Y, He K, Zhu Y, Han F, Xu Y, Matsuda I, Ishii Y, Cumings J and Wang C 2014 Expanded graphite as superior anode for sodium-ion batteries *Nat. Commun.* **5** 4033

[10] Tian Y *et al* 2023 Design of high-performance defective graphite-type anodes for sodium-ion batteries *ACS Appl. Energy Mater.* **6** 3854–61

[11] Cabello M, Bai X, Chyrka T, Ortiz G F, Lavela P, Alcántara R *et al* 2017 On the reliability of sodium Co-intercalation in expanded graphite prepared by different methods as anodes for sodium-ion batteries *J. Electrochem. Soc.* **164** A3804–13

[12] Stevens D A and Dahn J 2001 The mechanisms of lithium and sodium insertion in carbon materials *J. Electrochem. Soc.* **148** A803

[13] Komaba S, Murata W, Ishikawa T, Yabuuchi N, Ozeki T, Nakayama T, Ogata A, Gotoh K and Fujiwara K 2011 Electrochemical Na insertion and solid electrolyte interphase for hard-carbon electrodes and application to Na-ion batteries *Adv. Funct. Mater.* **21** 3859–67

[14] Bommier C, Surta T W, Dolgos M and Ji X 2015 New mechanistic insights on Na-ion storage in nongraphitizable carbon *Nano Lett.* **15** 5888–92

[15] Wang Y X, Chou S L, Liu H K and Dou S X 2013 Reduced graphene oxide with superior cycling stability and rate capability for sodium storage *Carbon* **57** 202–8

[16] Ding J *et al* 2013 Carbon nanosheet frameworks derived from peat moss as high-performance sodium ion battery anodes *ACS Nano* **7** 11004–15

[17] Datta D, Li J and Shenoy V B 2014 Defective graphene is a high-capacity anode material for Na- and Ca-ion batteries *ACS Appl. Mater. Interfaces* **6** 1788–95

[18] Kale S B, Chothe U P, Kale B B, Kulkarni M V, Pavitran S and Gosavi S W 2021 Synergetic strategy for the fabrication of self-standing distorted carbon nanofibers with heteroatom doping for sodium-ion batteries *ACS Omega* **6** 15686–97

[19] Li W, Zhou M, Li H, Wang K, Cheng S and Jiang K 2015 A high performance sulfur-doped disordered carbon anode for sodium-ion batteries *Energy Environ. Sci.* **8** 2916–21

[20] Yang H S, Kim S W, Kim K H, Yoon S H, Ha M J and Kang J 2021 S and P dual-doped carbon nanospheres as anode material for high rate performance sodium-ion batteries *Appl. Sci.* **11** 12007

[21] Li Z, Cai L, Chu K, Xu S, Yao G, Wei L and Zheng F 2021 Heteroatom-doped carbon materials with interconnected channels as ultrastable anodes for lithium/sodium ion batteries *Dalton Trans.* **50** 4335–44

[22] Alcántara R, Jaraba M, Lavela P and Tirado J L 2002 $NiCo_2O_4$ spinel: first report on a transition metal oxide for the negative electrode of sodium-ion batteries *Chem. Mater.* **14** 2847–8

[23] Zhang N, Han X, Liu Y, Hu X, Zhao Q and Chen J 2015 3D porous γ-Fe_2O_3@ C nanocomposite as high-performance anode material of Na-ion batteries *Adv. Energy Mater.* **5** 1401123

[24] Liu S, Wang Y, Dong Y, Zhao Z, Wang Z and Qiu J 2016 Ultrafine Fe_3O_4 quantum dots on hybrid carbon nanosheets for long-life, high-rate alkali-metal storage *ChemElectroChem.* **3** 38–44

[25] Yang J, Zhou T, Zhu R, Chen X, Guo Z, Fan J, Liu H K and Zhang W X 2016 Highly ordered dual porosity mesoporous cobalt oxide for sodium-ion batteries *Adv. Mater. Interfaces* **3** 1500464

[26] Lu Y C, Ma C, Alvarado J, Kidera T, Dimov N, Meng Y S and Okada S 2015 Electrochemical properties of tin oxide anodes for sodium-ion batteries *J. Power Sources* **284** 287–95

[27] Klein F, Jache B, Bhide A and Adelhelm P 2013 Conversion reactions for sodium-ion batteries *Phys. Chem. Chem. Phys.* **15** 15876–87

[28] Hu Z, Wang L, Zhang K, Wang J, Cheng F, Tao Z and Chen J 2014 MoS_2 nanoflowers with expanded interlayers as high-performance anodes for sodium-ion batteries *Angew. Chem. Int. Ed.* **53** 12794–8

[29] Wang J, Yue X, Xie Z, Abudula A and Guan G 2021 MOFs-derived transition metal sulfide composites for advanced sodium-ion batteries *Energy Storage Mater.* **41** 404–26

[30] Hou T, Liu B, Sun X, Fan A, Xu Z, Cai S, Zheng C, Yu G and Tricoli A 2021 Covalent coupling-stabilized transition-metal sulfide/carbon nanotube composites for lithium/sodium-ion batteries *ACS Nano* **15** 6735–46

[31] Zhu X, Liu D, Zheng D, Wang G, Huang X, Harris J, Qu D and Qu D 2018 Dual carbon-protected metal sulfides and their application to sodium-ion battery anodes *J. Mater. Chem.* A **6** 13294–301

[32] Wang X, Huang J, Li J, Cao L, Hao W and Xu Z 2016 Improved Na storage performance by incorporating nitrogen-doped conductive carbon into WS_2 nanosheets *ACS Appl. Mater. Interfaces* **8** 23899–908

[33] Deng M, Li S, Hong W, Jiang Y, Xu W, Shuai H, Li H, Wang W, Hou H and Ji X 2019 Natural stibnite ore (Sb_2S_3) embedded in sulfur-doped carbon sheets: enhanced electrochemical properties as anode for sodium ions storage *RSC Adv.* **9** 15210–6

[34] Yu Z, Li X, Yan B, Xiong D, Yang M and Li D 2017 Rational design of flower-like tin sulfide@ reduced graphene oxide for high-performance sodium-ion batteries *Mater. Res. Bull.* **96** 516–23

[35] Nam K H, Hwa Y and Park C M 2020 Zinc phosphides are outstanding sodium-ion battery anodes *ACS Appl. Mater. Interfaces* **12** 15053–62

[36] Liang J, Zhu G, Zhang Y, Liang H and Huang W 2022 Conversion of hydroxide into carbon-coated phosphide using plasma for sodium-ion batteries *Nano Res.* **15** 2023–9

[37] Lu Y, Zhou P, Lei K, Zhao Q, Tao Z and Chen J 2017 Selenium phosphide (Se_4P_4) as a new and promising anode material for sodium-ion batteries *Adv. Energy Mater.* **7** 1601973

[38] Kim Y, Kim Y, Choi A, Woo S, Mok D, Choi N S, Jung Y S, Ryu J H, Oh S M and Lee K T 2014 Tin phosphide is a promising anode material for Na-ion batteries *Adv. Mater.* **26** 4139–44

[39] Li W J, Yang Q R, Chou S L, Wang J Z and Liu H K 2015 Cobalt phosphide is a new anode material for sodium storage *J. Power Sources* **294** 627–32

[40] Jung S C, Jung D S, Choi J W and Han Y K 2014 Atom-level understanding of the sodiation process in silicon anode material *J. Phys. Chem. Lett.* **5** 1283–8

[41] Sharma V, Ghatak K and Datta D 2018 Amorphous germanium as a promising anode material for sodium-ion batteries: a first principle study *J. Mater. Sci.* **53** 14423–34

[42] Li Z, Ding J and Mitlin D 2015 Tin and tin compounds for sodium ion battery anodes: phase transformations and performance *Acc. Chem. Res.* **48** 1657–65

[43] Sayed S Y, Kalisvaart W P, Luber E J, Olsen B C and Buriak J M 2020 Stabilizing tin anodes in sodium-ion batteries by alloying with silicon *ACS Appl. Energy Mater.* **3** 9950–62
[44] Hu L, Zhu X, Du Y, Li Y, Zhou X and Bao J 2015 A chemically coupled antimony/multilayer graphene hybrid as a high-performance anode for sodium-ion batteries *Chem. Mater.* **27** 8138–45
[45] Song J, Yu Z, Gordin M L, Li X, Peng H and Wang D 2015 Advanced sodium ion battery anode constructed via chemical bonding between phosphorus, carbon nanotube, and a cross-linked polymer binder *ACS Nano* **9** 11933–41
[46] Yang F, Yu F, Zhang Z, Zhang K, Lai Y and Li J 2016 Bismuth nanoparticles embedded in carbon spheres as anode materials for sodium/lithium-ion batteries *Chem. Eur. J.* **22** 2333–8
[47] Delmas C, Fouassier C and Hagenmuller P 1980 Structural classification and properties of the layered oxides *Physica B+ c* **99** 81–5
[48] Peng B, Sun Z, Jiao S, Wang G and Zhang G 2020 Electrochemical performance optimization of layered P2-type $Na_{0.67}MnO_2$ through simultaneous Mn-site doping and nanostructure engineering *Batter. Supercaps* **3** 147–54
[49] Peng B, Sun Z, Zhao L, Zeng S and Zhang G 2021 Shape-induced kinetics enhancement in layered P2-$Na_{0.67}Ni_{0.33}Mn_{0.67}O_2$ porous microcuboids enables high energy/power sodium-ion full battery *Batter. Supercaps* **4** 456–63
[50] Zhao C *et al* 2020 Rational design of layered oxide materials for sodium-ion batteries *Science* **370** 708–11
[51] Zhang J, Yan Y, Wang X, Cui Y, Zhang Z, Wang S, Xie Z, Yan P and Chen W 2023 Bridging multiscale interfaces for developing ionically conductive high-voltage iron sulfate-containing sodium-based battery positive electrodes *Nat. Commun.* **14** 3701
[52] Guo Y J *et al* 2021 Boron-doped sodium layered oxide for reversible oxygen redox reaction in Na-ion battery cathodes *Nat. Commun.* **12** 5267
[53] Wang C *et al* 2021 Tuning local chemistry of P2 layered-oxide cathode for high energy and long cycles of sodium-ion battery *Nat. Commun.* **12** 2256
[54] Xu X *et al* 2017 Alkaline earth metal vanadates as sodium-ion battery anodes *Nat. Commun.* **8** 460
[55] Ma Z, Wang Y, Sun C, Alonso J A, Fernández-Díaz M T and Chen L 2014 Experimental visualization of the diffusion pathway of sodium ions in the $Na_3[Ti_2P_2O_{10}F]$ anode for sodium-ion battery *Sci. Rep.* **4** 7231
[56] Liu M, Zhang P, Qu Z, Yan Y, Lai C, Liu T and Zhang S 2019 Conductive carbon nanofiber interpenetrated graphene architecture for ultra-stable sodium-ion battery *Nat. Commun.* **10** 3917
[57] Willis T J, Porter D G, Voneshen D J, Uthayakumar S, Demmel F, Gutmann M J, Roger M, Refson K and Goff J P 2018 Diffusion mechanism in the sodium-ion battery material sodium cobaltate *Sci. Rep.* **8** 3210
[58] Lei D *et al* 2019 Cross-linked beta alumina nanowires with compact gel polymer electrolyte coating for ultra-stable sodium metal battery *Nat. Commun.* **10** 4244
[59] Ma X, An K, Bai J and Chen H 2017 $NaAlTi_3O_8$, a novel anode material for sodium-ion battery *Sci. Rep.* **7** 162
[60] Sahu T S and Mitra S 2015 Exfoliated MoS_2 sheets and reduced graphene oxide-an excellent and fast anode for sodium-ion battery *Sci. Rep.* **5** 12571
[61] Sun H *et al* 2019 A safe and non-flammable sodium metal battery based on an ionic liquid electrolyte *Nat. Commun.* **10** 3302

[62] Sun Y *et al* 2013 Direct atomic-scale confirmation of three-phase storage mechanism in $Li_4Ti_5O_{12}$ anodes for room-temperature sodium-ion batteries *Nat. Commun.* **4** 1870
[63] Lee H W, Wang R Y, Pasta M, Woo Lee S, Liu N and Cui Y 2014 Manganese hexacyanomanganate open framework as a high-capacity positive electrode material for sodium-ion batteries *Nat. Commun.* **5** 5280
[64] Sun J, Lee H W, Pasta M, Yuan H, Zheng G, Sun Y, Li Y and Cui Y 2015 A phosphorene–graphene hybrid material as a high-capacity anode for sodium-ion batteries *Nat. Nanotechnol.* **10** 980–5
[65] Ponrouch A, Marchante E, Courty M, Tarascon J M and Palacin M 2012 In search of an optimized electrolyte for Na-ion batteries *Energy Environ. Sci.* **5** 8572–83
[66] Wu F, Zhu N, Bai Y, Liu L, Zhou H and Wu C 2016 Highly safe ionic liquid electrolytes for sodium-ion battery: wide electrochemical window and good thermal stability *ACS Appl. Mater. Interfaces* **8** 21381–6
[67] Hagiwara R, Matsumoto K, Hwang J and Nohira T 2019 Sodium-ion batteries using ionic liquids as electrolytes *Chem. Rec.* **19** 758–70
[68] Lourenco T C, Dias L G and Da Silva J L 2021 Theoretical investigation of the Na^+ transport mechanism and the performance of ionic liquid-based electrolytes in sodium-ion batteries *ACS Appl. Energy Mater.* **4** 4444–58
[69] Di Lecce D, Minnetti L, Polidoro D, Marangon V and Hassoun J 2019 Triglyme-based electrolyte for sodium-ion and sodium–sulfur batteries *Ionics* **25** 3129–41
[70] Kim J I, Choi Y, Chung K Y and Park J H 2017 A structurable gel-polymer electrolyte for sodium-ion batteries *Adv. Funct. Mater.* **27** 1701768
[71] Gao H, Guo B, Song J, Park K and Goodenough J B 2015 A composite gel–polymer/glass–fiber electrolyte for sodium-ion batteries *Adv. Energy Mater.* **5** 1402235
[72] Wang X, Liu Z, Tang Y, Chen J, Wang D and Mao Z 2021 Low temperature and rapid microwave sintering of $Na_3Zr_2Si_2PO_{12}$ solid electrolytes for Na-ion batteries *J. Power Sources* **481** 228924
[73] Komaba S, Ishikawa T, Yabuuchi N, Murata W, Ito A and Ohsawa Y 2011 Fluorinated ethylene carbonate as electrolyte additive for rechargeable Na batteries *ACS Appl. Mater. Interfaces* **3** 4165–8
[74] Bouibes A, Takenaka N, Fujie T, Kubota K, Komaba S and Nagaoka M 2018 Concentration effect of fluoroethylene carbonate on the formation of solid electrolyte interphase layer in sodium-ion batteries *ACS Appl. Mater. Interfaces* **10** 28525–32
[75] Shi J, Ding L, Wan Y, Mi L, Chen L, Yang D, Hu Y and Chen W 2021 Achieving long-cycling sodium-ion full cells in ether-based electrolyte with vinylene carbonate additive *J. Energy Chem.* **57** 650–5
[76] Huang Z X, Zhang X L, Zhao X X, Zhao Y Y, Aravindan V, Liu Y H, Geng H and Wu X L 2021 Electrode/electrolyte additives for practical sodium-ion batteries: a mini-review *Inorg. Chem. Front.* **10** 37–48
[77] Guo Y J, Niu Y B, Wei Z, Zhang S Y, Meng Q, Li H, Yin Y X and Guo Y G 2021 Insights on electrochemical behaviors of sodium peroxide as a sacrificial cathode additive for boosting energy density of Na-ion battery *ACS Appl. Mater. Interfaces* **13** 2772–8
[78] Li R R, Yang Z, He X X, Liu X H, Zhang H, Gao Y, Qiao Y, Li L and Chou S L 2019 Binders for sodium-ion batteries: progress, challenges, and strategies *Chem. Commun.* **57** 12406–16

[79] Subramanian Y, Oh W, Choi W, Lee H, Jeong M, Thangavel R and Yoon W S 2021 Optimizing high voltage $Na_3V_2(PO_4)_2F_3$ cathode for achieving high rate sodium-ion batteries with long cycle life *Chem. Eng. J.* **403** 126291

[80] Callegari D, Colombi S, Nitti A, Simari C, Nicotera I, Ferrara C, Mustarelli P, Pasini D and Quartarone E 2021 Autonomous self-healing strategy for stable sodium-ion battery: A case study of black phosphorus anodes *ACS Appl. Mater. Interfaces* **13** 13170–82

[81] Malchik F, Shpigel N, Levi M D, Penki T R, Gavriel B, Bergman G, Turgeman M, Aurbach D and Gogotsi Y 2021 MXene conductive binder for improving the performance of sodium-ion anodes in water-in-salt electrolyte *Nano Energy* **79** 105433

[82] Suharto Y, Lee Y, Yu J S, Choi W and Kim K J 2018 Microporous ceramic coated separators with superior wettability for enhancing the electrochemical performance of sodium-ion batteries *J. Power Sources* **376** 184–90

[83] R A, Vijaya Kumar Saroja A P and Sundara R 2020 Barium titanate-based porous ceramic flexible membrane as a separator for room-temperature sodium-ion battery *ACS Appl. Mater. Interfaces* **11** 3889–96

[84] Janakiraman S, Khalifa M, Biswal R, Ghosh S, Anandhan S and Venimadhav A 2020 High-performance electrospun nanofiber coated polypropylene membrane as a separator for sodium-ion batteries *J. Power Sources* **460** 228060

[85] Zhou D *et al* 2020 Polyolefin-based Janus separator for rechargeable sodium batteries *Angew. Chem. Int. Ed.* **59** 16725–34

Chapter 9

Materials for solid-state batteries

Keith Sirengo, Aswathy Babu and Suresh C Pillai

Lithium-ion batteries have dominated as an energy storage technology for over three decades. In time, this technology will reach its maximum capacity. As a result, alternative options to enhance battery performance are needed. Among them, replacing the graphite anode with a lithium anode was projected to increase the energy density beyond the theoretical capacity of graphite. However, the lithium anode was unstable when exposed to convectional liquid electrolytes because the reaction between organic electrolytes and the lithium anode leads to dendritic growth. Moreover, organic electrolytes have low flash points and are highly flammable. These two challenges, especially at elevated temperatures, compromise battery life and to some extent lead to catastrophic battery failures that raise safety concerns. In this context, solid-state electrolytes are anticipated to address these challenges and accommodate the reactive lithium anode needed for high-energy-density performance. This chapter, therefore, presents the fundamentals of the ionic conductivity of solid-state electrolytes (SSEs), their types, and their expected contributions to battery technology. We also discuss the fascinating chemistry of their composites and the constructive interaction between them. Finally, we examine the challenges associated with the current SSEs and the strategies required to expedite their commercial recognition.

9.1 Introduction

Commercial alkali-based batteries such as lithium-ion, sodium-ion, and potassium-ion batteries are some of the current energy storage technologies [1]. Among them, lithium-ion technology has dominated the market because of its high energy density, superior lithium kinetics, and zero 'memory' effect [2]. On the other hand, sodium and potassium are potential substitutes for lithium because of their availability, affordability, and suitability for large-scale production [3]. The maturity of lithium-ion

doi:10.1088/978-0-7503-5749-4ch9

batteries has played vital role in repurposing their technology for use with other battery chemistries. For instance, the basic working principle is common: a single cell comprises electrodes, a separator, and an electrolyte system. Unlike the electrode and the separator, which are in the solid phase, the electrolyte is in the liquid phase. It is made from conductive salts and organic carbonates, which raises a safety concern at high temperatures [4] and high current densities [5]. These unstable organic solvents are flammable and support the thermal runaway process, which has recently led to catastrophic battery failures [6]. In addition to their poor mechanical stability and their emission of toxic gases, these solvents are associated with low cation transference numbers [7, 8]. This provides scant energy to meet the requirement for long-distance driving over distances of more than 300 km [9].

On the other hand, SSEs have high thermal stability, superior mechanical properties, and a wide electrochemical window and they operate in a wide range of temperatures from −50 °C to >200 °C [10]. Within this range, liquid electrolytes either freeze at extremely low temperatures or decompose at extremely high temperatures. In addition to facilitating ionic conductivity [11], SSEs will also have to meet the high safety standards needed to power hybrid cars, robots, and electric buses [12]. For these reasons, SSEs are potential replacements for traditional liquid-state electrolytes [13], as shown in figure 9.1. When SSEs are employed, all the cell components are solids, transforming the entire device into an all-solid-state battery.

All-solid-state battery technology is projected to improve battery capacity by 50% [14, 15]. This is because it will possibly replace graphite with high-energy-density lithium anodes (with ten times higher theoretical capacity). Second, a further increase in battery energy density is expected from the design of SSEs, which allows several unit cells to be stacked in one compartment instead of the single unit cell per compartment of typical liquid-based cells [16]. Finally, because higher-temperature operations are possible with SSEs, the kinetics of ions are elevated, and the costs of installing an advanced cooling system and operating it during battery cycling are reduced.

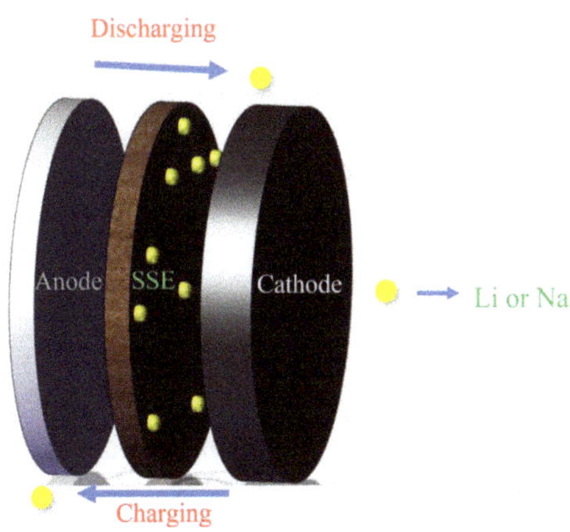

Figure 9.1. A schematic illustration of a working mechanism of a solid-state battery.

9.2 The history of solid-state electrolytes

The history of solid-state electrolytes dates back to 1883, when Michael Faraday discovered silver sulfide (Ag_2S) and lead fluoride (PbF) as the first inorganic materials to display ionic conductivity [17]. This discovery was followed by the first determination of a sodium transference number in 1888 after Warburg and Tegetmeier realized that sodium could be transported through a glass in 1884. Although both liquid- and solid-state electrolytes were developed at this time, the liquid state was dominant because of its easy preparation and high ionic conductivity. Further applications of SSEs in batteries were reported in the 1950s, which used $Pb/PbCl_2/AgCl/Ag$ and $Ag/AgI/I_2$, closely followed by research into β-alumina ($Na_2O{\cdot}11Al_2O_3$) in 1960 [18]. Progress was mainly hindered by a low voltage window (<1 V) until the 1970s, when a scientific leap occurred: lithium iodide was found to display a higher voltage window (>3 V) and energy densities between 100–200 Wh kg^{-1}, which were large enough to power lower-energy devices [19].

This interesting ionic conductivity coupled with the fuel crisis in the 1970s spiked an interest in other SSEs beyond inorganic electrolytes, leading to the first discovery of ionic conductivity in organic solid polymers in 1975 [20]. However, the superior liquid-state electrolytes overshadowed the existence of SSEs at the time. Moreover, the world was less developed, and most energy storage devices only had to satisfy the low energy demands of the technology used at the time. However, the current demand for high safety standards in batteries has prompted a review of SSEs. To this end, therefore, SSEs remain a feasible alternative for the realization of safer batteries. Toyota has already envisioned this, as they are expected to launch two solid-state batteries at the end of 2027, supplying enough energy for at least 1000 km with a very short charging time of 10 min.

9.3 The fundamentals of ionic conductivity in the solid state

In conventional electrolytes, alkali salts are dissolved in organic carbonates, forming an electrolyte system with free mobile ions. However, in solid-state electrolyte systems, polymers and inorganic materials act as solid solvents for lithium salts. In organic liquids, the ionic conductivity is controlled by the presence of either a chemical, electrochemical, or conventional gradient [21]. In contrast, the ionic conductivity of SSEs is primarily controlled by electrochemical gradients [21]. This is because solids have a low concentration potential and negligible convection gradients, which therefore do not significantly contribute to ionic conductivity at room temperature.

9.3.1 Ionic conductivity mechanisms in inorganic electrolytes

When a potential difference is applied across an SSE, ions diffuse with the aid of a hopping mechanism [15]. They acquire a small amount of energy known as activation energy [22] that allows them to move from one stable site to another (via wave-like motion) by overcoming a height of the energy barrier known as the bottleneck [23] via tunnels in the crystal lattice, grain boundaries, and defects/

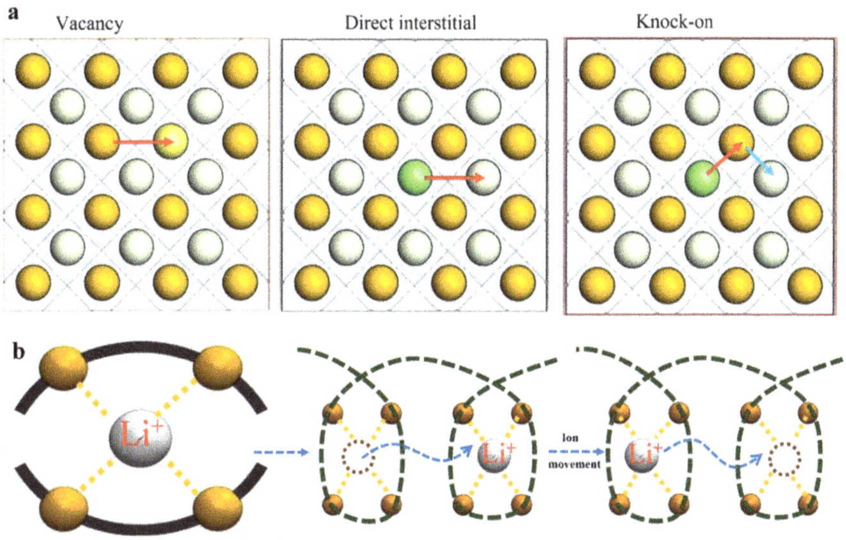

Figure 9.2. Ionic conductivity mechanisms in solid (a) inorganic and (b) polymer electrolytes.

imperfections [24]. Generally, ion diffusion in solids requires structural defects such as missing ions. Figure 9.2(a) illustrates how ionic conductivity is facilitated by three mechanisms: the interstitial mechanism, vacancies, and interstitial substitution. In the interstitial mechanism, an ion moves from one vacant site to another. The vacancy mechanism involves the movement of ions from an occupied site to a vacant site. Finally, interstitial substitution involves an ion in a vacant site that knocks another ion [10]. This is also known as a knock-on mechanism because it displaces an ion from an occupied site.

9.3.2 Ionic conductivity mechanisms in polymers

In solid polymer electrolytes, ionic conductivity is facilitated by interactions between the Lewis base units on the polymer structures and the cations on the conductive salts [25]. In these interactions, the mechanisms involve the formation and breakage of the bonds between cations and the Lewis base units (electron donors) or the available functional groups, such as ether, amide, carbonyl, or phosphine groups. For a typical electrolyte system composed of poly(ethylene oxide) (PEO) and a lithium salt, ionic conductivity involves the segmental amplitude-like motion of lithium in the PEO polymer structure, resulting in a constant process of the formation and breakage of oxygen–lithium (O–Li) bonds [26, 27] as illustrated in figure 9.2(b) [28].

Ionic conductivity is different for every salt–polymer electrolyte system. This is because of different physical properties such as the glass transition temperature as well as differences in Lewis functional groups that have varying electron donor abilities. On the other hand, the salts also contribute to these variations in ionic conductivity because of their different lattice energies and varying anion sizes. To sum up, a highly conductive polymer–salt electrolyte system can be achieved by focusing on amorphous polymers (high-dielectric-constant polymers) combined with low-lattice-energy salts. Therefore, several strategies have been proposed in this context, such as the use of plasticizers and cross-linkers, and it has attracted intensive research into the ionic conductivity mechanism, especially at a micro level.

9.4 Categories of solid-state electrolytes

There are three general classifications of SSEs. These include inorganic electrolytes, organic polymeric materials, and composites of inorganic materials and polymeric materials [29].

9.4.1 Inorganic materials

Inorganic electrolyte materials can be structurally classified as crystalline (ceramics), amorphous (glassy), or crystalline–amorphous [11]. Unlike liquids and polymers, in which anions are involved in ionic conductivity, inorganic materials have stationary anions whose ionic conductivity is primarily contributed by metal cations [30], translating to a high cation transference number [9]. Recent studies have shown that crystalline materials have higher bulk ionic conductivity that is close to and even better than that of organic electrolytes, but experience substantial resistance at the grain boundaries [31]. Based on these stationary anions and functional groups (oxygen, sulfide, and polyanionic moieties), which interact with alkali cations [10], ceramic electrolytes can be further classified as either oxide-based, sulfide-based, or halide-based electrolytes [1].

9.4.1.1 Oxide-based inorganic electrolytes

This category occurs in three structures: perovskite, superionic conductors, and the garnet type [32]. The garnet type has an overall formula of $A_2B_2(XO_4)_3$, where A is either Ca, Mg, Y, La, or a rare-earth metal, while B can be Al, Fe Ga, Ge, Mn, or Ni, and X represents either Si, Ge, or Al. An example of the garnet-type structure is $Li_7La_3Zr_2O_{12}$ (LLZO) [1], discovered in 2007 as a ceramic electrolyte with promising electrochemical stability in lithium reduction, especially for tantalum- and lanthanum-based materials [17]. It also demonstrates an ionic conductivity close to the practically acceptable range (1–0.1 mS cm^{-1}), which is attributed to low activation enthalpy, reliable thermal stability, and electrochemical stability when exposed to lithium metal [9, 33]. Despite this high bulk ionic conductivity, the overall battery performance is compromised and its progress is slow because of high interphase resistance [34]. In addition, it is unstable in the environment and susceptible to dendrite penetration along the grain boundaries [5]. For example, when lithium aluminum germanium phosphate (LAGP) is exposed to air, Ge^{4+} is reduced to Ge^{2+}, which subsequently produces GeO, leading to cracks and high resistance [35]. Moreover, this interphase can supply oxygen at elevated temperatures (>200 °C), which facilitates thermal runaway [35].

The second category is superionic conductors with an overall formula of $MA_2(BO_4)_3$, where M=Li, Na, K, or Ag, while A=Ti, Zr, Ge, or V, and B=P, Si, or Mo. They include natrium superionic conductor (NASICON)-type structures such as $Na_3Zr_2Si_2PO_{12}$ and lithium superionic conductors (LISICONs) such as Li_3PO_4 [36]. They demonstrate superior stability in air and water, respectively. However, they face a major obstacle because of their high resistance to lithium metal [37].

The perovskite-type is the third class, with a structure denoted by ABO_3 that is estimated to display an ionic conductivity of approximately 10^{-3} S cm^{-1} [32, 38]. The high ionic conductivity and high voltage stability of up to 8 V reported for some perovskite materials, such as $Li_{3x}La_{2/3-x}TiO_3$, make these materials attractive for use in energy applications [39]. However, performance is not guaranteed due to environmental instability, because the reduction of Ti^{4+} by lithium metal remains an unsolved challenge [39].

SSE electrolytes are also large compared with other options, which might have a detrimental effect on the overall energy density of batteries. Because of this challenge, the use of thin-film technology is important in solid-state batteries. One of the SSE electrolytes at the center of this technology is lithium phosphorus oxynitride (LiPON), represented by the general chemical formula $Li_xPO_yN_z$ [40]. It has a high operational voltage window (5 V) [41, 42] but low ionic conductivity in bulk, and the LiPON–cathode interface needs to be enhanced [34]. Some recently proposed strategies, namely doping with transition metals or elements such as fluorine, nitrogen, silicon, or boron, have been shown to improve the kinetics of LIPON [43]. For instance, LIPON doped with fluorine demonstrated higher ionic conductivity of $\sim 1.0 \times 10^{-6}$ S cm^{-1} compared to that of neat LiPON ($\sim 7.7 \times 10^{-7}$ S cm^{-1}) [44]. Unfortunately, this ionic conductivity is still below the recommended ionic conductivity for practical applications ($\sim 1.0 \times 10^{-3}$ S cm^{-1}). The replacement of oxygen anions with less electronegative sulfur anions could reduce the interaction of lithium with the anions. This might improve lithium-ion conductivity, which will be discussed in the next section.

9.4.1.2 Sulfide-based inorganic electrolytes

This is a class of SSEs with extremely high ionic conductivity [22]. They are among SSEs with the highest ionic conductivity, which is comparable to that of organic electrolytes [36]. This is because the substitution of oxygen with less electronegative sulfur reduces the interaction of the anion with lithium ions, prompting better lithium movement. Moreover, the Li–O bond is stronger than the Li–S bond, making the latter more conductive and the former more electrochemically stable [45]. A case in point is an investigation of $Li_{9.54}Si_{1.74}P_{1.44}S_{11.7}Cl_{0.3}$, which demonstrated an ionic conductivity of 25 mS cm^{-1} (far better than those of liquid electrolytes) and a subsequent high power delivery after 3 min of charging [8]. More sulfide inorganic electrolytes have been reported to have ionic conductivities higher than the 10 mS cm^{-1} of typical organic electrolytes [8, 46–50]. This exceptionally high ionic conductivity is attributed to concerted migrations of multiple ions and a comparatively lower activation energy experienced during hopping [28]. The main challenge facing this category of material is the instability of sulfur when exposed to sodium or the environment [47, 51]. Sulfur decomposes at low voltages and reacts with air and water to generate highly toxic hydrogen sulfide gas [36].

As mentioned above, the replacement of less electronegative anions and large sizes has a positive impact on the overall ionic conductivity. However, ion substitution should be done with great caution to maintain the thermodynamic stability of the electrode [45]. In addition, ionic conductivity may be reduced because of inductive effects in which the replacement of large cations with smaller ones may lower the ionic conductivity. This is because the cation attracts a large volume of charge, strengthening the bonds between the moving charges and the host lattice and reducing the ionic conductivity [52].

To sum up, inorganic electrolytes are promising electrolytes in terms of safety because of their thermal and mechanical stability. Besides, based on recent research, the gap between the ionic conductivity of commercial electrolytes and some SSEs has been reduced, and to some extent, SSEs have even superseded the former. For example, a combination of sulfide and oxygen demonstrates excellent ionic conductivity of 10 mS cm^{-1} and excellent electrochemical stability of 10 V. Unfortunately, a fully assembled device with an initial capacity of about 100 mAh g^{-1} fades quickly after 10 cycles (as shown in figure 9.3(a)) because of high resistance at the solid electrolyte interphase (SEI), which is about four times (figure 9.3(b)) that demonstrated in liquid electrolytes. This is attributed to breakage of the rigid inorganic materials during cycling and poor electrode wettability. In addition to this, their practical application is limited by the high cost linked to their complex production processes [37], expensive film technology [40], and high heat treatment [53]. This includes the melting of highly reactive metals such as lithium and sodium, which raises a safety concern [41].

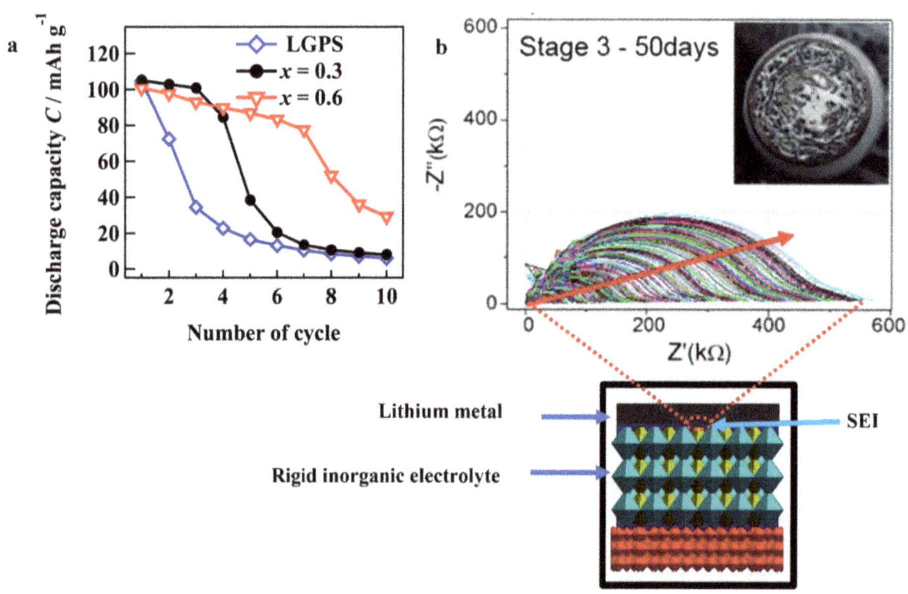

Figure 9.3. (a) The discharge capacity of a combined sulfide and oxide inorganic electrolyte (reprinted from [45], Copyright 2016, with permission from Elsevier). (b) The SEI of inorganic electrolytes (reprinted with permission from [35], Copyright 2017, American Chemical Society).

9.4.2 Solid polymer electrolytes

Polymers are large synthetic or organic compounds composed of numerous smaller building units known as monomers. The fundamental properties of polymers are determined by their composition, size, structure, morphology, and internal forces [30]. The interest in polymers as potential electrolytes dates back to 1973, when Wright published research work on PEO [54] in which the ionic conductivity was argued to be facilitated by coordination between the cation and the lone pairs on the oxygen atom in the polymer system. Since then, research into various categories of polymers has been reported, based on their advantages such as environmental friendliness, affordability [55], safety [56], and attractive flexibility [57]. They are also less reactive with lithium and have moderate interfacial contact compared to inorganic materials [58]. The quest to enhance the ionic conductivity of polymers remains an active research field.

9.4.3 Types of polymer electrolytes

Several types of polymers have made significant contributions to battery technology [59, 60], based on a wide range of properties attributed to their dual (both crystalline and amorphous) existence [61]. These two phases contribute directly to their electrochemical properties and indirectly to their other physical properties, such as their mechanical and thermal properties [60]. For instance, the amorphous phase improves ionic conductivity, while crystallinity is important for mechanical stability [7].

9.4.3.1 Poly (ethylene oxide)

PEO is a polymer compound derived from natural petroleum. It has the molecular formula ($H{-}(O{-}CH_2{-}CH_2)_n{-}OH$) and a molecular weight greater than 20 000 g mol^{-1}. Otherwise known as polyethylene glycol (PEG) when its molecular weight is below 20 000 g mol^{-1}, this polymer was among the first polymers to be investigated [62]. This is because of its capability to complex with a wide range of conductive salts, thus allowing a wide range of alkali salts to be investigated in a single polymer [63, 64]. However, low ionic conductivity (10^{-8}–10^{-4} S cm^{-1}) has halted its research trajectory due to low salt dissociation, high crystallinity, and low transference number [38, 57]. In addition to a narrow voltage window (<4 V) [7], its safety is also questionable based on a possible low transition melting temperature, making it vulnerable to possible short circuits [65].

A possible strategy to improve the performance of PEG is to lower the glass transition temperature and increase the dielectric constant by incorporating other polymeric materials and plasticizers into it [66]. For example, a combination of poly (arylene ether sulfone)-g-poly(ethylene glycol) (PAES-g-PEG) with functionalized PEG demonstrated better performance, as shown by the Fourier-transform infrared spectroscopy (FTIR) spectra in figure 9.4(a). While the functional groups (–CH_3, –OH, and –CN) increase the dielectric properties, ionic liquid brings a dual contribution as a plasticizer by suppressing crystallinity and introducing outstanding thermal stability (>430 °C) compared to that of neat PEG (220 °C) (figure 9.4(b)), accompanied by better ionic conductivity, a wide electrochemical window and

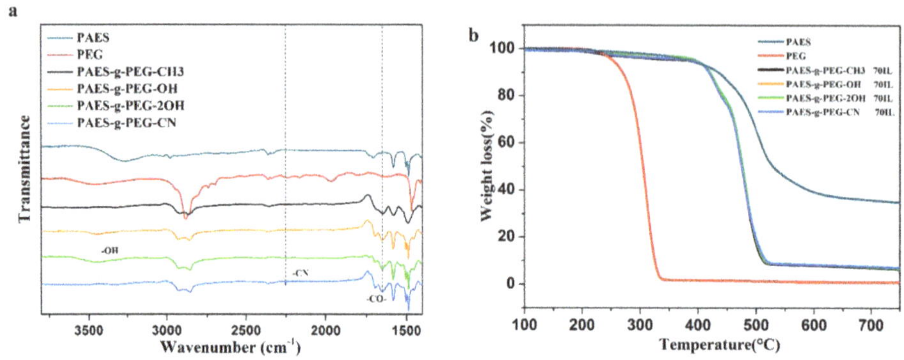

Figure 9.4. (a) The FTIR spectra of pure polymers and functionalized polymers and (b) thermogravimetric analysis (TGA) graphs of PEG with different functional groups. ((a) and (b) reprinted from [66], Copyright (2021), with permission from Elsevier.)

overall cycle stability for the nitrile functional group. The better performance of the nitrile group is attributed to the increased ion cluster dimension, which leads to higher lithium-ion concentration and free volume changes [27].

9.4.3.2 Poly (vinylidene fluoride)

Poly (vinylidene fluoride) (PVDF) is an inert polymer with the general chemical formula $(C_2H_2F_2)_n$ and good insulation properties ($\sim$14 Ω cm^{-1}). As a highly porous and flexible material characterized by high electrochemical stability, PVDF is very attractive for use in polymer batteries [60]. Its porosity allows the polymer to hold a high electrolyte load, while its electrochemical stability means that the electrolyte system can be coupled with high-voltage cathodes. In addition, recent reports have shown that it contributes to sulfur batteries by mitigating shuttle effects [67]. However, the room temperature ionic conductivity of this polymer is still low, and an attempt to improve ionic conductivity resulted in the loss of mechanical properties. Therefore, striking a balance between ionic conductivity and mechanical properties is still challenging. A possible strategy is to combine monomers from different species, a process known as copolymerization. A good example is poly(vinylidene fluoride-*co*-hexafluoropropylene) (P(VDF-*co*-HFP)), a copolymer which has demonstrated better mechanical stability attributed to the VDF part and good ionic conductivity attributed to the HFP amorphous phase [68]. Moreover, a highly electronegative fluorine atom improves anodic stability and contributes to salt dissociation [69]. However, more research is needed to address the safety challenges associated with the exothermic reactions related to the formation of lithium fluoride [70].

Combining P(VdF-*co*-HFP) with other polymers is also an alternative way to enhance the electrochemical properties of PVDF [60]. This minimizes the possible exothermic reaction associated with the high fluorine content in the PVDF-HFP. Figure 9.5(a) shows a combination of PVDF with PEG that forms a PVDF@PEG electrolyte. Composite electrolytes have high thermal (300 °C) and anodic stability above 5.5 V. Moreover, they have demonstrated promising flexibility with constant voltage output in bendable devices (figure 9.5(b)) and better cycle stability at a higher

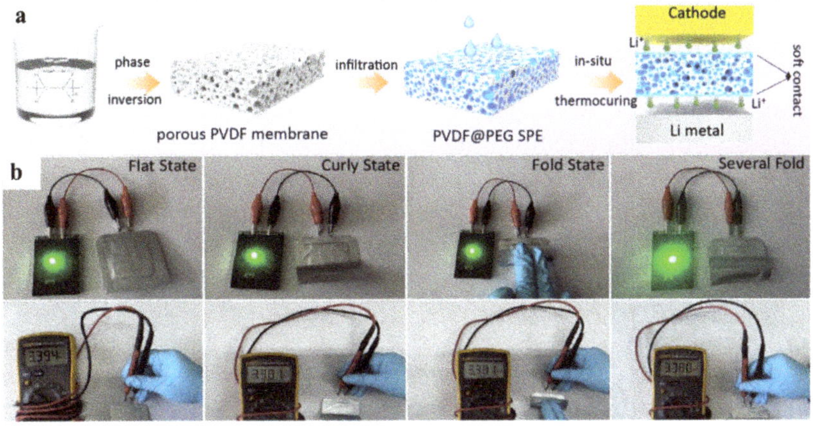

Figure 9.5. (a) The synthesis of PVDF@PEG and (b) a pouch cell in different topologies powering an LED. ((a) and (b) reprinted from [71], Copyright (2022), with permission from Elsevier.)

current density of 0.6 mA cm^{-2}. The promising structural integrity of PVDF is responsible for mitigating the growth of dendrites, improving a battery's cycle life [71].

9.4.3.3 Polyacrylonitrile

Polyacrylonitrile (PAN) is a synthetic, semi-amorphous organic polymer with the formula $(CH_2CHCN)_n$; it is composed of acrylonitrile monomers that form possible copolymers with methyl acrylate and methyl methacrylate [72]. High-energy-density batteries can be fabricated from this polymer material because of its good fiber-forming capabilities and suitability for use with thin-film technology [73]. It is also easy to modify and has superior morphology and a porosity of up to 90%, which is needed to host huge amounts of electrolytes [73]. As a result, it boasts a high ionic conductivity of about 1 uS cm^{-1} at 25 °C, which is suitable for practical applications [70].

Apart from this attractive ionic conductivity, it also provides safe battery operation based on its electrochemical stability [56], thermal stability (melting point of 319 °C) [72], and mechanical properties that suppress dendrites [74]. Unfortunately, the presence of the –CN group speeds up the growth of a resistive passivation layer which reduces the flexibility of the polymer [70] and has detrimental effects on the SEI chemistry and battery life [75, 76]. Combining PAN with other polymers such as poly(dimethylsiloxane) (PDMS), plasticizers, and filler materials is a suitable strategy that may lower its high glass transition point (107 °C) [72, 73].

9.4.3.4 Poly (methyl methacrylate)

Poly (methyl methacrylate) (PMMA), also known as acrylic, is a synthetic polymer composed of methyl methacrylate monomers [30, 77]. Given that the current hydrophobic polyethylene separator faces challenges with poor organic electrolyte wettability, hydrophilic PMMA is a suitable polymer with which to address this challenge. Its affinity for electrolytes improves its electrolyte uptake, which makes it suitable for combination with polymers that have a low affinity for organic electrolytes [78]. In addition to this, the polymer has a 5 V window that provides anodic stability and cathodic stability in high-performance lithium–sulfur batteries [79]. However, it has major disadvantages in terms of difficulties in processing and is also prone to mechanical deformation when filled with huge volumes of electrolyte [80]. It also has a high glass transition temperature ranging from 100 °C to 130 °C, limiting its ionic conductivity at room temperature [77]. The incorporation of other polymers such as PVDF is an important strategy for enhancing its ionic conductivity to as much as 1.31 mS cm^{-1} by contributing to better dielectric properties [80].

9.4.3.5 Poly(dimethylsiloxane)

Poly(dimethylsiloxane) (PDMS) is a silicon-based inorganic polymer with the chemical formula ($CH_3[Si(CH_3)_2O]_nSi(CH_3)_3$) [4]. It is also referred to as dimethylpolysiloxane or dimethicone and has beneficial low glass transition temperatures (−125 °C) and superior chemical stability compared to that of PEO [4, 65, 81]. It is a suitable candidate for cross-linking that meets high safety standards (high thermal stability) and has promising flexibility [82, 83] and a high cation transport number in lithium polymer chemistry [4, 83]. Inspired by this, recent studies have investigated both physical and electrochemical properties derived from cross-linked PDMS and other polymers such as PEO [61]. For instance, poly (ethylene glycol) diacrylate combined with methacrylate-terminated PDMS not only demonstrated better thermal stability (>200 °C) and tensile strength (2.74 MPa) [84] but also displayed a high level of flexibility, which makes it suitable for wearable technology [85]. Another vital contribution of PDMS as an additive is that it improves overall battery stability when incorporated into electrodes [86], organic electrolytes, and separator modifications [87]. It combines well with glass fiber separators and allows huge volume changes of lithium metal; it is therefore a superior candidate for separator and lithium anode modifications.

9.4.4 Hybrid polymer–inorganic composites

Hybrid SSE systems are composed of both polymers and inorganic materials [68, 88], either as a chemical composite [9] or as a design [6]. This combination reduces the interfacial resistance observed in inorganic electrolytes while improving ionic conductivity in polymers. Figure 9.6 shows how the polymer surface increases the contact between the brittle inorganic electrolyte and the electrolyte. In this case, the ionic conductivity of the inorganic electrolyte is maintained while the interfacial chemistry of the polymers is also utilized; the composite therefore realizes both high bulk ionic conductivity and mechanical stability [7]. The main disadvantage of this design is the increased volume of the SSE, which lowers the battery's energy density, especially for highly energy-demanding applications. This makes the above direct compositing of an inorganic electrolyte with a polymer a suitable option for achieving better electrochemical performance with high energy density if well optimized [75].

Inorganic electrolytes classified as active fillers can directly or indirectly contribute to ionic conductivity and electrochemical and cycle stability [68]. In contrast, inactive fillers indirectly affect ionic conductivity by suppressing the crystalline phases in polymers [89].

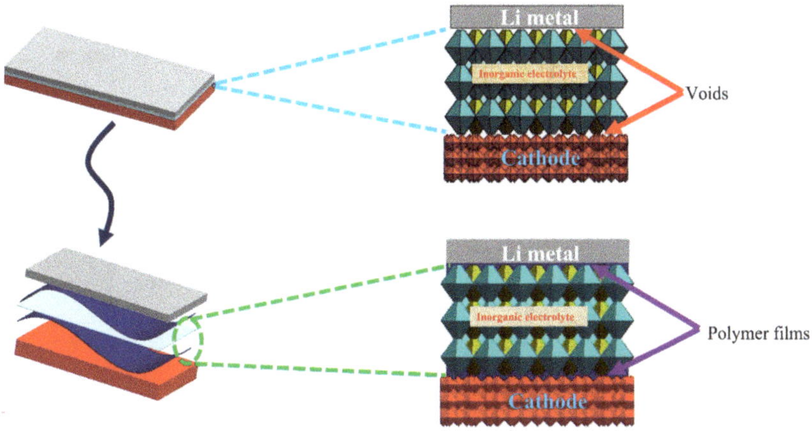

Figure 9.6. A schematic illustration of a hybrid solid-state battery design.

9.4.4.1 Inactive fillers

Inactive fillers or passive fillers are inorganic materials doped with polymers (also known as ceramic polymers) to indirectly enhance ionic conductivity [25, 90]. They include silicon oxide [63], titanium oxide (TiO_2) [90], aluminum oxide (Al_2O_3) [91], zirconium oxide (ZrO_2) [92], zinc oxide (ZnO) [90], boron nitride (BN) [73], and metal–organic frameworks (MOFs). In addition to improving ionic conductivity, they improve other physical properties of polymers [7]. For instance, silicon oxide (silica) and Al_2O_3 make significant contributions in improving the wettability and mechanical properties of PEO and PVDF [29, 91, 93], while TiO_2 and ZnO increase the ionic conductivity, lithium transference number, and mechanical stability of polymers [90]. Besides being involved in cross-linking, aluminum oxide fillers contribute ionic conductivity thermal stability [92] and improve stability when the filler is exposed to a lithium metal anode [24].

BN has dominated this field because of its outstanding and unique contributions ranging from liquid- and solid-state electrolytes [73] to separators [94] and electrodes [95]. BN is thermally stable as an electrolyte additive and has superb insulation and anti-corrosion properties [96]. It also prevents a reaction with lithium metal and boosts ionic conductivity in the grain boundary of ceramics up to a practically acceptable limit [97]. Similarly, it can be added to an electrode to improve the contact between the electrode and the electrolyte interphase. As shown in figure 9.7(a), this is due to better cohesive forces than those of pure lithium metal, contributing to more than 300 h of stable cycling. This cycling stability may also result from other superior boron nitride properties, such as its ability to relax grain boundaries (low activation energy) [98], increase the modulus constant (dendrite mitigation) [99], and interact with anions (high lithium transference number) [100].

In addition, boron nitride also improves the electrochemical stability of other inorganic fillers. For example, the electrochemical stability of $Li_{1.3}Al_{0.3}Ti_{1.7}(PO_4)_3$ (LATP) was extended by boron nitride from 2.17 vs Li/Li^+ up to 4.5 V vs Li/Li^+. Overall, a battery assembled from these electrolytes demonstrated higher coulombic

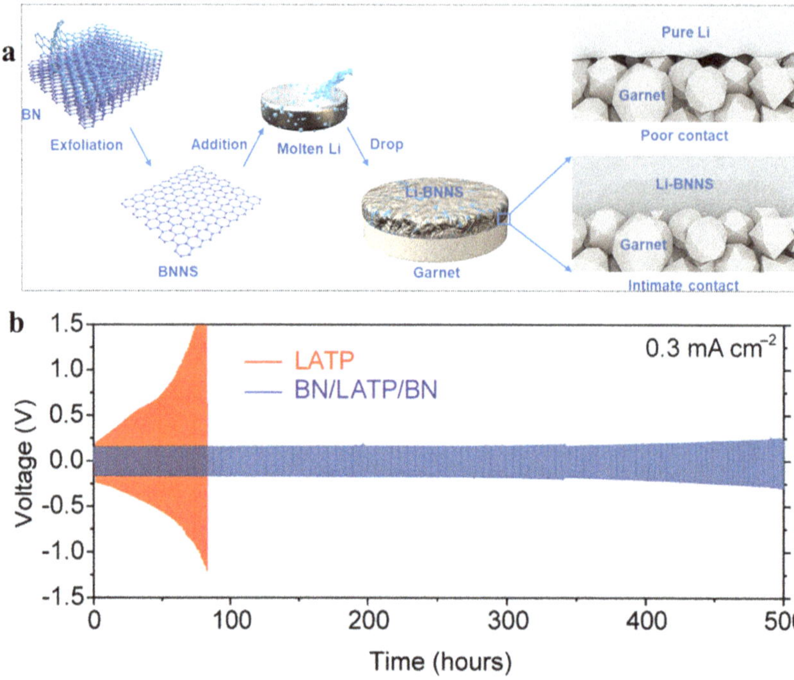

Figure 9.7. (a) A schematic illustration showing the preparation and modification of a lithium anode using boron nitride (reprinted with permission from [98], Copyright (2019), American Chemical Society). (b) The plating and stripping of LAP and a composite of BN/$Li_{1.3}Al_{0.3}Ti_{1.7}(PO_4)_3$/BN (reprinted from [95], Copyright (2019), with permission from Elsevier).

efficiency and a constant delivery capacity of 142 mAh g^{-1} and remained stable for 500 h (a performance better than that of pure LATP (<100 h)), as shown in figures 9.7(b) and (e), respectively [95].

In conclusion, inactive fillers contribute to electrolyte ionic conductivity and wettability, improving the SEI, mechanical, thermal, and electrochemical stability, and overall battery life. This has translated to better battery shelf life. In addition, they also have economic advantages over active fillers because of the absence of lithium ions and easily affordable synthetic processes [7]. However, their ionic conductivity is still low, prompting the next strategy of exploring active fillers that actively contribute to ion conductivity.

9.4.4.2 Active fillers

Active fillers are inorganic materials that contain lithium ions; therefore, they are directly involved in ionic conductivity [7, 101]. They are mostly composed of inorganic electrolytes added to polymers in small quantities. They include LATP [95], LAGP, LLZO [102], $Li_{10}GeP_2S_{12}$ (LGPS) [103], $Li_{3/8}Sr_{7/16}Zr_{1/4}Ta_{3/4}O_3$ (LSZT) [101] and Li–La–Ti–O (LLTO). Figures 9.8(a) and (b) clearly show increases in transference number and anodic stability, respectively, when LLZO was added to PEO [102]. This is because of the introduction of additional lithium ions from the active filler (LLZO). The mechanical

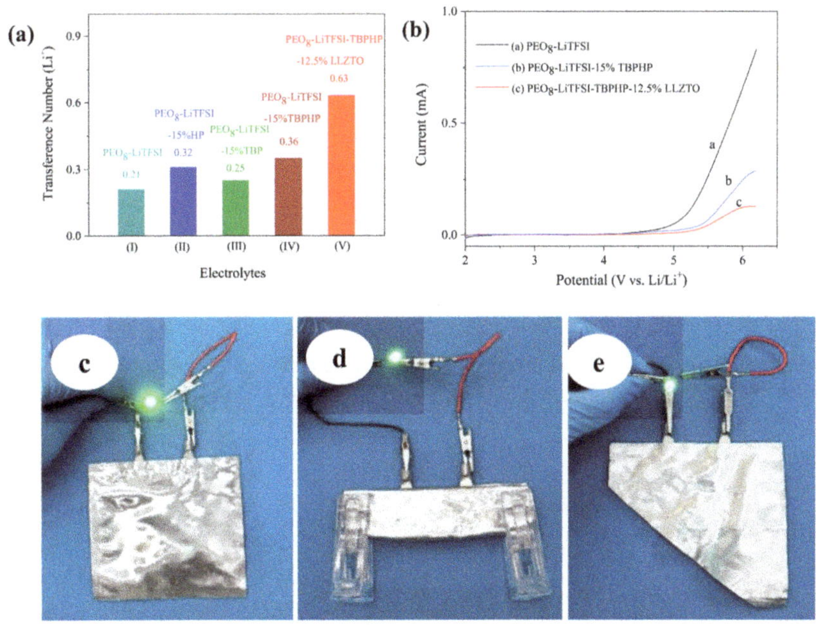

Figure 9.8. A comparison of (a) the lithium transference numbers, (b) the electrochemical stabilities of PEO and PEO/LLZO, the performance of a polymer in lighting an LED when the battery is (c) flat, (d) folded, and (e) halved. ((a)–(e) reprinted from [102], Copyright (2019), with permission from Elsevier.)

and thermal stability of polymers can also be improved by inorganic materials [89] while still maintaining their performance in flat, folded, and deformed topologies, as shown in figures 9.8(c), (d), and (e), respectively. Other polymers such as PVDF-HFP have been investigated; they benefited from a doubling in ionic conductivity and mechanical stability and improved mechanical stability (from 9.22 to 8.82 MPa) [68].

9.4.5 Metal–organic frameworks

Compared with classical nano inorganic additives, MOFs have a unique structural coordination in which the organic ligands are linked to either a central atom or an inorganic cluster. This network forms highly porous inorganic organic crystalline materials characterized by a uniform pore distribution. In addition to their electrical insulation properties, MOFs are potential additives in SSEs for three major reasons: (i) they exist at the nanoscale and have high surface areas which improve the contact at the SEI [38]; (ii) their adjustable pores control the motion of lithium ions, contributing to uniform lithium-ion deposition; (iii) the open metal sites limit anion movement, thus promoting salt dissociation and freeing lithium ions [104]. All these properties mitigate dendrites and improve the overall cycle life of electrolytes [105] and battery safety [106]. For example, as opposed to a neat polymer that shrinks after 40 s of flame exposure (figure 9.9(a)), PEO composited with MOF extended this time to 80 s (figure 9.9(b)) [106]. Furthermore, the MOF composite had better cycle stability (>1400 h) than PEO (<200 h) under the same conditions (figure 9.9(c)).

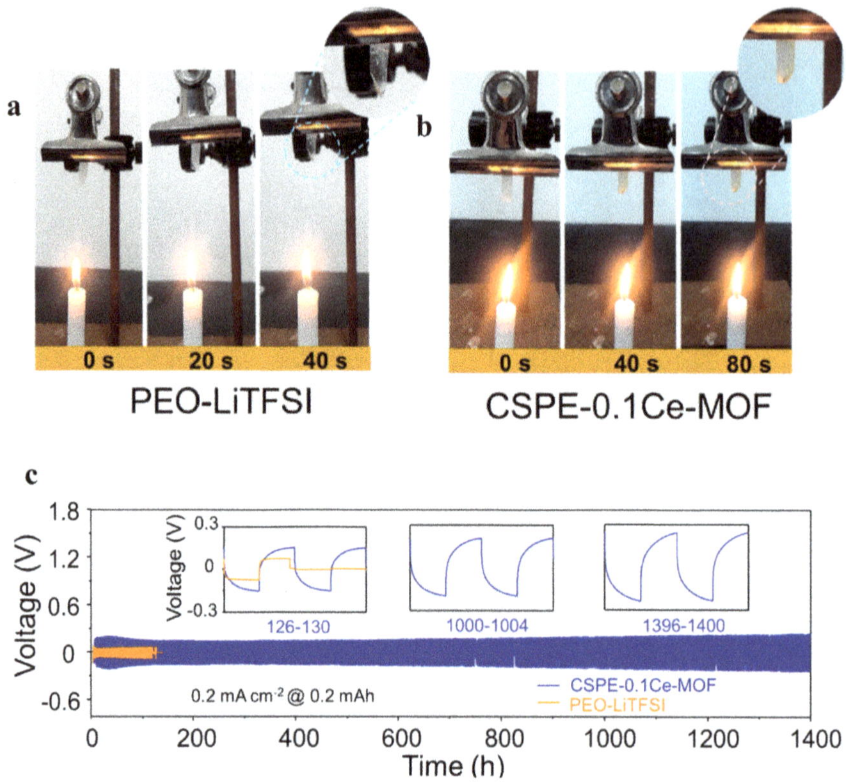

Figure 9.9. A flame test of (a) a PEO composite, (b) a PEO/MOF composite and (c) the cycle stability of PEO and PEO/MOF. ((a)–(c) reprinted from [106], Copyright (2021), with permission from Elsevier.)

9.4.6 Gel electrolytes

Gel electrolytes are liquid electrolyte systems stored in porous solid-state electrolytes (polymers, inorganic materials, and composites) [78]. This incorporates the advantages of liquid electrolytes and the advantages of SSEs. In this case, the liquid electrolyte primarily controls the ionic conductivity, while the SSE matrix contributes to mechanical, thermal, and electrochemical stabilities [64]. The higher the volume of the liquid electrolyte, the higher the resultant ionic conductivity; therefore, highly porous polymers are considered for gel electrolytes [75]. A case in point is a highly porous PVDF-HFP polymer with a honeycomb-like structure (a porosity of 78%) and an 86.2% electrolyte uptake. This porosity contributed to a high ionic conductivity of 1.03 mS cm^{-1} at room temperature while maintaining high thermal stability (350 °C) [69].

Even though porous polymers contribute to high ionic conductivity, this compromises the mechanical properties of the polymer, as this use is usually accompanied by swelling and distortion of the polymeric structure (low mechanical stability). A new strategy is to use special liquids such as N-methyl-2-pyrrolidine (NMP) that react with the polymer to improve its mechanical integrity [2]. A comparative analysis of NMP and tetrahydrofuran (THF) showed that a gel electrolyte prepared with NMP not only had better mechanical stability but also provided a stable SEI as compared to that of a gel electrolyte prepared with THF. In practical terms, when both electrolytes were charged and discharged at a constant current of 0.1 mA cm^{-1}, the NMP gel sustained 500 h of charge-discharge, while the THF cycles were limited to 300 h (figure 9.10(a)). Scanning electron microscopy (SEM) and x-ray photoelectron spectroscopy (XPS) elemental analyses of the lithium anode revealed the important components for stable SEI formation. The smooth anode surface in figure 9.10(b) is a clear indication of dendrite suppression, which was probably due to improved mechanical stability. The XPS C 1s spectra confirmed the presence of C–C/C–H at peaks at 284.6 eV and $-CF_3$ at 292.8 eV, which are associated with PVDF-HFP and LiTFSI, respectively. The presence of peaks at 288.7 and 285.95 eV represented CO= and C–N/C–O, respectively, from NMP, Li_2CO_3, and $ROCO_2Li$. Some of the most important components that contribute to a stable SEI, namely $-CF_3$ and LiF, are represented by the F 1s spectrum with peaks at 688.6 and 685 eV, respectively.

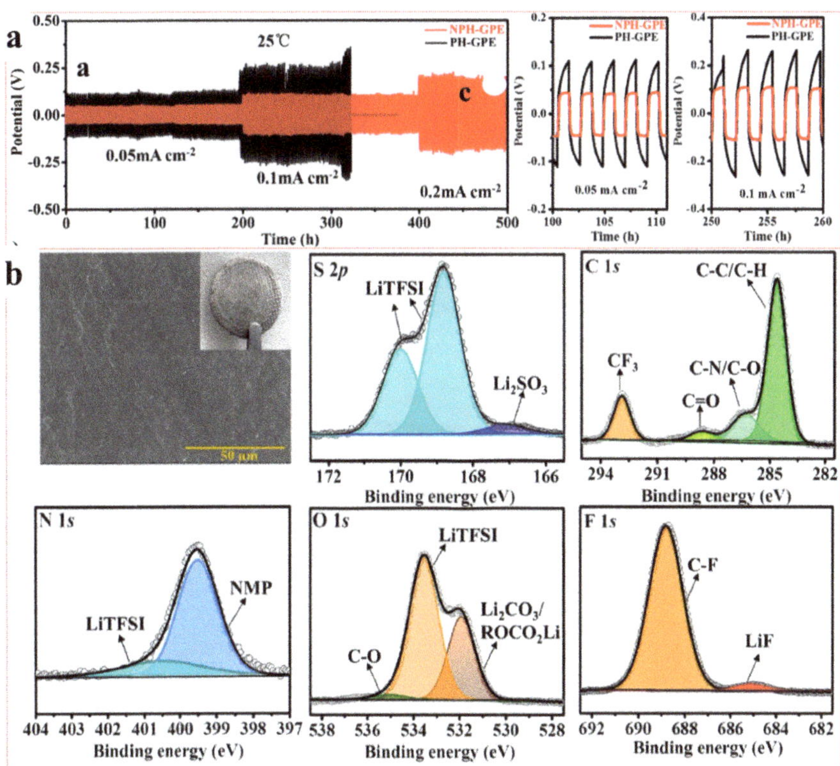

Figure 9.10. (a) The voltage profiles of NMP and THF-based gel electrolytes at increasing currents and (b) the SEM and XPS spectra of the lithium anode. ((a) and (b) reprinted from [2], Copyright (2020), with permission from Elsevier.)

Polymers lose their mechanical stability after they become swollen with huge volumes of electrolytes. An alternative approach would be to opt for porous inorganic fillers with better mechanical [107] and electrochemical properties [103]. For instance, modified montmorillonite (MMT) clay has a high surface area for electrolyte storage and a high dielectric constant to dissociate the same electrolyte. An optimized weight (4 wt%) clay improved the mechanical properties, ionic conductivity (2.3×10^{-3} S cm^{-1}), and wettability of PVDF [108].

In our opinion, gel electrolytes are the only group of electrolytes that can average all the important properties ranging from thermal properties to mechanical and electrochemical properties and ionic conductivity. Their main contribution has been reducing the interfacial resistance in polymers and inorganic electrolytes. However, the use of organic electrolytes as liquids in gels should be replaced with stable liquids, such as less viscous ionic liquids.

9.5 Challenges in solid-state batteries

Solid-state batteries are still in their infancy in many ways, and significant progress is needed to fully realize their practical applications. Initially, ionic conductivity was the biggest obstacle, but some solid-state ionic conductivity can match that of its liquid counterparts whose ionic conductivity is higher than 1 mS cm^{-1} [109], and other electrolytes have as much as 10 mS cm^{-1}, which is enough for high-energy and power-intensive applications [15]. The major challenges faced by solid-state batteries are summarized in figure 9.11, while minor challenges that are specific to certain electrolytes are discussed in this section.

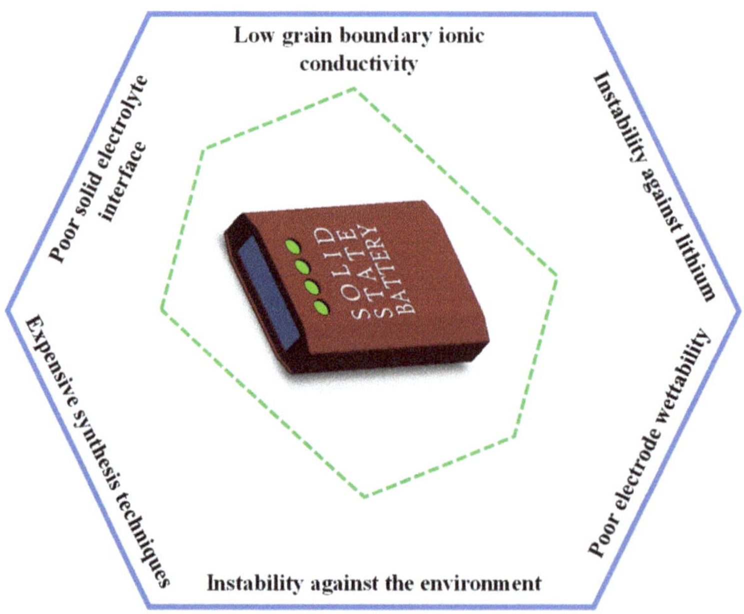

Figure 9.11. Major challenges in solid-state electrolytes.

In addition to better mechanical, chemical, electrochemical, and thermal properties, SSEs need to be affordable. Most solid-state batteries require high temperatures to improve their electrode–electrolyte interphases [14] or alternatively, expensive techniques such as thin-film technology. However, the high-temperature treatment process also raises a safety concern related to the handling of reactive lithium and sodium metals. As a result, safer approaches, such as the use of computational and theoretical studies (such as density functional theory analysis) to understand the fundamental thermodynamics between the electrode and the electrolyte, are highly welcome [14].

The discussion of mechanical stability is still an active research topic in SSEs because reports show that they are still vulnerable to dendritic penetration [110]. Therefore, it is essential to investigate filler materials such as boron nitride and MOFs because they have the chemistry to control lithium deposition by electrostatically interacting with anions [110].

9.6 Future research trajectories

Solid-state materials provide a safe alternative material to organic liquids because of their thermal stability. Stable materials are suitable for high-temperature operation and have an economic advantage over conventional high-temperature operation because they do not have the extra cost of a battery cooling system. Moreover, the advantage of placing SSE cells in one compartment gives SSEs an edge over the traditional liquid electrolytes because it enables a potential design that achieves high energy density. This is important in automobiles, where less volume and weight translate to covering a greater distance before recharging. In addition, they have a high modulus, which contributes to superior mechanical properties and can mitigate the growth of dendrites and/or slow down physical penetration in the case of a physical abuse condition. However, solid-state materials need improvement to compete favorably with conventional organic liquids and meet safety requirements. First, depending on the structure, the presence of grain boundaries in these materials still makes them vulnerable to dendrite penetration. Besides, the rigid nature of solid inorganic materials provides a poor interface with the electrode and is unable to accommodate huge volume changes during battery cycling. Due to this, solid inorganic materials constantly break during battery cycling, a process which results in high cell interfacial impedance, leading to poor battery cycle life.

Solid polymers offer a more flexible alternative to rigid inorganic material, as they can conform to and improve contact with the surface of the electrode, hence reducing the resistance at the interface. This ability can improve battery cycle life but causes a significant loss of mechanical properties. Once the mechanical stability is compromised, the chances of dendritic penetration across the polymer increase. Due to this, optimized composites of low-glass-transition-temperature polymers and inorganic electrolyte materials make up another active research area. This research field of composite electrolytes shows some promising advantages, such as improved interfacial chemistry and the reduced environmental footprints associated with polymer materials. While retaining the mechanical stability of solid-state materials, research targeting interfacial ionic conductivity mechanisms at the micro level can speed up the commercial recognition of these electrolytes.

Over the last few years, researchers have investigated hybrid energy storage devices that combine batteries and supercapacitors using a shared electrolyte. Despite this, there is a lack of thorough investigation into the use of hybrid energy storage devices with solid electrolytes in high-temperature applications. Polymeric hydrogels have emerged as promising solid electrolytes for such applications, offering advantages such as anti-freezing properties. Consequently, future research efforts should be directed toward comprehensive studies to assess the thermal safety characteristics of hybrid energy storage devices that employ a shared solid electrolyte.

On a commercial scale, the design of current low-power delivery energy storage is key to transforming this technology into high power delivery battery systems. Second, safe materials and short charging times that offer increased range without recharging are the future of the automobile industry. This goes hand in hand with cost reduction of the processes involved, as stated by automobile companies such as Toyota, Ford, and Hyundai. Based on the current situation, solid batteries will fully realize commercialization between 2025 and 2030, depending on individual companies. This means that it is only a matter of time before solid-state materials replace liquid electrolytes on a commercial scale.

9.7 Conclusions

This chapter discussed the basics of ionic conductivity in solid-state electrolytes, exploring their various types and their potential contributions to advancements in battery technology. In addition, we explored their chemistry and the chemical interactions among their composites. We discussed the challenges linked to existing solid-state electrolytes in detail and outlined trajectory strategies aimed at accelerating their recognition in the commercial sphere. It was concluded that these composites are suitable for low-powered portable devices. This means that it is only a matter of time before solid-state materials replace liquid electrolytes on a commercial scale. Based on their safety and high energy density, solid-state batteries have a future in electric mobility ranging from cold to low-temperature environments.

List of abbreviations

BN	Boron nitride
DFTB	Density-functional tight-binding
FTIR	Fourier-transform infrared spectroscopy
LiPON	Lithium phosphorus oxynitride
LISICON	Lithium superionic conductor
MMT	Modified montmorillonite.
MOF	Metal–organic framework
NASICON	Natrium superionic conductor
NMP	*N*-methyl-2-pyrrolidine

PAES-g-PEG	Poly(arylene ether sulfone)-g-poly(ethylene glycol)
PAN	Poly(acrylonitrile)
PDMS	Poly(dimethylsiloxane)
PEG	Poly(ethylene glycol)
PEO	Poly(ethylene oxide)
PMMA	Poly(methyl methacrylate)
PVDF	Poly(vinylidene fluoride)
PVDF-HFP	Poly(vinylidene fluoride-co-hexafluoropropylene)
SEI	Solid electrolyte interphase
SEM	Scanning electron microscopy
SSE	Solid-state electrolyte
TGA	Thermogravimetric analysis
THF	Tetrahydrofuran
VDF	Vinylidene fluoride
XPS	X-ray photoelectron spectroscopy

References

[1] Mandade P, Weil M, Baumann M and Wei Z 2023 Environmental life cycle assessment of emerging solid-state batteries: a review *Chem. Eng. J. Adv.* **13** 100439

[2] Jie J, Liu Y, Cong L, Zhang B, Lu W, Zhang X, Liu J, Xie H and Sun L 2020 High-performance PVDF-HFP based gel polymer electrolyte with a safe solvent in Li metal polymer battery *J. Energy Chem.* **49** 80–8

[3] Zhang M, Zhang A-M, Chen Y, Xie J, Xin Z-F, Chen Y-J, Kan Y-H, Li S-L, Lan Y-Q and Zhang Q 2011 Polyoxovanadate-polymer hybrid electrolyte in solid-state batteries *Energy Storage Mater.* **29** 172–81

[4] Wang H, Chen S, Li Y, Liu Y, Jing Q, Liu X, Liu Z and Zhang X 2021 Organosilicon-based functional electrolytes for high-performance lithium batteries *Adv. Energy Mater.* **11** 2101057

[5] Chen L, Liu Y and Fan L-Z 2017 Enhanced interface stability of polymer electrolytes using organic cage-type Cucurbit[6]uril for lithium metal batteries *J. Electrochem. Soc.* **164** A1834–40

[6] Ma X, Xu Y, Zhang B, Xue X, Wang C, He S, Lin J and Yang L 2020 Garnet Si–$Li_7La_3Zr_2O_{12}$ electrolyte with a durable, low resistance interface layer for all-solid-state lithium metal batteries *J. Power Sources* **453** 227881

[7] Didwal P N, Singhbabu Y N, Verma R, Sung B-J, Lee G-H, Lee J-S, Chang D R and Park C-J 2021 An advanced solid polymer electrolyte composed of poly(propylene carbonate) and mesoporous silica nanoparticles for use in all-solid-state lithium-ion batteries *Energy Storage Mater.* **37** 476–90

[8] Kato Y, Hori S, Saito T, Suzuki K, Hirayama M, Mitsui A, Yonemura M, Iba H and Kanno R 2016 High-power all-solid-state batteries using sulfide superionic conductors *Nat. Energy* **1** 16030

[9] Zhang W, Nie J, Li F, Wang Z L and Sun C 2018 A durable and safe solid-state lithium battery with a hybrid electrolyte membrane *Nano Energy* **45** 413–9

[10] Famprikis T, Canepa P, Dawson J A, Islam M S and Masquelier C 2019 Fundamentals of inorganic solid-state electrolytes for batteries *Nat. Mater.* **18** 1278–91

[11] Das A, Sahu S, Mohapatra M, Verma S, Bhattacharyya A J and Basu S 2022 Lithium-ion conductive glass-ceramic electrolytes enable safe and practical Li batteries *Mater. Today Energy* **29** 101118

[12] Choudhury S, Stalin S, Vu D, Warren A, Deng Y, Biswal P and Archer L A 2019 Solid-state polymer electrolytes for high-performance lithium metal batteries *Nat. Commun.* **10** 4398

[13] Ahmed D and Maraz K M 2023 Polymer electrolyte design strategies for high-performance and safe lithium-ion batteries: recent developments and future prospects *Mater. Eng. Res.* **5** 245–55

[14] Richards W D, Miara L J, Wang Y, Kim J C and Ceder G 2021 Interface stability in solid-state batteries *Chem. Mater.* **28** 266–73

[15] Ohno S, Banik A, Dewald G F, Kraft M A, Krauskopf T, Minafra N, Till P, Weiss M and Zeier W G 2020 Materials design of ionic conductors for solid state batteries *Prog. Energy* **2** 022001

[16] Zhao Q, Stalin S, Zhao C-Z and Archer L A 2020 Designing solid-state electrolytes for safe, energy-dense batteries *Nat. Rev. Mater.* **5** 229–52

[17] Wang C *et al* 2020 Garnet-type solid-state electrolytes: materials, interfaces, and batteries *Chem. Rev.* **120** 4257–300

[18] Li C, Wang Z, He Z, Li Y, Mao J, Dai K, Yan C and Zheng J 2021 An advance review of solid-state battery: challenges, progress and prospects *Sustain. Mater. Technol.* **29** e00297

[19] Chen R, Li Q, Yu X, Chen L and Li H 2020 Approaching practically accessible solid-state batteries: stability issues related to solid electrolytes and interfaces *Chem. Rev.* **120** 6820–77

[20] Wright P V 1975 Electrical conductivity in ionic complexes of poly(ethylene oxide) *J. Polym* **7** 319–27

[21] Hallinan D T and Balsara N P 2013 Polymer electrolytes *Annu. Rev. Mater. Res.* **43** 503–25

[22] Zhu L *et al* 2022 Enhancing ionic conductivity in solid electrolyte by relocating diffusion ions to under-coordination sites *Sci. Adv.* **8** 7698

[23] Krauskopf T, Culver S P and Zeier W G 2018 Bottleneck of diffusion and inductive effects in $Li_{10}Ge_{1-x}Sn_xP_2S_{12}$ *Chem. Mater.* **30** 1791–8

[24] Bertasi F, Pagot G, Vezzù K, Negro E, Sideris P J, Greenbaum S G, Ohno H, Scrosati B and Di Noto V 2018 Exotic solid state ion conductor from fluorinated titanium oxide and molten metallic lithium *J. Power Sources* **400** 16–22

[25] Jung S, Kim D W, Lee S D, Cheong M, Nguyen D Q, Cho B W and Kim H S 2009 Fillers for solid-state polymer electrolytes: highlight *Bull. Korean Chem. Soc.* **30** 2355–61

[26] Su Y, Xu F, Zhang X, Qiu Y and Wang H 2023 Rational design of high-performance PEO/ceramic composite solid electrolytes for lithium metal batteries *Nano-Micro Lett.* **15** 82

[27] Jung H Y, Mandal P, Jo G, Kim O, Kim M, Kwak K and Park M J 2017 Modulating ion transport and self-assembly of polymer electrolytes via end-group chemistry *Macromolecules* **50** 3224–33

[28] He X, Zhu Y and Mo Y 2017 Origin of fast ion diffusion in super-ionic conductors *Nat. Commun.* **8** 15893

[29] Arifeen W U, Abideen Z U, Guru Parakash N, Xiaolong L and Ko T J 2023 Effects of a high-performance, solution-cast composite electrolyte on the host electrospun polymer membrane for solid-state lithium metal batteries *Mater. Today Energy* **33** 101270

[30] Whba R A G, TianKhoon L, Su'ait M S, Rahman M Y A and Ahmad A 2020 Influence of binary lithium salts on 49% poly(methyl methacrylate) grafted natural rubber based solid polymer electrolytes *Arab. J. Chem.* **13** 3351–61

[31] Dawson J A, Canepa P, Famprikis T, Masquelier C and Islam M S 2018 Atomic-scale influence of grain boundaries on Li-ion conduction in solid electrolytes for all-solid-state batteries *J. Am. Chem. Soc.* **140** 362–8

[32] Kong Y, Li Y, Li J, Hu C, Wang X and Lu J 2018 Li ion conduction of perovskite $Li_{0.375}Sr_{0.4375}Ti_{0.25}Ta_{0.75}O_3$ and related compounds *Ceram. Int.* **44** 3947–50

[33] Thangadurai V, Kaack H and Weppner W J F 2004 Novel fast lithium ion conduction in garnet-type $Li_5La_3M_2O_{12}$ (M=Nb, Ta) *J. Am. Ceram. Soc.* **86** 437–40

[34] Miao X, Wang H, Huang J, Meng K, Lai Y and Lin Z 2018 Interface engineering on inorganic solid state electrolytes for high performance lithium metal batteries *Energy Environ. Sci.* **11** 772–99

[35] Chung H and Kang B 2017 Mechanical and thermal failure induced by contact between a $Li_{1.5}Al_{0.5}Ge_{1.5}(PO_4)_3$ solid electrolyte and Li metal in an all solid-state Li cell *Chem. Mater.* **29** 8611–9

[36] Deng Y, Eames C, Fleutot B, David R, Chotard J-N, Suard E, Masquelier C and Islam M S 2017 Enhancing the lithium ion conductivity in lithium superionic conductor (LISICON) solid electrolytes through a mixed polyanion effect *ACS Appl. Mater. Interfaces* **9** 7050–8

[37] Huang J, Liang F, Hou M, Zhang Y, Chen K and Xue D 2020 Garnet-type solid-state electrolytes and interfaces in all-solid-state lithium batteries: progress and perspective *Appl. Mater. Today* **20** 100750

[38] Zuo X, Chang K, Zhao J, Xie Z, Tang H, Li B and Chang Z 2016 Bubble-template-assisted synthesis of hollow fullerene-like MoS_2 nanocages as a lithium ion battery anode material *J. Mater. Chem.* A **4** 51–8

[39] Bohnke O 2008 The fast lithium-ion conducting oxides $Li_{3x}La_{2/3-x}TiO_3$ from fundamentals to application *Solid State Ionics* **179** 9–15

[40] LaCoste J D, Zakutayev A and Fei L 2021 A review on lithium phosphorus oxynitride *J. Phys. Chem.* C **125** 3651–67

[41] Lee S, Jung S, Yang S, Lee J-H, Shin H, Kim J and Park S 2022 Revisiting the LiPON/Li thin film as a bifunctional interlayer for NASICON solid electrolyte-based lithium metal batteries *Appl. Surf. Sci.* **586** 152790

[42] Li J and Lai W 2020 Structure and ionic conduction study on Li_3PO_4 and LiPON (lithium phosphorous oxynitride) with the density-functional tight-binding (DFTB) method *Solid State Ionics* **351** 115329

[43] Song X, Yu W, Zhou S, Zhao L, Li A, Wu A, Li L and Huang H 2023 Enhancement of Mn-doped LiPON electrolyte for higher performance of all-solid-state thin film lithium battery *Mater. Today Phys.* **33** 101037

[44] He X, Ma Y, Liu J, Wang J, Hu X, Dong H and Huang X 2023 Improved electrochemical performance and chemical stability of thin-film lithium phosphorus oxynitride electrolyte by appropriate fluorine plasma treatment *Electrochim. Acta* **454** 142411

[45] Sun Y, Suzuki K, Hara K, Hori S, Yano T, Hara M, Hirayama M and Kanno R 2016 Oxygen substitution effects in $Li_{10}GeP_2S_{12}$ solid electrolyte *J. Power Sources* **324** 798–803

[46] Jung W D *et al* 2020 Superionic halogen-rich Li-argyrodites using *in situ* nanocrystal nucleation and rapid crystal growth *Nano Lett.* **20** 2303–9

[47] Lu P *et al* 2023 Realizing long-cycling all-solid-state Li–In||TiS_2 batteries using $Li_{6+x}M_xAs_{1-x}S_5$ (M=Si, Sn) sulfide solid electrolytes *Nat. Commun.* **14** 4077

[48] Seino Y, Ota T, Takada K, Hayashi A and Tatsumisago M 2014 A sulphide lithium super ion conductor is superior to liquid ion conductors for use in rechargeable batteries *Energy Environ. Sci.* **7** 627–31
[49] Kamaya N *et al* 2011 A lithium superionic conductor *Nat. Mater.* **10** 682–6
[50] Zhou L, Assoud A, Zhang Q, Wu X and Nazar L F 2019 New family of argyrodite thioantimonate lithium superionic conductors *J. Am. Chem. Soc.* **141** 19002–13
[51] Zhao Y, Liu Z, Xu J, Zhang T, Zhang F and Zhang X 2019 Synthesis and characterization of a new perovskite-type solid-state electrolyte of $Na_{1/3}La_{1/3}Sr_{1/3}ZrO_3$ for all-solid-state sodium-ion batteries *J. Alloys Compd.* **783** 219–25
[52] Etourneau J, Portier J and Ménil F 1992 The role of the inductive effect in solid state chemistry: how the chemist can use it to modify both the structural and the physical properties of the materials *J. Alloys Compd.* **188** 1–7
[53] Li X *et al* 2019 Air-stable Li_3InCl_6 electrolyte with high voltage compatibility for all-solid-state batteries *Energy Environ. Sci.* **12** 2665–71
[54] Fenton D E, Parker J M and Wright P V 1973 Complexes of alkali metal ions with poly (ethylene oxide) *Polymer* **14** 589
[55] Sirengo K, Babu A, Brennan B and Pillai S C 2023 Ionic liquid electrolytes for sodium-ion batteries to control thermal runaway *J. Energy Chem.* **81** 321–38
[56] Ludwig K B, Correll-Brown R, Freidlin M, Garaga M N, Bhattacharyya S, Gonzales P M, Cresce A V, Greenbaum S, Wang C and Kofinas P 2023 Highly conductive polyacrylonitrile-based hybrid aqueous/ionic liquid solid polymer electrolytes with tunable passivation for Li-ion batteries *Electrochim. Acta* **453** 142349
[57] Abraham K M, Jiang Z and Carroll B 1997 Highly conductive PEO-like polymer electrolytes *Chem. Mater.* **9** 1978–88
[58] Zhang Q, Liu K, Liu K, Zhou L, Ma C and Du Y 2020 Imidazole containing solid polymer electrolyte for lithium ion conduction and the effects of two lithium salts *Electrochim. Acta* **351** 136342
[59] Gou J, Liu W and Tang A 2021 To improve the interfacial compatibility of cellulose-based gel polymer electrolytes: a cellulose/PEGDA double network-based gel membrane designed for lithium-ion batteries *Appl. Surf. Sci.* **568** 150963
[60] Liang Y F, Xia Y, Zhang S Z, Wang X L, Xia X H, Gu C D, Wu J B and Tu J P 2019 A preeminent gel blending polymer electrolyte of poly(vinylidene fluoride-hexafluoropropylene)–poly(propylene carbonate) for solid-state lithium ion batteries *Electrochim. Acta* **296** 1064–9
[61] Kalybekkyzy S, Kopzhassar A-F, Kahraman M V, Mentbayeva A and Bakenov Z 2021 Fabrication of UV-crosslinked flexible solid polymer electrolyte with PDMS for Li-ion batteries *Polymers* **13** 15
[62] Xue Z, He D and Xie X 2015 Poly(ethylene oxide)-based electrolytes for lithium-ion batteries *J. Mater. Chem.* A **3** 19218–53
[63] Liu Y, Lee J Y and Hong L 2004 *In situ* preparation of poly(ethylene oxide)–SiO_2 composite polymer electrolytes *J. Power Sources* **129** 303–11
[64] Bocharova V and Sokolov A P 2020 Perspectives for polymer electrolytes: a view from fundamentals of ionic conductivity *Macromolecules* **53** 4141–57
[65] Grewal M S, Tanaka M and Kawakami H 2019 Free-standing polydimethylsiloxane-based cross-linked network solid polymer electrolytes for future lithium ion battery applications *Electrochim. Acta* **307** 148–56

[66] Tian Z and Kim D 2021 Solid electrolyte membranes prepared from poly(arylene ether sulfone)-g-poly(ethylene glycol) with various functional end groups for lithium-ion battery *J. Membr. Sci.* **621** 119023
[67] Yu J, Liu S, Duan G, Fang H and Hou H 2020 Dense and thin coating of gel polymer electrolyte on sulfur cathode toward high performance Li–sulfur battery *Compos. Commun.* **19** 239–45
[68] Lu J, Liu Y, Yao P, Ding Z, Tang Q, Wu J, Ye Z, Huang K and Liu X 2019 Hybridizing poly(vinylidene fluoride-co-hexafluoropropylene) with $Li_{6.5}La_3Zr_{1.5}Ta_{0.5}O_{12}$ as a lithium-ion electrolyte for solid state lithium metal batteries *Chem. Eng. J.* **367** 230–8
[69] Zhang J, Sun B, Huang X, Chen S and Wang G 2014 Honeycomb-like porous gel polymer electrolyte membrane for lithium ion batteries with enhanced safety *Sci. Rep.* **4** 6007
[70] Raghavan P, Manuel J, Zhao X, Kim D-S, Ahn J-H and Nah C 2011 Preparation and electrochemical characterization of gel polymer electrolyte based on electrospun polyacrylonitrile nonwoven membranes for lithium batteries *J. Power Sources* **196** 6742–9
[71] Wang Z, Guo Q, Jiang R, Deng S, Ma J, Cui P and Yao X 2022 Porous poly(vinylidene fluoride) supported three-dimensional poly(ethylene glycol) thin solid polymer electrolyte for flexible high temperature all-solid-state lithium metal batteries *Chem. Eng. J.* **435** 135106
[72] Jyothi N K, Venkataratnam K K, Murty P N and Kumar K V 2016 Preparation and characterization of PAN–KI complexed gel polymer electrolytes for solid-state battery applications *Bull. Mater. Sci.* **39** 1047–55
[73] Aydın H, Çelik S Ü and Bozkurt A 2017 Electrolyte loaded hexagonal boron nitride/polyacrylonitrile nanofibers for lithium ion battery application *Solid State Ionics* **309** 71–6
[74] Tatsuma T, Taguchi M, Iwaku M, Sotomura T and Oyama N 1999 Inhibition effects of polyacrylonitrile gel electrolytes on lithium dendrite formation *J. Electroanal. Chem.* **472** 142–6
[75] Wang X, Hao X, Xia Y, Liang Y, Xia X and Tu J 2019 A polyacrylonitrile (PAN)-based double-layer multifunctional gel polymer electrolyte for lithium–sulfur batteries *J. Membr. Sci.* **582** 37–47
[76] Chai S, Zhang Y, Wang Y, He Q, Zhou S and Pan A 2022 Biodegradable composite polymer as advanced gel electrolyte for quasi-solid-state lithium-metal battery *eScience* **2** 494–508
[77] Ali U, Karim K J B A and Buang N A 2015 A review of the properties and applications of poly (methyl methacrylate) (PMMA) *Polym. Rev.* **55** 678–705
[78] Chiu L-L and Chung S-H 2022 Composite gel-polymer electrolyte for high-loading polysulfide cathodes *J. Mater. Chem.* A **10** 13719–26
[79] Yang D, He L, Liu Y, Yan W, Liang S, Zhu Y, Fu L, Chen Y and Wu Y 2019 An acetylene black modified gel polymer electrolyte for high-performance lithium–sulfur batteries *J. Mater. Chem.* A **7** 13679–86
[80] Zhang J, Chen S, Xie X, Kretschmer K, Huang X, Sun B and Wang G 2014 Porous poly (vinylidene fluoride-co-hexafluoropropylene) polymer membrane with sandwich-like architecture for highly safe lithium ion batteries *J. Membr. Sci.* **472** 133–40
[81] Tsao C H and Kuo P L 2015 Poly(dimethylsiloxane) hybrid gel polymer electrolytes of a porous structure for lithium ion battery *J. Membr. Sci.* **489** 36–42
[82] Boaretto N, Joost C, Seyfried M, Vezzù K and Di Noto V 2016 Conductivity and properties of polysiloxane-polyether cluster-LiTFSI networks as hybrid polymer electrolytes *J. Power Sources* **325** 427–37

[83] Ren C, Liu M, Zhang J, Zhang Q, Zhan X and Chen F 2018 Solid-state single-ion conducting comb-like siloxane copolymer electrolyte with improved conductivity and electrochemical window for lithium batteries *J. Appl. Polym. Sci.* **135** 45848

[84] Grewal M S, Tanaka M and Kawakami H 2020 Fabrication and characterizations of soft and flexible Poly(dimethylsiloxane)-incorporated network polymer electrolyte membranes *Polymer* **186** 122045

[85] Huang X, Guo J-Y, Yang J, Xia Y, Zhang Y-F, Fu P and Du F-P 2022 High mechanical properties and ionic conductivity of polysiloxane sulfonate via tuning ionization degree with clicking chemical reaction *Polymer* **254** 125066

[86] Gao S *et al* 2021 Glass-fiber-reinforced polymeric film as an efficient protecting layer for stable Li metal electrodes *Cell Rep. Phys. Sci.* **2** 100534

[87] Kim K M, Ly N V, Won J H, Lee Y-G, Cho W I, Ko J M and Kaner R B 2014 Improvement of lithium-ion battery performance at low temperature by adopting polydimethylsiloxane-based electrolyte additives *Electrochim. Acta* **136** 182–8

[88] Sumathipala H H, Hassoun J, Panero S and Scrosati B 2008 Li–$LiFePO_4$ rechargeable polymer battery using dual composite polymer electrolytes *J. Appl. Electrochem.* **38** 39–42

[89] Zheng J, Tang M and Hu Y-Y 2016 Lithium-ion pathway within $Li_7La_3Zr_2O_{12}$-polyethylene oxide composite electrolytes *Angew. Chem. Int. Ed.* **55** 12538–42

[90] Banitaba S N, Semnani D, Heydari-Soureshjani E, Rezaei B and Ensafi A A 2019 Effect of titanium dioxide and zinc oxide fillers on morphology, electrochemical and mechanical properties of the PEO-based nanofibers, applicable as an electrolyte for lithium-ion batteries *Mater. Res. Express* **6** 0850d6

[91] Wei N, Hu J, Zhang M, He J and Ni P 2019 Cross-linked porous polymer separator using vinyl-modified aluminum oxide nanoparticles as cross-linker for lithium-ion batteries *Electrochim. Acta* **307** 495–502

[92] Xiao W, Wang Z, Zhang Y, Fang R, Yuan Z, Miao C, Yan X and Jiang Y 2018 Enhanced performance of P(VDF-HFP)-based composite polymer electrolytes doped with organic-inorganic hybrid particles PMMA-ZrO_2 for lithium-ion batteries *J. Power Sources* **382** 128–34

[93] Zhang S S and Tran D T 2013 How a gel polymer electrolyte affects performance of lithium/sulfur batteries *Electrochim. Acta* **114** 296–302

[94] Ma T, Wang R, Jin S, Zheng S, Li L, Shi J, Cai Y, Liang J and Tao Z 2021 Functionalized boron nitride-based modification layer as ion regulator toward stable lithium anode at high current densities *ACS Appl. Mater. Interfaces* **13** 391–9

[95] Cheng Q *et al* 2019 Stabilizing solid electrolyte–anode interface in Li-metal batteries by boron nitride-based nanocomposite coating *Joule* **3** 1510–22

[96] Gautam C and Chelliah S 2021 Methods of hexagonal boron nitride exfoliation and its functionalization: covalent and non-covalent approaches *RSC Adv.* **11** 31284–327

[97] Duan S *et al* 2020 Selective doping to relax glassified grain boundaries substantially enhances the ionic conductivity of $LiTi_2(PO_4)_3$ glass-ceramic electrolytes *J. Power Sources* **449** 227574

[98] Wen J *et al* 2019 Highly adhesive Li–BN nanosheet composite anode with excellent interfacial compatibility for solid-state Li metal batteries *ACS Nano* **13** 14549–56

[99] Hyun W J, De Moraes A C M, Lim J-M, Downing J R, Park K-Y, Tan M T Z and Hersam M C 2019 High-modulus hexagonal boron nitride nanoplatelet gel electrolytes for solid-state rechargeable lithium-ion batteries *ACS Nano* **13** 9664–72

[100] Chen Y, Kang Q, Jiang P and Huang X 2021 Rapid, high-efficient and scalable exfoliation of high-quality boron nitride nanosheets and their application in lithium–sulfur batteries *Nano Res.* **14** 2424–31

[101] Lu J, Li Y and Huang W 2022 Study on structure and electrical properties of PVDF/$Li_{3/8}Sr_{7/16}Zr_{1/4}Ta_{3/4}O_3$ composite solid polymer electrolytes for quasi-solid-state Li battery *Mater. Res. Bull.* **153** 111880

[102] Xie Z, Wu Z, An X, Yoshida A, Wang Z, Hao X, Abudula A and Guan G 2019 Bifunctional ionic liquid and conducting ceramic co-assisted solid polymer electrolyte membrane for quasi-solid-state lithium metal batteries *J. Membr. Sci.* **586** 122–9

[103] Zhao Y, Wu C, Peng G, Chen X, Yao X, Bai Y, Wu F, Chen S and Xu X 2016 A new solid polymer electrolyte incorporating $Li_{10}GeP_2S_{12}$ into a polyethylene oxide matrix for all-solid-state lithium batteries *J. Power Sources* **301** 47–53

[104] Chu Z, Gao X, Wang C, Wang T and Wang G 2021 Metal–organic frameworks as separators and electrolytes for lithium–sulfur batteries *J. Mater. Chem.* A **9** 7301–16

[105] Lu G *et al* 2022 Bifunctional MOF doped PEO composite electrolyte for long-life cycle solid lithium ion battery *ACS Appl. Mater. Interfaces* **14** 45476–83

[106] Wu X, Chen K, Yao Z, Hu J, Huang M, Meng J, Ma S, Wu T, Cui Y and Li C 2021 Metal organic framework reinforced polymer electrolyte with high cation transference number to enable dendrite-free solid state Li metal conversion batteries *J. Power Sources* **501** 229946

[107] Yao P *et al* 2018 PVDF/Palygorskite nanowire composite electrolyte for 4 V rechargeable lithium batteries with high energy density *Nano Lett.* **18** 6113–20

[108] Deka M and Kumar A 2011 Electrical and electrochemical studies of poly(vinylidene fluoride)–clay nanocomposite gel polymer electrolytes for Li-ion batteries *J. Power Sources* **196** 1358–64

[109] Xia S, Wu X, Zhang Z, Cui Y and Liu W 2019 Practical challenges and future perspectives of all-solid-state lithium-metal batteries *Chem.* **5** 753–85

[110] Huo H, Wu B, Zhang T, Zheng X, Ge L, Xu T, Guo X and Sun X 2019 Anion-immobilized polymer electrolyte achieved by cationic metal–organic framework filler for dendrite-free solid-state batteries *Energy Storage Mater.* **18** 59–67

Chapter 10

Electrocatalysts for hybrid water electrolysis

Shaista Jabeen, Sreedhanya Pallilavalappil and Suresh C Pillai

The implementation of hybrid water splitting techniques to generate hydrogen plays a pivotal role in advancing sustainable energy solutions. This chapter reviews recent advances in electrocatalysts for hydrogen generation via hybrid water electrolysis. Electrocatalysts, comprising noble and non-noble metals, nonmetals, transition metals, and alloys, are evaluated for their selectivity, activity, and durability.

Enhancing electrocatalyst performance involves exploring structure–activity relationships by adopting different strategies such as electronic structure optimization, crystal phase orientation, ionic manipulation, and doping. The chapter also covers anodic electrocatalysts that oxidize various chemicals (e.g. biomass, alcohol, urea) to produce high-purity H_2. Noble metals such as Pt, Pd, Ru, and Rh are promising, while non-noble metal-based catalysts gain attention because of their cost-effectiveness and activity. Ongoing research focuses on developing monofunctional, bifunctional, and multifunctional catalysts, driving their practical use in hybrid water electrolysis. These advancements contribute to a cleaner and more sustainable energy future.

10.1 Introduction

10.1.1 The fundamentals of hybrid water electrolysis

The ever-expanding global demand for energy is closely linked to the scarcity of fossil fuels and the adverse environmental consequences resulting from this depletion. Considering these circumstances, the development of green and renewable energy sources follows an inexorable trend toward the establishment of a low-carbon energy system. Hydrogen, characterized by its high gravimetric energy density, lack of carbon emissions, and potential to be produced using renewable energy sources such as wind- and sun-driven water electrolysis, is emerging as a worthwhile alternative fuel to support a sustainable energy model [1].

doi:10.1088/978-0-7503-5749-4ch10

Hybrid water electrolysis is basically an advanced technological approach that integrates the fundamental principles of the proton exchange membrane (PEM) and alkaline electrolysis methodologies to achieve proficient and environmentally sustainable hydrogen generation. This approach strategically harnesses the advantages of both systems to optimize the performance metrics of environmentally sustainable energy solutions, overcome their limitations, and in turn, make a substantial contribution to their progress. Pure water electrolysis (PWE) entails a dual half-reaction configuration, namely, the cathodic process of the hydrogen evolution reaction (HER) and the anodic process of the oxygen evolution reaction (OER). However, in conventional water splitting procedures, the production of hydrogen (H_2) via the HER is accompanied by the OER, a process characterized by slow kinetics that occur at a potential value of 1.23 V under standard conditions. Consequently, the unfavorable OER leads to an increase in the electrical energy input required for water electrolyzers [2].

Moreover, the concurrent production of hydrogen and oxygen gases leads to gas crossover across the membrane, giving rise to safety concerns. Conversely, the O_2 produced at the anode is sparse and typically lacks significant market value, often leading to its venting rather than capture for potentially valuable applications.

10.1.1.1 Anodic oxidation reactions

To address these persistent challenges, researchers have turned to hybrid water electrolysis, an innovative approach in which the OER is replaced by an alternative oxidation process. This strategy has earned increasing research attention due to its potential to enhance overall efficiency and effectiveness in H_2 production processes. Efforts to develop efficient electrocatalysts for hydrogen production have made substantial progress in addressing several key objectives and challenges such as scalability, safety, efficiency, and cost improvement. However, achieving satisfactory levels of hydrogen production efficiency remains a challenge.

In essence, at the cathode, hybrid water electrolysis shares fundamental similarities with conventional water electrolysis. At the cathode, two-electron transfers (the HER) dominate, which can be described either by equation (10.1) or equation (10.2) in alkaline or acidic conditions, respectively [3]. However, in hybrid water electrolysis, in contrast to the OER in conventional water splitting, a set of complex anodic oxidation reactions (AORs) involving the oxidation of organic compounds takes place on the opposite side from the cathode, as shown in figure 10.1.

The basic reactions involved in water electrolysis are:

$$2H_2O + 2e^- \rightarrow H_2 + 2OH^- \text{ (Alkaline)} \tag{10.1}$$

$$2H^+ + 2e^- \rightarrow H_2 \qquad \text{(Acidic)} \tag{10.2}$$

10.1.1.2 The criteria for anodic oxidation reactions

Hybrid water electrolysis systems have witnessed significant and swift advancements in recent years, as they present numerous advantages over traditional water electrolysis systems. When assessing the prospective application of an oxidation

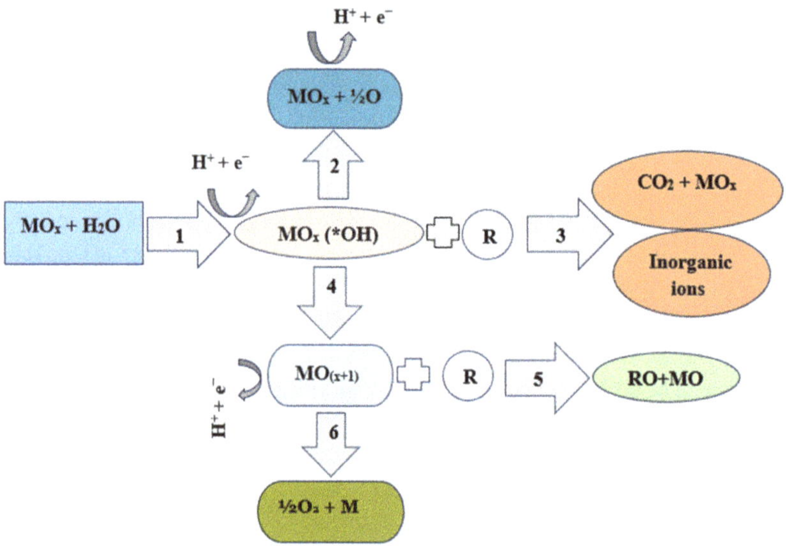

Figure 10.1. The mechanism of the anodic oxidation reaction (AOR).

reaction for an optimal commercial output, it is crucial to consider both qualitative and quantitative criteria. These criteria play a pivotal role in determining the feasibility and success of such systems. The following are some criteria that can maximize the commercial output of hybrid water electrolysis:

Environmental impact: the chosen reaction should not yield toxic or polluting by-products. This is critical to maintain environmental sustainability and ensure that the process aligns with clean energy goals.

Cost-effective electrocatalysts: it is essential to employ electrocatalysts that are both cost-effective and highly efficient. These catalysts play a central role in driving the chosen oxidation reaction and must be optimized for maximum performance.

Cost-effective reductive additives: one approach used to enhance the effectiveness of hybrid water electrolysis involves the incorporation of cost-effective reductive chemical additives, such as ammonia, urea, alcohol, and carbonyl compounds. These substances act as sacrificial agents, effectively replacing the slow kinetics of the OER. This substitution reduces the energy input required for the process, enhancing overall energy efficiency, and making the system more cost-effective [4].

Minimal competition with the HER: the selected oxidation reaction should minimally interfere with the HER. This ensures the production of hydrogen remains efficient and uncontested.

Optimization of reaction conditions: fine-tuning reaction conditions is crucial. Factors such as electrolyte selection, pH value, and kinetics must be optimized to achieve efficient electrocatalysis.

Simply, hybrid water electrolysis offers benefits such as reduced voltage input, the avoidance of gas mixture formation, and the mitigation of reactive species of oxygen. These advantages contribute to the overall efficiency, safety, and effectiveness of the system [5].

10.2 Strategies for tailored electrocatalysts in hybrid water electrolysis: design and synthesis

Electrocatalysts play a pivotal role in diminishing the energy barrier of reactions, expediting reaction kinetics and controlling reaction selectivity by enhancing energy utilization. Consequently, considerable endeavors have been dedicated to the development of efficient electrocatalysts with the objective of reducing the energy requirements for H_2 production [6]. This section introduces a rational framework for the design and optimization of electrocatalysts as well as different synthetic techniques for electrocatalysts.

10.2.1 A rational framework for electrocatalyst design and optimization

In contrast to conventional water splitting, hybrid water electrolysis is considerably more intricate due to the variability of the intermediates and the active sites of catalysts in response to different substrates. Therefore, it is essential to focus on understanding the mechanisms involved and obtain theoretical guidance for catalyst design to navigate these complexities effectively. The following are some aspects on which the design principles of efficient electrocatalysts are mainly based.

(i) Increasing the surface area and number of active sites to enhance the density of active sites
(ii) Improving the electrical conductivity to increase the *m*/*e* (mass-to-charge ratio) transfer rate and the utilization of active sites
(iii) Tailoring electronic properties through integrating foreign species with the aim of boosting electrocatalytic performance.

Considering the rational framework for the design and optimization of electrocatalysts, there are two parts to discuss: (1) surface engineering and (2) ionic manipulation.

10.2.1.1 Surface engineering

The surface engineering of electrocatalysts is a vital strategy for advancing hybrid water electrolysis. By tailoring the catalyst's surface at the nanoscale, introducing active sites, employing protective coatings, and leveraging advanced characterization techniques, researchers can enhance catalytic activity, durability, and selectivity. These surface engineering strategies are pivotal in realizing the full potential of hybrid water electrolysis as a clean and sustainable hydrogen production technology. Surface engineering methodologies can be categorized into four distinct groups, which are discussed below.

10.2.1.2 Defect and vacancy tailoring

In general, nanomaterials can be designed to incorporate interstitial atoms, lattice imperfections, and voids within their atomic structures. By introducing defects into electrocatalysts, their electronic structures can be modified to produce additional active sites which have the potential to change the charge distribution [7, 8]. For example, transition metals possess an abundance of electrocatalytic active sites, which significantly enhance their electrocatalytic activity.

Another approach to tailoring electrocatalytic performance at the atomic level is to introduce dopants with different electronegativities and sizes, but this is a major challenge. The combined heterostructures induce a synergistic effect with the base material whose strong electronic interactions result in enhanced electrocatalytic activity and electronic conductivity compared to those of the individual components. The distinctive physicochemical characteristics exhibited by heterostructured electrocatalysts render them exceptionally well-suited for electrochemical water splitting, as they have improved conductivity, selectivity, stability, and mass transport. Hence, electrocatalysts incorporating transition-metal elements and subjected to heteroatom doping, i.e. doping with oxygen, nitrogen, phosphorous, carbon, sulfur, etc., have been used extensively to modulate the electronic configuration of electrocatalysts to make them more efficient [9].

10.2.1.3 Surface morphology tailoring

The modulation of surfaces as part of the design of 1D, 2D, and 3D nanomaterials with different sizes and morphologies has proven essential in augmenting electrocatalytic performance in hybrid water electrolysis. One-dimensional nanostructures, i.e. nanorods, nanofibers, nanowires, and nanotubes, exhibit exceptional physicochemical characteristics, such as a high electron transport rate due to direct transfer routes, which make them a very attractive approach for electrocatalysis. Like 1D nanostructures, 2D nanostructures in the form of nanoflakes, nanofilms, nanosheets (NSs), and nanowalls have shown potential for enhanced electrochemical activity due to their high catalytic surface area and fast reaction kinetics. Regarding 3D nanostructures, in addition to their significant specific surface area, they exhibit rapid mass–charge transport pathways and an abundance of active sites, making them exceptional materials for water splitting [10].

10.2.1.4 Crystallographic tailoring

Crystallographic tailoring is a powerful approach for fine-tuning the atomic structure of electrocatalysts for hybrid water electrolysis. By systematically engineering the catalyst's crystal lattice through doping, alloying, and facet exposure, researchers can optimize catalytic activity, selectivity, and stability [11]. This precise control over catalyst materials is central to advancing the efficacy and viability of hybrid water electrolysis, thus promoting its role as a clean and sustainable technology for hydrogen production.

10.2.1.5 Interface engineering

Surface interface engineering is a fundamental aspect of optimizing electrocatalyst performance [12]. By tailoring the interaction between a catalyst and a solid-state electrolyte, researchers can enhance ion and electron transfer, reduce overpotentials, and improve stability. Surface interface engineering involves different strategies such as adding surface coatings of different materials, i.e. graphene, metal carbides, oxides, and different functional groups, to reduce energy losses. Recently, self-supported electrocatalysts have garnered significant research attention, primarily because of their highly advantageous interfaces [13]. During overall charge transport within a self-supported electrocatalyst, the process interfaces involved are: (1) the electrolyte–catalyst interface, (2) the substrate–catalyst interface, and (3) the inner electrocatalyst interface.

10.2.1.6 Ionic manipulation/tuning

In the process of increasing the active sites of electrocatalysts and free carrier density, cation/anion exchange strategies and tensile strain are powerful tools for optimizing the atomic structures of electrocatalysts. This can be done through surface faceting, heteroatom doping, and strain modulation [14]. Anions, such as nitrogen, oxygen, sulfur, boron, and selenium-based nanomaterials, exhibit a notable capacity for flexible modulation of electronic structures. Their elevated electronegativity plays a pivotal role in positioning them as promising electrocatalysts, characterized by enhanced performance and reduced reliance on precious-metal content. Similarly, in cation regulation strategies, the different atomic sizes and the electronegativity of cations boost the intrinsic electrochemical activities of catalysts; this approach shows great potential for use in the design of energy conversion devices [15].

10.2.2 The synthesis of electrocatalysts

The characteristics and potential applications of diverse metal nanostructures are profoundly influenced by factors such as size, shape, structure, and the presence of defects within their configurations. To obtain the expected and amplified performance increases, precise management of defect density, shape, and size in these materials is therefore a crucial and key factor. The realization of optimum functionality across a range of applications requires this strategic control over the material properties. Nanomaterial synthesis methods are categorized into two principal approaches: the top-down method and the bottom-up method. The top-down method starts off with a bulk material and gradually reduces it to the nanoscale using methods such as milling, etching, or lithography to produce precise structures. The bottom-up strategy, on the other hand, entails assembling nanoscale structures from separate atoms, molecules, or smaller parts. This method depends on the controlled manipulation of building blocks, chemical reactions, or self-assembly to produce the required nanomaterials. The bottom-up approach is preferred for its effectiveness and environmental friendliness because it produces less waste during synthesis.

Various methods are employed to prepare electrocatalyst materials for electrochemical studies, including solution-phase methods such as the sol–gel, hydrothermal, microwave-assisted synthetic, precipitation, and electrochemical methods. Solid-phase methods such as high-temperature techniques, flux growth, solid-phase combustion, and MOF-derived approaches, as well as vapor phase methods such as spray pyrolysis, chemical vapor deposition, and magnetron sputtering, are also utilized for synthesis, as represented in figure 10.2.

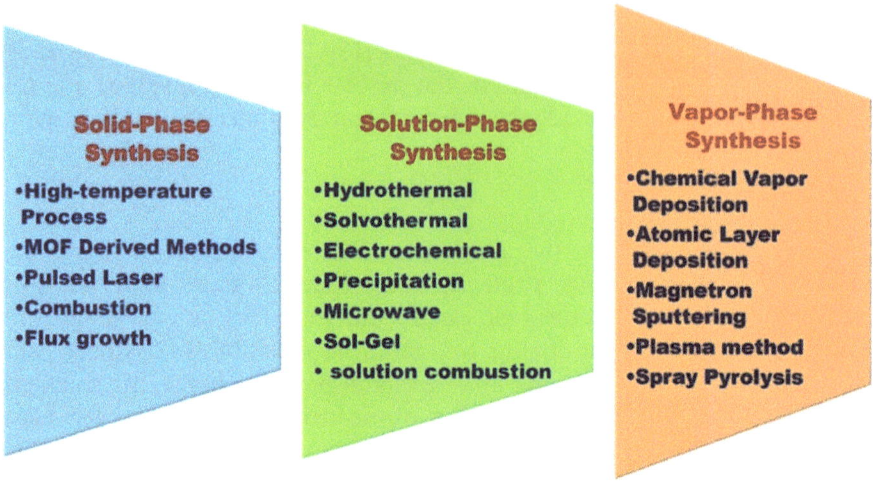

Figure 10.2. Different synthetic approaches and strategies for electrocatalysts (reprinted from [16], Copyright (2021), with permission from Elsevier).

Hydrothermal methods are effective techniques for synthesizing a variety of materials, including simple oxides, nanosized spinels, and perovskites, which are renowned for their chemical homogeneity, purity, and optimized nanostructure [17]. Notably, this method's extraordinary resilience is due to its capacity to change the morphology of the synthesized material by varying reaction conditions. Metal salt precursors or catalyst solutions such as KOH or NaOH are stirred together to form specific materials before being heated in an autoclave. The production of the required products with the desired qualities results from this process.

Gao and colleagues focused on both active sites and electrode conductivity in their thorough investigation of Ni-based electrocatalysts for 5-hydroxymethylfurfural (HMF) oxidation [18]. They suggested a novel strategy in response to the constraints posed by oxidized Ni species that have low electrical conductivity. They created a NiSe@NiO_x structure using a hydrothermal selenization reaction, which was then subjected to cyclic voltammetric activation, resulting in the development of a NiO_x shell over NiSe nanowires. This novel manufacturing method improved electrode conductivity as well as the active sites, showing promise for enhanced HMF oxidation performance. A hierarchical nanostructured Mo-doped Ni_3S_2 nanoforest was produced by Xu *et al* using a hydrothermal technique, which effectively controlled the nanostructure and increased the density of electroactive sites for improved urea oxidation reaction performance [19].

Although solvothermal and hydrothermal processes are comparable, the type of solvent employed varies. Aqueous solutions are used in the hydrothermal process, while nonaqueous solvents are used in the solvothermal approach. An *in situ* hydrothermal selenization method was employed to synthesize a Ni_3Se_4 nanorod array using dimethylformamide (DMF) as a solvent, which exhibited exceptional activity for anodic urea oxidation reactions [20]. A heterogeneously structured material with a hierarchical nanoporous MoS_2/Ni_3S_2 catalyst was synthesized in two steps: the electrodeposition of a macroporous nickel layer on Ni foam, followed by a solvothermal reaction with $(NH_4)_2MoS_4$ to create a MoS_2/Ni_3S_2 heterostructure [21].

In summary, the precise manipulation of variables such as defect density, size, and shape are imperative to optimize the efficacy of metal nanostructures in a multitude of applications. For the synthesis of nanomaterials, there are two main methods: top-down and bottom-up. The bottom-up method is preferred because of its effectiveness and environmental friendliness. Hydrothermal techniques stand out among the various synthetic techniques discussed for their capacity to produce controlled nanostructures and materials with favorable electrocatalytic properties, providing promising ways to improve performance in processes such as HMF oxidation and urea oxidation reactions.

10.3 Crucial benchmarks for analyzing electrocatalyst effectiveness

Certainly, evaluating the effectiveness of an electrocatalyst and its activity in the context of hybrid water electrolysis requires specific benchmarks to be assessed to determine the performance and suitability of the electrocatalyst for the process. The following sections discuss some crucial standards used to analyze electrocatalyst effectiveness in this process.

10.3.1 Stability and durability

Stability and durability are the key parameters used to ensure the reliable and long-term operation of the electrolysis system. An efficient electrocatalyst is expected to maintain its activity and efficacy over an extended period. To assess stability, the η value, derived from the polarization curve for a given current density (j), is compared before and after each operational cycle. The lower the difference between the obtained values, the greater and more enduring the stability of the electrocatalyst [22]. In parallel to stability, durability relates to the electrocatalyst's capacity to endure the rigorous conditions of the electrolysis process without experiencing substantial degradation or a decline in catalytic activity. In addition, it is imperative for electrocatalysts to prevent the crossover of reactants or products between the anode and cathode compartments, as such crossover can have significant implications for overall process efficiency.

10.3.2 Overpotential (η)

Overpotential denotes the extra voltage required to drive the electrolysis reaction compared to the thermodynamically expected voltage. This is the fundamental aspect employed to assess the electrocatalytic performance of the HER. In other words, the value of the overpotential represents the energy barrier that needs to be overcome to initiate and sustain the electrochemical reaction.

A lower overpotential corresponds to a lower activation energy, enhances the reaction kinetics, suppress competing reactions, and improves the selectivity and specificity of the electrocatalyst, thus enabling the reaction to occur at a faster rate with improved catalytic activity. Modifying the η values allows for the adjustment of the overall work potential. An increase in temperature can be employed to reduce η

and facilitate ion diffusion [23]. The lower the overpotential, the more efficiently the electrocatalyst facilitates the reaction [24].

10.3.3 Faradaic efficiency

Faradaic efficiency quantifies the ratio of the actual amount of product generated during electrolysis to the theoretically expected amount [25]. In essence, faradic efficiency represents the electron transfer efficiency controlled by the external circuit used to drive the electrochemical reaction. In industrial applications, it is imperative for the electrocatalyst to exhibit a high faradaic efficiency, indicating the optimal utilization of external energy. A high faradaic efficiency indicates minimal side reactions and the effective use of electrical energy for hydrogen and oxygen production [26].

10.3.4 The turnover frequency

The turnover frequency (TOF) quantifies the intrinsic electrocatalytic activity of the catalyst. It is the rate of conversion of active reactant molecules per active site in a given time. So, the greater its value, the more effective the catalyst. The TOF can be determined as follows:

$$\mathrm{TOF} = \frac{jA}{2}Fn, \tag{10.3}$$

where F, n, j, and A signify the faradaic constant, the number of moles per active site of the electrode, the current density, and the surface area of the electrode, respectively [27, 28].

10.3.5 The Tafel slope

The Tafel slope provides an insight into the rate at which reaction kinetics change with voltage. The Tafel equation establishes a linear relationship between the logarithm of the current density and the overpotential to elucidate the rate of a reaction.

A lower Tafel slope on the plot corresponds to faster reaction kinetics, suggesting better catalytic performance, whereas a higher slope indicates the opposite [29]. Evaluating electrocatalyst effectiveness using these benchmarks helps researchers and engineers to select the most suitable catalysts for hybrid water electrolysis, thereby contributing to the progress of efficient and sustainable hydrogen production methods.

10.4 Combining organic electrocatalytic oxidation with the hydrogen evolution reaction

In recent years, there has been a notable surge in the development of hybrid water electrolysis techniques. These methods have primarily focused on the electro-oxidation of a wide range of oxidative substances, including alcohols, urea, hydrazine (NH_2NH_2), ammonia, nitrogen, glucose, and other biomass, as exemplified in figure 10.3. In this process, the anode contributes to the production of value-

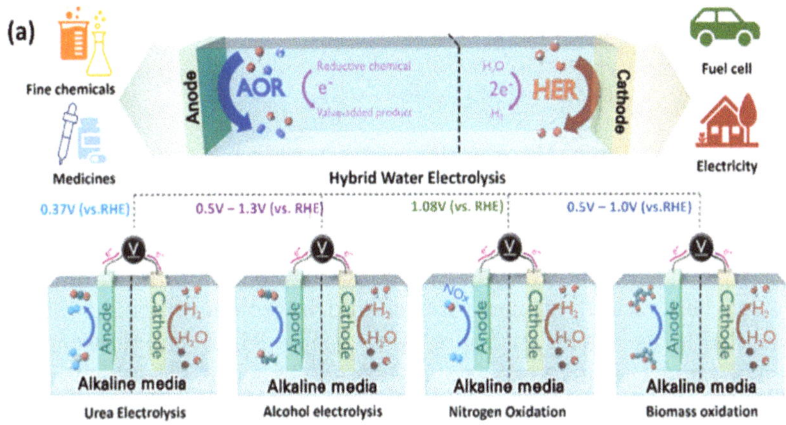

Figure 10.3. Hybrid water electrolysis systems (reprinted from [30], Copyright (2023), with permission from Elsevier).

added chemicals. In contrast, the cathode output consists of hydrogen fuel, which holds practical relevance for electricity production and fuel cells.

To carry out these AORs, a variety of combinations of metals, nonmetals and their oxides, carbides, nitrides, and phosphides are used as electrocatalysts to enhance the hydrogen production efficiency. Depending upon the choice of reaction, hybrid water electrolysis is divided into three categories (discussed next).

10.4.1 Reagent-sacrificing reactions

An AOR can replace the OER, but results in the transformation of a valuable substrate into a less valuable or worthless product; therefore, the substrate participating in this process can be classified as a sacrificial reagent. This type of reaction is then defined as a reagent-sacrificing mechanism. While costs may rise as a result of consuming valuable substrates, the electricity input required for such a reaction can, nevertheless, be diminished. Examples of this type of reaction are the electrochemical oxidation of hydrazine (N_2H_4) and ammonia (NH_3).

In the case of hydrazine oxidation, different noble and non-noble based electrocatalysts have been used for the activation of N_2H_4 and its subsequent dehydrogenation to N_2 in order to boost hydrogen (H_2) production. The thermodynamic potential for hydrazine oxidation, as shown in equation (10.5), is −0.38 V, significantly lower than the 1.23 V of the OER. Hydrazine oxidation as an anode reaction allows a remarkable current density of 500 mA cm^{-2} to be achieved within a cell voltage of 1 V, and notably, without the generation of oxygen (O_2) [31]. According to density functional theory (DFT) calculations, the oxidation process of hydrazine (N_2H_4) can be broken down into six elementary steps. The initial step involves the adsorption of hydrazine, while the last step is the desorption of nitrogen gas (N_2). In between, there are four intermediate steps, each of which involves the formation of both one electron and one proton. This sequence of six steps collectively describes the complex process of hydrazine oxidation.

$$\text{Cathode: } 4H_2O + 4e^- \rightarrow 2H_2 + 4OH^- \quad (10.4)$$

$$\text{Anode: } N_2H_4 + 4OH^- \rightarrow N_2 + 4H_2O + 4e^- \quad (10.5)$$

A study conducted by Du *et al* observed that the use of a bifunctional electrocatalyst of nickel and palladium alloy NSs decreased the required voltage from 1.97 to 0.63 V at 15 mA cm^{-2} [32]. Similarly, in the case of RhPb nanoflowers, the amorphous form showed better catalytic performance and lower overpotential than those of crystalline RhPb, as shown in figure 10.4. Using RhPb nanoflowers at current densities of 10 and 100 mA cm^{-2}, voltages of 0.095 and 0.321 mV were achieved.

Ammonia oxidation is another good illustration of a reagent-sacrificing reaction, as expressed in equation (10.6). Ammonia oxidation is thermodynamically more favorable than the OER. Notably, as ammonia is more abundant and nontoxic than hydrazine, it is a promising candidate for replacing the OER. Platinum has long been acknowledged as the benchmark catalyst for the oxidation of NH_3. However, its limitations of being expensive and rare have prompted research efforts to explore the development of electrocatalysts based on non-noble metals [34].

$$2NH_3 + 6OH^- \rightarrow N_2 + 6H_2O + 6e^- \quad (10.6)$$

An electrocatalyst composed of layered hydroxides (LHs) of Ni and Cu achieved a significant current density of 35 mA cm^{-2} at a relatively low potential value of 0.55 V (versus Ag/AgCl), as shown in figure 10.5. The enhanced catalytic performance was due to the presence of numerous active sites and the synergistic effect resulting

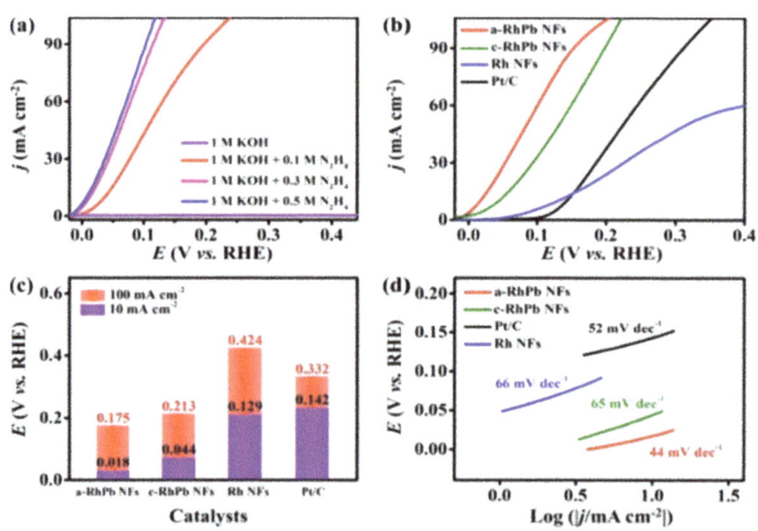

Figure 10.4. (a) Linear sweep voltammetry (LSV) curves of a-RhPb/nanoflowers used for N_2H_4 oxidation in 1 M KOH at different N_2H_4 concentrations, (b) in 1 M KOH and 0.1 M N_2H_4, (c) the corresponding overpotentials, and (d) Tafel plots. ((a)–(d) reprinted from [33], Copyright (2022), with permission from Elsevier.)

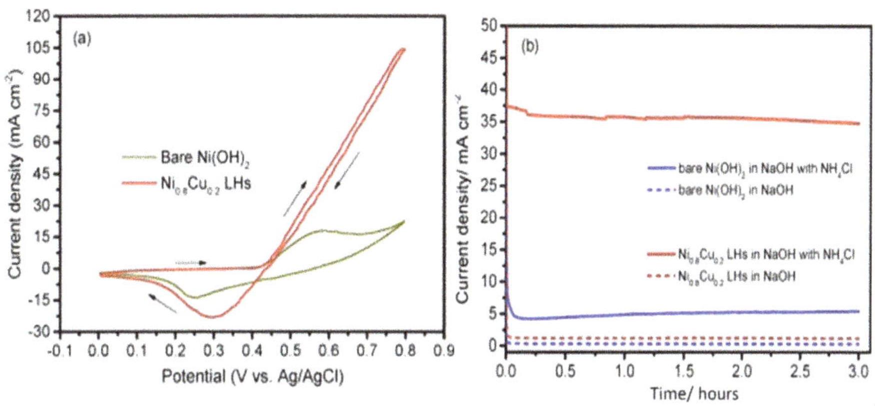

Figure 10.5. (a) Constant voltammetry results for $Ni_{0.8}Cu_{0.2}$ layered hydroxides (LHs) and bare $Ni(OH)_2$, (b) chronoamperometry curves. ((a) and (b) reprinted from [35], CC BY 4.0.)

from the combination of Ni and Cu in the catalyst. Notably, this mixed Ni–Cu/LHs nanowire catalyst beat Pt/C in terms of both current and stability, suggesting its potential as a Pt alternate.

In summary, while the utilization of sacrificial reagents may incur additional costs, the significantly reduced energy input guarantees the reliable operation of water splitting at higher current densities without generating oxygen. Consequently, this type of reaction has emerged as an attractive choice for hybrid water electrolysis.

10.4.2 Pollutant-degrading reactions

The incorporation of sacrificial reagents tends to elevate the overall cost of the system; therefore, alternative substrates such as environmental contaminants, may serve as viable candidates for oxidative degradation. In the context of hybrid water electrolysis integration, this form of anode oxidation bears a resemblance to the processes commonly employed in wastewater treatment, hence it is denoted as a pollutant-degrading approach. We now discuss examples of some electrocatalysts used for such oxidation reactions, including the electrochemical oxidation of urea, organic dyes, and exhaust gases.

Urea is a common material. Despite its relatively low concentration in wastewater, it significantly contributes to eutrophication. Numerous studies of urea oxidation have consistently observed that it has an approximate onset potential of 1.3 V. This behavior can be attributed to the slow kinetics associated with a $6e^-$ transfer process, as given in equation (10.7). Notably, urea oxidation follows both diffusion and kinetic control mechanisms. Consequently, integrating urea oxidation with water electrolysis offers a promising avenue for simultaneously achieving environmental preservation and clean energy generation. This interaction can effectively yield hydrogen (H_2) while refining urea waste, thus advancing ecological conservation and sustainable energy production [36].

$$CO(NH_2)_2 + 6OH^- \rightarrow N_2 + 5H_2O + CO_2 + 6e^- \tag{10.7}$$

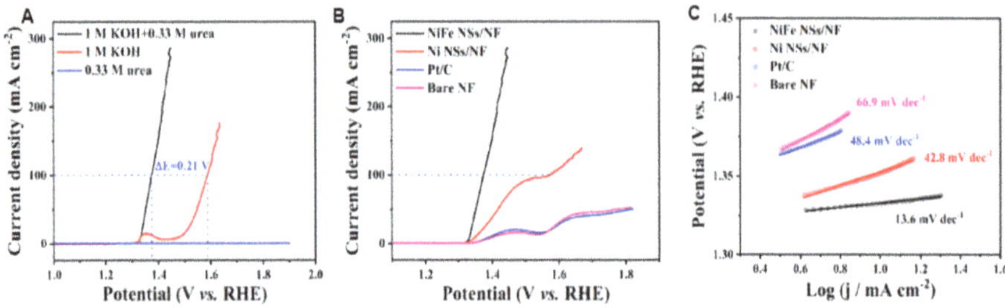

Figure 10.6. (a) LSV curves, (b) Tafel slope curves, and (c) EIS spectra of NiFe NSs/NF. ((a)–(c) reprinted from [39], Copyright (2022), with permission from Elsevier).

Nickel-based catalysts are currently considered to be superior electrocatalysts for urea oxidation, even compared with precious metals [37, 38]. An electrocatalyst made from ultrathin NSs of NiFe decorated on nickel foam (NiFe NSs/NF) was reported to alleviate the sluggish kinetics even at a low onset potential, as shown in figure 10.6. The results indicated a 15.2% reduction in cell voltage compared to the voltage of standard water splitting. In addition, the urea concentration decreased by 55.6% following 36 h of electrolysis at 1.70 V.

The fusion of the HER with pollutant degradation undeniably presents a promising trajectory for hydrogen production.

10.4.3 Value-added reactions

In an ideal hybrid system, the coupling of a valuable oxidation reaction with the HER is required. This greatly boosts the benefits of electrolysis by producing valuable products at the anode and the cathode while reducing the electricity needed. A key way to do this is to use biomass oxidation (glucose, alcohol, HMF) at the anode [40]. This process is crucial for upgrading biomass-based materials into valuable chemicals such as plastics and medicines [41]. The specific products formed through oxidation are strongly influenced by the electrocatalysts and electrolytes used. For instance, when glycerol undergoes oxidation with gold (Au) as an electrocatalyst, it yields products such as mesoxalic acid, tartronic acid, 1,3-dihydroxyacetone, carbon dioxide (CO_2), and glyoxylic acid. In contrast, when platinum (Pt) serves as the catalyst in an alkaline electrolyte, the oxidation of glycerol creates different products, including formic acid, glyoxylic acid, CO_2, glycolic acid, etc [42]. Hence, this mechanism of biomass oxidation is desirable due to its reaction selectivity. In the latest developments in the electrooxidation of additional biomass and biomass-derived compounds such as glucose and glycerol, different electrocatalysts are being used [43, 44].

For the oxidation of CH_3OH, a ruthenium- and cobalt-based alloy supported on carbon black (RhCo/CB) exhibited composition-dependent enhanced catalytic activity compared to that of platinum supported on carbon (Pt/C), as illustrated in figure 10.7.

Similarly, another electrocatalyst used for the oxidation of ethanol is a carbon-based alloy of platinum and molybdenum (PtMo/C), which showed higher catalytic activity

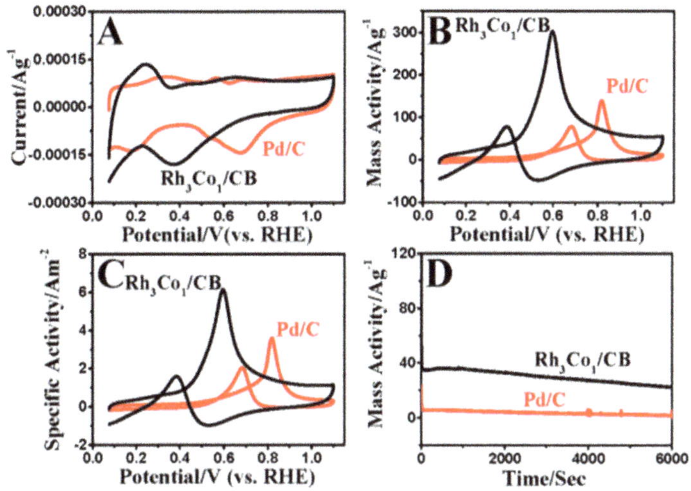

Figure 10.7. (A) Cyclic voltammetry (CV) curves of Rh_3Co_1/CB nanohybrids and Pd/C in 1 M KOH. (B) Metal mass-normalized, (C) electrochemical active surface area (ECSA) normalized CV curves, and (D) chronoamperometric curves of Rh_3Co_1/CB nanohybrids and Pd/C in 1 M KOH+1 M CH_3OH. ((A)–(C) reprinted from [45], Copyright (2017), with permission from Elsevier.)

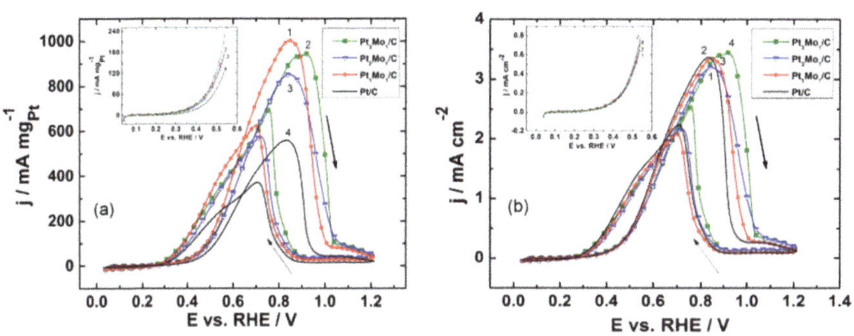

Figure 10.8. (a) The mass activity of the ethanol oxidation reaction on PtMo/C and Pt/C. (b) Specific activity in 0.5 mol l^{-1} KOH+0.5 mol l^{-1} C_2H_5OH. ((a) and (b) reprinted from [46], Copyright (2017), with permission from Elsevier.)

and CO oxidation at a negative potential than those of Pt/C. In terms of catalytic performance, Pt_3Mo_1/C showed the highest activity, as illustrated in figure 10.8.

Cost and profitability are the predominant motivators behind the ongoing research efforts. Among the three types of oxidation reactions, the value-added reactions hold the most promise for practical implementation.

10.5 Electrocatalysts for hybrid water electrolysis

10.5.1 Metal- and nonmetal-based electrocatalysts

Metals and alloys, particularly noble, non-noble, and rare-earth metals, are outstanding electrocatalysts for both cathodic and anodic reactions. This can be attributed to their high electrocatalytic activity and durability [47].

Electrocatalysts based on noble metals are characterized by high values of current density and potential. These exceptional catalytic properties render them attractive candidates for optimizing efficiency and performance in hybrid water electrolysis. The noble-metal-based electrocatalysts most frequently used for hybrid water electrolysis are platinum (Pt), palladium (Pd), ruthenium (Ru), gold (Au), and iridium (Ir) [48]. Noble metals are usually employed in various sizes and configurations of nanoparticles to harness the benefits of their high surface area and defect-rich surface structures. It is obvious from numerous research results that both the size and the morphology of the nanoparticles exert significant influence on electrocatalytic performance [49]. Also, noble-metal electrocatalysts show optimal binding interactions, as suggested by the Sabatier principle, making them promising catalysts for different electrocatalytic applications.

Unfortunately, at the industrial scale, H_2 production by water splitting using electrocatalysis based on noble metals is somewhat challenging because of their high cost, limited availability, susceptibility to dissolution, and reduced tolerance for catalytic poisoning. To date, electrocatalysts that facilitate the oxygen reduction reaction under basic conditions have seen significant improvements. These improvements extend to various key parameters of catalytic performance, including surface area, faradaic efficiency, and specific mass activity. Such progress is instrumental in lowering overpotential and costs, making it a crucial step in the pursuit of a H_2-based economy.

Noble metals such as palladium nanocrystals exhibit significant performance as catalysts for methanol and ethanol electrooxidation in alkaline environments. They provide a notable combination of high current/mass efficiency and elevated catalytic performance when used in oxidative processes. This phenomenon can be attributed to the inhibition of nanocrystal aggregation, facilitated by their two-dimensional structural alignment along with their substantial catalytically active surface area. The observed mass activities of the electrochemical oxidation of methanol and ethanol at their respective anodic oxidation peaks are approximately 1040 and 5500 A g^{-1}, respectively, as shown in figure 10.9.

One possibility for enhancing catalysis in hybrid water electrolysis involves creating an oxygen-enriched environment. In water dissociation, to achieve a robust

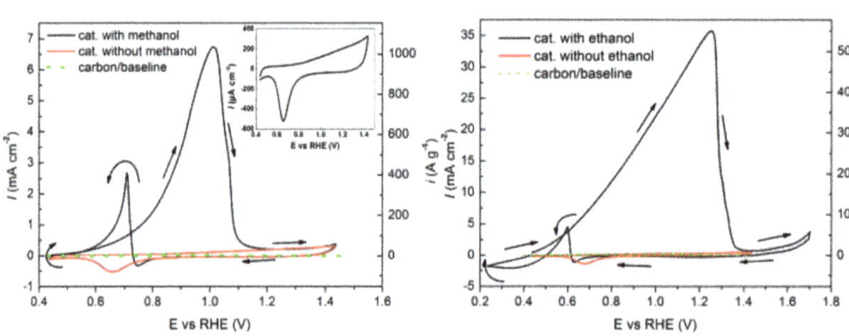

Figure 10.9. CV of the methanol oxidation reaction and the ethanol oxidation reaction in 0.5 M NaOH with 1.0 M ethanol. The base line represents oxidation in the absence of the Pd electrocatalyst (reprinted from [50], Copyright (2016), with permission from Elsevier).

electrocatalytic output at minimal overpotentials, non-precious metals such as copper, nickel, tin, ruthenium, etc. are alloyed with noble metals. For instance, in the case of ethanol oxidation, the combination of ruthenium (Ru) with tin dioxide (SnO_2) subsequently coated with platinum (Pt) or palladium (Pd) nanoparticles exhibited improved catalytic properties and stability under acidic conditions. The most interesting results were the twofold higher mass activity exhibited by Pt/Ru–SnO_2 compared to that of commercial Pt/C [51]. Another investigation focused on platinum-modified tungsten carbide revealed that bilayer platinum supported on tungsten carbide, whether in a face-centered or hexagonal structure, serves as an economically advantageous and high-performance electrocatalyst for the electro-oxidation of methanol. Pt–WC catalysts display onset potentials similar to those observed for pure platinum (Pt) but demonstrate reactivity levels up to 2.4 times higher than that of pure platinum [52]. A nanoarray of cobalt and phosphorus anchored to a titania mesh (CoP/TiM) electrode required a low potential of −6 mV to run at 100 mA cm^{-2} in a solution of 1 M KOH+0.1 M N_2H_4. This performance surpassed that of Pt/C/TiM. In addition, when employed in a N_2H_4-supported electrolyzer, CoP-based TiM produced 10 mA cm^{-2} at 0.2 V while maintaining 100% faradic efficiency [53].

Transition metals such as cobalt, nickel, and iron have gained significant attention for use in electrocatalysis because of their theoretically predicted high electrocatalytic activity, excellent conductivity, favorable electronic structure, cost-effectiveness, and widespread availability. In recent decades, a lot of research effort has been invested in the engineering and modification of electrocatalytically active surfaces and interfaces. The resulting designs aim to optimize defect concentrations, atomic arrangements, and electronic structures that work synergistically to boost specific electrocatalytic reactions.

Studies have indicated that the replacement of anodic reactions with non-precious-metal catalysts in hybrid water electrolysis is a promising approach for the development of alternative environmental remediation and energy conversion processes. Zhang and colleagues conducted a study in which they synthesized nickel selenide nanorods (Ni_3Se_4-NRs) and highlighted their remarkable catalytic competence in anodic reactions. An electrode modified with Ni_3Se_4 nanorods displayed exceptional electrochemical performance in oxygen evolution, attaining 10 mA cm^{-2} of anodic current density at 1.47 V. This voltage requirement could be lowered to 1.38 V and further to 0.32 V by substituting oxygen evolution with hydrazine oxidation and urea, underscoring the versatility and efficiency of the catalyst [20].

The most prevailing approach for enhancing the conductivity of nickel-based materials involves cultivating an active and conductive catalytic base such as Ni foam (NF), Cu foam, carbon paper, carbon cloth, or graphene. An alternative strategy that is gaining considerable interest is the utilization of bifunctional electrocatalysts in water electrolysis. These catalysts offer more favorable thermodynamics for hydrogen production while demonstrating outstanding activity and stability, making them particularly appealing for this application. For example, a nanoarray of nickel phosphide (Ni_2P/NF) functioned as a robust electrocatalyst for hydrazine oxidation under alkaline conditions. This catalyst exhibited a remarkable faradaic efficiency of 100% for hydrogen evolution, observed in both 1.0 and 0.5 M solutions of KOH and hydrazine, respectively [54].

Likewise, for the electrochemical oxidation of benzyl alcohol and hydrogen (H_2) production, a bifunctional electrocatalyst composed of copper and cobalt was produced in the form of 3D arrays of CuCoNx grown on a carbon fiber substrate encased within a nitrogen-doped carbon shell. This bifunctional electrocatalyst demonstrated exceptional performance by achieving high conversion rates and selective outcomes. The exceptional electrocatalytic activity may be attributed to its hierarchical architecture and its synergistic interactions between the nitride compounds of copper and cobalt. These interactions contribute to the creation of more catalytically active sites and enhance mass transport [55].

Other bifunctional electrocatalysts such as 3D-Ni_2P nanoparticle arrays on a Ni foam and tubular cobalt perselenide Ni_xCoSe_2 nanosheet electrode, nanoflowers of nickel–copper oxide [56], and zinc molybdate [57] have also been reported to couple hydrogen production with hydroxymethylfurfural oxidation, hydrazine oxidation, and the electrooxidation of glucose, respectively. All have proved the water splitting at low voltage while exhibiting promising stability and effective faradaic efficiency [58]. A 3D nanostructure composed of NiFe oxide and nitride, as reported by Yu and Huber, was found to exhibit excellent activity for HER and glucose oxidation. The $NiFeN_x$/NF electrode required remarkably low overpotentials of 40.6 and 104 mV to achieve 10 and 100 mA cm^{-2} in the HER. In addition, when used as anodes and cathodes in glucose-supported electrolysis with 1 M KOH and 0.1 M glucose, it demonstrated a high j value of 200 mA cm^{-2} at 1.48 V, and good stability for over 24 h for hydrogen production [59].

The choice of metal-based electrocatalyst depends on different factors such as cost, availability, activity, stability, and the specific requirements of the hybrid water electrolysis process. Researchers continue to investigate and develop novel materials and strategies to boost the output of these electrocatalysts. Some of the electrocatalysts studied for hybrid water electrolysis are listed in table 10.1.

Table 10.1. Electrocatalysts for hybrid water electrolysis.

Electrocatalysts	Electrolyte	Onset potential (V vs. reversible hydrogen electrode (RHE))	Current density (mA cm^{-2})	References
PtAuRu-supported rGO	1 M KOH + 1 M CH_3OH	0.42	—	[60]
Pt/Ni/C	1 M KOH + 1 M CH_3OH	0.42	11.5	[61]
Cyclic penta twinned Rh nanobranches	1 M NaOH + 1 M C_2H_5OH	0.13	0.361	[62]

(Continued)

Table 10.1. (*Continued*)

Electrocatalysts	Electrolyte	Onset potential (V vs. reversible hydrogen electrode (RHE))	Current density (mA cm^{-2})	References
Co–S–P/carbon cloth	1.0 M KOH + 1.0 M C_2H_5OH	1.63 V	10	[63]
Co/Ni alloy/carbon cloth	1.0 M KOH + 0.1 mM $C_6H_{12}O_6$	1.172 V	20	[64]
Ce/CoP NSs (Ce doping + P vacancies)	10 mM HMF + 1.0 M KOH	1.54 V	20	[65]
Ni–Cu alloy/nickel foam	1 M KOH + 0.5 M N_2H_4	0.41 V	100	[66]
Mo–Ni_3N/Ni/nickel foam	1.0 M KOH + 0.5 M N_2H_4	−0.3 mV	10	[67]
RuP_2/carbon micro-sheets (porous)	1.0 M KOH + 0.3 M N_2H_4	0.023 V	10	[68]
Pt/Cu nanoalloy	1 M KOH + 1 M N_2H_4	668 mV	200	[69]
Pt–Ru nano-cube	1 M KOH + 0.1 M NH_3	0.5 V	—	[70]
Platinum nanoparticles (cubic)	0.2 M NaOH + 0.1 M NH_3	0.5 V	1.8	[71]
Ni/carbon nanotubes (CNTs)	600 ppm		Ni/CNTs = 1.9	[72]
Ni/carbon nanospheres (CNSs)	ammonia + 10 mM potassium sulfate	~1.3 V	Ni/CNSs = 3.3	
Ni/Mo alloy (nanotubes)	1.0 M KOH + 0.1 M $CO(NH_2)_2$	1.43 V	10	[73]
Ni–Fe–Mo hybrid film	1 M KOH with 0.33 M $CO(NH_2)_2$	1.38 V	10	[74]
Nanoporous Ni–Fe	1 M KOH with 0.33 M $CO(NH_2)_2$	1.55 V	10	[75]

10.5.2 Other metal-based electrocatalysts for hybrid water electrolysis

Hybrid water electrolysis offers a fascinating context for the investigation of electrocatalysts beyond noble and non-noble metals. Alternative metal-based electrocatalysts are being investigated by researchers to address the drawbacks of conventional ones. Non-noble metals can lack activity and stability, while noble metals are effective but costly. Novel candidates such as metal oxides, hydroxides, nitrides, sulfides, carbides, and phosphides are attracting attention in this area due to their fascinating electrical and catalytic properties. By enabling both the OER and the HER, these materials show potential for improving efficiency and sustainability in hybrid water electrolysis. This wide variety of metal-based catalysts presents interdisciplinary strategies, opening fresh opportunities for improved water splitting and the promotion of environmentally friendly hydrogen production.

10.5.2.1 Metal oxides and hydroxides

Electrocatalysts rooted in transition-metal (hydr)oxides are gaining increasing prominence because of their adaptable chemical structure, potent catalytic activity, and economic viability. In particular, metal (hydr)oxides based on nickel (Ni) [76], cobalt (Co), and iron (Fe) have shown impressive electrocatalytic performances in electrooxidation reactions [77]. Due to their tuneable characteristics, which enable tailored catalyst design, their ability to efficiently accelerate electrochemical reactions, and their low price, these materials have attracted attention and are expected to make significant improvements to environmentally friendly energy conversion and storage systems.

A promising method for the energy-efficient production of organic acids with added value and hydrogen is presented by innovative electrolytic cell design. For example, electrodeposition and sulfurization were used by Xiang *et al* to produce cobalt hydroxide@hydroxysulfide NSs [78], which served as a robust bifunctional and stable electrocatalyst for the HER and oxidative methanol refining. Due to advantageous methanol oxidation dynamics, these NSs achieved 10 mA cm^{-2} at an incredibly low potential of 1.497 V. Energy conversion efficiency was maximized due to the simultaneous production of high-purity hydrogen and valuable formate compounds at the cathode and anode sites, respectively. The core–shell structure of the NSs and their partially sulfurized surfaces provided plentiful active sites, optimal electronic structures, robust heterointerfaces, and effective interaction between reactants and intermediates, which were the sources of this extraordinary electrocatalytic activity.

For the urea oxidation process, conventional catalysts such as Pt, Rh, and Ir have proven to be highly effective electrocatalysts. However, their limited availability, high costs, and enhanced oxidation potentials inhibit their use, imposing serious restrictions on their suitability for wider application [79, 80]. The development of electrocatalysts made from plentiful, non-precious materials is therefore urgently needed to provide long-term stability and meet the demand for sustainable urea oxidation reaction (UOR) applications. Numerous low-cost catalysts have recently attracted a lot of interest for use in the electrochemical oxidation of urea. Metals

[81], metal oxides [82], metal hydroxides [83], layered double hydroxides [84], nitrides [85], phosphides [86], and carbides [87] are only a few examples of the diverse materials that can be used as catalysts. This vast investigation, which has lasted for several decades, highlights the growing interest in less expensive, more promising, and technologically wise alternatives for urea oxidation.

In a study by Zhang *et al*, Co_3O_4 NSs were effectively doped with several transition metals [88]. Significant changes in the properties of the material were brought about by the addition of Ni as a dopant [89], which was particularly noteworthy. The potential of the doped Co_3O_4 NSs to be used in a variety of applications was further increased by the fact that Ni is cost-effective and has desirable electrochemical characteristics.

A significant development in the technology of urea electrolytic cells was reported by Qiao and colleagues in 2016 [90]. They used MnO_2 nanolayer hybrid electrodes as anodes and CoPx–Ni foam as cathodes to produce H_2 while simultaneously purifying urea-rich effluent. These MnO_2 nanolayer hybrid electrodes outperformed the Pt/C standard in urea oxidation due to their superior catalytic activity. They were made of finely dispersed MnO_2 nanocrystals within a 3D graphene–Ni foam hybrid framework. For effective hydrogen evolution, the CoPx–NF cathodes were created utilizing an electrodeposition approach.

The effectiveness of nickel hydroxide as an electrocatalyst for the oxidation of urea was studied by Boggs and colleagues [91]. Their investigation showed that this catalyst made it possible for pure hydrogen gas to be produced at the cathode. The potential of $Ni(OH)_2$ for effective urea oxidation and hydrogen synthesis in an alkaline environment was revealed by this study.

By potentially substituting alcohol oxidation for the OER during electrochemical H_2 evolution, electrochemical alcohol oxidation offers a potential alternative to thermal catalytic processes. Using abundant molecules such as HMF and glycerol, this innovative approach may produce beneficial products rather than less useful by-products. This strategy is gaining attention because it has the potential to enhance energy efficiency and thus lower the overall cost. The conversion of HMF to 2,5-furandicarboxylic acid (FDCA), in particular, has demonstrated astounding current density and efficiency when carried out using a variety of electrode materials, including oxides/hydroxides such as NiFe layered double hydroxides [92], meso-structured nickel oxide [93], CoOOH [94], Mno_2 [95], $Fe(NO_3)_3$ [96], and CuO [97]. This novel idea is ideally suited to the materials commonly employed in alkaline water electrolyzers, potentially eliminating stability issues and allowing for less expensive anode materials like iron.

Cobalt spinel oxides (Co_3O_4) were investigated by Jin and colleagues for their potential as electrooxidation catalysts for HMF [98]. They discovered that the Co^{2+} ions in the spinel oxide structure's tetrahedral positions have an affinity for chemisorbing acidic organic compounds. On the other hand, HMF oxidation is influenced by Co^{3+} ions in octahedral locations. Based on that understanding, they replaced Co^{2+} with Cu^{2+} to simultaneously increase acidic adsorption and increase the exposure of Co^{3+} sites. The electrooxidation efficiency of HMF was improved because of this tactical change.

The HER overpotential for the MoO_2–FeP@C heterojunction was surprisingly low at 103 mV at 10 mA cm^{-2}, which was attributed to extensive electron transport between MoO_2 and FeP at the active interface. As a result of this transfer, the water and hydrogen adsorption energies on FeP were adjusted, increasing HER efficiency, while hole accumulation on MoO_2 increased HMF oxidation activity [99]. The design of the heterojunction successfully captured these synergistic effects, improving electrocatalytic performance in both the HER and HMF oxidation processes. In recent developments, several transition-metal-oxide catalysts such as NiOOH [100] and $CuCo_2O_4$ have emerged as promising candidates for the electrooxidation of HMF [101]. These catalysts have highlighted significant activity levels comparable to those of precious-metal catalysts reported in previous studies.

Metal oxides and carbon are commonly used supports for heterogeneous catalysts in glycerol oxidation reactions. The selection of a suitable metal-oxide support is influenced by the strong metal–support interaction and unique reversible interactions with reagents. Notably, TiO_2 is favored as a metal oxide support for glycerol oxidation. Dimitratos *et al* studied the way in which factors such as the PVA/Au weight ratio, Au concentration, heat treatment, and support choice (TiO_2 or C) affected catalytic activity [102]. Among their samples, the Au/C catalyst with the lowest PVA/Au ratio showed the highest activity, underscoring the importance of precise parameter control for optimal catalytic performance.

In the case of methanol-based water electrolysis, Xiang *et al* introduced a bifunctional catalyst, Pt–Co_3O_4 on carbon paper, which served as both anode and cathode in methanol electrolysis [103]. The electrolyzer functioned successfully, requiring just 0.555 V to produce useful formate and H_2 while achieving a current density of 10 mA cm^{-2}. This voltage demand was noticeably 1022 mV lower than the voltages used in the conventional strategies for water dissociation. Li *et al* used a nanoneedle array of $Ni_xCo_{1-x}(OH)_2$ as an electrocatalyst for the efficient coelectrolysis of methanol and water, achieving highly selective methanol oxidation and hydrogen evolution [104]. This process exclusively generated valuable formate as the oxidation product and achieved a calculated faradaic efficiency close to 100%, indicating its promise. For the electrocatalytic oxidation of biomass, numerous composite catalysts have been reported, the majority of which comprise transition metals. In addition to the previously listed catalysts, we list some further catalysts in table 10.2 along with their primary products.

10.5.2.2 Metal–organic frameworks

Metal–organic frameworks (MOFs) belong to a novel category of crystalline porous materials which are made up of organic linkers and metallic nodes. These materials have drawn the interest of researchers because of their noteworthy properties, including their substantial surface area, exceptional structural stability, and tuneable electrical structures [116, 117]. However, a drawback of MOFs is their limited conductivity, which reduces their potential for use in electrocatalysis [118]. As a result, efforts have been focused on designing and synthesizing 2D MOFs (conductive or semiconductive), with the aim of improving their conductivity and boosting electrochemical reactions.

Table 10.2. Some additional metal-oxide-based catalysts with their primary products.

Electrocatalyst	Electrolyte	Faradaic efficiency	Final anodic product(s)	Reference
$Co_{0.26}$–$Ni(OH)_2$ NPs/CF	0.5 M $CO(NH_2)_2$	100	CO_2 and N_2	[105]
Ni–WO_2@C/N	0.5 M $CO(NH_2)_2$	—	CO_2 and N_2	[106]
β-$Co_{0.1}Ni_{0.9}(OH)_2$/NF	0.05 M C_2H_5OH	95		[107]
$Eu_3O_4/WO_{2.72}$	0.1 M glycerol	—	Glucaric acid	[108]
$NiCo_2O_4$	5	87.5	FDCA	[109]
Ru–NiO	50	43.3	FDCA	[110]
Pt–$Ni(OH)_2$	50	98.7	FDCA	[111]
Mo–Co_3O_4	50	92	FDCA	[112]
NiO– Co_3O_4	10	96	FDCA	[113]
PbO–CuO	50	93.7	FDCA	[114]
CoP–CoOOH	150	96.3	FDCA	[115]

* FDCA—2,5-furandicarboxylic acid.

Recently, a novel electrocatalyst for the HER and the oxidation of ethylene glycol was developed using Pt nanoclusters rooted within Fe–MOF [119]. The Pt nanoclusters enhanced interfacial connections, electron transport, and catalytic activity by forming strong Pt–O interactions with the oxygen centers on the surface of the Fe–MOF. Due to its superior electrocatalytic performance (low overpotential and enhanced stability and current density) in the HER and the ethylene glycol oxidation reaction compared to that of commercial Pt/C, this groundbreaking method demonstrated the potential of Fe–MOF combinations in the design of effective dual-function electrocatalysts (figure 10.10).

Xiang *et al* utilized a derived Ni/Co MOF electrode to efficiently drive primary amine dehydrogenation oxidation at a low potential [120]. Similarly, Cu-MOF with Ni and Co are used for preparation of a highly efficient electrocatalyst for methanol oxidation reaction [121]. In addition to enhancing the hydrogen evolution process in a membrane-free electrolyzer setup, the combination of the Ni/Co bimetal catalyst and the MOF structure allowed for the conversion of primary amines into nitriles with good yields and substrate adaptability. This electrode showed great promise as a multifunctional, high-performance electrocatalyst. In a research study, Qiao *et al* proposed two-dimensional MOF made of benzenedicarboxylic acid and nickel species (Ni^{2+}/Ni^{3+}). They demonstrated the excellent electrocatalytic properties of these materials in the UOR [122]. Their results showed how the favorable combination of quick charge transfer properties and a significant concentration of active Ni sites residing on the surfaces of these 2D NSs considerably increased the catalytic efficiency of nickel-based MOFs in the UOR [123].

Similarly, Jiang *et al* created two-dimensional Ni–MOFs by solvothermal synthesis using Ni^{2+} and an organic ligand of 4-dimethylaminopyridine (Ni-DMAP-*t*, where *t* is the time required for synthesis) [124]. They developed self-supported Ni–DMAP-2 NSs on Ni foam by adjusting the nanosheet thickness. This material had excellent electrocatalytic output in the urea oxidation process, attaining a

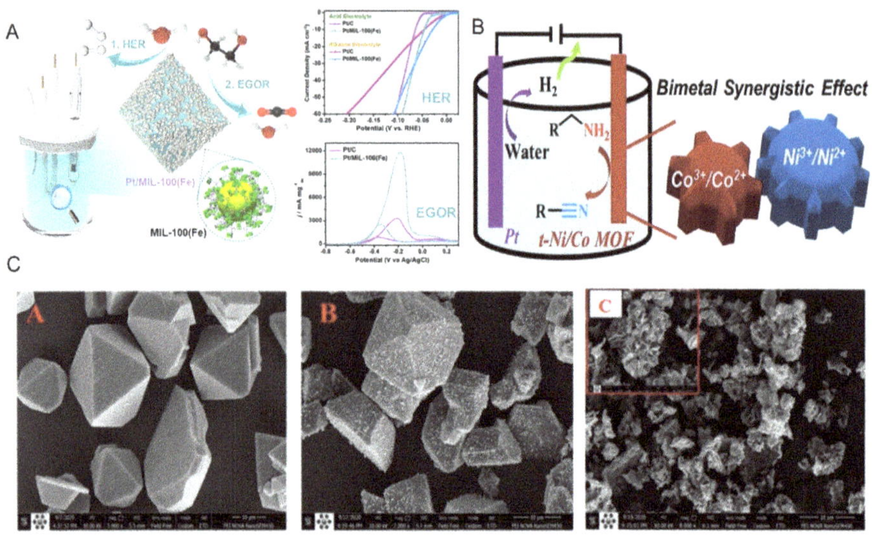

Figure 10.10. (A) Pt/MIL-100(Fe) catalyst for electrocatalytic hydrogen evolution and ethylene glycol oxidation (reprinted from [119], Copyright (2022), with permission from Elsevier). (B) Ni/Co MOF-derived anode for the UOR (reprinted from [120], Copyright (2022), with permission from Elsevier), (C) SEM images depicting (a) Cu–MOF, (b) CuO–C, and (c) CuO–C/$NiCo_2O_4$ composites (reprinted from [121], Copyright (2022), with permission from Elsevier).

remarkably low voltage of 1.45 V at 100 mA cm^{-2}. A potential direction for ultrathin Ni–MOFs in urea-assisted energy applications was provided by the superior urea oxidation activity in Ni–DMAP-2/NF, which was attributed to the abundance of exposed Ni active sites and the ease of electron/ion transfer. Using a regulated spontaneous galvanic replacement reaction, Wang *et al* effectively produced a three-dimensional Ru/NiFe MOF array on Ni foam [125]. The Ru modification significantly improved the ability of intermediate catalyst species to adsorb materials, and self-driven electronic reconstruction at the interface produced many active high-valence Ni sites. Both the water and urea electrolysis procedures were significantly more effective because of these synergistic effects.

10.5.2.3 Metal chalcogenides

The immense potential of transitional-metal chalcogenides, particularly sulfides and selenides, in a variety of applications involving the electrocatalytic oxidation of biomass has drawn the attention of researchers beyond transition-metal oxides. These materials have several benefits, including simple synthetic processes, adjustable constituent composition, outstanding electrical conductivity, and controlled electronic structures. Several examples have shown intriguing bifunctional possibilities for hybrid water electrolysis in accessible heterostructures. Notable instances are Cu_2Se/Co_3Se_4 as reported by Zhao *et al* [126], h-NiSe/CNTs exhibited by Zhao *et al* [127], nitrogen-doped NiS/NiS_2 elucidated by Liu *et al* [128], Co_3S_4-NSs/Ni–F described by Ding *et al* [129], and $Ni_2P/Ni_{0.96}S$ featured in a study by He *et al* [130]. Due to their distinctive compositions and architectures, these diverse combinations demonstrate excellent catalytic activities for water electrolysis, advancing effective and affordable electrochemical processes.

Liu *et al* synthesized Fe-doped CoS_2 NSs, which exhibited adaptable bifunctional electrocatalytic activity, allowing for the integration of direct hydrazine fuel cells and complete hydrazine splitting units [131]. This innovative method created a self-driven H_2 production system that exclusively used hydrazine for fuel. It achieved remarkable hydrogen evolution rates (9.95 mmol h^{-1}), 98% faradaic efficiency, and noteworthy stability (20 h), comparable to the top self-powered water splitting technologies. These improvements were attributed to the Fe doping (as supported by theoretical calculations), which reduced the amount of energy needed for hydrazine dehydrogenation and H adsorption on CoS_2. Using a bifunctional $CoSe_2$ nanosheet electrode in anodic hydrazine oxidation [132], Xia's research team demonstrated energy-efficient H_2 evolution. They created tubular $CoSe_2$ NSs on a Ni foam substrate using a two-step hydrothermal technique, resulting in a dense network of interconnected structures on the tubes' exteriors. In three-electrode system setup, this electrode displayed considerable electrocatalytic capabilities for both the HER and the hydrazine oxidation reaction (HzOR).

Huang *et al* demonstrated the successful selective synthesis of nitrile from primary amines with acceptable yield, substrate compatibility, and olefinic bond tolerance utilizing a NiSe anode [133]. These conversions were greatly aided by the use of redox-active Ni(II)/Ni(III) species. The more advantageous thermodynamics of primary amine electrooxidation over oxygen evolution was harnessed using CoP as a cathode for the HER, substantially reducing the cell voltage and permitting concurrent nitrile production at the anode and the production of H_2 at the cathode.

Among various modification approaches, the utilization of Mott–Schottky heterojunction engineering has garnered significant attention due to its capability to facilitate controlled electron movement within catalysts [30]. The following are some notable examples: NiS/MoS_2 [134], CoS_2/MoS_2 [135], and $CoMn/CoMn_2O_4$ [136]. This method offers two advantages: the abundance of active sites produced by the electrochemical urea oxidation reaction and the efficient electron transfer and optimized hydrogen adsorption energy provided by the catalyst's inherent electric field. For instance, by hydrothermally synthesizing a Mo-doped Ni_3S_2 nanoforest [19], Xu *et al* produced a regulated nanostructure with enhanced electroactive sites. The introduction of Mo induced electronic structure modifications in Ni_3S_2, enhancing its Gibbs free adsorption energy and consequently elevating catalytic activity and endurance. In a similar way, Hao *et al* discovered that the inclusion of electrodeposition-derived Ni in Co_9S_8 NSs [137] improved UOR performance due to improved electrical conductivity, effective charge transfer, and an increase in electroactive sites. These results highlight the potential of cation doping in the progress of efficient, reliable, and financially viable electrocatalysts for the UOR.

Zhong *et al* created NSs of nickel- and iron-based layered double hydroxide on the surface of a conductive Ni_xSe_y nanowire array in an effort to enhance its electrochemically active surface area and electrical conductivity [138]. Using this novel method, a core–shell nanostructured nanoarray catalyst was created. In contrast to individual NiFe–LDH nanosheet arrays and Ni_xSe_y nanowire arrays, this unique catalyst design impressively displayed greater activity in the HMF oxidation reaction. Using a 3D nickel-foam-supported Ni_3S_2/Ni [139] heterostructure nanobelt array resulted in a few beneficial outcomes. This structural arrangement notably increased the effectiveness of electron transfer, redistributed charges, generated nickel species lacking in electrons, raised the adsorption energy of OH– groups, and made urea breakdown simpler. As a specific example, a study of NF@Mo–$Ni_{0.85}$Se [140] showed that doping with Mo can be used to precisely modify the d-band center of Ni, opening a potentially fruitful path for improving the electrochemical oxidation of HMF. Through the introduction of oxygen vacancies via Se doping, Song's research team improved the performance of a CoO electrocatalyst [141]. As a result, CoO–$CoSe_2$ was produced, which demonstrated outstanding selectivity, stability, and performance in the oxidation of HMF through the hydroxymethyl-2-furan–carboxylic acid pathway. Notably, in a mixture of KOH and HMF, this catalyst achieved 97.9% efficiency at 1.43 vs RHE, a 99% yield of FDCA, and an onset potential of 1.3 vs RHE. A crucial element that increased the active surface area, decreased the resistance of charge transfer, and promoted electrocatalytic activity in the oxidation of HMF to FDCA was found to be the abundance of oxygen vacancies in CoO–$CoSe_2$.

In conclusion, a wide variety of metal-based materials, including oxides, hydroxides, nitrides, sulfides, carbides, and phosphides, have been investigated in the search for efficient and reasonably priced electrocatalysts for a variety of applications involving hybrid water electrolysis. These materials have special electrical and catalytic capabilities that increase productivity and sustainability in hydrogen synthesis and electrocatalytic biomass reactions. With their variable topologies and conductivity enhancements, metal chalcogenides and MOFs have also emerged as significant competitors and are appealing options for electrocatalysis. To further enhance the performance of these materials, cutting-edge techniques such as heterojunction engineering and controlled doping have been used, paving the way for improvements in renewable energy conversion and storage technologies. Table 10.3 briefly compares different metal-chalcogenide-based electrocatalysts.

Table 10.3. A comparative analysis of different metal-chalcogenide-based electrocatalysts.

Electrocatalyst	Electrolyte	Faradaic efficiency	Final anodic product	Reference
h-NiSe/CNTs	1 M methanol	98	Formate	[127]
MoS_2/Ni_3S_2/NF	0.3 M $CO(NH_2)_2$	—	CO_2 and N_2	[21]
NiCoS	10 mmol HMF	96.4	FDCA	[142]
NiS_x–Ni_2P	10 mmol HMF	95.1	FDCA	[143]
Cu_2S@Ni_3Se_2/CF	0.5 M urea	92	CO_2 and N_2	[144]
Ni–MoS_2	0.3 M glucose	—	—	[145]

10.6 Summary and outlook

The development of hybrid water splitting for energy-efficient H_2 production holds substantial importance in extending the green energy agenda worldwide. In contrast to conventional water electrolysis, this novel approach offers a range of benefits. These advantages encompass the reduction of essential cell voltages through thermodynamically favorable oxidation reactions, the prevention of explosive gas mixture formation, and the mitigation of catalyst poisoning, all the while yielding value-added by-products.

This chapter provided a comprehensive overview of recent advancements in electrocatalyst development for anodic reactions aimed at hydrogen (H_2) production. These electrocatalyst developments must fulfill the requirements of high selectivity, activity, and durability. These electrocatalysts encompass a wide range of materials, including noble metals, non-noble metals, nonmetals, transition metals, and their various alloys. To enhance the electrochemical activity of these catalysts, considerable attention has been devoted to investigating the relationships between catalyst structure and activity. This exploration encompasses variations in composition and morphology and reconstruction activities that aim to increase the catalytic active sites. In pursuit of these goals, diverse synthetic strategies are employed to optimize the electronic structure, crystal phase orientation, ionic manipulation, and interface engineering through techniques such as doping. These strategies address various atomic-level properties, including electronic and strain effects, with the aim of augmenting electrocatalytic performance.

Some qualitative and quantitative decision criteria such as stability, selectivity, overpotential value, turnover frequency, and Tafel slope have been discussed. These are used to determine the performance and suitability of electrocatalysts for hybrid water electrolysis. Furthermore, this chapter gave a complete and refined explanation of hybrid water electrolysis focused on the various anodic electrocatalysts used for the electrochemical oxidation of different chemicals (such as biomass, amine, alcohol, etc.) to produce high-purity H_2.

Among the noble metals, Pt, Pd, Ru, and Rh are promising catalysts. Non-noble metal-based electrocatalysts are gaining attention for use as potential alternatives to noble metals due to their abundance, cost-effectiveness, impressive activity, and robust stability. However, their electrocatalytic activities are still unsatisfactory. To bridge this performance gap, ongoing research efforts are focused on the development of various compositions of monofunctional, bifunctional, and multifunctional catalysts. These catalysts are based on transition metals, rare-earth metal oxides, hydroxides, MOFs, and metal chalcogenides with different structural and morphological optimizations. The goal is to harness their enhanced electrocatalytic properties for practical applications in hybrid water electrolysis. Continued advancements in catalyst design and materials science are expected to drive further improvements in this technology, making it a key player in the transition to clean and sustainable energy sources.

Acknowledgments

A part of this work was supported by Science Foundation Ireland (SFI-21/FFP-A/9161) under the SFI Frontiers for the Future 2021—Frontiers for Partnership Awards.

References

[1] You B and Sun Y 2018 Innovative strategies for electrocatalytic water splitting *Acc. Chem. Res.* **51** 1571–80

[2] Suen N T, Hung S F, Quan Q, Zhang N, Xu Y J and Chen H M 2017 Electrocatalysis for the oxygen evolution reaction: recent development and future perspectives *Chem. Soc. Rev.* **46** 337–65

[3] Shang X, Dong B, Chai Y M and Liu C G 2018 In-situ electrochemical activation designed hybrid electrocatalysts for water electrolysis *Sci. Bull.* **63** 853–76

[4] Li Y, Wei X, Han S, Chen L and Shi J 2021 MnO_2 electrocatalysts coordinating alcohol oxidation for ultra-durable hydrogen and chemical productions in acidic solutions *Angew. Chem.* **133** 21634–42

[5] Bender M T, Yuan X and Choi K S 2020 Alcohol oxidation as alternative anode reactions paired with (photo)electrochemical fuel production reactions *Nat. Commun.* **11** 4594

[6] Liu J, Duan S, Shi H, Wang T, Yang X, Huang Y and Li Q 2022 Rationally designing efficient electrocatalysts for direct seawater splitting: challenges, achievements, and promises *Angew. Chem.* **134** e202210753

[7] Ji L, Wang J, Teng X, Meyer T J and Chen Z 2020 CoP Nanoframes as bifunctional electrocatalysts for efficient overall water splitting *ACS Catal.* **10** 412–9

[8] Wang P and Wang B 2021 Designing self-supported electrocatalysts for electrochemical water splitting: surface/interface engineering toward enhanced electrocatalytic performance *ACS Appl. Mater. Interfaces* **13** 59593–617

[9] Zhang L, Xiao J, Wang H and Shao M 2017 Carbon-based electrocatalysts for hydrogen and oxygen evolution reactions *ACS Catal.* **7** 7855–65

[10] Zhan S, Zhou Z, Liu M, Jiao Y and Wang H 2019 3D NiO nanowalls grown on Ni foam for highly efficient electro-oxidation of urea *Catal. Today* **327** 398–404

[11] Hosseini H and Roushani M 2020 Rational design of hollow core-double shells hybrid nanoboxes and nanopipes composed of hierarchical Cu–Ni–Co selenides anchored on nitrogen-doped carbon skeletons as efficient and stable bifunctional electrocatalysts for overall water splitting *J. Chem. Eng.* **402** 126174

[12] Jiao S, Fu X, Wang S and Zhao Y 2021 Perfecting electrocatalysts via imperfections: towards the large-scale deployment of water electrolysis technology *Energy Environ. Sci.* **14** 1722–70

[13] Yang H, Driess M and Menezes P W 2021 Self-supported electrocatalysts for practical water electrolysis *Adv. Energy Mater.* **11** 2102074

[14] Li B Q, Zhang S Y, Tang C, Cui X and Zhang Q 2017 Anionic regulated NiFe (oxy)sulfide electrocatalysts for water oxidation *Small* **13** 1700610

[15] Tian L, Pang X, Xu H, Liu D, Lu X, Li J *et al* 2022 Cation–anion dual doping modifying electronic structure of hollow cop nanoboxes for enhanced water oxidation electrocatalysis *Inorg. Chem.* **61** 16944–51

[16] Vazhayil A, Vazhayal L, Thomas J, Ashok C S and Thomas N 2021 A comprehensive review on the recent developments in transition metal-based electrocatalysts for oxygen evolution reaction *Appl. Surf. Sci. Adv.* **6** 100184

[17] Naz S, Durrani S K, Mehmood M and Nadeem M 2016 Hydrothermal synthesis, structural and impedance studies of nanocrystalline zinc chromite spinel oxide material *J. Saudi Chem. Soc.* **20** 585–93

[18] Gao L, Liu Z, Ma J, Zhong L, Song Z, Xu J *et al* 2020 NiSe@NiO$_x$ core–shell nanowires as a non-precious electrocatalyst for upgrading 5-hydroxymethylfurfural into 2,5-furandicarboxylic acid *Appl. Catal.* B **261** 118235
[19] Xu H, Liao Y, Gao Z, Qing Y, Wu Y and Xia L 2021 A branch-like Mo-doped Ni_3S_2 nanoforest as a high-efficiency and durable catalyst for overall urea electrolysis *J. Mater. Chem. A Mater.* **9** 3418–26
[20] Zhang J Y, Tian X, He T, Zaman S, Miao M, Yan Y *et al* 2018 *In situ* formation of Ni_3Se_4 nanorod arrays as versatile electrocatalysts for electrochemical oxidation reactions in hybrid water electrolysis *J. Mater. Chem. A Mater.* **6** 15653–8
[21] Li F, Chen J, Zhang D, Fu W F, Chen Y, Wen Z *et al* 2018 Heteroporous MoS_2/Ni_3S_2 towards superior electrocatalytic overall urea splitting *Chem. Commun.* **54** 5181–4
[22] Zeradjanin A R, Masa J, Spanos I and Schlögl R 2021 Activity and stability of oxides during oxygen evolution reaction—from mechanistic controversies toward relevant electrocatalytic descriptors *Front. Energy Res.* **8** 613092
[23] Niu S, Li S, Du Y, Han X and Xu P 2020 How to reliably report the overpotential of an electrocatalyst *ACS Energy Lett.* **5** 1083–7
[24] Wang H, Sun M, Ren J and Yuan Z 2023 Circumventing challenges: design of anodic electrocatalysts for hybrid water electrolysis systems *Adv. Energy Mater.* **13** 2203568
[25] Görlin M, Chernev P, Ferreira de Araújo J, Reier T, Dresp S, Paul B *et al* 2016 Oxygen evolution reaction dynamics, faradaic charge efficiency, and the active metal redox states of Ni–Fe oxide water splitting electrocatalysts *J. Am. Chem. Soc.* **138** 5603–14
[26] Kweon D H, Okyay M S, Kim S J, Jeon J P, Noh H J, Park N *et al* 2020 Ruthenium anchored on carbon nanotube electrocatalyst for hydrogen production with enhanced Faradaic efficiency *Nat. Commun.* **11** 1278
[27] Anantharaj S, Karthik P E and Noda S 2021 The significance of properly reporting turnover frequency in electrocatalysis research *Angew. Chem. Int. Ed.* **60** 23051–67
[28] Costentin C, Drouet S, Robert M and Savéant J M 2012 Turnover numbers, turnover frequencies, and overpotential in molecular catalysis of electrochemical reactions. cyclic voltammetry and preparative-scale electrolysis *J. Am. Chem. Soc.* **134** 11235–42
[29] Shinagawa T, Garcia-Esparza A T and Takanabe K 2015 Insight on tafel slopes from a microkinetic analysis of aqueous electrocatalysis for energy conversion *Sci. Rep.* **5** 13801
[30] Veeramani K, Janani G, Kim J, Surendran S, Lim J, Jesudass S C *et al* 2023 Hydrogen and value-added products yield from hybrid water electrolysis: a critical review on recent developments *Renew. Sustain. Energy Rev.* **177** 113227
[31] Hu S, Tan Y, Feng C, Wu H, Zhang J and Mei H 2020 Synthesis of N doped NiZnCu-layered double hydroxides with reduced graphene oxide on nickel foam as versatile electrocatalysts for hydrogen production in hybrid-water electrolysis *J. Power Sources* **453** 227872
[32] Du M, Sun H, Li J, Ye X, Yue F, Yang J *et al* 2019 Integrative Ni@Pd–Ni alloy nanowire array electrocatalysts boost hydrazine oxidation kinetics *ChemElectroChem.* **6** 5581–7
[33] Tian W, Zhang X, Wang Z, Cui L, Li M, Xu Y *et al* 2022 Amorphization activated RhPb nanflowers for energy-saving hydrogen production by hydrazine-assisted water electrolysis *J. Chem. Eng* **440** 135848
[34] Peng W, Xiao L, Huang B, Zhuang L and Lu J 2011 Inhibition effect of surface oxygenated species on ammonia oxidation reaction *J. Phys. Chem.* C **115** 23050–6

[35] Xu W, Lan R, Du D, Humphreys J, Walker M, Wu Z *et al* 2017 Directly growing hierarchical nickel–copper hydroxide nanowires on carbon fibre cloth for efficient electro-oxidation of ammonia *Appl. Catal.* B **218** 470–9
[36] Gnana kumar G, Farithkhan A and Manthiram A 2020 Direct urea fuel cells: recent progress and critical challenges of urea oxidation electrocatalysis *Adv. Energy Sustain. Res.* **1** 2000015
[37] Han W K, Li X P, Lu L N, Ouyang T, Xiao K and Liu Z Q 2020 Partial S substitution activates $NiMoO_4$ for efficient and stable electrocatalytic urea oxidation *Chem. Commun.* **56** 11038–41
[38] Chen Z, Wei W, Song L and Ni B J 2022 Hybrid water electrolysis: a new sustainable avenue for energy-saving hydrogen production *Sustain. Horizons* **1** 100002
[39] Diao Y, Liu Y, Hu G, Zhao Y, Qian Y, Wang H *et al* 2022 NiFe nanosheets as urea oxidation reaction electrocatalysts for urea removal and energy-saving hydrogen production *Biosens. Bioelectron.* **211** 114380
[40] Yang X, Gao Y, Zhao Z, Tian Y, Kong X, Lei X *et al* 2021 Three-dimensional spherical composite of layered double hydroxides/carbon nanotube for ethanol electrocatalysis *Appl. Clay Sci.* **202** 105964
[41] Li K and Sun Y 2018 Electrocatalytic upgrading of biomass-derived intermediate compounds to value-added products *Chem. Eur. J.* **24** 18258–70
[42] Gomes J F and Tremiliosi-Filho G 2011 Spectroscopic studies of the glycerol electro-oxidation on polycrystalline au and pt surfaces in acidic and Alkaline media *Electrocatalysis* **2** 96–105
[43] Luo H, Barrio J, Sunny N, Li A, Steier L, Shah N *et al* 2021 Progress and perspectives in photo- and electrochemical-oxidation of biomass for sustainable chemicals and hydrogen production *Adv. Energy Mater.* **11** 2101180
[44] Martínez N P, Isaacs M and Nanda K K 2020 Paired electrolysis for simultaneous generation of synthetic fuels and chemicals *New J. Chem.* **44** 5617–37
[45] Zhai Y N, Li Y, Zhu J Y, Jiang Y C, Li S N and Chen Y 2017 The electrocatalytic performance of carbon ball supported RhCo alloy nanocrystals for the methanol oxidation reaction in alkaline media *J. Power Sources* **371** 129–35
[46] Pech-Rodríguez W J, González-Quijano D, Vargas-Gutiérrez G, Morais C, Napporn T W and Rodríguez-Varela F J 2017 Electrochemical and *in situ* FTIR study of the ethanol oxidation reaction on PtMo/C nanomaterials in alkaline media *Appl. Catal.* B **203** 654–62
[47] Bullock R M 2013 Abundant metals give precious hydrogenation performance *Science (1979)* **342** (6162)
[48] Li C and Baek J B 2020 Recent advances in noble metal (Pt, Ru, and Ir)-based electrocatalysts for efficient hydrogen evolution reaction *ACS Omega* **5** 31–40
[49] Rodríguez-Gómez A, Lepre E, Sánchez-Silva L, López-Salas N and de la Osa A R 2022 PtRu nanoparticles supported on noble carbons for ethanol electrooxidation *J. Energy Chem.* **66** 168–80
[50] Davi M, Keßler D and Slabon A 2016 Electrochemical oxidation of methanol and ethanol on two-dimensional self-assembled palladium nanocrystal arrays *Thin Solid Films* **615** 221–5
[51] Marčeta Kaninski M P, Šaponjić Z V, Mudrinić M D, Milovanović D S, Rajčić B M, Radulović A M *et al* 2021 Comparison of Pt and Pd anode catalysts supported on nanocrystalline Ru–SnO_2 for ethanol oxidation in fuel cell applications *Int. J. Hydrogen Energy* **46** 38270–80

[52] Sheng T, Lin X, Chen Z Y, Hu P, Sun S G, Chu Y Q *et al* 2015 Methanol electro-oxidation on platinum modified tungsten carbides in direct methanol fuel cells: a DFT study *Phys. Chem. Chem. Phys.* **17** 25235–43
[53] Liu H, Liu Y, Li M, Liu X and Luo J 2020 Transition-metal-based electrocatalysts for hydrazine-assisted hydrogen production *Mater. Today Adv.* **7** 100083
[54] Tang C, Zhang R, Lu W, Wang Z, Liu D, Hao S *et al* 2017 Energy-saving electrolytic hydrogen generation: Ni_2P nanoarray as a high-performance non-noble-metal electrocatalyst *Angew. Chem. Int. Ed.* **56** 842–6
[55] Zheng J, Chen X, Zhong X, Li S, Liu T, Zhuang G *et al* 2017 Hierarchical porous NC@CuCo nitride nanosheet networks: highly efficient bifunctional electrocatalyst for overall water splitting and selective electrooxidation of benzyl alcohol *Adv. Funct. Mater.* **27** 1704169
[56] Cao M, Cao H, Meng W, Wang Q, Bi Y, Liang X *et al* 2021 Nickel–copper oxide nanoflowers for highly efficient glucose electrooxidation *Int. J. Hydrogen Energy* **46** 28527–36
[57] Khoobi A and Salavati-Niasari M 2019 High performance of electrocatalytic oxidation in direct glucose fuel cell using molybdate nanostructures synthesized by microwave-assisted method *Energy* **178** 50–6
[58] You B, Jiang N, Liu X and Sun Y 2016 Simultaneous H_2 generation and biomass upgrading in water by an efficient noble-metal-free bifunctional electrocatalyst *Angew. Chem. Int. Ed.* **55** 9913–7
[59] Liu W J, Xu Z, Zhao D, Pan X Q, Li H C, Hu X *et al* 2020 Efficient electrochemical production of glucaric acid and H_2 via glucose electrolysis *Nat. Commun.* **11** 265
[60] Ren F, Wang C, Zhai C, Jiang F, Yue R, Du Y *et al* 2013 One-pot synthesis of a RGO-supported ultrafine ternary PtAuRu catalyst with high electrocatalytic activity towards methanol oxidation in alkaline medium *J. Mater. Chem. A Mater.* **1** 7255
[61] Lu S, Li H, Sun J and Zhuang Z 2018 Promoting the methanol oxidation catalytic activity by introducing surface nickel on platinum nanoparticles *Nano Res.* **11** 2058–68
[62] Zhang J, Ye J, Fan Q, Jiang Y, Zhu Y, Li H *et al* 2018 Cyclic penta-twinned rhodium nanobranches as superior catalysts for ethanol electro-oxidation *J. Am. Chem. Soc.* **140** 11232–40
[63] Sheng S, Ye K, Sha L, Zhu K, Gao Y, Yan J *et al* 2020 Rational design of Co–S–P nanosheet arrays as bifunctional electrocatalysts for both ethanol oxidation reaction and hydrogen evolution reaction *Inorg. Chem. Front* **7** 4498–506
[64] Lin C, Zhang P, Wang S, Zhou Q, Na B, Li H *et al* 2020 Engineered porous Co–Ni alloy on carbon cloth as an efficient bifunctional electrocatalyst for glucose electrolysis in alkaline environment *J. Alloys Compd.* **823** 153784
[65] Bi J, Ying H, Xu H, Zhao X, Du X, Hao J *et al* 2022 Phosphorus vacancy-engineered Ce-doped CoP nanosheets for the electrocatalytic oxidation of 5-hydroxymethylfurfural *Chem. Commun.* **58** 7817–20
[66] Sun Q, Wang L, Shen Y, Zhou M, Ma Y, Wang Z *et al* 2018 Bifunctional copper-doped nickel catalysts enable energy-efficient hydrogen production via hydrazine oxidation and hydrogen evolution reduction *ACS Sustain. Chem. Eng* **6** 12746–54
[67] Liu Y, Zhang J, Li Y, Qian Q, Li Z and Zhang G 2021 Realizing the synergy of interface engineering and chemical substitution for Ni_3N enables its bifunctionality toward hydrazine oxidation assisted energy-saving hydrogen production *Adv. Funct. Mater.* **31** 2103673

[68] Li Y, Zhang J, Liu Y, Qian Q, Li Z, Zhu Y *et al* 2020 Partially exposed RuP_2 surface in hybrid structure endows its bifunctionality for hydrazine oxidation and hydrogen evolution catalysis *Sci. Adv.* **6** eabb4197
[69] Ge S, Zhang L, Hou J, Liu S, Qin Y, Liu Q *et al* 2022 Cu_2O-derived PtCu nanoalloy toward energy-efficient hydrogen production via hydrazine electrolysis under large current density *ACS Appl. Energy Mater.* **5** 9487–94
[70] Xue Q, Zhao Y, Zhu J, Ding Y, Wang T, Sun H *et al* 2021 PtRu nanocubes as bifunctional electrocatalysts for ammonia electrolysis *J. Mater. Chem. A Mater.* **9** 8444–51
[71] Martínez-Rodríguez R A, Vidal-Iglesias F J, Solla-Gullón J, Cabrera C R and Feliu J M 2014 Synthesis of Pt nanoparticles in water-in-oil microemulsion: effect of HCl on their surface structure *J. Am. Chem. Soc.* **136** 1280–3
[72] Gonzalez-Reyna M, Luna-Martínez M S and Perez-Robles J F 2020 Nickel supported on carbon nanotubes and carbon nanospheres for ammonia oxidation reaction *Nanotechnology* **31** 235706
[73] Zhang J Y, He T, Wang M, Qi R, Yan Y, Dong Z *et al* 2019 Energy-saving hydrogen production coupling urea oxidation over a bifunctional nickel-molybdenum nanotube array *Nano Energy* **60** 894–902
[74] Lv Z, Li Z, Tan X, Li Z, Wang R, Wen M *et al* 2021 One-step electrodeposited NiFeMo hybrid film for efficient hydrogen production via urea electrolysis and water splitting *Appl. Surf. Sci.* **552** 149514
[75] Cao Z, Zhou T, Ma X, Shen Y, Deng Q, Zhang W *et al* 2020 Hydrogen production from urea sewage on NiFe-based porous electrocatalysts *ACS Sustain. Chem. Eng* acssuschemeng **8** 11007–15
[76] Xie Y, Wang X, Tang K, Li Q and Yan C 2018 Blending Fe_3O_4 into a Ni/NiO composite for efficient and stable bifunctional electrocatalyst *Electrochim. Acta* **264** 225–32
[77] Taitt B J, Nam D H and Choi K S 2018 A comparative study of nickel, cobalt, and iron oxyhydroxide anodes for the electrochemical oxidation of 5-hydroxymethylfurfural to 2,5-furandicarboxylic acid *ACS Catal.* **9** 660–70
[78] Xiang K, Wu D, Deng X, Li M, Chen S, Hao P *et al* 2020 Boosting H_2 generation coupled with selective oxidation of methanol into value-added chemical over cobalt hydroxide@-hydroxysulfide nanosheets electrocatalysts *Adv. Funct. Mater.* **30** 1909610
[79] Chakrabarty S, Offen-Polak I, Burshtein T Y, Farber E M, Kornblum L and Eisenberg D 2021 Urea oxidation electrocatalysis on nickel hydroxide: the role of disorder *J. Solid State Electrochem.* **25** 159–71
[80] Feng Y, Wang X, Dong P, Li J, Feng L, Huang J *et al* 2019 Boosting the activity of Prussian-blue analogue as efficient electrocatalyst for water and urea oxidation *Sci. Rep.* **9** 15965
[81] Yan W, Wang D and Botte G G 2012 Electrochemical decomposition of urea with Ni-based catalysts *Appl. Catal.* B **127** 221–6
[82] Pham T N T and Yoon Y S 2020 Development of nanosized Mn_3O_4–Co_3O_4 on multiwalled carbon nanotubes for cathode catalyst in urea fuel cell *Energies (Basel)* **13** 2322
[83] Zhu X, Dou X, Dai J, An X, Guo Y, Zhang L *et al* 2016 Metallic nickel hydroxide nanosheets give superior electrocatalytic oxidation of urea for fuel cells *Angew. Chem. Int. Ed.* **55** 12465–9
[84] Zeng M, Wu J, Li Z, Wu H, Wang J, Wang H *et al* 2019 Interlayer effect in NiCo layered double hydroxide for promoted electrocatalytic urea oxidation *ACS Sustain. Chem. Eng* **7** 4777–83

[85] Hu S, Wang S, Feng C, Wu H, Zhang J and Mei H 2020 Novel MOF-derived nickel nitride as high-performance bifunctional electrocatalysts for hydrogen evolution and urea oxidation *ACS Sustain. Chem. Eng* **8** 7414–22
[86] Zheng J, Wu K, Lyu C, Pan X, Zhang X, Zhu Y *et al* 2020 Electrocatalyst of two-dimensional CoP nanosheets embedded by carbon nanoparticles for hydrogen generation and urea oxidation in alkaline solution *Appl. Surf. Sci.* **506** 144977
[87] Fan J, Dou Y, Jiang R, Du K, Deng B and Wang D 2021 Electro-synthesis of tungsten carbide containing catalysts in molten salt for efficiently electrolytic hydrogen generation assisted by urea oxidation *Int. J. Hydrogen Energy* **46** 14932–43
[88] Zhang S L, Guan B Y, Lu X F, Xi S, Du Y and Lou X W 2020 Metal atom-doped Co_3O_4 hierarchical nanoplates for electrocatalytic oxygen evolution *Adv. Mater.* **32** 2002235
[89] Cardenas Flechas L J, Raba Paéz A M, Rincón and Joya M 2020 Synthesis and evaluation of nickel doped Co_3O_4 produced through hydrothermal technique *Dyna* **87** 184–91
[90] Chen S, Duan J, Vasileff A and Qiao S Z 2016 Size fractionation of two-dimensional sub-nanometer thin manganese dioxide crystals towards superior urea electrocatalytic conversion *Angew. Chem. Int. Ed.* **55** 3804–8
[91] Boggs B K, King R L and Botte G G 2009 Urea electrolysis: direct hydrogen production from urine *Chem. Commun.* **32** 4859–61
[92] Qi Y F, Wang K Y, Sun Y, Wang J and Wang C 2022 Engineering the electronic structure of NiFe layered double hydroxide nanosheet array by implanting cationic vacancies for efficient electrochemical conversion of 5-hydroxymethylfurfural to 2,5-furandicarboxylic acid *ACS Sustain. Chem. Eng.* **10** 645–54
[93] Holzhäuser F J, Janke T, Öztas F, Broicher C and Palkovits R 2020 Electrocatalytic oxidation of 5-hydroxymethylfurfural into the monomer 2,5-furandicarboxylic acid using mesostructured nickel oxide *Adv. Sustain. Syst.* **4** 1900151
[94] Zhu B, Chen C, Huai L, Zhou Z, Wang L and Zhang J 2021 2,5-Bis(hydroxymethyl)furan: a new alternative to HMF for simultaneously electrocatalytic production of FDCA and H_2 over CoOOH/Ni electrodes *Appl. Catal.* B **297** 120396
[95] Hayashi E, Komanoya T, Kamata K and Hara M 2017 Heterogeneously-catalyzed aerobic oxidation of 5-hydroxymethylfurfural to 2,5-furandicarboxylic acid with MnO_2 *ChemSusChem.* **10** 654–8
[96] Fang C, Dai J J, Xu H J, Guo Q X and Fu Y 2015 Iron-catalyzed selective oxidation of 5-hydroxylmethylfurfural in air: a facile synthesis of 2,5-diformylfuran at room temperature *Chin. Chem. Lett.* **6** 1265–8
[97] Ren J, Song K he, Li Z, Wang Q, Li J, Wang Y *et al* 2018 Activation of formyl CH and hydroxyl OH bonds in HMF by the CuO (1 1 1) and Co_3O_4 (1 1 0) surfaces: A DFT study *Appl. Surf. Sci.* **456** 174–83
[98] Lu Y, Liu T, Dong C L, Yang C, Zhou L, Huang Y C *et al* 2022 Tailoring competitive adsorption sites by oxygen-vacancy on cobalt oxides to enhance the electrooxidation of biomass *Adv. Mater.* **34** 2107185
[99] Yang G, Jiao Y, Yan H, Xie Y, Wu A, Dong X *et al* 2020 Interfacial engineering of MoO_2–FeP heterojunction for highly efficient hydrogen evolution coupled with biomass *Electrooxidation. Adv. Mater.* **32** e2000455
[100] Latsuzbaia R, Bisselink R, Anastasopol A, Van der Meer H, Van Heck R, Yagüe M S *et al* 2018 Continuous electrochemical oxidation of biomass derived 5-(hydroxymethyl) furfural into 2,5-furandicarboxylic acid *J. Appl. Electrochem.* **48** 611–26

[101] Nam D H, Taitt B J and Choi K S 2018 Copper-based catalytic anodes to produce 2,5-furandicarboxylic acid, a biomass-derived alternative to terephthalic acid *ACS Catal.* **8** 1197–206
[102] Dimitratos N, Villa A, Prati L, Hammond C, Chan-Thaw C E, Cookson J *et al* 2016 Effect of the preparation method of supported Au nanoparticles in the liquid phase oxidation of glycerol *Appl. Catal. A Gen* **514** 267–75
[103] Xiang K, Song Z, Wu D, Deng X, Wang X, You W *et al* 2021 Bifunctional Pt–Co_3O_4 electrocatalysts for simultaneous generation of hydrogen and formate via energy-saving alkaline seawater/methanol co-electrolysis *J. Mater. Chem. A Mater.* **9** 6316–24
[104] Li M, Deng X, Xiang K, Liang Y, Zhao B, Hao J *et al* 2020 Value-added formate production from selective methanol oxidation as anodic reaction to enhance electrochemical hydrogen cogeneration *ChemSusChem.* **13** 914–21
[105] Sun C B, Guo M W, Siwal S S and Zhang Q B 2020 Efficient hydrogen production via urea electrolysis with cobalt doped nickel hydroxide-riched hybrid films: cobalt doping effect and mechanism aspect *J. Catal.* **381** 454–61
[106] Shen F, Jiang W, Qian G, Chen W, Zhang H, Luo L *et al* 2020 Strongly coupled carbon encapsulated Ni–WO_2 hybrids as efficient catalysts for water-to-hydrogen conversion via urea electro-oxidation *J. Power Sources* **458** 228014
[107] Chen W, Xie C, Wang Y, Zou Y, Dong C L, Huang Y C *et al* 2020 Activity origins and design principles of nickel-based catalysts for nucleophile electrooxidation *Chem.* **6** 2974–93
[108] Yang Z, Niu H, Xia L, Li L, Xiang M, Yu C *et al* 2023 Rare-earth europium heterojunction electrocatalyst for hydrogen evolution linking to glycerol oxidation *Int. J. Hydrogen Energy* **48** 32304–12
[109] Kang M J, Park H, Jegal J, Hwang S Y, Kang Y S and Cha H G 2019 Electrocatalysis of 5-hydroxymethylfurfural at cobalt based spinel catalysts with filamentous nanoarchitecture in alkaline media *Appl. Catal.* B **242** 85–91
[110] Ge R, Wang Y, Li Z, Xu M, Xu S, Zhou H *et al* 2022 Selective electrooxidation of biomass-derived alcohols to aldehydes in a neutral medium: promoted water dissociation over a nickel-oxide-supported ruthenium single-atom catalyst *Angew. Chem.* **134** 22908–14
[111] Zhou B, Li Y, Zou Y, Chen W, Zhou W, Song M *et al* 2021 Platinum modulates redox properties and 5-hydroxymethylfurfural adsorption kinetics of $Ni(OH)_2$ for biomass upgrading *Angew. Chem. Int. Ed.* **60** 22908–14
[112] Xia B, Wang G, Cui S, Guo J, Xu H, Liu Z *et al* 2023 High-valance molybdenum doped Co_3O_4 nanowires: origin of the superior activity for 5-hydroxymethyl-furfural oxidation *Chin. Chem. Lett.* **34** 107810
[113] Lu Y, Dong C L, Huang Y C, Zou Y, Liu Y, Li Y *et al* 2020 Hierarchically nanostructured NiO–Co_3O_4 with rich interface defects for the electro-oxidation of 5-hydroxymethylfurfural *Sci. China Chem.* **63** 980–6
[114] Zhou P, Lv X, Tao S, Wu J, Wang H, Wei X *et al* 2022 Heterogeneous-interface-enhanced adsorption of organic and hydroxyl for biomass electrooxidation *Adv. Mater.* **34** 2204089
[115] Wang H, Zhou Y and Tao S 2022 CoP–CoOOH heterojunction with modulating interfacial electronic structure: a robust biomass-upgrading electrocatalyst *Appl. Catal.* B **315** 121588
[116] Ding M, Flaig R W, Jiang H L and Yaghi O M 2019 Carbon capture and conversion using metal–organic frameworks and MOF-based materials *Chem. Soc. Rev.* **48** 2783–828

[117] Guo C, Jiao Y, Zheng Y, Luo J, Davey K and Qiao S Z 2019 Intermediate modulation on noble metal hybridized to 2D metal–organic framework for accelerated water electrocatalysis *Chem.* **5** 2429–41

[118] Yang M, Zhou Y N, Cao Y N, Tong Z, Dong B and Chai Y M 2020 Advances and challenges of Fe-MOFs based materials as electrocatalysts for water splitting *Appl. Mater. Today* **20** 100692

[119] He Z L, Huang X, Chen Q, Zhai C, Hu Y and Zhu M 2022 Pt nanoclusters embedded Fe-based metal-organic framework as a dual-functional electrocatalyst for hydrogen evolution and alcohols oxidation *J. Colloid Interface Sci.* **616** 279–86

[120] Xiang M, Xu Z, Wu Q, Wang Y and Yan Z 2022 Selective electrooxidation of primary amines over a Ni/Co metal–organic framework derived electrode enabling effective hydrogen production in the membrane-free electrolyzer *J. Power Sources* **535** 231461

[121] Sheikhi S and Jalali F 2022 Hierarchical $NiCo_2O_4$/CuO-C nanocomposite derived from copper-based metal organic framework and Ni/Co hydroxides: excellent electrocatalytic activity towards methanol oxidation *J. Alloys Compd.* **907** 164510

[122] Zhu D, Guo C, Liu J, Wang L, Du Y and Qiao S Z 2017 Two-dimensional metal–organic frameworks with high oxidation states for efficient electrocatalytic urea oxidation *Chem. Commun.* **53** 10906–9

[123] Zheng S, Zheng Y, Xue H and Pang H 2020 Ultrathin nickel terephthalate nanosheet three-dimensional aggregates with disordered layers for highly efficient overall urea electrolysis *Chem. Eng. J.* **395** 125166

[124] Jiang H, Bu S, Gao Q, Long J, Wang P, Lee C S *et al* 2022 Ultrathin two-dimensional nickel-organic framework nanosheets for efficient electrocatalytic urea oxidation *Mater. Today Energy* **27** 101024

[125] Wang Y, Wang C, Shang H, Yuan M, Wu Z, Li J *et al* 2022 Self-driven Ru-modified NiFe MOF nanosheet as multifunctional electrocatalyst for boosting water and urea electrolysis *J. Colloid Interface Sci.* **605** 779–89

[126] Zhao B, Liu J, Xu C, Feng R, Sui P, Luo J X *et al* 2021 Interfacial engineering of Cu_2Se/Co_3Se_4 multivalent hetero-nanocrystals for energy-efficient electrocatalytic co-generation of value-added chemicals and hydrogen *Appl. Catal.* B **285** 119800

[127] Zhao B, Liu J, Xu C, Feng R, Sui P, Wang L *et al* 2021 Hollow NiSe nanocrystals heterogenized with carbon nanotubes for efficient electrocatalytic methanol upgrading to boost hydrogen co-production *Adv. Funct. Mater.* **31** 2008812

[128] Liu H, Liu Z, Wang F and Feng L 2020 Efficient catalysis of N doped NiS/NiS_2 heterogeneous structure *Chem. Eng. J.* **397** 125507

[129] Ding Y, Xue Q, Hong Q L, Li F M, Jiang Y C, Li S N *et al* 2021 Hydrogen and potassium acetate Co-production from electrochemical reforming of ethanol at ultrathin cobalt sulfide nanosheets on nickel foam *ACS Appl. Mater. Interfaces* **13** 4026–33

[130] He M, Feng C, Liao T, Hu S, Wu H and Sun Z 2020 Low-cost Ni_2P/$Ni_{0.96}S$ heterostructured bifunctional electrocatalyst toward highly efficient overall urea-water electrolysis *ACS Appl. Mater. Interfaces* **12** 2225–33

[131] Liu X, He J, Zhao S, Liu Y, Zhao Z, Luo J *et al* 2018 Self-powered H_2 production with bifunctional hydrazine as sole consumable *Nat. Commun.* **9** 4365

[132] Zhang J Y, Wang H, Tian Y, Yan Y, Xue Q, He T *et al* 2018 Anodic hydrazine oxidation assists energy-efficient hydrogen evolution over a bifunctional cobalt perselenide nanosheet electrode *Angew. Chem. Int. Ed.* **57** 7649–53

[133] Huang Y, Chong X, Liu C, Liang Y and Zhang B 2018 Boosting hydrogen production by anodic oxidation of primary amines over a NiSe nanorod electrode *Angew. Chem. Int. Ed.* **57** 13163–6

[134] Gu C, Zhou G, Yang J, Pang H, Zhang M, Zhao Q *et al* 2022 NiS/MoS_2 Mott-Schottky heterojunction-induced local charge redistribution for high-efficiency urea-assisted energy-saving hydrogen production *Chem. Eng. J.* **443** 136321

[135] Li C, Liu Y, Zhuo Z, Ju H, Li D, Guo Y *et al* 2018 Local charge distribution engineered by Schottky heterojunctions toward urea electrolysis *Adv. Energy Mater.* **8** 1801775

[136] Wang C, Lu H, Mao Z, Yan C, Shen G and Wang X 2022 Bimetal Schottky heterojunction boosting energy-saving hydrogen production from alkaline water via urea electrocatalysis *Adv. Funct. Mater.* **30** 2000556

[137] Hao P, Zhu W, Li L, Tian J, Xie J, Lei F *et al* 2020 Nickel incorporated Co_9S_8 nanosheet arrays on carbon cloth boosting overall urea electrolysis *Electrochim. Acta* **338** 135883

[138] Zhong Y, Ren R Q, Wang J B, Peng Y Y, Li Q and Fan Y M 2022 Grass-like Ni_xSe_y nanowire arrays shelled with NiFe LDH nanosheets as a 3D hierarchical core–shell electrocatalyst for efficient upgrading of biomass-derived 5-hydroxymethylfurfural and furfural *Catal. Sci. Technol.* **12** 201–11

[139] Zhuo X, Jiang W, Qian G, Chen J, Yu T, Luo L *et al* 2021 Ni_3S_2/Ni heterostructure nanobelt arrays as bifunctional catalysts for urea-rich wastewater degradation *ACS Appl. Mater. Interfaces* **13** 35709–18

[140] Yang C, Wang C, Zhou L, Duan W, Song Y, Zhang F *et al* 2021 Refining d-band center in $Ni_{0.85}Se$ by Mo doping: a strategy for boosting hydrogen generation via coupling electro-catalytic oxidation 5-hydroxymethylfurfural *Chem. Eng. J.* **422** 130125

[141] Huang X, Song J, Hua M, Xie Z, Liu S, Wu T *et al* 2020 Enhancing the electrocatalytic activity of CoO for the oxidation of 5-hydroxymethylfurfural by introducing oxygen vacancies *Green Chem.* **22** 843–9

[142] Zhao Z, Guo T, Luo X, Qin X, Zheng L, Yu L *et al* 2022 Bimetallic sites and coordination effects: electronic structure engineering of NiCo-based sulfide for 5-hydroxymethylfurfural electrooxidation *Catal. Sci. Technol.* **12** 3817–25

[143] Zhang B, Fu H and Mu T 2022 Hierarchical NiSx/Ni_2P nanotube arrays with abundant interfaces for efficient electrocatalytic oxidation of 5-hydroxymethylfurfural *Green Chem.* **24** 877–84

[144] Lv L, Li Z, Wan H and Wang C 2021 Achieving low-energy consumption water-to-hydrogen conversion via urea electrolysis over a bifunctional electrode of hierarchical cuprous sulfide@nickel selenide nanoarrays *J. Colloid Interface Sci.* **592** 13–21

[145] Liu X, Cai P, Wang G and Wen Z 2020 Nickel doped MoS_2 nanoparticles as precious-metal free bifunctional electrocatalysts for glucose assisted electrolytic H_2 generation *Int. J. Hydrogen Energy* **45** 32940–8

IOP Publishing

Chapter 11

Hydrogen production and storage: fundamentals and recent advances

Nikoleta D Nikolova and Priyanka Ganguly

In recent years, hydrogen has garnered considerable attention as an eco-friendly alternative to fossil fuels. The increasing demand for energy, coupled with growing concerns about climate change, has compelled stakeholders to explore new horizons. The integration of renewable energy sources such as solar and wind power with hydrogen production is a promising combination that can ensure a steady electricity supply to the grid. Consequently, there has been a significant focus on the generation of environmentally friendly hydrogen and the enhancement of storage efficiency. This chapter offers a thorough summary of both the fundamental principles and recent advancements in methods for producing renewable hydrogen. In addition, it reviews the utilization of nanomaterials (NMs) for hydrogen storage, comparing the characteristics of various NMs used, including carbon nanotubes, other carbon-based NMs, and metal–organic frameworks (MOFs). It also outlines other materials such as complex hydrides. It concludes that the integration of wind and solar energy for hydrogen production represents a pivotal step in the transition to a sustainable and resilient energy landscape. This synergy is a testament to the power of innovation and collaboration in the pursuit of a greener and more sustainable future.

11.1 Introduction

Hydrogen is one of the key green alternatives to conventional carbon-based nonrenewable energy sources [1]. Although the combustion of hydrogen does not produce harmful greenhouse gases such as CO_2, it is, however, essential to consider the method used to produce the hydrogen. The common routes for hydrogen production are steam reforming and the use of fossil fuels such as coal to generate H_2. The overall carbon footprints of these processes are quite high and do not meet the requirements of the climate pledges made at COP26 [2].

doi:10.1088/978-0-7503-5749-4ch11

The reaction of hydrogen (H_2) through combustion results in the formation of water vapor and boasts a calorific value of 141.9 kJ g^{-1}, which is three times greater than the energy content of gasoline (47 kJ g^{-1}) and 2.6 times higher than that of natural gas (54 kJ g^{-1}). [3]. In comparison with fossil fuels such as ethanol and natural gas, hydrogen is lighter; it has a density of 0.089 88 g l^{-1} at standard temperature and pressure (STP) and a molar mass of 1.008 g mol^{-1} [4]. However, the energy density per volume of hydrogen (0.09 kg m^{-3}) is relatively low at STP when compared with those of other fuels. As a result, the storage of hydrogen in vehicles requires a sizable tank at an elevated pressure compared to the pressures used for other gaseous fuels. Thus, it is crucial to advance energy storage technologies to enhance the energy density per unit volume. Therefore, optimizing safe operational storage solutions for hydrogen is critical for the everyday usage of hydrogen in forthcoming years [5].

In this chapter, we discuss the fundamentals of hydrogen production such as thermochemical water splitting, photoelectrochemical hydrogen generation, etc. We also discuss several conventional and some unconventional storage solutions. Metal hydrides and ceramic components for hydrogen storage are discussed. Moreover, we detail some of the more advanced materials, such as MOFs, used for hydrogen storage.

11.2 Hydrogen production methods

There are numerous methods for producing hydrogen from gaseous or liquid fuels. These vary depending on the primary fuel type (ammonia, methanol, ethanol, gaseous or liquid hydrocarbons, or water) and the chemical reactions carried out (decomposition, steam reforming, partial oxidation, electrolysis, or gasification). Nowadays, most H_2 is created through the steam reforming of fossil fuels, primarily methane-containing natural gas. Methods of manufacturing hydrogen without the emission of CO_2 will be required as concerns rise regarding the mitigation of potential climate change and the reduction of greenhouse gas emissions. Therefore, it is crucial to consider alternative sources for hydrogen production [6]. To increase production capacity while reducing or eliminating net greenhouse gas emissions without the use of carbon sequestration technologies, it is possible to produce hydrogen from renewable sources obtained from agricultural or alternative waste streams, which allows for increased flexibility. To produce hydrogen locally and independently, electrolysis and bioproduction can be simply adapted, eliminating the need to build a substantial and costly distribution network [7]. Unfortunately, the low efficiency of using biomass as fuel is among its main disadvantages, as it achieves a thermal efficiency of only 10%–30% [8]. A comparative overview of the major renewable hydrogen production methods has been carried out, together with a discussion of the advancements in hydrogen storage techniques, analyzing the basic processes through which hydrogen is produced and stored and the major challenges and difficulties faced when considering mass production. Figure 11.1 displays a summary of the different primary energy sources and methods for H_2 production [9].

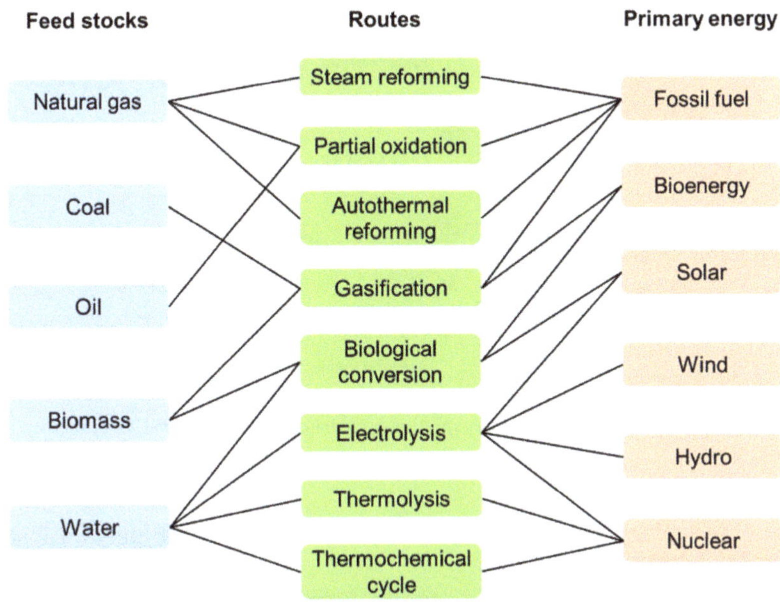

Figure 11.1. The primary energy sources, feedstocks, and routes for H_2 production (reprinted from [9], Copyright (2021), with permission from Elsevier).

11.2.1 Thermochemical water splitting

A variety of methods are used to produce renewable hydrogen, including biomass processing; biological, solar, and thermal water splitting; and water electrolysis [10]. Hydrogen and oxygen can be produced from water through thermochemical water splitting, a process initially developed to use waste heat from nuclear power reactions, which later evolved to use concentrated solar energy. This is a long-term technological route that could produce zero or very few greenhouse gas emissions. In thermochemical cycles, elevated temperatures (500 °C–2000 °C) are employed in thermochemical water splitting systems, facilitating a sequence of chemical reactions that produce hydrogen. The chemical inputs for each cycle are recycled, creating a closed loop that utilizes only water and generates hydrogen and oxygen. The simplest thermochemical water splitting cycle consists of a two-step process in which a metal oxide used as a catalyst undergoes one endothermic reaction and one exothermic reaction. The metal oxide is initially reduced and reacted with water to yield hydrogen [5]. The reaction is as follows:

$$MO_{ox} \rightarrow MO_{red} + \frac{1}{2}O_2 \tag{11.1}$$

$$MO_{red} + H_2O \rightarrow MO_{ox} + H_2. \tag{11.2}$$

Hydrogen production using solar thermochemical water splitting has been studied using a variety of distinct cycles, each with its own operating circumstances,

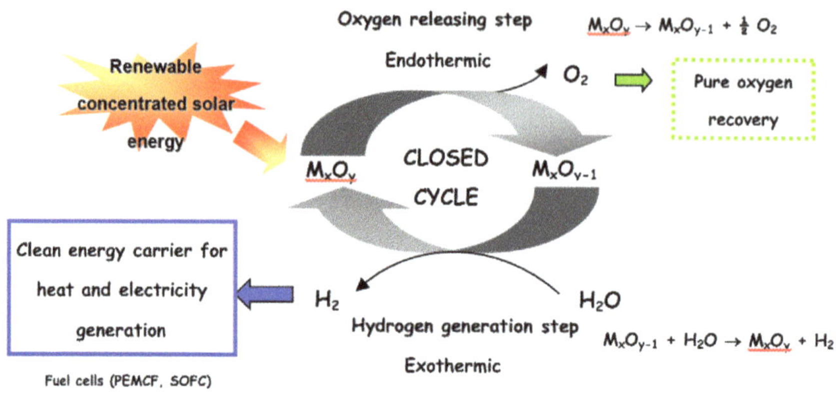

Figure 11.2. A schematic of the two-step water splitting cycle based on the M_xO_y/M_xO_{y-1} system (reproduced from [11], CC BY 4.0).

engineering difficulties, and hydrogen production prospects. Figure 11.2 schematically summarizes the process [11].

The research literature describes over 300 distinct water splitting cycles, examples of which include the 'direct' two-step cerium oxide thermal cycle and the 'hybrid' copper chloride cycle, as explained in a document published by the US Office of Energy Efficiency and Renewable Energy [12]. In contrast to the more complex hybrid cycles, direct cycles are typically simpler and involve fewer phases, but they also need greater operating temperatures [13].

Concentrated solar power plants are highly efficient and do not need expensive catalysts [14]. To raise the temperature to a level that can be used for hydrogen generation, incident solar radiation is concentrated over a smaller area using mirrors, parabolic dishes, or power towers. Although water thermolysis is the cleanest method and hence the most popular, the very high operational temperatures (2500 K–3000 K) required restrict the materials that can endure these extreme conditions and accessibility at a reasonable price, which are the main obstacles to its widespread adoption [15]. In addition, to prevent the recombination of oxygen and hydrogen, an effective separation mechanism is also necessary.

11.2.2 Photocatalytic hydrogen production

Another strategy used to convert solar energy into chemical energy is photocatalytic splitting. The word 'photo' refers to photons, whilst a catalyst is a chemical that affects the speed at which a process takes place [16]. Hence, photocatalysts are materials that, upon exposure to light, modify the pace of a chemical reaction. All photocatalysts can be regarded as semiconductors. In the phenomenon called photocatalysis, a semiconducting substance generates an electron–hole pair when exposed to light. [17]. Depending on the physical appearance of the reactants, two types of photocatalytic reactions are identified:

- *Homogeneous photocatalysis* occurs when the semiconductor and reactant both exist in the same phase (gas, solid, or liquid).

- *Heterogeneous photocatalysis* occurs when the semiconductor and reactant are in distinct phases.

The bandgap (E_g) represents the energy difference between the valence band (highest occupied molecular orbital, HOMO) and the conduction band (lowest unoccupied molecular orbital, LUMO). Materials are categorized into three fundamental groups based on their bandgaps: metals or conductors (E_g < 1.0 eV), semiconductors (1.5–3.0 eV), and insulators (E_g > 5.0 eV). Semiconductors, due to their capacity to conduct electricity in the presence of light even at ambient temperatures, serve as effective photocatalysts. [18]. The energy of photons is absorbed by a ground-state electron (e), which is promoted to the excited state when a photocatalyst is exposed to light of the appropriate wavelength (i.e. carrying enough energy). A hole (h^+) is made in the valence band during this process, in which an e–h^+ pair is produced, thus creating a photoexcited state. The excited electron is utilized to reduce an acceptor, while the donor molecules are oxidized by the h^+ left in the acceptor. A photocatalyst's ability to simultaneously supply an oxidation environment and a reduction environment is what gives photocatalysis its significance [19].

Natural photosynthesis offers a model for converting solar energy into chemical fuels (such as glucose and cellulose) and storing it. In the natural photosynthetic process, the initial step is the light reaction, in which water is split. Subsequently, the photogenerated holes are captured by the oxygen-evolving complex (Mn_4CaO_5) to oxidize water into dioxygen and protons. This process directly drives the energy-storing chemical reactions for further chemical synthesis. In a broad sense, the transformation of solar energy into chemical energy can be seen as a form of artificial photosynthesis [20].

Photocatalytic hydrogen production from water is seen as a highly promising method for developing a hydrogen economy. Figure 11.3 schematically illustrates photocatalytic H_2 production on the surface of a semiconductor. The main challenge encountered has been that of finding a photocatalyst that is chemically stable, resistant to corrosion, and capable of harvesting visible light whilst creating a suitable bandgap [21]. Advances in nanoscience and nanotechnology have accelerated the improvement of current photocatalysts as well as the identification and creation of new candidate materials [21].

For a nanocrystal to reduce water, four electrons must be generated. Therefore, the nanocrystal surface needs to be affected by four photons of sufficient energy within a brief period to produce four electrons. The lifespan of the charge carriers is expected to be 1 s, and the photons require at least 4 ms for absorption by the nanoparticles (NPs), making it quite difficult to obtain a solar energy flux that completes the water splitting reaction. Two-dimensional (2D) architectural NMs can meet these needs by decreasing recombination and cutting down on the charge carriers' transit time. Multiple layers facilitate light absorption at low flux density and reduce the distance traveled by the charge carriers [5].

The overall water splitting reaction encompasses two fundamental reactions: the hydrogen evolution reaction (HER) and the oxygen evolution reaction (OER).

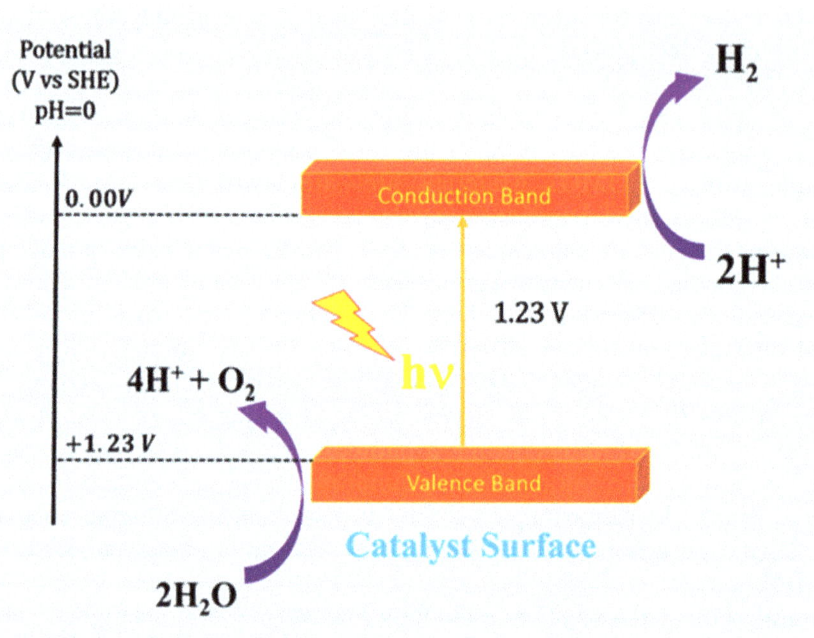

Figure 11.3. A schematic illustration of photocatalytic hydrogen generation on the catalytic surface sites of two-dimensional (2D) NMs (reprinted with permission from [5], Copyright (2019), American Chemical Society).

The overall goal is to identify photoreactive surfaces that have the capacity to capture light and generate electron–hole pairs. After that, the HER and OER mechanisms can proceed. To achieve this goal, it is crucial that the conduction band edge of the surface has a lower potential than the reduction potential of H^+ to H_2 (0.00 V), while the valence band edge of the catalyst has a higher potential than the oxidation potential of H_2O to O_2 (1.23 V). [5]. The photocatalytic conversion of H_2 and O_2 is a thermodynamically unfavorable reaction:

$$H_2O \rightarrow H_2 + \frac{1}{2}O_2 \Delta G_0 = 237\ \text{kJ mol}^{-1}, \qquad (11.3)$$

and it requires the successful completion of multiple difficult processes, including:

- activation of the semiconductor photocatalyst with highly energetic photons
- the movement of the induced electrons and holes to the reaction sites on the outermost layer
- the use of these charge carriers in the oxidation/reduction reactions
- the desorption of the resulting compounds from the photocatalyst's surface to the liquid/gas material

Recombination of the e–h^+ pair in the bulk or on the surface occurs more rapidly than the chemical oxidation/reduction reactions, since the timescale of these processes varies. Therefore, recombination is thought to be one of the primary factors limiting photocatalytic activity. Natural photosynthesis has better charge carrier and mass transfer characteristics than artificial water splitting; therefore, it has significantly greater rates of O_2 evolution. This comparison shows that more work needs to be done on photocatalytic devices if they are to compete with nature's sophisticated structure [22].

Numerous studies have emphasized the utilization of various ranges of semiconductor NMs, such as metal oxides, metal sulfides, and metal-doped metal oxides/sulfides and the formation of different heterojunctions. Such NMs and heterojunctions are some of the common types of composite structures utilized for photocatalytic hydrogen generation. A recent study examined an sp^2-carbon-linked triazine-based covalent organic framework (COF-JLU100). The material showed exceptional characteristics, including high crystallinity, a large surface area, robust durability, and enhanced carrier mobility, which are crucial to achieving an efficient photon-mediated HER. Notably, the introduction of Pt into COF-JLU100, also known as terephthalaldehyde-covalent organic framework, resulted in an impressive rate of hydrogen evolution exceeding 1 00 000 $\mu mol\ g^{-1}\ h^{-1}$ during water splitting under visible light illumination ($\lambda > 420$ nm). The authors reported that empirical and theoretical studies supported the idea that the cyano-vinylene segments within COF-JLU100 promoted extended π-delocalization, facilitating rapid charge transfer and separation rates along with favorable dispersion in water. Furthermore, the use of inexpensive and readily available monomers in the preparation of COF-JLU100, coupled with its remarkable stability, underscored its viability for large-scale solar-driven hydrogen production [23]. Another attempt to *in situ* photodeposit Pt on COF (PY-DHBD, PY-1,4-dihydroxybenzidine) used the adsorption of the Pt precursor $PtCl_6^{2-}$ in proximity to specific sites near the –OH group and the –C=NR group to initiate *in situ* photodeposition of well-dispersed Pt clusters on a 2D COF surface layer. This localized photoreduction process resulted in a highly dispersed deposition of the Pt cocatalyst, consequently leading to an impressive hydrogen evolution rate of 42 432 $\mu mol\ g^{-1}\ h^{-1}$ at 1 wt% Pt loading. This research not only demonstrates a precise approach for controlling the deposition of photocatalytic cocatalysts at the atomic level but also highlights the advantages of leveraging the customizable and adjustable pore morphology of a COF to engineer an efficient photocatalyst. The use of multicomponent structures as composites leads to interesting properties. Four distinct types of synthesized ternary chalcogenides ($AgInS_2/TiO_2$, $AgBiS_2/TiO_2$, $AgInSe_2/TiO_2$, $AgBiSe_2/TiO_2$) exhibited distinctive structural and electronic properties. These materials, characterized by narrow bandgaps, demonstrated significant potential when hybridized with 2D counterparts, presenting a promising avenue for diverse multifunctional applications. Table 11.1 summarizes the composites fabricated, the type of heterojunction created, and the hydrogen produced.

The intimate interfaces generated by this amalgamation contribute to delayed recombination pathways. Furthermore, the incorporation of an additional element, such as a cocatalyst, or the use of multiple fabrication techniques, has notably enhanced charge carrier transport. The innovative production process of 2D/2D compound structures has led to improved visible light absorption and enhanced material stability, effectively mitigating concerns related to photocorrosion.

Table 11.1. A summary of the composites fabricated, the type of heterojunction created, and the hydrogen produced.

Composites	Heterojunction	Hydrogen generation (μ mol m^{-1})
$AgBiS_2$–TiO_2	Type II	1300
$AgInS_2$–TiO_2	Type II	300
$AgBiSe_2$–TiO_2	*p–n*	180
$AgInSe_2$–TiO_2	Type II	250

11.2.3 Electrochemical hydrogen generation

Another similar method for producing hydrogen is based on electrochemistry and consists of the electrolytic decomposition of water in saline, acidic, or alkaline solutions. The method of water decomposition differs in acidic and alkaline media, although the overall reaction is the same [24]. During an electrochemical reaction, one molecule of the substrate is first changed into an intermediate with an unpaired electron; the electron is then moved to an electrode. A series of bond-forming or bond-breaking reactions, as well as an additional single-electron transfer processes, are required to convert the intermediate into the final product. An ionized salt solution occupies the region between the anode and the cathode, and charge travels through the solution between the electrodes due to ion migration. The cathode and anode are typically split into separate cells, facilitating ion diffusion but hindering the movement of precursors and products across compartments, which avoids undesired reactions [25]. Water electrolysis is generally separated into two half-cell reactions: the HER, in which water is reduced at the negative electrode to form H_2, and the OER, in which water is oxidized at the positive electrode to produce O_2. The delayed reaction times of the OER and the HER are due to elevated overpotentials. This means that the potential difference between the thermodynamically calculated half-cell reaction voltage and the experimentally determined value is overlarge and therefore requires a higher amount of energy than the thermodynamically expected amount to drive the reaction. Catalysts play a key role in optimizing this process but are dependent on the conditions under which the cell functions. As mentioned above, three main types of electrolysis technologies are used nowadays:

- proton exchange membrane (PEM) electrolysis
- alkaline electrolysis
- high-temperature solid oxide water electrolysis

PEM electrolysis is performed under acidic conditions and presents low gas transmission and high proton conductivity, although it limits the choice of OER electrocatalysts to precious metals and noble-metal oxides. This leads to high production costs for the cells. On the other hand, alkaline electrolysis makes it possible to broaden the range of electrocatalysts used. Still, it lowers the efficiency of the HER by two to three orders of magnitude in contrast to an acidic environment. A positive value of ΔG (the Gibbs free energy) indicates that electrolytic water splitting is a reaction that requires an input of energy; in other words, it moves against the thermodynamic gradient and must overcome a large kinetic barrier. Catalysts are critical in decreasing the kinetic barrier and are assessed according to their activity, stability, and effectiveness. The overpotential, Tafel gradient, and the exchange current density are used to express the activity. Fluctuations in the overpotential or current over time characterize the stability. Finally, efficiency is defined by faradaic efficiency and turnover frequency (figure 11.4) [26].

As the synthetic process is complex, creating bifunctional catalysts remains difficult for both the OER and the HER. Until now, many catalysts have been created that perform better than noble metals in only a single area; however, the challenge is to create a catalyst that can outperform noble metals in all areas. Ultimately, future research areas will solve current difficulties in the creation of successful and widely used methods for non-noble-metal-based catalysts for water electrolysis [27].

Platinum-based catalysts have a pivotal role in the electrochemical HER, a sustainable and environmentally friendly method for hydrogen production. The authors of one study reported the successful synthesis of ultrafine PtNiP nanowires (NWs) through a two-step hydrothermal technique. This resulted in the production of PtNiP NWs with a diameter of less than 5 nm. Notably, the PtNiP NWs

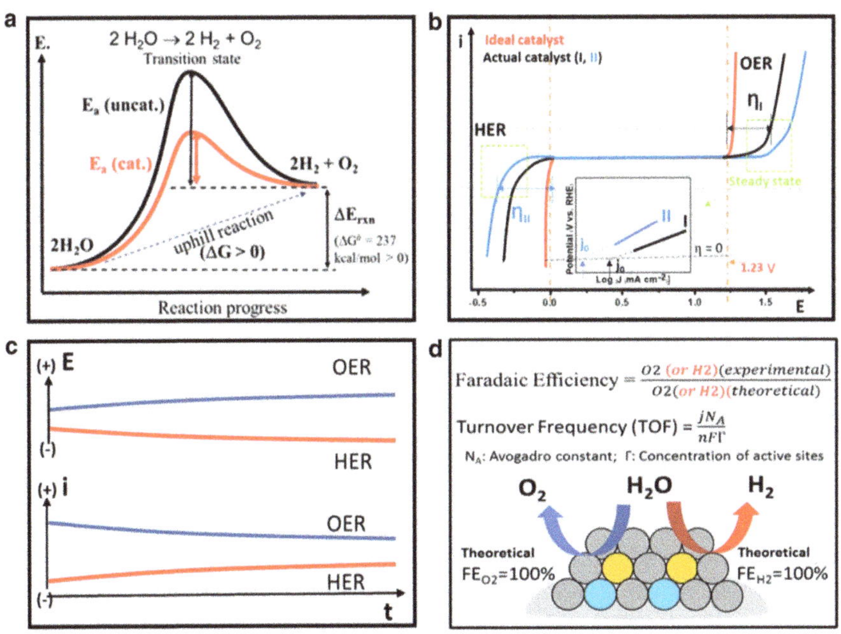

Figure 11.4. (a) A visual representation of the catalyst's function in reducing the activation energy barrier; (b–d) diagrammatic representations of the parameters used to assess electrocatalysts, encompassing (b) effectiveness, measured by overpotential, Tafel slope, and exchange current density; (c) durability, indicated by current– and potential–time curves, and (d) efficacy gauged using faradaic efficiency and turnover frequency (reproduced from reference [26], CC BY 4.0).

demonstrated exceptional electrocatalytic activity for the HER under alkaline conditions, achieving a remarkable current density of 500 mA cm^{-2} at a mere −0.153 V vs a reversible hydrogen electrode (RHE), with a Tafel slope of only 30 mV dec^{-1}. The phosphorization of PtNi NWs resulted in an even distribution of phosphorus atoms across the NW structure. Theoretical studies (density functional theory, DFT) highlighted that the presence of Ni atoms promoted the chemisorption of H_2O molecules, while P atoms facilitated the binding of hydroxyl species. This assisted in the rapid diffusion of H* to Pt atoms and subsequently enhanced the production of hydrogen. The remarkable HER performance of PtNiP NWs under alkaline conditions, which surpassed that of commercial Pt/C, was attributed to the synergistic effects of Pt, Ni, and P [28]. The incorporation of Pt on carbon nanofiber (CNF) was also reported to improve the HER. These structures were designed to enable an unbiased, exceptionally effective, and enduring HER. Carbon dots served as foundational units, aiding in the creation of graphene on the CNF surface. The incorporation of graphene improved the electronic conductivity of the CNFs to approximately 3013.5 S m^{-1}, while also promoting the uniform growth of Pt clusters, facilitating effective electron transfer during the HER. Notably, an electrode with a low Pt loading of 3.4 μg cm^{-2} exhibited outstanding mass activity for the HER in both acidic and alkaline media, outperforming commercial Pt/C (with 31 μg cm^{-2} of Pt loading). Moreover, a solar cell coupled with a luminescent solar concentrator provided the necessary voltage, resulting in a bias-free water splitting system with a solar-to-hydrogen (STH) efficiency of 0.22% under one-solar illumination (standard illumination at air mass (AM) 1.5, or 1 kW m^{-2}) [29]. The authors of another recent report reported the fabrication of a ternary composite of Co_3O_4–CeO_2/g-C_3N_4 (CC–CNx%). These composites were shown to be efficient electrode materials that improved the performance of hydrogen storage at a current of 1 mA. Notably, amongst the various additives used, a ternary nanocompound with 50.0 wt% CC NPs demonstrated a higher discharge capacity of approximately 1020.53 mAh g^{-1} in a three-electrode cell. The observed synergistic interaction of the interfacial architecture offers valuable insights into the construction of uniformly distributed nanocomposites designed for exceptionally efficient energy storage applications [30].

11.2.4 Photoelectrochemical hydrogen generation

Photoelectrochemistry (PEC) is the study of the interplay between photons and electrochemical systems. The basic reaction mechanism consists of redox reactions between electrochemically active species in solution and photoexcited materials (figure 11.5). Today, photoactive materials are made from hybrid semiconductors. They are fabricated using NMs and organic semiconductors and consist of small photoactive molecules such as metal complexes and polymers [31]. The mechanism of photoelectrochemical hydrogen generation is based on photosynthetic cells, in which two redox systems in the electrolyte react with photogenerated holes and electrons at the surfaces of the semiconductors [32]. As mentioned in previous sections, the conductivity is dependent on the bandgap between the conduction band and the valence band. The peculiarity, in this case, is the presence of extra charge

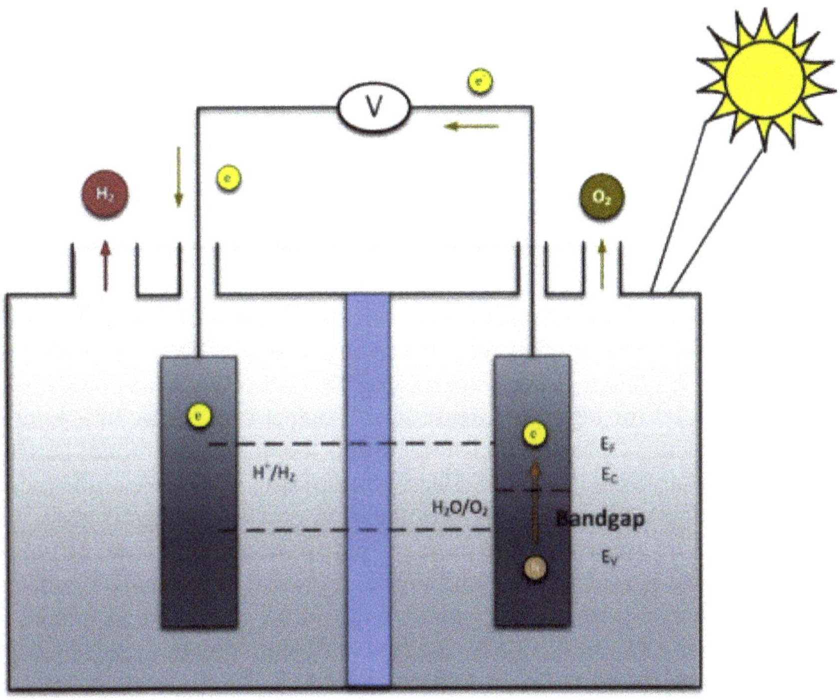

Figure 11.5. A general schematic of the water splitting photoelectrochemical cell (reprinted from [33], Copyright (2013), with permission from Elsevier).

carriers in both *n*-type (electrons) and *p*-type semiconductors (holes), which allows the n-type to be easily oxidized and the p-type reduced, creating the foundation for energy transformation in photoelectrochemistry. When light with sufficient energy is incident on the semiconductors, it generates holes and electrons which move to the junction between the semiconducting material and the electrolyte, promoting oxidation and reduction reactions until equilibrium is reached. The photoelectrochemical methodology is inherently uncomplicated; however, its efficacy hinges upon intricate synergies among solar radiation, semiconductors, and liquid solutions [2, 5].

Several challenges are therefore faced: the energy absorbed by the semiconducting materials should be high enough to split water and maintain both H_2 and O_2 evolution reactions at the semiconductor–liquid interface. Furthermore, to adopt PEC for large-scale deployments whilst maximizing its performance, it is crucial to select the correct materials and system settings, together with consistent working conditions at reasonable costs. Materials ought to have suitable energy gaps and energy edges, high optical absorption coefficients, little reflectance, and low carrier recombination rates in order to reduce the excess voltage associated with the movement of electrons to protons in solution. Different designs are currently being reviewed to address the main challenges, such as single cell photoelectrodes, dual cell photoelectrodes, and photoelectrochemical cells [34].

Silicon is one of the most abundantly available elements, and it therefore makes sense to use it to fabricate cathode surfaces. The authors of a recent work reported the deposition of *n*-type TiO_2 on a Si surface to fabricate a p–n heterojunction and a protective layer, which effectively enhanced charge separation and augmented the stability of water splitting [35]. Under AM1.5G illumination, the optimized Si/TiO_2 photocathode demonstrated a remarkable photocurrent density of -15.53 mA cm^{-2} at 0 V vs the RHE, a substantial onset potential of 0.60 V vs the RHE, and a remarkable efficiency in converting photons to current under applied bias of 2.22% for hydrogen production. The modification of the surface to have a nanoporous structure and the establishment of the p–n heterojunction provided valuable knowledge for the development of effective photoelectrodes for solar energy conversion. [36]. A unique three-dimensional (3D) SiNW@MoS_2/NiS_2 photocathode was developed through the integration of defect-rich MoS_2/NiS_2 nanosheets onto silicon nanowires (SiNWs). This integration facilitated additional active sites, enhancing charge movement and the effective separation of electron–hole pairs. The 3D architecture of the photocathode promoted an efficient charge transfer mechanism and provided a smooth pathway for rapid hydrogen emission. Notably, the SiNW@MoS_2/NiS_2 photocathode exhibited exceptional performance, achieving a maximum photocurrent density of 21.4 mA·cm^{-2} at 0.9 V vs RHE, a maximum H_2 production rate of 183 μmol·h^{-1}, a minimal diffusion resistance of 34.7 Ω, and remarkable catalytic stability for over 10 h at pH = 7 [35].

Similarly, plasmon-mediated photoelectrocatalysis is a favorable strategy for enhancing photomediated energy conversion. However, its dependence on costly noble metals such as gold (Au) and silver (Ag) and the localized spatial distribution of surface plasmon resonance (SPR) limit the enhancement of photoelectrochemical production and hinder practical applications. The authors of a recent work reported the design of a near-infrared (NIR) light-active periodic plasmonic heterostructure that combined semimetallic Bi NPs and $Bi_3(Se_nTe_{1-n})_2$ ternary alloy NWs to expand the light absorption range and maximize the SPR effect. Unlike noble metals, metallic Bi enabled the stimulation of the SPR effect across the entire ultraviolet-to-NIR spectrum. The periodic heterostructure efficiently alleviated localized SPR, promoting charge transfer and redox kinetics by effectively harnessing local electromagnetic fields and photothermal heating. As a result, a photoanode made using the plasmonic heterostructure achieved an impressive incident photon-to-current conversion efficiency of 22% at 800 nm and a significant photocurrent density of 13.8 mA cm^{-2} at 0.85 V vs RHE under visible light, without the need for any cocatalysts [37].

11.2.5 Green hydrogen generation

The definition of green hydrogen has been approached in a variety of ways, depending on factors such as whether there is a necessity to produce green hydrogen using sustainable energy, the boundaries of the carbon accounting procedure, the pollution standards at which hydrogen is designated green, and the resources and manufacturing processes that are used. As in the case of electricity, different feedstocks (biological or not) and energy paths (renewable or not) can be used to make hydrogen. Green hydrogen does not yet have a market, unlike renewable electricity. However, large amounts of hydrogen are already utilized in industry, most of which are derived from fossil fuels and associated with heavy CO_2 emissions [38].

Hydrogen is regarded as the optimal environmentally friendly, secure, and renewable energy source due to its cost-effectiveness and minimal environmental footprint. Furthermore, the abundance of water and sunlight are excellent resources for producing hydrogen fuel; therefore, after the creation of nanotechnology, further research was conducted on water splitting. The numerous chemical processes used to create NPs were deemed hazardous to both people and the environment, so the focus, in recent years, was to create NPs through environmentally friendly processes that use plants and other microorganisms. The design of NPs with good performance represents a complicated challenge due to a series of requirements; the NPs need to be stable under both UV and visible radiation, and they need to have an extended surface region with significant surface energy and high crystallinity. The sustainable methods of NP production employ plant species, microbes, and templates to provide the enzymes, polyphenols, and other molecules used in the redox reactions of metallic ions. The development of *in vitro* methods also allows NPs' size and morphology to be controlled, thanks to changes in the acidity of the system. Several other factors control the formation of NPs, such as temperature, which increases the reaction rate; plant chemical constituents, present at different concentrations for different plant species; and the electrochemical potentials of metal ions, which influence the plant extraction ability [39]. Figure 11.6 is a schematic outline of different routes for green hydrogen production.

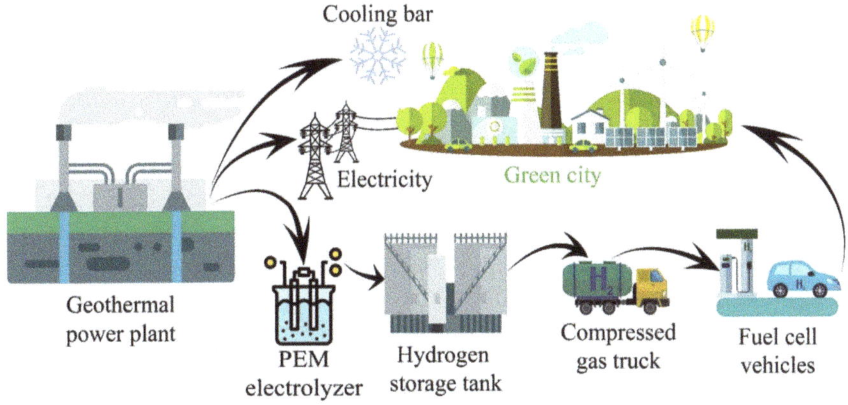

Figure 11.6. An outline displaying routes for green hydrogen production from eco-friendly power generation and other electricity sources such as electrolyzers and distribution vessels (reprinted from [40], Copyright (2022), with permission from Elsevier).

11.3 Hydrogen storage solutions

Hydrogen is one of the major renewable alternatives, as it can be converted into electricity by fuel cells. However, a major hurdle arises from the storage of hydrogen, as a light fuel-cell vehicle would need 4 kg of H_2 to travel 400 km. Nevertheless, each kilogram of H_2 requires 11 m^3 of volumetric space for storage at normal temperature and pressure [41, 42]. Therefore, hydrogen storage mechanisms have become essential and play a major role in storing energy sources for time-delayed end uses. While it is an excellent medium for renewable energy storage, hydrogen is extremely hard to store. The key issues demanding attention revolve around the effective, secure, and economically viable storage of hydrogen. The difficulties arise due to hydrogen's low volumetric energy density compared to other gasses, and its requirement for greater storage space. It also has a boiling point close to absolute zero, needing cryogenic storage. Ultimately, under certain conditions, hydrogen can also cause cracks in metals, which can lead to dangerous leaks [43]. Today, there are four main technologies used to store hydrogen: high-pressure gas compression, liquefaction, metal hydride storage, and carbon nanotube absorption. The systems and vessels used differ, as explained in the following [44]. The vessels must endure and guard against H_2 permeation and corrosion, factors that have led to the development of new materials such as ceramic composites, improved resins, and engineered fibers that meet financial targets without compromising on functionality [45].

Compressed storage is the commonest and simplest technique utilized across the field, in which hydrogen is compressed to pressures ranging from 20 to 100 MPa in carbon fiber composite or metal pressure vessels [44]. Although compression can be performed at room temperature, its energy density is low in comparison with those of other methods. Liquefied cryogenic hydrogen storage is a method used when abundant amounts of H_2 need to be moved, requiring high gravimetric performance. Cylindrical tanks are normally vacuum insulated and kept at temperatures of –253 °C, which leads to higher energy densities [44]. Due to the low temperatures that hydrogen needs to be kept at, the energy necessary for the liquefaction of H_2 gas is exceedingly high.

Finally, hydrogen can be stored on the surfaces of solids (by adsorption) or within solids (by absorption), resulting in higher hydrogen moles per unit volume storage in comparison to those of other storage systems. Different types of materials are currently used to store H_2 in solid forms, such as metal hydrides, nanostructured porous compounds, zeolites, MOFs, and carbon nanotubes. Hydrogen can be reversibly absorbed by solid compounds at room temperature and atmospheric pressure [45, 46]. Figure 11.7 displays a comparative overview of different varieties of materials employed for the storage of hydrogen.

11.3.1 Liquified H_2 storage

Liquefied hydrogen simply refers to the liquid form of H_2, which is obtained by cooling gaseous hydrogen to temperatures below –253 °C at ambient pressure. This method of storage is normally used when high-volume transport is required in the absence of pipelines and in high-tech fields such as space travel. The major challenges

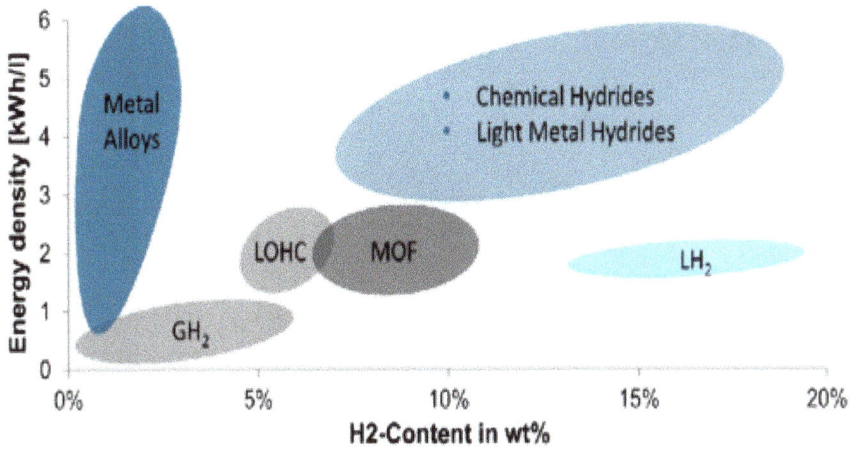

Figure 11.7. A comparison of different hydrogen storage technologies (reproduced from [47], CC BY 4.0). LOHC—liquid organic hydrogen carriers; MOF—metal–organic frameworks; GH_2—gaseous hydrogen; LH_2—liquid hydrogen

faced in this case are the fact that it takes energy to reduce hydrogen to cryogenic temperatures, consuming over 30% of the energy content of hydrogen, resulting in relatively high costs [48]. The reason for liquefying hydrogen is that its energy density is improved when it is stored in a liquid state, allowing the storage of much larger masses and resulting in more economical long-distance transport. The Joule–Thomson cycle is the simplest liquefaction cycle, which consists of cooling compressed hydrogen gas in a heat exchanger, facilitating isenthalpic Joule–Thomson expansion (meaning that the enthalpy of the fluid is constant during the process) and producing some liquid. After cooling, the gas is isolated from the liquid and subsequently reintroduced into the compressor to repeat the cycle (figure 11.8) [42].

The low temperature is sustained through the application of liquid helium/nitrogen cylinders, and the storage vessels are equipped with optimal vacuum insulation to reduce the boil-off rate. Boil-off losses are a result of thermal inefficiency; they depend on the configuration and the heat conservation properties of the containers and scale in direct proportion with the surface-to-volume ratio. The remaining issues with liquid hydrogen storage revolve around the hydrogen boil-off point, together with the energy required for hydrogen liquefaction, which reduce the total efficiency of the storage setup. In addition, liquid hydrogen tank systems are generally quite complex, consisting of vacuum-insulated tanks covered with over 40 individual layers of radiation shields to reduce heat influx. The costs associated with manufacturing such tanks are quite high [45]. However, hydrogen liquefaction has achieved a relatively solid foundation, boasting a worldwide capability of 355 tons per day (tpd), and the most extensive facility currently functions at a capacity of 34 tpd [49].

11.3.2 Gaseous H_2 storage

The most established storage method for hydrogen involves compressing gaseous hydrogen and storing it in high-pressure containers, relying on the physical

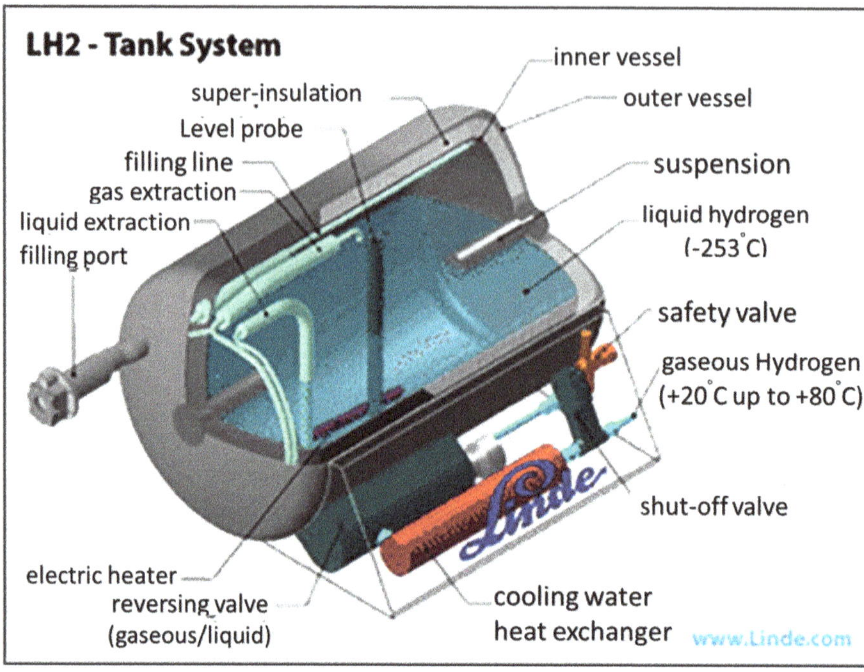

Figure 11.8. A liquid hydrogen storage tank system, horizontally mounted with a double gasket and a dual seal (reproduced from [45], CC BY 4.0).

compression of the H_2 gas. This method is divided into four different types, depending on the design and the materials used to assemble it. Figure 11.9 is a diagrammatic depiction of the four varieties of pressure vessels used for storing compressed hydrogen. [50]. Type I vessels are normally steel vessels and are mainly adopted for stationary use due to their large weight. Type II vessels consist of a metal-lined hoop-wrapped composite cylinder. However, both types of vessels have low hydrogen storage density and face the challenges of metal degradation. Type III vessels consist of a composite cylinder surrounded by an aluminum layer to prevent hydrogen permeation and embrittlement. Type III vessels allow the storage of between 25% and 75% more mass compared to type I and II vessels, rendering them more apt for transport purposes.

Similarly, type IV vessels are made of a composite cylinder with a high-density polyethylene liner. They are the lightest of the pressure vessels, capable of withstand pressures up to 1000 bar. Changes in the materials used to produce pressure vessels can drastically improve the weight of the vessels, together with the storage density capacity, but their production costs increase [51].

The main challenges associated with compressed hydrogen gas storage lie in hydrogen's low specific gravity, which is approximately only 7% of that of air. (The specific gravity is the ratio of the density of a substance to that of a reference substance at STP.) This means that H_2 gas has a lower density than air and tends to ascend and diffuse rapidly if dispersed in the environment, constituting a safer alternative to

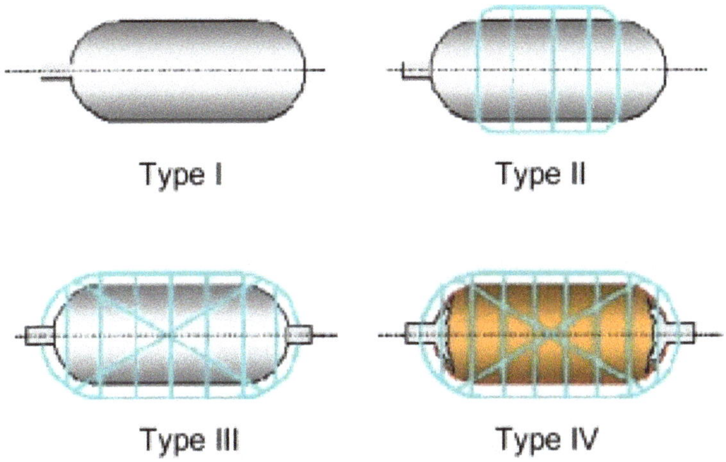

Figure 11.9. A schematic representation of the four pressure vessel types for compressed hydrogen storage (reprinted from [50], Copyright (2022), with permission from Elsevier).

hydrocarbon fuels from a safety standpoint [50]. This also implies that higher amounts of energy are required to compress hydrogen gas and that storing large quantities of compressed hydrogen at over 200 bar introduces multiple problems [52].

Hydrogen embrittlement is another factor to be considered; this refers to the ability of hydrogen to reduce the mechanical strength of certain materials. In addition, the heat transfer capability of hydrogen is greater than that of other gasses, meaning that its temperature increases upon expansion in a Joule–Thomson process [50]. Currently, great attention is being given to the development and manufacture of alloys that can prevent hydrogen embrittlement.

The geological storage of hydrogen is another alternative widely considered because it offers unrestricted storage capacity. The theory behind this idea is that of injecting hydrogen gas underground under pressure, to be extracted in future. Its advantages lie in the fact that it provides secure and safe storage, that can be easily integrated with an urban environment, and that it results in a more economical alternative compared to other modern methods. Depleted natural gas reservoirs, salt and rock caverns, and abandoned mines can all be used as storage vessels. Cushion gas constitutes the base gas of the storage reservoir, maintaining the pressure at adequate levels, while the actual hydrogen stored is referred to as the working gas. This type of storage uses input and output wells together with a confining layer. It consists of a series of injections and withdrawals from the reservoir (as shown in figure 11.10), processes that might eventually lead to a disturbance in the equilibrium of residual gasses with the rocks, risking potential hydrogen contamination. To avoid this, studies are being conducted on the number of cycles, rate, and intervals of injections and withdrawals, aiming to minimize the degree of contamination. Ultimately, several scientific challenges need to be addressed before geological hydrogen storage can be implemented as a global mechanism, but research continues due to the high potential of this alternative [52, 53].

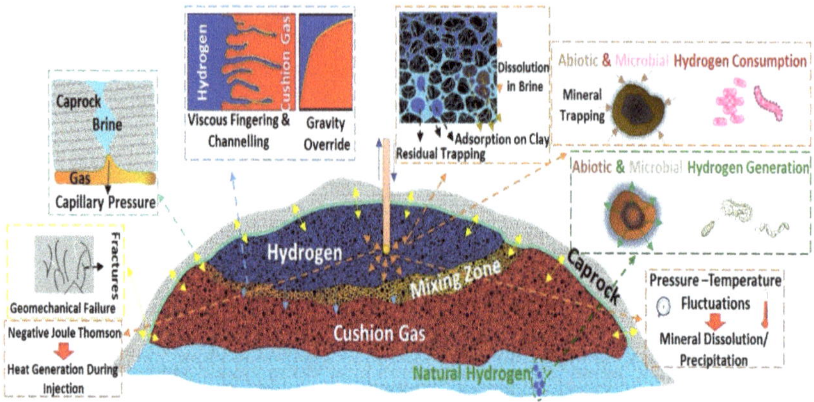

Figure 11.10. A graphical illustration of various physical/geochemical/microbial reactions that lead to hydrogen contamination (reprinted with permission from [53], Copyright (2021), American Chemical Society).

11.3.3 Metal hydrides and ceramics for hydrogen storage

Hydrides have a very rich materials chemistry. 'Metal hydride' is the term used to refer to compounds formed of metal cations and hydride anions, and this family of materials has gained increased popularity for hydrogen storage and compression applications. Metal hydrides have the advantages of requiring no moving parts, allowing the use of simple designs, being compact and safe, and having high volumetric hydrogen storage densities. They are normally formed by the following reaction:

$$M(s) + \frac{x}{2}H_2 \rightarrow MH_x(s) + Q \tag{11.4}$$

As the metal hydride is formed, energy is released in the form of heat (figure 11.11).

This is a promising alternative for storing hydrogen at ambient pressure and temperature [54]. Historically, the first synthesis of a hydride was reported in the 1860s, following the discovery that palladium had the ability to absorb large amounts of H_2. The last two decades have seen a huge volume of research focus on this area of materials chemistry due to the opportunity to effectively store hydrogen reversibly and in solid form. Hydrides with high hydrogen storage capacities are formed from light metals in the periodic table and are referred to as complex hydrides. Due to their high thermodynamic stability, they are suited for thermal storage, electrochemical batteries, hydrogen compression, hydrogen sensing, and power generation. Metal hydrides are also divided into ionic, covalent, metallic, and/or polymeric compounds. Covalent hydrides are not usually sufficiently stable under ambient conditions and are not deemed efficient. In addition, hydrides can also be grouped based on their operational temperatures (high or low) for hydrogen release/uptake [55].

For example, MgH_2 is one of the metal hydrides recently reported to be used for hydrogen storage, primarily due to its safety, efficiency, substantial storage capacity,

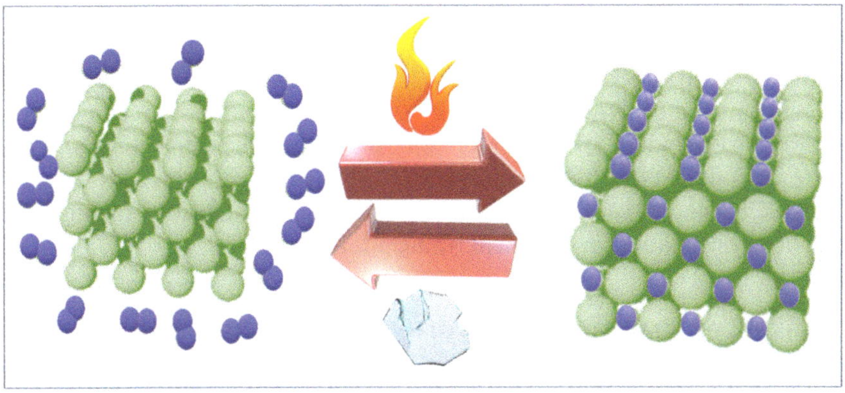

Figure 11.11. Metal hydrides store hydrogen in metal powder: the absorption of hydrogen by hydrides is an exothermic process, while the desorption of H_2 from hydrides is an endothermic process. Thermal management is essential to reverse the reaction.

and cost-effectiveness. However, it faces some drawbacks, including the need for elevated temperatures and slow H_2 absorption dynamics, hindering its widespread implementation. To address these challenges, porous microspheres of $Ni_3ZnC_{0.7}$/Ni-loaded carbon nanotubes (NZC/Ni@CNT) were prepared through a simple purification and calcination method. Subsequently, various proportions of NZC/Ni@CNT (2.5, 5.0, and 7.5 wt%) were added to MgH_2 via ball milling. Among a range of samples exhibiting varied quantities of NZC/Ni@CNT, a composite containing 5 wt% NZC/Ni@CNT exhibited the most promising H_2 storage performance. Notably, the MgH_2–5 wt% NZC/Ni@CNT composite initiated hydrogen discharge at approximately 110 °C, and its hydrogen absorption capacity reached 2.34 wt% H_2 at 80 °C within 60 min. Furthermore, the compound demonstrated a hydrogen emission of approximately 5.36 wt% H_2 at 300 °C. The *in situ*-formed Mg_2NiH_4/Mg_2Ni was reported to function as a hydrogen pump by providing additional reaction centers and enhancing H_2 diffusion channels, facilitating splitting during the absorption process. Moreover, the uniform distribution of Zn and $MgZn_2$ within Mg and MgH_2 provided nucleation sites for Mg/MgH_2 and facilitated H_2 diffusion channels. This study underscores the effectiveness of the bimetallic carbide $Ni_3ZnC_{0.7}$ as an additive for enhancing the hydrogen storage performance of MgH_2, while also highlighting the potential for the facile synthesis of $Ni_3ZnC_{0.7}$/Ni@CNT to guide the design of high-performance carbide catalysts for improving MgH_2 performance [56].

Another group of materials widely considered for hydrogen storage is ceramics. A ceramic material is an inorganic, nonmetallic solid, commonly formed from an oxide, nitride, boride, or carbide and forged at high temperatures. This process leads to many porous compounds that can store hydrogen through physisorption. Most ceramics, like metal compounds, form crystalline structures in which atoms are arranged in 3D patterns that repeat regularly [57]. For optimal storage, these materials need to meet certain requirements, such as: high hydrogen storage capacity, low thermodynamic stability, high kinetics for hydrogen absorption/desorption, high stability when exposed to oxygen and moisture, and good thermal conductivity. Various ceramic materials have been developed, including activated carbon, carbon nanotubes, graphene-based structures, alanates, Mg-based compounds, and so on. Normally, material systems with complex structures and reaction paths are considered for storage purposes [55]. Although numerous studies have been conducted, most were either simulation-based or performed at high pressure and therefore unsuitable for practical applications.

11.4 Advanced materials for hydrogen production and storage

It has become clear that one of the major impediments to hydrogen-based energy systems arises from the lack of satisfactory hydrogen production and storage materials. As previously mentioned, the focus of research is currently on zeolites, advanced carbon nanostructured materials, and MOFs [58]. The scientific community has shown significant interest in nanoporous materials owing to their notable properties, including a large surface area, substantial pore volume, and selective surface chemistry.

Amongst the above materials, zeolites are the most popular thus far, due to unique attributes such as superior resistance to chemicals, excellent mechanical properties, and good adsorption and catalytic properties, highlighting their potential for gas storage and separation. Zeolites are a family of open-framework aluminosilicate materials, neatly arranged in three-dimensional tetrahedral structures, leading to a variety of zeolite frameworks. Until now, a total of 235 unique zeolites have been recognized and are referred to by three-letter codes. To synthesize a zeolite, hydrothermal, sonochemical, and/or microwave-assisted methods can be used. The aim is to obtain zeolites of superior quality in terms of their form, dimensions, porosity (figure 11.12), uniform structure, and crystallinity. *Ex situ* methods are normally preferred to create a better membrane on the support surface, since they permit control of the membrane microstructure and the rate of reaction [59].

Carbon nanostructured materials, commonly referred to as carbon nanotubes, have also gathered significant interest in the past few years, due to their simple synthesis. They also present high hydrogen storage capacities at room temperatures. Single-wall nanotubes are formed by rolling a single graphene sheet, which allows all carbon atoms to be exposed to the surface, resulting in materials with high surface area. Furthermore, the density of the nanotubes is a lot lower than those of metals and MOFs. Carbon nanotubes absorb hydrogen by both physisorption and chemisorption (figure 11.13), and their adsorption capacity is mainly dependent on the diameter of the tubes. Two main methods are used to store hydrogen in nanotubes: gas-phase storage and electrochemical storage. In gas-phase hydrogen storage methods, a sample of the nanotube is exposed to pure hydrogen and the amount of hydrogen absorbed is measured gravimetrically using a microbalance. In electrochemical hydrogen storage, carbon nanotubes are used to create an electrochemical cell, where hydrogen is directly stored following water reduction processes [61].

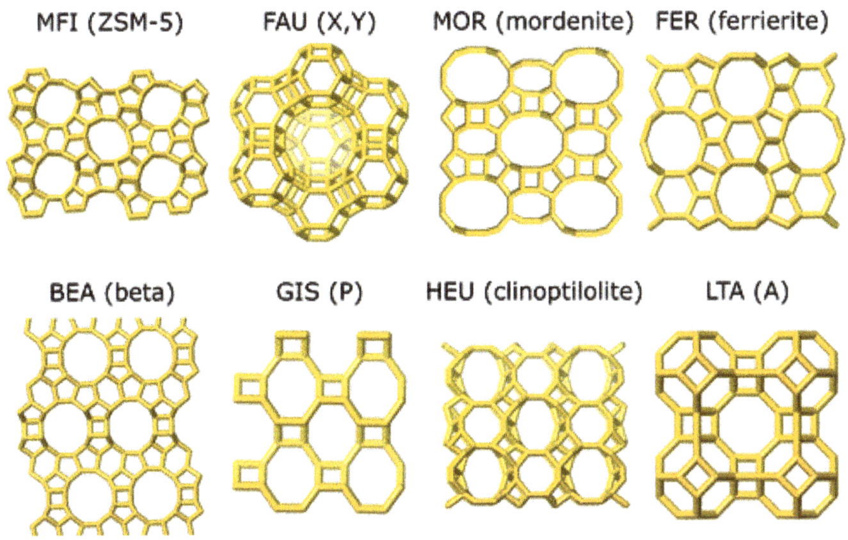

Figure 11.12. Front views of the main pores of selected zeolites. The three-letter code written above each picture is the official name given to each structure, while the common name of the zeolite is indicated in parentheses (reproduced from [60], CC BY 4.0).

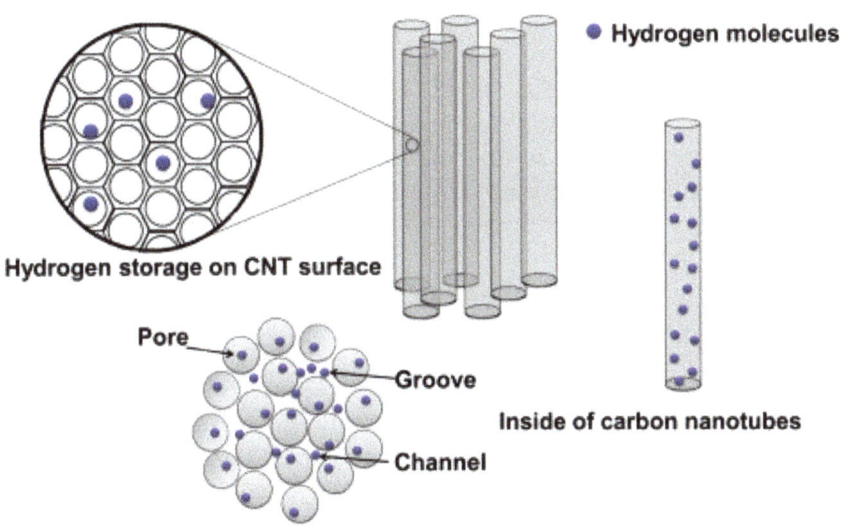

Figure 11.13. A structural representation of the different storage sites of carbon nanotubes. Reprinted from [62], Copyright (2018), with permission from Elsevier).

Finally, MOFs are crystalline materials with infinite lattices, high porosity, and high surface area, which allow them to absorb hydrogen efficiently through weak van der Waals attractive forces. MOFs consist of secondary building units (metal atoms) and organic linkers (carbon chains), forming three-dimensional networks

Figure 11.14. Representative MOFs. Reprinted from [64], Copyright (2019), with permission from Elsevier.

with a defined pore size. Numerous highly porous structures of MOFs are shown in figure 11.14. The presence of strong bonds within the structure allows for frameworks that do not collapse even after the removal of solvent molecules remaining after synthesis. The great variety of organic linkers and metal ions used provides a wide range of different materials [63]. The main benefits of MOFs lie in their bidirectional and rapid H_2 adsorption process, although it can only be achieved at relatively low temperatures. Several challenges need to be overcome, such as improving the interaction between H_2 and the framework to enhance storage capacities at room temperature, as well as optimizing both gravimetric and volumetric functionality.

Several studies have highlighted the synthesis of novel MOFs and their potential usage for hydrogen storage. In one such study, the authors described the synthesis of MOF-76(Tb) for hydrogen storage and humidity-sensing applications. The MOF-76 (Tb) crystal structure is made up of terbium(III) and benzene-1,3,5-tricarboxylate (III) ions, one coordinated aqua ligand, and one crystallization *N,N′*-dimethylformamide molecule. The synthesized material displayed good heat resistance up to 600 °C, as observed using thermogravimetric analysis (TGA). The MOF sample displayed a hydrogen storage capacity of 0.6 wt.% at 77 K and pressures of up to 20 bar [65]. The authors of another recent work studied the H_2 storage capacity of an MOF ($V_2Cl_{2.8}$(btdd)) at ambient temperature. The MOF featured exposed vanadium(II) sites capable of forming back bonds with weak π acids. Notably, it was observed that the adsorbents bonded to H_2 with a heat of adsorption (Q_{st}) of 21 kJ mol^{-1}, enabling the available hydrogen capacity to surpass that of compressed storage, reaching 10.7 g l^{-1}. This study suggests that MOFs containing high-density and weak-π alkali metal sites have the potential to significantly exceed the storage capacity achieved by compressed gas storage under ambient conditions [66]. However, studies to achieve improved storage efficiency under ambient conditions have been a prolonged dream. Long and his colleagues explored the possibility of hydrogen retention at close-to-room temperature conditions. In particular, MOFs such as Ni_2(m-dobdc) demonstrated maximum adsorption capacity at temperatures ranging from −75 °C to 100 °C and pressures from 0 to 100 bar. Remarkably, the H_2 adsorption capacities at 75 °C and below surpassed the pure compression capacity of H_2 at 25 °C. The storage of H_2 in Ni_2(m-dobdc) reached 11.9 g l^{-1} at 25 °C and 100 bar. Findings obtained using neutron diffraction and infrared spectroscopy experiments suggested that the notable H_2 volumetric uptake of Ni_2(m-dobdc) is linked to the presence of highly polarized Ni^{2+} sites with a significant binding enthalpy. This phenomenon leads to the dense packaging of H_2 within the framework [67].

11.5 Outlook and perspectives

The wide-reaching production of sun-powered hydrogen is a critical milestone in the journey toward decarbonizing society. However, enhancing the STH conversion efficiency, ensuring prolonged durability, and achieving affordability in the HER remain areas where improvement is imperative. Moreover, the sluggish kinetic rate, mass transfer constraints, and overall formation technology are some of the major factors to be considered [5].

While NMs have exhibited significant potential in the realm of biological H_2 manufacture, the precise process underlying NM-mediated hydrogen production through organic pathways remains inadequately understood. Furthermore, integrating NMs in the context of biological hydrogen production represents a relatively unexplored area of study. Notably, the combination of photofragmentation and dark fermentation, facilitated by NMs, represents a promising research direction that remains largely uninvestigated. The amalgamation of diverse biological pathways holds the potential to yield fresh insights and foster the growth of optimized and affordable processes for H_2 production.

Further research into hydrogen storage is also critical. The use of physisorbent materials such as MOFs and activated carbon in hydrogen (H_2) storage involves creating advanced functional materials with increased surface area, pore volume, and stability. Carbon nanotubes' effectiveness for H_2 storage depends on their purity, requiring research to improve synthetic protocols. Investigating the impact of oxygen-rich carbon groups on the reversibility of carbon-based nanocomposites, especially in ammonia-borane variants, is crucial. In addition, identifying carbon-based NMs that enhance adsorption kinetics and contribute to hydrogen storage reversibility is a key research area.

Fundamentally, more efforts need to be made in the integration of wind and solar energy to produce hydrogen, which can also be used to mitigate the individual disadvantages associated with each renewable energy-generation system. For example, instability in wind speed leads to unstable power generation from wind turbines; on the other hand, photovoltaic (PV) panels remain inactive during the night. The use of a combined system has been shown to have the capacity to produce hydrogen at a rate of 130–140 ml min^{-1} under solar radiation intensities ranging from 200 to 800 W m^{-2} and wind speeds between 2 and 5 m s^{-1} [68]. At the same time, the initial investment in the hybrid system leads to higher production costs for hydrogen [69]. However, the overall efficiency of hydrogen production surpasses that of relying solely on solar panels or wind turbines. A case study conducted in the Netherlands estimated the manufacturing cost to be 8.7 € kg^{-1}, which is less than the end-user price of 10 € kg^{-1} [70]. Incorporating an inland windmill into the hybrid setup, coupled with an alkaline electrolyzer, further reduced the operational outlay to 4.33 \$ kg^{-1} [71].

11.6 Conclusions

Hydrogen plays a pivotal role in transitioning towards a fossil-fuel-free world, necessitating careful consideration of optimal solutions for its generation, transmission, and storage. This chapter provided a concise overview of existing and emerging technologies in hydrogen production and storage. Overcoming challenges in physical infrastructure remains a key focus in technology development. Strict adherence to safety measures is crucial for the secure transportation and storage of hydrogen to mitigate the risk of accidents. The aim of achieving a specific energy consumption of approximately 5–6 $kWh_{el}/kg\text{-}H_2$ in the near future aligns with technologies such as zero boil-off storage and transportation, offering a significant reduction in the overall cost of liquid hydrogen production, storage, and transportation.

References and further reading

[1] Ishaq H, Dincer I and Crawford C 2022 A review on hydrogen production and utilization: challenges and opportunities *Int. J. Hydrogen Energy* **47** 26238–64

[2] Pillai S C and Ganguly P 2021 2D nanomaterials and composites for energy storage and conversion *2D Materials for Energy Storage and Conversion* (Bristol: IOP Publishing)

[3] Hübert T, Boon-Brett L, Black G and Banach U 2011 Hydrogen sensors—a review *Sens. Actuators* B **157** 329–52

[4] Epelle E I *et al* 2022 A comprehensive review of hydrogen production and storage: a focus on the role of nanomaterials *Int. J. Hydrogen Energy* **47** 20398–431

[5] Ganguly P *et al* 2019 2D Nanomaterials for photocatalytic hydrogen production *ACS Energy Lett.* **4** 1687–1709

[6] Kothari R, Buddhi D and Sawhney R 2008 Comparison of environmental and economic aspects of various hydrogen production methods *Renew. Sustain. Energy Rev.* **12** 553–63

[7] Levin D B and Chahine R 2010 Challenges for renewable hydrogen production from biomass *Int. J. Hydrogen Energy* **35** 4962–9

[8] Ni M, Leung D Y, Leung M K and Sumathy K 2006 An overview of hydrogen production from biomass *Fuel Process. Technol.* **87** 461–72

[9] Ji M and Wang J 2021 Review and comparison of various hydrogen production methods based on costs and life cycle impact assessment indicators *Int. J. Hydrogen Energy* **46** 38612–35

[10] Pérez-Herranz V, Pérez-Page M and Beneito R 2010 Monitoring and control of a hydrogen production and storage system consisting of water electrolysis and metal hydrides *Int. J. Hydrogen Energy* **35** 912–9
[11] Abanades S 2019 Metal oxides applied to thermochemical water-splitting for hydrogen production using concentrated solar energy *ChemEngineering* **3** 63
[12] Mehrpooya M and Habibi R 2020 A review on hydrogen production thermochemical water-splitting cycles *J. Clean. Prod.* **275** 123836
[13] Boretti A, Nayfeh J and Al-Maaitah A 2021 Hydrogen production by solar thermochemical water-splitting cycle via a beam down concentrator *Front. Energy Res.* **9** 116
[14] Ngoh S K and Njomo D 2012 An overview of hydrogen gas production from solar energy *Renew. Sustain. Energy Rev.* **16** 6782–92
[15] Villafán-Vidales H I, Arancibia-Bulnes C A, Valades-Pelayo P J, Romero-Paredes H, Cuentas-Gallegos A K and Arreola-Ramos C E 2019 Hydrogen from solar thermal energy *Solar Hydrogen Production* (Amsterdam: Elsevier) pp 319–63
[16] Mathew S *et al* 2018 Cu-doped TiO_2: visible light assisted photocatalytic antimicrobial activity *Appl. Sci.* **8** 2067
[17] Ganguly P, Panneri S, Hareesh U, Breen A and Pillai S C 2019 Recent advances in photocatalytic detoxification of water *Nanoscale Materials in Water Purification* (Amsterdam: Elsevier) pp 653–88
[18] Padmanabhan N T *et al* 2021 Graphene coupled TiO_2 photocatalysts for environmental applications: a review *Chemosphere* **271** 129506
[19] Ameta R, Solanki M S, Benjamin S and Ameta S C 2018 Photocatalysis *Advanced Oxidation Processes for Waste Water Treatment* (Amsterdam: Elsevier) pp 135–75
[20] Li R and Li C 2017 Photocatalytic water splitting on semiconductor-based photocatalysts *Advances in Catalysis* **60** 1–57 (Amsterdam: Elsevier)
[21] Zhu J and Zäch M 2009 Nanostructured materials for photocatalytic hydrogen production *Curr. Opin. Colloid Interface Sci.* **14** 260–9
[22] Ipek B and Uner D 2019 On the limits of photocatalytic water splitting *Water Chemistry* (London: Intechopen) 1–23
[23] Ma S *et al* 2022 Photocatalytic hydrogen production on a sp^2-carbon-linked covalent organic framework *Angew. Chem.* **134** e202208919
[24] Nefedov V *et al* 2023 Electrochemical production of hydrogen in reactors with reduced energy costs *IOP Conf. Ser.: Earth Environ. Sci.* **1156** 012034
[25] Grimshaw J 2000 *Electrochemical Reactions and Mechanisms in Organic Chemistry* (Amsterdam: Elsevier)
[26] Wang S, Lu A and Zhong C-J 2021 Hydrogen production from water electrolysis: role of catalysts *Nano Convergence* **8** 23
[27] Khan M A *et al* 2018 Recent progresses in electrocatalysts for water electrolysis *Electrochem. Energy Rev.* **1** 483–530
[28] Chen Q *et al* 2022 Synergistic effect in ultrafine PtNiP nanowires for highly efficient electrochemical hydrogen evolution in alkaline electrolyte *Appl. Catal.* B **301** 120754
[29] Wang X *et al* 2022 Platinum cluster/carbon quantum dots derived graphene heterostructured carbon nanofibers for efficient and durable solar-driven electrochemical hydrogen evolution *Small Methods* **6** 2101470

[30] Heydariyan Z, Monsef R and Salavati-Niasari M 2022 Insights into impacts of Co_3O_4–CeO_2 nanocomposites on the electrochemical hydrogen storage performance of g-C_3N_4: pechini preparation, structural design and comparative study *J. Alloys Compd.* **924** 166564
[31] Selvolini G and Marrazza G 2023 Sensor principles and basic designs *Fundamentals of Sensor Technology* (Amsterdam: Elsevier) pp 17–43
[32] Wijayantha K, Auty D and Bhuiyan M S H 2016 Twin cell technology for hydrogen generation *Materials Science and Materials Engineering* (Amsterdam: Elsevier)
[33] Dincer I and Ishaq H 2022 Solar energy-based hydrogen production *Renewable Hydrogen Production* (Amsterdam: Elsevier) pp 91–122
[34] Ahmed M and Dincer I 2019 A review on photoelectrochemical hydrogen production systems: challenges and future directions *Int. J. Hydrogen Energy* **44** 2474–507
[35] Lin F *et al* 2023 Defect-rich MoS_2/NiS_2 nanosheets loaded on SiNWs for efficient and stable photoelectrochemical hydrogen production *J. Colloid Interface Sci.* **631** 133–42
[36] Jian J-X *et al* 2022 Surface engineering of nanoporous silicon photocathodes for enhanced photoelectrochemical hydrogen production *Catal. Sci. Technol.* **12** 5640–8
[37] Liu G-Q *et al* 2023 Near-infrared-active periodic plasmonic heterostructures enable high-efficiency photoelectrochemical hydrogen production *Chem. Mater.* **35** 5822–31
[38] Abad A V and Dodds P E 2020 Green hydrogen characterisation initiatives: definitions, standards, guarantees of origin, and challenges *Energy Policy* **138** 111300
[39] Ali I 2019 Water photo splitting for green hydrogen energy by green nanoparticles *Int. J. Hydrogen Energy* **44** 11564–73
[40] Alirahmi S M, Assareh E, Pourghassab N N, Delpisheh M, Barelli L and Baldinelli A 2022 Green hydrogen and electricity production via geothermal-driven multi-generation system: thermodynamic modeling and optimization *Fuel* **308** 122049
[41] Schlapbach L and Züttel A 2001 Hydrogen-storage materials for mobile applications *Nature* **414** 353–8
[42] Züttel A 2003 Materials for hydrogen storage *Mater. Today* **6** 24–33
[43] Willige A 2022 4 ways of storing hydrogen from renewable energy (https://spectra.mhi.com/4-ways-of-storing-hydrogen-from-renewable-energy)
[44] Ni M 2006 An overview of hydrogen storage technologies *Energy Explor. Exploit.* **24** 197–209
[45] Krishna R *et al* 2012 Hydrogen storage for energy application *Hydrogen Storage* (London: IntechOpen)
[46] Barthélémy H, Weber M and Barbier F 2017 Hydrogen storage: recent improvements and industrial perspectives *Int. J. Hydrogen Energy* **42** 7254–62
[47] Reuß M, Grube T, Robinius M, Preuster P, Wasserscheid P and Stolten D 2017 Seasonal storage and alternative carriers: a flexible hydrogen supply chain model *Appl. Energy* **200** 290–302
[48] Yanxing Z, Maoqiong G, Yuan Z, Xueqiang D and Jun S 2019 Thermodynamics analysis of hydrogen storage based on compressed gaseous hydrogen, liquid hydrogen and cryo-compressed hydrogen *Int. J. Hydrogen Energy* **44** 16833–40
[49] Schüth F 2009 Challenges in hydrogen storage *Eur. Phys. J. Spec. Top.* **176** 155–66
[50] Langmi H W, Engelbrecht N, Modisha P M and Bessarabov D 2022 Hydrogen storage *Electrochemical Power Sources: Fundamentals, Systems, and Applications* (Amsterdam: Elsevier) pp 455–86

[51] Rambert O and Febvre L 2021 The challenges of hydrogen storage on a large scale (https://hysafe.info/uploads/papers/2021/189.pdf)
[52] 2022 H2 safety #1: understanding hydrogen. Physical and chemical properties of the fuel of the future (https://seshydrogen.com/en/safety-of-hydrogen-systems/)
[53] Hassanpouryouzband A, Joonaki E, Edlmann K and Haszeldine R S 2021 Offshore geological storage of hydrogen: is this our best option to achieve net-zero? *ACS Energy Lett.* **6** 2181–6
[54] Liu L *et al* 2023 Metal hydride composite structures for improved heat transfer and stability for hydrogen storage and compression applications *Inorganics* **11** 181
[55] Salman M S *et al* 2022 The power of multifunctional metal hydrides: a key enabler beyond hydrogen storage *J. Alloys Compd.* **920** 165936
[56] Zhang B *et al* 2022 *In situ* formation of multiple catalysts for enhancing the hydrogen storage of MgH_2 by adding porous $Ni_3ZnC_{0.7}$/Ni loaded carbon nanotubes microspheres *J. Magnes. Alloy.* **12** 1227–38
[57] Guo Z, Shang C and Aguey–Zinsou K 2008 Materials challenges for hydrogen storage *J. Eur. Ceram. Soc.* **28** 1467–73
[58] 2023 Metal Hydrides: The Key to Sustainable and Infinite Hydrogen Storage (https://mincatec-energy.com/en/technology/)
[59] Hirscher M and Panella B 2007 Hydrogen storage in metal–organic frameworks *Scr. Mater.* **56** 809–12
[60] Pérez-Botella E, Valencia S and Rey F 2022 Zeolites in adsorption processes: state of the art and future prospects *Chem. Rev.* **122** 17647–95
[61] Das N and Das J K 2020 Zeolites: an emerging material for gas storage and separation applications *Zeolites-New Challenges* (London: IntechOpen)
[62] Rajaura R S *et al* 2018 Structural and surface modification of carbon nanotubes for enhanced hydrogen storage density *Nano-Struct. Nano-Objects* **14** 57–65
[63] David E 2005 An overview of advanced materials for hydrogen storage *J. Mater. Process. Technol.* **162** 169–77
[64] Safaei M, Foroughi M M, Ebrahimpoor N, Jahani S, Omidi A and Khatami M 2019 A review on metal-organic frameworks: synthesis and applications *TrAC, Trends Anal. Chem.* **118** 401–25
[65] Garg A *et al* 2023 A highly stable terbium (III) metal–organic framework MOF-76 (Tb) for hydrogen storage and humidity sensing *Environ. Sci. Pollut. Res.* **30** 98548–62
[66] Jaramillo D E *et al* 2021 Ambient-temperature hydrogen storage via vanadium (II)-dihydrogen complexation in a metal–organic framework *J. Am. Chem. Soc.* **143** 6248–56
[67] Kapelewski M T *et al* 2018 Record high hydrogen storage capacity in the metal–organic framework Ni_2 (m-dobdc) at near-ambient temperatures *Chem. Mater.* **30** 8179–89
[68] Sopian K, Ibrahim M Z, Daud W R W, Othman M Y, Yatim B and Amin N 2009 Performance of a PV–wind hybrid system for hydrogen production *Renew. Energy* **34** 1973–8
[69] Nasser M, Megahed T F, Ookawara S and Hassan H 2022 Techno-economic assessment of clean hydrogen production and storage using hybrid renewable energy system of PV/wind under different climatic conditions *Sustain. Energy Technol. Assessments* **52** 102195
[70] van der Roest E, Snip L, Fens T and van Wijk A 2020 Introducing power-to-H_3: combining renewable electricity with heat, water and hydrogen production and storage in a neighbourhood *Appl. Energy* **257** 114024

[71] Schnuelle C, Wassermann T, Fuhrlaender D and Zondervan E 2020 Dynamic hydrogen production from PV and wind direct electricity supply–modeling and techno-economic assessment *Int. J. Hydrogen Energy* **45** 29938–52
[72] Li Y *et al* 2022 *In situ* photodeposition of platinum clusters on a covalent organic framework for photocatalytic hydrogen production *Nat. Commun.* **13** 1355
[73] Ganguly P *et al* 2019 Theoretical and experimental investigation of visible light responsive $AgBiS_2$–TiO_2 heterojunctions for enhanced photocatalytic applications *Appl. Catal.* B **253** 401–18
[74] Ganguly P *et al* 2020 Ternary metal chalcogenide heterostructure ($AgInS_2$–TiO_2) nanocomposites for visible light photocatalytic applications *ACS Omega* **5** 406–21
[75] Ganguly P *et al* 2020 New insights into the efficient charge transfer of ternary chalcogenides composites of TiO_2 *Appl. Catal.* B **282** 119612
[76] Ganguly P 2021 Ternary chalcogenides and their composites with titanium dioxide for photocatalytic applications *PhD* Atlantic Technological University http://research.thea.ie/handle/20.500.12065/4033

www.ingramcontent.com/pod-product-compliance
Ingram Content Group UK Ltd.
Pitfield, Milton Keynes, MK11 3LW, UK
UKHW051907100926
8556IPUK00005B/792